75—

Pitman Research Notes in Mathematics Series

AF711375

Submission of proposals for consideration

Suggestions for publication, in the form of outlines and representative samples, are invited by the Editorial Board for assessment. Intending authors should approach one of the main editors or another member of the Editorial Board, citing the relevant AMS subject classifications. Alternatively, outlines may be sent directly to the publisher's offices. Refereeing is by members of the board and other mathematical authorities in the topic concerned, throughout the world.

Preparation of accepted manuscripts

On acceptance of a proposal, the publisher will supply full instructions for the preparation of manuscripts in a form suitable for direct photo-lithographic reproduction. Specially printed grid sheets can be provided and a contribution is offered by the publisher towards the cost of typing. Word processor output, subject to the publisher's approval, is also acceptable.

Illustrations should be prepared by the authors, ready for direct reproduction without further improvement. The use of hand-drawn symbols should be avoided wherever possible, in order to maintain maximum clarity of the text.

The publisher will be pleased to give any guidance necessary during the preparation of a typescript, and will be happy to answer any queries.

Important note

In order to avoid later retyping, intending authors are strongly urged not to begin final preparation of a typescript before receiving the publisher's guidelines. In this way it is hoped to preserve the uniform appearance of the series.

Longman Scientific & Technical
Longman House
Burnt Mill
Harlow, Essex, CM20 2JE
UK
(Telephone (0279) 426721)

Titles in this series. A full list is available on request from the publisher.

100 Optimal control of variational inequalities
V Barbu
101 Partial differential equations and dynamical systems
W E Fitzgibbon III
102 Approximation of Hilbert space operators Volume II
C Apostol, L A Fialkow, D A Herrero and D Voiculescu
103 Nondiscrete induction and iterative processes
V Ptak and F-A Potra
104 Analytic functions – growth aspects
O P Juneja and G P Kapoor
105 Theory of Tikhonov regularization for Fredholm equations of the first kind
C W Groetsch
106 Nonlinear partial differential equations and free boundaries. Volume I
J I Díaz
107 Tight and taut immersions of manifolds
T E Cecil and P J Ryan
108 A layering method for viscous, incompressible L_p flows occupying R^n
A Douglis and E B Fabes
109 Nonlinear partial differential equations and their applications: Collège de France Seminar. Volume VI
H Brezis and J L Lions
110 Finite generalized quadrangles
S E Payne and J A Thas
111 Advances in nonlinear waves. Volume II
L Debnath
112 Topics in several complex variables
E Ramírez de Arellano and D Sundararaman
113 Differential equations, flow invariance and applications
N H Pavel
114 Geometrical combinatorics
F C Holroyd and R J Wilson
115 Generators of strongly continuous semigroups
J A van Casteren
116 Growth of algebras and Gelfand–Kirillov dimension
G R Krause and T H Lenagan
117 Theory of bases and cones
P K Kamthan and M Gupta
118 Linear groups and permutations
A R Camina and E A Whelan
119 General Wiener–Hopf factorization methods
F-O Speck
120 Free boundary problems: applications and theory. Volume III
A Bossavit, A Damlamian and M Fremond
121 Free boundary problems: applications and theory. Volume IV
A Bossavit, A Damlamian and M Fremond
122 Nonlinear partial differential equations and their applications: Collège de France Seminar. Volume VII
H Brezis and J L Lions
123 Geometric methods in operator algebras
H Araki and E G Effros
124 Infinite dimensional analysis–stochastic processes
S Albeverio
125 Ennio de Giorgi Colloquium
P Krée
126 Almost-periodic functions in abstract spaces
S Zaidman
127 Nonlinear variational problems
A Marino, L Modica, S Spagnolo and M Degliovanni
128 Second-order systems of partial differential equations in the plane
L K Hua, W Lin and C-Q Wu
129 Asymptotics of high-order ordinary differential equations
R B Paris and A D Wood
130 Stochastic differential equations
R Wu
131 Differential geometry
L A Cordero
132 Nonlinear differential equations
J K Hale and P Martinez-Amores
133 Approximation theory and applications
S P Singh
134 Near-rings and their links with groups
J D P Meldrum
135 Estimating eigenvalues with *a posteriori/a priori* inequalities
J R Kuttler and V G Sigillito
136 Regular semigroups as extensions
F J Pastijn and M Petrich
137 Representations of rank one Lie groups
D H Collingwood
138 Fractional calculus
G F Roach and A C McBride
139 Hamilton's principle in continuum mechanics
A Bedford
140 Numerical analysis
D F Griffiths and G A Watson
141 Semigroups, theory and applications. Volume I
H Brezis, M G Crandall and F Kappel
142 Distribution theorems of L-functions
D Joyner
143 Recent developments in structured continua
D De Kee and P Kaloni
144 Functional analysis and two-point differential operators
J Locker
145 Numerical methods for partial differential equations
S I Hariharan and T H Moulden
146 Completely bounded maps and dilations
V I Paulsen
147 Harmonic analysis on the Heisenberg nilpotent Lie group
W Schempp
148 Contributions to modern calculus of variations
L Cesari
149 Nonlinear parabolic equations: qualitative properties of solutions
L Boccardo and A Tesei
150 From local times to global geometry, control and physics
K D Elworthy

151 A stochastic maximum principle for optimal control of diffusions
U G Haussmann

152 Semigroups, theory and applications. Volume II
H Brezis, M G Crandall and F Kappel

153 A general theory of integration in function spaces
P Muldowney

154 Oakland Conference on partial differential equations and applied mathematics
L R Bragg and J W Dettman

155 Contributions to nonlinear partial differential equations. Volume II
J I Díaz and P L Lions

156 Semigroups of linear operators: an introduction
A C McBride

157 Ordinary and partial differential equations
B D Sleeman and R J Jarvis

158 Hyperbolic equations
F Colombini and M K V Murthy

159 Linear topologies on a ring: an overview
J S Golan

160 Dynamical systems and bifurcation theory
M I Camacho, M J Pacifico and F Takens

161 Branched coverings and algebraic functions
M Namba

162 Perturbation bounds for matrix eigenvalues
R Bhatia

163 Defect minimization in operator equations: theory and applications
R Reemtsen

164 Multidimensional Brownian excursions and potential theory
K Burdzy

165 Viscosity solutions and optimal control
R J Elliott

166 Nonlinear partial differential equations and their applications: Collège de France Seminar. Volume VIII
H Brezis and J L Lions

167 Theory and applications of inverse problems
H Haario

168 Energy stability and convection
G P Galdi and B Straughan

169 Additive groups of rings. Volume II
S Feigelstock

170 Numerical analysis 1987
D F Griffiths and G A Watson

171 Surveys of some recent results in operator theory. Volume I
J B Conway and B B Morrel

172 Amenable Banach algebras
J-P Pier

173 Pseudo-orbits of contact forms
A Bahri

174 Poisson algebras and Poisson manifolds
K H Bhaskara and K Viswanath

175 Maximum principles and eigenvalue problems in partial differential equations
P W Schaefer

176 Mathematical analysis of nonlinear, dynamic processes
K U Grusa

177 Cordes' two-parameter spectral representation theory
D F McGhee and R H Picard

178 Equivariant K-theory for proper actions
N C Phillips

179 Elliptic operators, topology and asymptotic methods
J Roe

180 Nonlinear evolution equations
J K Engelbrecht, V E Fridman and E N Pelinovski

181 Nonlinear partial differential equations and their applications: Collège de France Seminar. Volume IX
H Brezis and J L Lions

182 Critical points at infinity in some variational problems
A Bahri

183 Recent developments in hyperbolic equations
L Cattabriga, F Colombini, M K V Murthy and S Spagnolo

184 Optimization and identification of systems governed by evolution equations on Banach space
N U Ahmed

185 Free boundary problems: theory and applications. Volume I
K H Hoffmann and J Sprekels

186 Free boundary problems: theory and applications. Volume II
K H Hoffmann and J Sprekels

187 An introduction to intersection homology theory
F Kirwan

188 Derivatives, nuclei and dimensions on the frame of torsion theories
J S Golan and H Simmons

189 Theory of reproducing kernels and its applications
S Saitoh

190 Volterra integrodifferential equations in Banach spaces and applications
G Da Prato and M Iannelli

191 Nest algebras
K R Davidson

192 Surveys of some recent results in operator theory. Volume II
J B Conway and B B Morrel

193 Nonlinear variational problems. Volume II
A Marino and M K V Murthy

194 Stochastic processes with multidimensional parameter
M E Dozzi

195 Prestressed bodies
D Iesan

196 Hilbert space approach to some classical transforms
R H Picard

197 Stochastic calculus in application
J R Norris

198 Radical theory
B J Gardner

199 The C^*-algebras of a class of solvable Lie groups
X Wang

200 Stochastic analysis, path integration and dynamics
K D Elworthy and J C Zambrini

201 Riemannian geometry and holonomy groups
S Salamon
202 Strong asymptotics for extremal errors and polynomials associated with Erdös type weights
D S Lubinsky
203 Optimal control of diffusion processes
V S Borkar
204 Rings, modules and radicals
B J Gardner
205 Two-parameter eigenvalue problems in ordinary differential equations
M Faierman
206 Distributions and analytic functions
R D Carmichael and D Mitrovic
207 Semicontinuity, relaxation and integral representation in the calculus of variations
G Buttazzo
208 Recent advances in nonlinear elliptic and parabolic problems
P Bénilan, M Chipot, L Evans and M Pierre
209 Model completions, ring representations and the topology of the Pierce sheaf
A Carson
210 Retarded dynamical systems
G Stepan
211 Function spaces, differential operators and nonlinear analysis
L Paivarinta
212 Analytic function theory of one complex variable
C C Yang, Y Komatu and K Niino
213 Elements of stability of visco-elastic fluids
J Dunwoody
214 Jordan decomposition of generalized vector measures
K D Schmidt
215 A mathematical analysis of bending of plates with transverse shear deformation
C Constanda
216 Ordinary and partial differential equations. Volume II
B D Sleeman and R J Jarvis
217 Hilbert modules over function algebras
R G Douglas and V I Paulsen
218 Graph colourings
R Wilson and R Nelson
219 Hardy-type inequalities
A Kufner and B Opic
220 Nonlinear partial differential equations and their applications: Collège de France Seminar. Volume X
H Brezis and J L Lions
221 Workshop on dynamical systems
E Shiels and Z Coelho
222 Geometry and analysis in nonlinear dynamics
H W Broer and F Takens
223 Fluid dynamical aspects of combustion theory
M Onofri and A Tesei
224 Approximation of Hilbert space operators. Volume I. 2nd edition
D Herrero
225 Operator theory: proceedings of the 1988 GPOTS–Wabash conference
J B Conway and B B Morrel
226 Local cohomology and localization
J L Bueso Montero, B Torrecillas Jover and A Verschoren
227 Nonlinear waves and dissipative effects
D Fusco and A Jeffrey
228 Numerical analysis 1989
D F Griffiths and G A Watson
229 Recent developments in structured continua. Volume II
D De Kee and P Kaloni
230 Boolean methods in interpolation and approximation
F J Delvos and W Schempp
231 Further advances in twistor theory. Volume I
L J Mason and L P Hughston
232 Further advances in twistor theory. Volume II
L J Mason and L P Hughston
233 Geometry in the neighborhood of invariant manifolds of maps and flows and linearization
U Kirchgraber and K Palmer
234 Quantales and their applications
K I Rosenthal
235 Integral equations and inverse problems
V Petkov and R Lazarov
236 Pseudo-differential operators
S R Simanca
237 A functional analytic approach to statistical experiments
I M Bomze
238 Quantum mechanics, algebras and distributions
D Dubin and M Hennings
239 Hamilton flows and evolution semigroups
J Gzyl
240 Topics in controlled Markov chains
V S Borkar
241 Invariant manifold theory for hydrodynamic transition
S Sritharan
242 Lectures on the spectrum of $L^2(\Gamma\backslash G)$
F L Williams
243 Progress in variational methods in Hamiltonian systems and elliptic equations
M Girardi, M Matzeu and F Pacella
244 Optimization and nonlinear analysis
A Ioffe, M Marcus and S Reich
245 Inverse problems and imaging
G F Roach
246 Semigroup theory with applications to systems and control
N U Ahmed
247 Periodic-parabolic boundary value problems and positivity
P Hess
248 Distributions and pseudo-differential operators
S Zaidman
249 Progress in partial differential equations: the Metz surveys
M Chipot and J Saint Jean Paulin
250 Differential equations and control theory
V Barbu

251 Stability of stochastic differential equations with respect to semimartingales
X Mao
252 Fixed point theory and applications
J Baillon and M Théra
253 Nonlinear hyperbolic equations and field theory
M K V Murthy and S Spagnolo
254 Ordinary and partial differential equations. Volume III
B D Sleeman and R J Jarvis
255 Harmonic maps into homogeneous spaces
M Black
256 Boundary value and initial value problems in complex analysis: studies in complex analysis and its applications to PDEs 1
R Kühnau and W Tutschke
257 Geometric function theory and applications of complex analysis in mechanics: studies in complex analysis and its applications to PDEs 2
R Kühnau and W Tutschke
258 The development of statistics: recent contributions from China
X R Chen, K T Fang and C C Yang
259 Multiplication of distributions and applications to partial differential equations
M Oberguggenberger
260 Numerical analysis 1991
D F Griffiths and G A Watson
261 Schur's algorithm and several applications
M Bakonyi and T Constantinescu
262 Partial differential equations with complex analysis
H Begehr and A Jeffrey
263 Partial differential equations with real analysis
H Begehr and A Jeffrey
264 Solvability and bifurcations of nonlinear equations
P Drábek
265 Orientational averaging in mechanics of solids
A Lagzdins, V Tamuzs, G Teters and A Kregers
266 Progress in partial differential equations: elliptic and parabolic problems
C Bandle, J Bemelmans, M Chipot, M Grüter and J Saint Jean Paulin
267 Progress in partial differential equations: calculus of variations, applications
C Bandle, J Bemelmans, M Chipot, M Grüter and J Saint Jean Paulin
268 Stochastic partial differential equations and applications
G Da Prato and L Tubaro
269 Partial differential equations and related subjects
M Miranda
270 Operator algebras and topology
W B Arveson, A S Mishchenko, M Putinar, M A Rieffel and S Stratila
271 Operator algebras and operator theory
W B Arveson, A S Mishchenko, M Putinar, M A Rieffel and S Stratila
272 Ordinary and delay differential equations
J Wiener and J K Hale
273 Partial differential equations
J Wiener and J K Hale
274 Mathematical topics in fluid mechanics
J F Rodrigues and A Sequeira
275 Green functions for second order parabolic integro-differential problems
M G Garroni and J F Menaldi
276 Riemann waves and their applications
M W Kalinowski
277 Banach C(K)-modules and operators preserving disjointness
Y A Abramovich, E L Arenson and A K Kitover
278 Limit algebras: an introduction to subalgebras of C*-algebras
S C Power
279 Abstract evolution equations, periodic problems and applications
D Daners and P Koch Medina
280 Emerging applications in free boundary problems
J M Chadam and H Rasmussen
281 Free boundary problems involving solids
J M Chadam and H Rasmussen
282 Free boundary problems in fluid flow with applications
J M Chadam and H Rasmussen
283 Asymptotic problems in probability theory: stochastic models and diffusions on fractals
K D Elworthy and N Ikeda
284 Asymptotic problems in probability theory: Wiener functionals and asymptotics
K D Elworthy and N Ikeda
285 Dynamical systems
R Bamon, R Labarca, J Lewowicz and J Palis
286 Models of hysteresis
A Visintin
287 Moments in probability and approximation theory
G Anastassiou
288 Mathematical aspects of penetrative convection
B Straughan
289 Ordinary and partial differential equations. Vol ume IV
B D Sleeman and R J Jarvis
290 *K*-theory for real *C**-algebras
H Schröder
291 Recent developments in theoretical fluid mechanics
G P Galdi and J Necas
292 Propagation of a curved shock and nonlinear ray theory
P Prasad
293 Non-classical elastic solids
M Ciarletta and D Ieşan
294 Multigrid methods
J Bramble
295 Entropy and partial differential equations
W A Day
296 Progress in partial differential equations: the Metz surveys 2
M Chipot
297 Nonstandard methods in the calculus of variations
C Tuckey
298 Barrelledness, Baire-like- and (LF)-spaces
M Kunzinger
299 Nonlinear partial differential equations and their applications. Collège de France Seminar. Volume XI
H Brezis and J L Lions
300 Introduction to operator theory
T Yoshino

301 Generalized fractional calculus and applications
V Kiryakova

Virginia Kiryakova

Bulgarian Academy of Sciences

Generalized fractional calculus and applications

Longman Scientific & Technical

Copublished in the United States with
John Wiley & Sons, Inc., New York

Longman Scientific & Technical
Longman Group UK Limited
Longman House, Burnt Mill, Harlow
Essex CM20 2JE, England
and Associated companies throughout the world.

Copublished in the United States with
John Wiley & Sons Inc., 605 Third Avenue, New York, NY 10158

First published 1994

AMS Subject Classifications: (Main) 26A33
(Subsidiary) 33CXX, 44XX, 45XX, 30XX

ISSN 0269-3674

ISBN 0 582 21977 9

British Library Cataloguing in Publication Data

A catalogue record for this book is
available from the British Library

Library of Congress Cataloging-in-Publication Data

Kiryakova, V. S. (Virginia S.)
Generalized fractional calculus and applications / V.S. Kiryakova.
p. cm. -- (Pitman research notes in mathematics series, ISSN 0269-3674)
Includes bibliographical references and indexes.
1. Calculus. I. Title. II. Series.
QA303.K62 1993
515--dc20 93-4705
CIP

Printed and bound in Great Britain
by Biddles Ltd, Guildford and King's Lynn

Contents

Preface

These notes are devoted to *a systematic and unified development of a new generalized fractional calculus*, closely related to *special functions* of a rather general nature. Generalized operators of integration and differentiation of arbitrary (not necessarily integer) multiorders $\delta = (\delta_1 \geq 0, ..., \delta_m \geq 0), m \geq 1,$ are introduced by means of kernel-functions, being $G^{m,0}_{m,m}$- and $H^{m,0}_{m,m}$-functions. Due to this *peculiar choice of Meijer's G-functions (Fox's H-functions)* in the single integral representations of the operators that are introduced here, a decomposition into commuting Erdélyi-Kober fractional operators holds, under suitable conditions. Thus, complicated multiple integrals or differintegral expressions can be represented alternatively by means of single integrals involving generalized hypergeometric functions. The beauty and succinctness of the notations and properties of the latter functions allow the development of a *full chain of operational rules, mapping properties and convolutional structure of the generalized (m-tuple) fractional integrals as well as an appropriate explicit definition of the corresponding generalized derivatives.* On the other hand, the frequent appearance of compositions of classical Riemann-Liouville and Erdélyi-Kober fractional operators in various problems of applied analysis is the key to the great number of *applications and known special cases* of our *generalized fractional differintegrals* (i.e. generalized operators of integration, differentiation or integro-differentiation), some of which are established in the present notes.

In the *Introduction* we give some historical background and present some of the main points of the theory presented here, from the author's point of view. *Chapter 1* deals with the basic definitions of the generalized (multiple) Erdélyi-Kober fractional integrals and derivatives, their properties and various special cases. In *Chapter 2* we go deeper into some less known aspects and details concerning the classical Erdélyi-Kober operators. *Chapter 3* is specially devoted to the important class of the so-called hyper-Bessel integral and differential operators, Bessel type equations of arbitrary order, their explicit solutions in terms of Meijer's G-functions, Poisson-Sonine-Dimovski transmutations and Obrechkoff integral transforms. Special cases of all these elements investigated by different authors are mentioned. In *Chapter 4* we find new integral and differintegral (integral, differential or integro-differential) formulas for the ${}_pF_q$-functions, which are generalizations both of Poisson type integrals and Rodrigues differential formulas for particular special functions. On the basis of these representations, a new classification of the generalized hypergeometric functions is proposed. *Chapter 5* deals with other applications of the generalized fractional calculus: to Abel integral equations, dual integral equations, univalent function theory and generalized Laplace type transformations. More general fractional integration operators involving Fox's $H^{m,0}_{m,m}$-function are studied there in different functional spaces. To make our presentation self-contained, in the *Appendix* we have collected most of the definitions and basic properties of the special functions that are used in these notes. The *References* include 519 titles and a *Citation Index* is provided, showing the articles referred to in the sections. The notes also contain some *open problems*, aimed at stimulating further research on this topic.

ACKNOWLEDGEMENTS

The author is obliged to *Prof. I. Dimovski* under whose guidance M.Sc. and Ph.D. theses were accomplished, leading finally to these notes, as well as to *Prof. P. Rusev* and the colleagues of the Seminar advised by him, for many useful discussions and support. The beginning of a deeper investigation of the problems of fractional calculus was inspired 13 years ago by the papers [163]-[164] of *Prof. S. Kalla* and I am especially grateful to him for the permanent interest and moral support in accomplishing these notes. Special thanks are due to the *Editorial Board of Pitman Research Notes in Mathematics Series* for including this book in the Series and in particular, to *Prof. G. Roach* who acted as initiator for this.

It is a pleasure to thank also *Acad. B. Sendov* for suggesting the idea about writing a book and the *other Bulgarian mathematicians* from the Complex Analysis Section and the Institute of Mathematics for supporting the author in various ways.

This notes have benefited from the many responsive *colleagues and friends over the world* who sent valuable and rare articles, discussed open problems and paid attention to various results related to this work: Profs. V. Adamchik, M. Al-Bassam, B. Al-Saqabi, L. M. B. S. Campos, R. Carroll, M. Džrbashjan, N. Hayek, J. Heading, J. Hubbell, I.Khachatrjan, A. Kilbas, E. Koh, E. Krätzel, M. Lapidus, J. Lighthill, O. Marichev, A. McBride, E. Meister, M. Mikolas, K. Nishimoto, F. Olver, S. Owa, R. S. Pathak, A. Prudnikov, R. K. Raina, G. L. N. Rao, J. Rodriguez, P. G. Rooney, B. Ross, M. Saigo, R. K. Saxena, I. Sneddon, H. M. Srivastava, V. Tkachenko, N. Virchenko, S. Yakubovich and Z. Yakubowski.

The author would also like to thank the following organizations: the *British Council* and the Mathematics Department of the *University of Strathclyde* (Glasgow) for ensuring the conditions for a scholarship in the UK in 1992, during which a time part of this work was done; the *Ministry of Education and Science, Bulgaria* and the *Union of Scientists in Bulgaria*, for the partial financial support during preparation this work.

Last but not least, thanks are due to *my son Atanas Kiryakov*, a student in Informatics at the Sofia University, for the accurate and really heavy work of TEX-ing all the text and formulas and thus preparing the camera-ready copy of the manuscript. I am grateful also to the *staff of Longman Group UK Ltd.* for arranging the final editing and publication of these notes.

Sofia, July 1993

Dr. Virginia Kiryakova, Asso. Prof.
Institute of Mathematics - Bulgarian Academy of Sciences,
P.O. Box 373, Sofia - 1090, Bulgaria
E-mail: virginia@bgearn.bitnet

Introduction

The fractional calculus, like many other mathematical disciplines and ideas, has its origin in the striving for extension of meaning. Well known examples are the extensions of the integers to the rational numbers, of the real numbers to the complex numbers, of the factorials of integers to the notion of the Γ-function. In differential and integral calculus the question of extension of meaning is: Can the derivatives $\frac{d^n y}{dx^n}$ of integer order $n > 0$, respectively the n-fold integrals, be extended when n is any number – fractional, irrational or complex? The affirmative answer has led to the so-called *fractional calculus, a misnomer for the theory of operators of integration and differentiation of arbitrary (fractional) order and their applications.*

This is a discipline, "*as old as mathematical analysis. It can be categorized, briefly, as applicable mathematical analysis. The properties and theory of fractional operators are proper objects of study in their own right. In recent decades, they have been found useful in various fields: rheology, quantitative biology, electrochemistry, scattering theory, diffusion, transport theory, probability, statistics, potential theory and elasticity ... . However, many mathematicians and scientists are unfamiliar with this topic ... and thus, while the theory has developed, its use has lagged behind ...* " ([404]).

These reasons made Prof. B. Ross, the author of the above-cited phrase, backed by other famous *fractional analysts*: Profs. A. Erdélyi, I. Sneddon and A. Zigmund, to organize in 1974 in New Haven the *historical first international conference on fractional calculus*. It lasted only two days but with 22 participants and 72 attendees, it had an exceptional turnout. This conference had *several "singular" purposes*: to popularize the topic in the hope that it would induce scientists to include it in their research and curricula; to exchange and impart information serving to suggest new areas of research; and to encourage attempts in discovering new formal methods of representing physical phenomena by mathematical models that could be treated with the elegance of fractional calculus. The lectures were published in the Proceedings [404] (ed. B. Ross). Ten years later, a *second international conference on fractional calculus* took place in Scotland, at Ross Priory – the Strathclyde University's stunning property on the shores of Loch Lomond. Later on, in 1989, the *third international conference* was organized in Tokyo, on the occasion of the hundredth anniversary of Nihon University. While the participants at the second conference might have had some trouble with the *nine (and a half?) hole golf course*, the attendees of the third conference were posed quite a hard problem: *What is the fractional dimension of Tokyo's subway net?* The contributions to more serious problems of the recent theory of fractional calculus were published in the Proceedings [291] (eds A. McBride and G.Roach) and [399] (ed. K. Nishimoto).

Going back to the history, it seems that Leibnitz was the first to try to extend the meaning of a derivative $\frac{d^n y}{dx^n}$ of integer order n. Perhaps, it was a naive toying with symbols that prompted l'Hospital to ask: *"What if n be $\frac{1}{2}$?"*. Leibnitz, in 1695, replied, *"It will lead to a paradox"* but added prophetically: *"From this apparent paradox, one day useful consequences will be drawn"*. In 1819 the first mentioning of a derivative of arbitrary order appeared in a published article by Lacroix. Later on, Euler and Fourier gave meaning to derivatives of arbitrary order but still without examples and applications. So, the honour of devising the first application belonged to Abel in 1823. He applied fractional calculus techniques to the soluiton of an integral equation, related to the so-called isochrone (tautochrone) problem. Abel's solution was so elegant that it probably attracted the attention of Liouville to make the first major attempt in giving a logical definition of the fractional derivative, starting from two different points of view. In 1847, Riemann, while a student, wrote a paper which was eventually published posthumously and gave the definition, whose modification is now known as the Riemann-Liouville integral. Like other mathematical ideas, the development of fractional calculus has passed through various errors, absurdities, controversies, etc. that sometimes made mathematicians distrusting in the general concept of fractional operators. Thus, it has taken 279 years, since l'Hospital first raised the question, for a text to appear entirely devoted to the topic of fractional calculus: the book of Oldham and Spanier [340]. The intermediate period (1695-1974) is thouroughly described by Ross in his *"Chronological Bibliography of the Fractional Calculus with Commentary"* [405], including more than 150 items. Among the many authors contributing to the topic we can mention the names of Lagrange, Laplace, de Morgan, Letnikov, Lourant, Hadamard, Heaviside, Hardy, Littlewood, Weyl, Post, Erdélyi, Kober, Widder, Osler, Sneddon, Mikolas, Al-Bassam and the other mathematicians cited in the References. The reader could find a more detailed exposition in the historical remarks of Ross [406] and Mikolas [303]-[305] as well as in the surveys of Sneddon [454]-[455] and Al-Bassam [19] and the huge encyclopaedic book of Samko, Kilbas and Marichev [434]. Let us emphasize *the essence of the mathematical problem for defining integrals and derivatives of fractional order.* It consists in the following: for each function $f(z)$, $z = x + iy$, of sufficiently large class and for each number δ (rational, irrational, complex), set up a correspondence with a function $g(z) = D^\delta f(z)$ satisfying the conditions:

1) If $f(z)$ is an analytic function of z, the derivative $D^\delta f(z)$ *is an analytic function* of z and δ.

2) The operation D^δ gives *the same result* as the usual differentiation of order n, when $\delta = n$ is a positive integer, and the same effect as n-fold integration, if $\delta = -n$ is a negative integer. Moreover, $D^\delta f(z)$ should vanish at the initial point $z = 0$ (or $z = c$) together with its first $(n-1)$ derivatives.

3) The operator of order $\delta = 0$ is the *identity operator*.

4) Fractional operators are *linear*:

$$D^\delta \{af(z) + bg(z)\} = aD^\delta f(z) + bD^\delta g(z).$$

5) For fractional integrations of arbitrary orders $\alpha > 0$, $\beta > 0$ ($\Re\alpha > 0, \Re\beta > 0$) the *additive index law (semigroup property)* holds:

$$D^{-\alpha} D^{-\beta} f(z) = D^{-(\alpha+\beta)} f(z).$$

It is well known that the so-called *Riemann-Liouville operator* of integration of fractional order δ ($\Re\delta > 0$), defined as

$$R^{\delta} f(x) = D^{-\delta} f(x) = \frac{1}{\Gamma(\delta)} \int_0^x (x-t)^{\delta-1} f(t) dt, \tag{0.1}$$

satisfies all the above conditions (see Ross [406]). In particular, by means of the Dirichlet formula one can establish that for integer $\delta = n > 0$, operator (0.1) coincides with the repeated (n-fold) integration:

$$\begin{aligned} R^n f(x) = D^{-n} f(x) &= \frac{1}{(n-1)!} \int_0^x (x-t)^{n-1} f(t) dt \\ &= \int_0^x dt_1 \int_0^{t_1} dt_2 \ldots \int_0^{t_{n-2}} dt_{n-1} \int_0^{t_{n-1}} f(t_n)\, dt_n = l^n f(x). \end{aligned} \tag{0.2}$$

If $z = x + iy$ is a complex variable, the above representations are modified to the Cauchy integral formula (see Section 5.5). Definition (0.1) concerns integrations of positive order δ (or $\Re\delta > 0$). For differentiation it could not be used directly. However, a little trick can help to find suitable expression. If $\delta > 0$ is non integer: $\delta = n - p$, where $n := [\delta] + 1$, $p := 1 - \{\delta\}$, then we can define

$$\begin{aligned} D^{\delta} f(x) = D^n D^{-p} f(x) &= \left(\frac{d}{dx}\right)^n R^p f(x) \\ &= \left(\frac{d}{dx}\right)^n \left\{ \frac{1}{\Gamma(1-\{\delta\})} \int_0^x (x-t)^{-\{\delta\}} f(t) dt \right\}. \end{aligned} \tag{0.3}$$

To compare this with the classical calculus, having in mind the formula

$$D^{\delta}\{x^{\alpha}\} = \frac{\Gamma(\alpha+1)}{\Gamma(\alpha+1-\delta)} x^{\alpha-\delta}, \quad \delta > 0, \ \alpha > -1, \tag{0.4}$$

we obtain for $\alpha = 0$:

$$D^{\delta}\{c\} = c \frac{x^{-\delta}}{\Gamma(1-\delta)},$$

i.e. a fractional derivative of a constant is zero only for positive integer values $\delta = n = 1, 2, 3, \ldots$. Also, for suitable functional spaces

$$D^{\delta} R^{\delta} f(x) = f(x), \quad \text{i.e.} \ \left\{R^{\delta}\right\}^{-1} = D^{\delta}.$$

The so-called Weyl fractional integral is also known:

$$W^{\delta} f(x) = \frac{1}{\Gamma(\delta)} \int_{x}^{\infty} (t-x)^{\delta-1} f(t)dt, \ \delta > 0, \tag{0.5}$$

and both fractional integrals (0.1), (0.5) are related via the *Parseval equality (fractional integration by parts)*:

$$\int_{0}^{\infty} f(x) \left(R^{\delta} g\right)(x)dx = \int_{0}^{\infty} \left(W^{\delta} f\right)(x)g(x)dx. \tag{0.6}$$

Besides the above Riemann-Liouville definitions of fractional operators, many other modifications and their generalizations have been used in the literature. The most useful classical fractional integrals however seem to be the *Erdélyi-Kober operators* (see e.g. Sneddon [452]-[455]), whose generalizations in the form

$$\begin{aligned} I_{\beta}^{\gamma,\delta} f(x) &= \int_{0}^{1} \frac{(1-\sigma)^{\delta-1}\sigma^{\gamma}}{\Gamma(\delta)} f(x\sigma^{\frac{1}{\beta}})\, d\sigma \\ &= x^{-\beta(\gamma+\delta)} \int_{0}^{x} \frac{(x^{\beta}-\tau^{\beta})^{\delta-1}}{\Gamma(\delta)} \tau^{\beta\gamma} f(\tau)\, d(\tau^{\beta}) = I_{\beta,1}^{\gamma,\delta}, \\ K_{\beta}^{\gamma,\delta} f(x) &= \int_{1}^{\infty} \frac{(\sigma-1)^{\delta-1}\sigma^{-(\gamma+\delta)}}{\Gamma(\delta)} f(x\sigma^{\frac{1}{\beta}})\, d\sigma \\ &= x^{\beta\gamma} \int_{x}^{\infty} \frac{(\tau^{\beta}-x^{\beta})^{\delta-1}}{\Gamma(\delta)} \tau^{-\beta(\gamma+\delta)} f(\tau)\, d(\tau^{\beta}) = W_{\beta,1}^{\gamma,\delta} f(x) \end{aligned} \tag{0.7}$$

are *used essentially in the present work.*

Several authors, like Love [262], Saxena [435], Kalla and Saxena [177]-[178], Saigo [415]-[418], McBride [289], etc. have studied and used different modifications of the so-called *hypergeometric operators of fractional integration* (cf. (1.1.19))

$$\begin{aligned} If(x) &= \frac{\mu x^{-\gamma-1}}{\Gamma(1-\delta)} \int_{0}^{x} {}_2F_1\left(\delta, \beta+m;\ \eta; a\left(\frac{t}{x}\right)^{\mu}\right) t^{\gamma} f(t)dt, \\ Kf(x) &= \frac{\mu x^{\gamma}}{\Gamma(1-\delta)} \int_{x}^{\infty} {}_2F_1\left(\delta, \beta+m;\ \eta; a\left(\frac{x}{t}\right)^{\mu}\right) t^{-\gamma-1} f(t)dt, \end{aligned} \tag{0.8}$$

involving the Gauss hypergeometric function.

An example of *fractional integration operators involving other special functions*, is given by the operators of Lowndes [268]-[269]:

$$I_\lambda(\eta,\nu+1)f(x) = \frac{2^{\nu+1}}{\lambda^\nu}x^{-(\nu+\eta+1)}\int_0^x t^{2\eta+1}\left(x^2-t^2\right)^{\frac{\nu}{2}} J_\nu\left(\lambda\sqrt{x^2-t^2}\right)f(t)dt,$$
$$K_\lambda(\eta,\nu+1)f(x) = \frac{2^{\nu+1}}{\lambda^\nu}x^{2\eta}\int_x^\infty t^{-2\eta-2\nu-1}\left(t^2-x^2\right)^{\frac{\nu}{2}} J_\nu\left(\lambda\sqrt{t^2-x^2}\right)f(t)dt, \tag{0.9}$$

related to the Bessel type differential operator $B_\eta = x^{-2\eta-1}\frac{d}{dx}x^{2\eta+1}\frac{d}{dx}$.

The most general fractional integration operators of type (0.1) are obtained when the *kernel-function is a Maijer G-function* (see Appendix, Sections A, C), as in Parashar [355], Kalla [159]-[160], [164], Rooney [403], etc.:

$$R_{\mathfrak{G}}f(x) = x^{-\gamma-1}\int_0^x G^{m,n}_{p,q}\left[a\left(\frac{t}{x}\right)^r \middle| \begin{matrix}(a_j)_1^p\\(b_k)_1^q\end{matrix}\right]t^\gamma f(t)dt, \tag{0.10}$$

or its further generalization, *the Fox H-function* (Appendix, Section E), as in Kalla [157]-[158], [164], Srivastava and Buschman [463], Saxena and Kumbhat [440], Saigo, Saxena and Ram [425], etc.:

$$R_{\mathfrak{H}}f(x) = x^{-\gamma-1}\int_0^x H^{m,n}_{p,q}\left[a\left(\frac{t}{x}\right)^r \middle| \begin{matrix}(a_j,A_j)_1^p\\(b_k,B_k)_1^q\end{matrix}\right]t^\gamma f(t)dt. \tag{0.11}$$

It is suggested then that all the above operators of Riemann-Liouville type (0.1) have the general form

$$Rf(x) = x^{-\gamma-1}\int_0^x \Phi\left(\frac{t}{x}\right)t^\gamma f(t)dt, \tag{0.12}$$

and respectively those of Weyl type (0.5) have the form:

$$Wf(x) = x^\gamma\int_x^\infty \Phi\left(\frac{x}{t}\right)t^{-\delta-1}f(t)dt, \tag{0.13}$$

where the kernel $\Phi(x)$ is an arbitrary continuous function so that integrals (0.12)-(0.13) make sense in sufficiently large functional spaces. The above forms for the *generalized operators of fractional integration* have been proposed by Kalla [161], [163]-[164] who has established a series of their formal properties, analogous to those of classical fractional integrals. By suitable choices of the kernel-function Φ, operators (0.12)-(0.13) can be shown to include all the other known fractional integrals as special cases.

Nowadays, there exist a great number of articles, surveys and several books, entirely devoted to fractional calculus, its generalizations and applications. Among the separate books and volumes, considered as *competitive titles* to the present work, we can mention: Olbham and Spanier [340], Ross (ed.) [404], McBride and Roach (eds) [291], Nishimoto [324], [326], [330], Nishimoto (ed.) [329], Samko, Kilbas and Marichev [434], Al-Bassam [21], three of them being the above-mentioned Proceedings. Most of the articles and surveys before 1987 have been referred to in the encyclopaedic book of Samko, Kilbas and Marichev [434]. There, an overall exposition of the classical theory of fractional calculus, its various applications and recent generalizations is given.

The *present book is of a slightly different nature.* It is devoted to a *systematic development of a new generalized fractional calculus*, with interesting special cases and applications to different topics of analysis. The exposition of this theory is self-contained and unified.

The basic operators of this calculus, the so-called *generalized (multiple) operators of fractional integration* are defined by means of convolutional type single integrals involving the Meijer $G^{m,0}_{m,m}$-functions as kernels. In the more general case, the H-functions of Fox appear as kernel-functions. The same operators can be considered also as compositions of an arbitrary number ($m > 1$) of commuting Erdélyi-Kober fractional integrals written by means of repeated integrals without any use of special functions. Since such compositions arise quite often in different problems and branches of analysis, mathematical physics and other areas of applied mathematics, this relationship proves to be *the key to the various applications* of the generalized fractional calculus presented here. The corresponding generalized (multiple) fractional derivatives are also introduced by means of explicit *differintegral* (e.g. differential, integral or integro-differential) expressions. A full chain of operational rules, mapping properties in the spaces of Lebesgue integrable, continuous and analytic functions, is proposed, based exclusively on the simple properties of the G- and H-functions. These functions of a generalized hypergeometric nature include the basic elementary functions, classical orthogonal polynomials and almost all of the special functions but the corresponding theory is not very popular, so we also include Appendix dealing with the basic facts that are scattered in the books about the G- and H-functions.

Most of the known fractional integration and differentiation operators as well as some other generalized integrations, differentiations and differintegrations are included in our scheme as *quite special cases. Examples:* the classical Riemann-Liouville fractional integrals and derivatives; Erdélyi-Kober operators of fractional integration; Hardy-Littlewood-Cesaro integral operators; Džrbashjan-Gelfond-Leontiev generalized integrations and differentiations with respect to Mittag-Leffler functions; Love's hypergeometric fractional integrals involving a Gauss function; the operators of Saigo, Sprinkhuisen-Kuiper, Marichev; Ruscheweyh derivatives; Euler, Poisson, Sonine, Uspensky, Darboux-Weinstein transformations; Delsarte transmutations; the Bessel differential operator and the so-called hyper-Bessel integral and differential operators of arbitrary order $m > 1$ as well as their fractional powers considered by Dimovski and McBride; all the transmutation operators in the theory of hyper-Bessel operators; and a great number of linear integral and differential operators used recently in univalent function theory like the operators of Libera, Bernardi, Biernacki, Komatu, Srivastava, Owa, Saigo, Carlson and Shaffer, Hohlov, etc. Acting as transmutation operators, the generalized fractional integrals generate a family of new G- and H-function integral transformations, generalizing the Laplace transform: the Borel-Džrbashjan transform, Meijer transform, Obrechkoff transform, and their properties follow easily from Laplace transform theory by the method of transmutations.

Various applications of this theory are shown: to differential and integral equations, operational calculi, convolutional calculi, integral transforms, univalent functions, special functions and other branches of analysis. In particular, for the generalized hypergeometric functions, a new classification and approach to them is proposed: each of them is shown to be nothing but a generalized fractional differintegal of one of three basic elementary functions. On this basis, new integral and differential representations of the ${}_pF_q$-functions are found as analogues of the well-known Poisson integral for the Bessel functions and Rodrigues differential formulas for the classical orthogonal polynomials. Also, the solutions of ordinary differential equations of Bessel type and arbitrary order are expressed by means of G-functions. Generalizations of the Abel integral equation and Titchmarsh dual integral equations are solved explicitly by means of the generalized fractional calculus techniques.

Some *open problems and relations to other topics*, like discrete calculus, moduli of smoothness, fractals, statistics, etc. are discussed.

Our aim has been to combine (at long last):

- *the European traditions in the fractional calculus and special functions*, having their origins in Great Britain and Germany (to mention Sir I. Newton and G. Leibnitz and later: Erdélyi, Kober, Wright, Sneddon, Lighthill, Slater, Heading, etc.) and recently developed intensively also in Russia, Byelorussia, Scotland and Bulgaria;
- the *extensive use of the G- and H-functions* of Meijer and Fox, a tool that has been used exclusively by the mathematicians from India;
- the *new trends in the integral transforms and special functions* that have arisen in the USA, Japan, Canada, Venezuela, etc.

Some of *the special features*, the author hopes to have achieved in the present work are:

- it is not a reference book but an exposition of a *self-contained new theory*;
- it is the only book solely devoted to fractional calculus, other than reference books and proceedings, that *uses special functions as kernels*;
- it provides a *uniform approach* to rather different operators, problems and topics via the introduciton of the *common notion of the generalized fractional differintegrals*;
- *includes as special cases* most of the known generalized and classical integrations and differentiations and *many particular examples of applications*;
- together with the *Appendix*, it offers a self-contained exposition.

To record the *contributions of Bulgarian mathematicians* to applied mathematical analysis and other topics related to fractional calculus, we list here these authors separately: Bainov and Shopolov ([26]), Bozhinov ([34]-[35]), Dimkov ([63]), Dimovski ([35],

[64]-[83]); Dryanov ([93]-[95]), Hristova ([139]-[141]), Ignatov and Kaishev ([144]-[145]), Iliev ([146]-[147]), Ivanov ([149]), Kaishev ([54]), Karanikoloff ([184]), Koprinski ([234]), Obrechkoff ([337]-[339]), Pashkuleva ([346]), Popov ([364], [448]), Raichinov ([372]-[383]), Rusev ([413]-[414]), Sendov ([445]-[448]), Shulev ([449]), Spirova ([213]), Takev ([486]) and Vasilev ([83], [500]-[501]).

The main results in the present work have been published in 42 articles by this author (some of them in collaboration): [74]-[82], [118]-[119], [174]-[175], [190]-[214], [320], [390], [438]-[439].

The book consists of this Introduction, five chapters, an Appendix and a list of References. It is intended for graduate and post-graduate students, lecturers and researchers *in applied mathematical analysis* in general, *and particularly* in the areas of fractional calculus, special functions, integral transforms, operational calculus, differential and integral equations, mathematical physics and engineering.

1 Generalized operators of fractional integration and differentiation

1.1. Definitions and examples of the generalized fractional integration operators

1.1.i. Functional spaces

First let us specify the main functional spaces $\mathcal{X}$ where the generalized operators of integration and differentiation of arbitrary fractional multiorder will be considered.

a) For the purposes of the practical applications and unity of the exposition, in Chapter 1 we confine our considerations mainly to spaces of functions which are continuous in the interval $(0, \infty)$ and subject to some growth conditions at zero and infinity. In this way, we follow the pattern of Erdélyi [104]-[105], Dimovski [64]-[71], Saigo [415]-[417], [420], Lowndes [268]-[269] and others.

Let α be an arbitrary real number, k a non negative integer and let $C^{(k)}[0, \infty)$ denote the set of real-valued functions with continuous k-th derivatives in $[0, \infty)$. Denote by $C_\alpha^{(k)}$ the linear space of functions

$$C_\alpha^{(k)} := \left\{ f(x) = x^p \tilde{f}(x); p > \alpha, \tilde{f} \in C^{(k)}[0, \infty) \right\}. \tag{1.1.1}$$

The space

$$\mathcal{X} = C_\alpha := C_\alpha^{(0)} \tag{1.1.1'}$$

is the basic one in which the Riemann-Liouville type generalized fractional integrals will be considered. This definition of the spaces C_α, $C_\alpha^{(k)}$ is due to Dimovski [64]-[71].

When dealing with the Weyl type fractional integrals whose upper limit of integration is infinity, we have to impose some additional conditions on the growth of the functions in a neighbourhood of this point. For instance, we shall consider the linear spaces of functions

$$C_\alpha^{*(k)} = \left\{ f(x) = x^q \hat{f}(x); q < \alpha^*, \hat{f} \in C^{(k)}[0, \infty), |\hat{f}| \le A_{\hat{f}} \right\} \tag{1.1.2}$$

with α^* being a real number, a constant $A_{\hat{f}}$ and $C_{\alpha^*}^* := C_{\alpha^*}^{*(0)}$. The following inclusions are easily verified:

$$C_{\alpha_1}^{(k_1)} \subseteq C_{\alpha_2}^{(k_2)}, \ C_{\alpha_1}^{*(k_1)} \subseteq C_{\alpha_2}^{*(k_2)} \text{ for } \alpha_1 \ge \alpha_2, \ \alpha_1^* \le \alpha_2^*, \ k_1 \ge k_2.$$

b) Fractional integrals and derivatives are often considered also in *Hölder (Lipschitz) classes* $\mathcal{X} = H^\lambda(\Omega)$ of functions satisfying conditions of the form:

$$|f(x_1) - f(x_2)| \le A|x_1 - x_2|^\lambda; \; |f(x_1) - f(x_2)| \le \frac{|x_1 - x_2|^\lambda}{(1 + |x_1|)^\lambda (1 + |x_2|)^\lambda} \tag{1.1.3}$$

in Ω being a finite interval $[a, b]$ or the half-line $(0, \infty)$. Weighted versions of such spaces (with respect to the points $x = a$, $x = b$ or $x = 0$, $x = \infty$) like $\mathcal{X} = H^\lambda((0, \infty), x^\mu)$, $\mathcal{X} = H^\lambda(\Omega, \rho(x))$ are also used. For details see Samko, Kilbas annd Marichev [434, §1], Saigo [485], Kilbas [188], etc.

c) It is most natural however, to consider the Riemann-Liouville, Erdélyi-Kober fractional integrals and their generalizations for functions which are locally integrable in $[0, \infty)$ (i.e. Lebesgue integrable in $[0, \mathcal{X}]$ for every $\mathcal{X} \in [0, \infty)$). Then, appropriate functional spaces are: $\mathcal{X} = L(0, \infty)$ (see e.g. Love [265]); more generaly: $\mathcal{X} = L^p(0, \infty)$, $1 \le p < \infty$ (see e.g. Hardy, Littlewood and Polya [125], Parashar [355], Kalla [163]-[164], etc.); or its weighted versions like: $\mathcal{X} = L_{\mu,p}$ (Rooney [399]-[400], Kalla and Kiryakova [174]-[175]; $\mathcal{X} = F_{p,\mu}$ (McBride [288]-[290], Raina and Saigo [394], Saigo and Glaeske [422]); $\mathcal{X} = L((0, \infty), x^\mu)$ (Love [262], Marichev [274]); $\mathcal{X} = L^p((0, \infty), x^\mu)$ (Samko, Kilbas and Marichev [434]), etc.

For example, we deal (see Chapter 5) with the spaces $L_{\mu,p}, 1 \le p < \infty, \mu \in \mathbb{R}$. This is a brief denotation for *the weighted space $L_{\mu,p}(0, \infty)$ of Lebesgue measurable functions* $f(x)$ defined almost everywhere on $(0, \infty)$ and such that

$$||f||_{\mu,p} = \left[\int_0^\infty x^{\mu-1} |f(x)| dx \right]^{\frac{1}{p}} < \infty. \tag{1.1.4}$$

d) Fractional integrals and differintegrals have been considered also for complex-valued functions $f(z)$ of a complex variable $z \in \mathbb{C}$, e.g. in the space $\mathfrak{H}(\Omega)$ of analytic functions in a domain Ω (or especially in a disk $|z| < R$, $R > 0$), see Hadamard [124], Džrbashjan [101], Nishimoto [324]-[326], Srivastava and Owa [471]-[476], Gelfond and Leontiev [120], Tkachenko [488]-[492], Dimovski [72]-[73], etc.

In particular, we shall consider the following classes of functions:

$$\mathfrak{H}_\mu(\Omega) = \left\{ f(z) = z^\mu \tilde{f}(z); \tilde{f}(z) \in \mathfrak{H}(\Omega) \right\}, \quad \mathfrak{H}_0(\Omega) := \mathfrak{H}(\Omega), \tag{1.1.5}$$

where $\mu \ge 0$ is a real number (usually $\mu = 0, 1, 2, \ldots$) and $\mathfrak{H}(\Omega)$ is the space of functions, holomorphic (analytic and single-valued) in a domain Ω, starlike with respect the origin $z = 0$. See Chapters 2 and 5.

1.1.ii. Definitions

We introduce the following *generalizations of the Riemann-Liouville and Erdélyi-Kober operators* concerning *integration of fractional multiorder* of functions associated with weights of the form $\rho_k(x) = x^{\gamma_k}$, $k = 1, \ldots, m$. The meaning of this characterization will become clear after we establish the composition structure of these operators in Section 1.2. The definitions and the main results of this chapter have appeared first in Kiryakova [194], [196], [201]-[202], [208] and Dimovski and Kiryakova [79].

Definition 1.1.1. Let $m \geq 1$ be an integer, $\beta > 0$, $\gamma_1, \ldots, \gamma_m$ and $\delta_1 > 0, \ldots, \delta_m > 0$ be arbitrary real numbers. Consider the set $\gamma = (\gamma_1, \ldots, \gamma_m)$ as a *multiweight* and $\delta = (\delta_1, \ldots, \delta_m)$ as a positive *multiorder of integration*. For functions $f \in C_\alpha$, $\alpha \geq \max_k[-\beta(\gamma_k + 1)]$, we define the *multiple Erdélyi-Kober (multi-E.-K.) operators* in the following way:

$$I_{\beta,m}^{(\gamma_k),(\delta_k)} f(x) = \int_0^1 G_{m,m}^{m,0} \left[\sigma \left| \begin{matrix} (\gamma_k + \delta_k)_1^m \\ (\delta_k)_1^m \end{matrix} \right. \right] f(x\sigma^{\frac{1}{\beta}})\, d\sigma. \tag{1.1.6}$$

Then, each operator of the form

$$Rf(x) = x^{\beta\delta_0} I_{\beta,m}^{(\gamma_k),(\delta_k)} f(x) \text{ with arbitrary } \delta_0 \geq 0, \tag{1.1.7}$$

is said to be a *generalized (m-tuple) operator of fractional integration of Riemann-Liouville type*, or briefly: a *generalized R.-L. fractional integral.*

Definition 1.1.2. The corresponding *generalized operators of fractional integration of Weyl type* are introduced by the representation

$$Wf(x) = x^{\beta\delta_0} W_{\beta,m}^{(\gamma_k),(\delta_k)} f(x), \quad \delta_0 \geq 0, \tag{1.1.8}$$

where

$$W_{\beta,m}^{(\gamma_k),(\delta_k)} f(x) = \int_1^\infty G_{m,m}^{m,0} \left[\frac{1}{\sigma} \left| \begin{matrix} (\gamma_k + \delta_k + 1)_1^m \\ (\gamma_k + 1)_1^m \end{matrix} \right. \right] f(x\sigma^{\frac{1}{\beta}})\, d\sigma \tag{1.1.9}$$

are the *multi-E.-K. operators of Weyl type* defined for functions $f \in C_{\alpha^*}^*$, $\alpha \leq \min_k(\beta\gamma_k)$.

Note. Other representations of the multi-E.-K. operators (1.1.6), (1.1.9), close to the commonly used forms of the Riemann-Liouville, Weyl, Erdélyi-Kober and other known fractional integrals, are:

$$I_{\beta,m}^{(\gamma_k),(\delta_k)} f(x) = x^{-\beta} \int_0^x G_{m,m}^{m,0} \left[\left(\frac{\tau}{x}\right)^\beta \left| \begin{matrix} (\gamma_k + \delta_k)_1^m \\ (\gamma_k)_1^m \end{matrix} \right. \right] f(\tau)\, d(\tau^\beta), \tag{1.1.6'}$$

$$W_{\beta,m}^{(\gamma_k),(\delta_k)} f(x) = x^{-\beta} \int_x^\infty G_{m,m}^{m,0} \left[\left(\frac{x}{\tau}\right)^\beta \left| \begin{matrix} (\gamma_k + \delta_k + 1)_1^m \\ (\gamma_k + 1)_1^m \end{matrix} \right. \right] f(\tau)\, d(\tau^\beta). \tag{1.1.9'}$$

However, we prefer using the representations of Definitions 1.1.1, 1.1.2 as more convenient, especially in the case of a complex variable $z = x$. Then, the choice of a single-valued branch in (1.1.6) depends on the fixed point $z = 0$ only, while in representation

(1.1.6′) the branch-point $z = x$ is variable. This approach, descending from Hadamard [124], allows us to consider functions defined in starlike complex domains even if they are only integrable there.

Note. One can suitably modify Definitions 1.1.1, 1.1.2 for multiorders of integrations δ consisting also of arbitrary complex components with $\Re\delta_k > 0$, $k = 1, \ldots, m$ as well as for purely imaginary multiorders with $\Re\delta_k = 0$, $k = 1, \ldots, m$. In this case the pattern from the classical fractional calculus should be followed; see Mikolas [302], Love [263], Ross [407], Marichev [275], Samko, Kilbas and Marichev [434, §4], etc.

The *kernel-function* of integral operators (1.1.6), (1.1.9) is a special case of the *Meijer G-function*, suitably chosen for our purposes. In the general case, this special function is defined by means of the contour Mellin-Barnes type integral

$$G_{p,q}^{m,n}\left[\sigma \,\middle|\, \begin{matrix}(a_k)_1^p\\(b_k)_1^q\end{matrix}\right] = G_{p,q}^{m,n}\left[\sigma \,\middle|\, \begin{matrix}a_1,\ldots,a_p\\b_1,\ldots,b_q\end{matrix}\right]$$
$$= \frac{1}{2\pi i}\int\limits_{\mathcal{L}} \frac{\prod_{k=1}^{m}\Gamma(b_k - s)\prod_{j=1}^{n}\Gamma(1 - a_j + s)}{\prod_{k=m+1}^{q}\Gamma(1 - b_k + s)\prod_{j=n+1}^{p}\Gamma(a_j - s)}\sigma^s\, ds. \tag{1.1.10}$$

Here $\sigma \neq 0$ is a complex variable, the integers $0 \leq m \leq q$, $0 \leq n \leq p$ define the order $(m, n; p, q)$ of the G-function, the parameters $(a_j)_1^p$, $(b_k)_1^q$ are such that none of the poles of $\Gamma(b_k - s)$, $k = 1, \ldots, m$ coincide with any of the poles of $\Gamma(1 - a_j + s)$, $j = 1, \ldots, n$ and the infinite contour $\mathcal{L}$ is situated in the complex plane in such a manner that it separates the poles of these two sets of Γ-functions in the numerator of the integrand. Depending on the values of all these parameters, Meijer's G-function is analytic in the whole complex plane, only in the unit disk $|\sigma| < 1$ or outside it. More details about the three possible contours $\mathcal{L} = \mathcal{L}_{\pm\infty}$, $\mathcal{L}_{i\infty}$ in (1.1.10), the historical origin of this function and its excellent (simple but quite general and useful) properties and numerous special cases (among them most of the known special functions and the basic elementary functions) can be found in any of the handbooks of Erdélyi [106, I], Luke [272], Marichev [276], Mathai and Saxena [286], Prudnikov, Brychkov and Marichev [369] etc. *A brief exposition of all the properties of the G-function is proposed in the Appendix* of this book.

Here we mention only some characteristic properties of the chosen subclass of G-functions of order $(m, 0; m, m)$ (with $m = p = q, n = 0$):

$$G_{m,m}^{m,0}(\sigma) = G_{m,m}^{m,0}\left[\sigma \,\middle|\, \begin{matrix}(a_k)_1^m\\(b_k)_1^m\end{matrix}\right] = \frac{1}{2\pi i}\int\limits_{\mathcal{L}}\left(\prod_{k=1}^{m}\frac{\Gamma(b_k - s)}{\Gamma(a_k - s)}\right)\sigma^s\, ds. \tag{1.1.11}$$

For $m = 1$ and $m = 2$ functions (1.1.11) are well-known (see the examples in 1.2.2), namely: they are the kernel-functions of the Erdélyi-Kober operators and of Love's hypergeometric fractional integrals, respectively. For $m > 2$ the function $G_{m,m}^{m,0}$ cannot be identified with any known elementary or special functions (except for a number of quite particular cases). It is interesting that *these kinds of functions have been used by Kabe [153] in statistics as density functions of a random variable.* Naturally, for $m = 1$ and $m = 2$ the "beta-distribution" and the "hypergeometric"-distribution respectively are obtained (see also [468], p. 262-263). It is worth mentioning that in 1982 McBride [289] used the $G_{m,m}^{m,0}$-function as a kernel of the integral representation of the fractional powers of Bessel type operators. After establishing the role of another Meijer's G-function (the $G_{0,m}^{m,0}$-function) in the theory of Bessel type operators, in 1983 Dimovski and Kiryakova [79] also used the $G_{m,m}^{m,0}$-function for these purposes. This was a starting point for the author to choose function (1.1.11) as a kernel of the generalized fractional integrals $I_{\beta,m}^{(\gamma_k)(\delta_k)}$ of a more general nature.

Meijer's $G_{m,m}^{m,0}$-function has *three regular singular points*: $\sigma = 0, \sigma = 1$ and $\sigma = \infty$. In the unit disk and outside it, this function is representable by two different analytic functions, namely:

For $|\sigma| < 1$ this is the uniformly convergent integral (1.1.11), where $\mathcal{L} = \mathcal{L}_{+\infty}$ is a loop starting and ending at infinity and encircling all the poles $s_{k,l} = b_k + l$, $l = 0, 1, 2, \ldots$, $k = 1, 2, \ldots, m$ of the functions $\Gamma(b_k - s)$, once in the negative direction.

For $|\sigma| > 1$, if we choose the contour $\mathcal{L} = \mathcal{L}_{-\infty}$ to be a loop inside which the integrand has no poles, then we obtain that

$$G_{m,m}^{m,0}\left[\sigma \left| \begin{matrix} (a_k)_1^m \\ (b_k)_1^m \end{matrix} \right.\right] \equiv 0, \quad |\sigma| > 1. \tag{1.1.12}$$

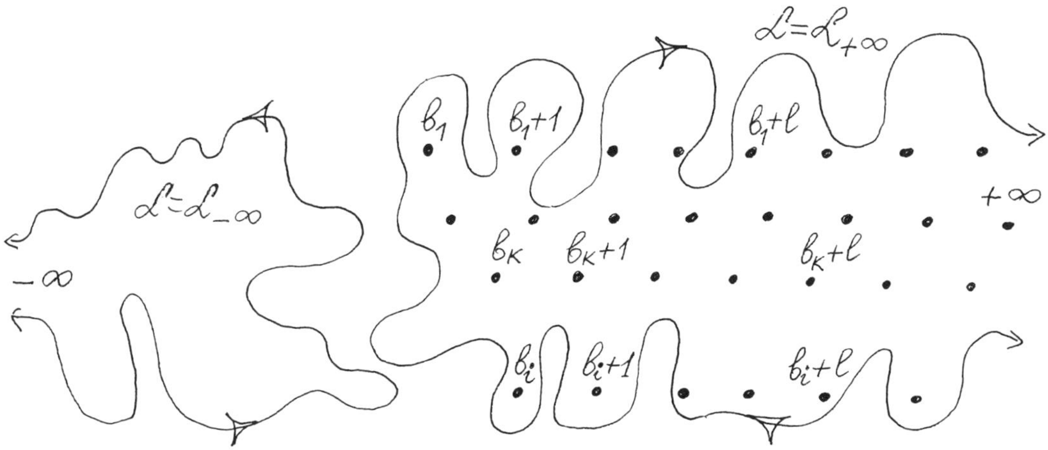

The asymptotic behaviour of the $G_{m,m}^{m,0}$-function near the origin $\sigma = 0$ follows easily by the general result (A.21), viz.:

$$G_{m,m}^{m,0}(\sigma) = \mathcal{O}\left(|\sigma|^{\mu}\right) \text{ as } \sigma \to 0, \text{ where } \mu = \min_{1 \le k \le m} (b_k). \tag{1.1.13}$$

The problem of the asymptotics of the $G_{m,m}^{m,0}$-function (and more generally of the $G_{p,p}^{m,n}$-functions) in a neighbourhood of the singular point $\sigma = 1$ has been obscure until recently (see [106, I, p. 221] : " ... In this case $(-1)^{p-m-n}$ is also a regular singular point, however no fundamental system is known in the literature in a neighbourhood of this point"). In 1981 Marichev [280] proposed asymptotic formulas for $G_{m,m}^{m,0}(\sigma)$ as

$\sigma \to 1$, in a few variants, depending on the values of $\nu_m^* = -\nu - 1 = \sum\limits_{k=1}^{m} (a_k - b_k) - 1$. For example if $\nu_m^* \neq 0, \pm 1, \pm 2, \dots$, then

$$G_{m,m}^{m,0}\left[\sigma \left| \begin{matrix} (a_k)_1^m \\ (b_k)_1^m \end{matrix} \right.\right] \sim \frac{(1-\sigma)^{\nu_m^*}}{\Gamma(\nu_m^* + 1)} \quad \text{as } \sigma \to 1, \quad \sigma < 1. \tag{1.1.14}$$

For $\nu_m^* > 0$ this function is continuous at the point $\sigma = 1$. In our case $a_k = \gamma_k + \delta_k$, $b_k = \gamma_k$ and for $\delta_k > 0$, $k = 1, \dots, m$:

$$-\nu = \sum_{k=1}^{m} \delta_k > 0, \qquad \text{therefore } \nu_m^* = \sum_{k=1}^{m} \delta_k - 1 > -1, \tag{1.1.15}$$

i.e. $\nu_m^* \neq -1, -2, \dots$. If $\sum\limits_{k=1}^{m} \delta_k \neq 1, 2, \dots$, then $\nu_m^* \neq 0, 1, 2, \dots$ and (1.1.14) is valid. Otherwise, if $\sum\limits_{k-1}^{m} \delta_k$ is a natural number, then the function $G_{m,m}^{m,0}(\sigma)$ has a logarithmic singularity at the point $\sigma = 1$. For more details see Marichev [280, p. 390].

In this book *attention is paid mainly* to the study and applications of *the Riemann-Liouville type generalized fractional integrals*. The corresponding results for Weyl type counterparts (1.1.9) are mentioned in Section 1.4 and used in some applications.

1.1.iii. Examples

We consider separately the cases $m = 1, 2$ (for which many generalized integration and differentiation operators introduced and used by various authors are included as special cases) and $m > 2$ (for which operators (1.1.7), (1.1.8) are comparatively unknown and used only in their alternative, repeated integrals representations; see further (1.2.27), (1.2.31)).

i) $m = 1$. The kernel-functions of the generalized fractional integrals are the simple elementary functions (see [286, p. 37] or [369, p. 631, (3)–(4)]):

$$G_{1,1}^{1,0}\left[\sigma \left| \begin{matrix} \gamma + \delta \\ \gamma \end{matrix} \right.\right] = \begin{cases} \frac{(1-\sigma)^{\delta-1}\sigma^{\gamma}}{\Gamma(\delta)}, & 0 < \sigma < 1, \\ 0, & \sigma > 1, \end{cases} \tag{1.1.16}$$

analogously,

$$G_{1,1}^{1,0}\left[\frac{1}{\sigma} \left| \begin{matrix} \gamma + \delta + 1 \\ \gamma + 1 \end{matrix} \right.\right] = G_{1,1}^{0,1}\left[\sigma \left| \begin{matrix} -\gamma \\ -\gamma - \delta \end{matrix} \right.\right] = \begin{cases} \frac{(\sigma-1)^{\delta-1}}{\Gamma(\delta)}\sigma^{-(\gamma+\delta)}, & \sigma > 1 \\ 0, & 0 < \sigma < 1. \end{cases} \tag{1.1.16*}$$

Therefore, for arbitrary $\beta > 0$, γ and $\delta > 0$ *the generalized fractional integrals* $I_{\beta,1}^{\gamma,\delta}$ (1.1.6) *and* $W_{\beta,1}^{\gamma,\delta}$ (1.1.9) *coincide with the well-known Erdélyi-Kober fractional integrals* (see [404, p. 55], [452], [454]-[455], [164], [289], etc.), namely:

$$I_\beta^{\gamma,\delta} f(x) = \int_0^1 \frac{(1-\sigma)^{\delta-1}\sigma^\gamma}{\Gamma(\delta)} f(x\sigma^{\frac{1}{\beta}})\,d\sigma$$
$$= x^{-\beta(\gamma+\delta)} \int_0^x \frac{(x^\beta-\tau^\beta)^{\delta-1}}{\Gamma(\delta)} \tau^{\beta\gamma} f(\tau)\,d(\tau^\beta) = I_{\beta,1}^{\gamma,\delta} \tag{1.1.17}$$

and

$$K_\beta^{\gamma,\delta} f(x) = \int_1^\infty \frac{(\sigma-1)^{\delta-1}\sigma^{-(\gamma+\delta)}}{\Gamma(\delta)} f(x\sigma^{\frac{1}{\beta}})\,d\sigma$$
$$= x^{\beta\gamma} \int_x^\infty \frac{(\tau^\beta-x^\beta)^{\delta-1}}{\Gamma(\delta)} \tau^{-\beta(\gamma+\delta)} f(\tau)\,d(\tau^\beta) = W_{\beta,1}^{\gamma,\delta} f(x). \tag{1.1.17*}$$

Here the following *particular cases* are listed:

a) *Ordinary n-fold integrations* (integer $n > 0$):

$$l^n f(x) = \int_0^x dx_1 \int_0^{x_1} dx_2 \ldots \int_0^{x_{n-1}} f(x_n)\,dx_n$$
$$= \frac{1}{(n-1)!} \int_0^x (x-\tau)^{n-1} f(\tau)\,d\tau = x^n I_{1,1}^{0,n} f(x) \tag{1.1.a}$$

and

$$l_*^n f(x) = \frac{1}{(n-1)!} \int_x^\infty (\tau-x)^{n-1} f(\tau)d\tau = x^n W_{1,1}^{-n,n} f(x). \tag{1.1.a*}$$

b) *The Riemann-Liouville (R.-L.) fractional integral of order $\delta > 0$*:

$$R^\delta f(x) = \frac{1}{\Gamma(\delta)} \int_0^x (x-\tau)^{\delta-1} f(\tau)\,d\tau = x^\delta I_{1,1}^{0,\delta} f(x) \tag{1.1.b}$$

and the *Weyl fractional integral* (actually introduced by Liouville in 1834):

$$W^\delta f(x) = \frac{x^\delta}{\Gamma(\delta)} \int_x^\infty (\tau-x)^{\delta-1} f(\tau)d\tau = x^\delta W_{1,1}^{-\delta,\delta} f(x), \tag{1.1.b*}$$

including the operators of a) for integer $\delta = n > 0$.

c) *The Hardy-Littlewood (Cesaro) integration operator* (see e.g. [240])

$$L_{1,0}f(x) = \frac{1}{x}\int_0^x f(\tau)\,d\tau = I_{1,1}^{0,1}f(x)$$

and its generalization for integers m, n; $n > m-1$ (for its commutant see Hristova [139]-[141], also Section 2.3):

$$L_{m,n}f(x) = x^{-m}\int_0^x \tau^n f(\tau)\,d\tau = x^{n-m+1}I_{1,1}^{n,1}f(x). \tag{1.1.c}$$

d) The operators, usually referred to as *Erdélyi-Kober operators* $I_{\gamma,\delta} = I_2^{\gamma,\delta}$, $K_{\gamma,\delta} = K_2^{\gamma,\delta}$:

$$I_{\gamma,\delta}f(x) = \frac{2x^{-2(\gamma+\delta)}}{\Gamma(\delta)}\int_0^x (x^2-\tau^2)^{\delta-1}\tau^{2\gamma+1}f(\tau)d\tau = I_{2,1}^{\gamma,\delta}f(x), \tag{1.1.d}$$

$$K_{\gamma,\delta}f(x) = \frac{2x^{2\gamma}}{\Gamma(\delta)}\int_x^\infty (\tau^2-x^2)\tau^{-2(\gamma+\delta)+1}f(\tau)d\tau = W_{2,1}^{\gamma,\delta}f(x), \tag{1.1.d*}$$

introduced by Sneddon [452], [454]-[455]. They follow from (1.1.17), (1.1.17*) for $\beta = 2$, while those originally considered by Kober [220] and Erdélyi [104] have $\beta = 1$.

e) *The Uspensky integral transform* ([499])

$$P^{(\alpha)}f(z) = \frac{1}{\Gamma(\alpha+\frac{1}{2})}\int_0^1 (1-\sigma^2)^{\alpha-\frac{1}{2}}f(z\sigma)d\sigma = \frac{1}{2}I_{2,1}^{-\frac{1}{2},\alpha+\frac{1}{2}}f(z), \tag{1.1.e}$$

used successfully by Rusev [413], [414] in solving the problem of the representation of analytic functions by series in Laguerre polynomials $L_n^\alpha(z)$, arbitrary real $\alpha \neq -1, -2, \ldots$, by referring to the corresponding results for the Hermite polynomials. To this end, the *transmutation formula* Uspensky([499]) is used:

$$L_n^{(\alpha)}\left(z^2\right) = \frac{2(-1)^n\Gamma\left(n+\alpha+1\right)}{\sqrt{\pi}(2n)!\Gamma\left(\alpha+\frac{1}{2}\right)}\int_0^1 \left(1-\sigma^2\right)^{\alpha-\frac{1}{2}}H_{2n}(z\sigma)d\sigma,\ n = 0, 1, 2, \ldots .$$

f) *An example of a Weyl type operator* $W_{\beta,1}^{\gamma,\delta}$ is the transmutation operator considered for functions, analytic in a starlike neighbourhood of infinity $z = \infty$:

$$Tf(z^3) = z\int_1^\infty \frac{(\sigma-1)^{-\frac{1}{6}}}{\Gamma(\frac{5}{6})}\sigma^{-\frac{1}{3}}f(z\sigma^{\frac{1}{3}})d\sigma = zW_{3,1}^{-\frac{1}{3},\frac{5}{6}}f(z). \tag{1.1.f}$$

This operator is a counterpart of the Riemann-Liouville type generalized Sonine transformation corresponding to the Bessel type operator $B = \frac{1}{z}\frac{d^2}{dz^2}$, proposed by Dimovski [44], [52]. The operator (1.1.f) is used in Kiryakova [191] to give another *explanation of the Stokes phenomenon for the Airy equation* $\frac{d^2u}{dz^2} - zu = 0$. For details see Section 3.7.iii.

g) A *generalized differential operator* D_α was used by *Iliev* [146], [147, p. 41-42] in *the theory of Laguerre entire functions.* For power series $f(z) = \sum_{k=0}^{\infty} a_k z^k$ (i.e. $f \in \mathfrak{H}(\Delta_R : |z| < R)$) this operator D_α, $\alpha > 0$ is defined by

$$D_\alpha f(z) = \sum_{k=1}^{\infty} a_k \frac{\Gamma(\alpha(k+1))}{\Gamma(\alpha k)} z^{k-1}, \quad D_\alpha z^0 = 0. \tag{1.1.g$'$}$$

Its right inverse operator L_α, considered by the author in [77], [196] is a generalized fractional integral (see Section 2.2), viz.

$$\begin{aligned} L_\alpha f(z) &= \sum_{k=0}^{\infty} a_k \frac{\Gamma(\alpha(k+1))}{\Gamma(\alpha(k+2))} z^{k+1} \\ &= z \int_0^1 \frac{(1-\sigma)^{\alpha-1}}{\Gamma(\alpha)} \sigma^{\alpha-1} f(z\sigma^\alpha)\, d\sigma = z I_{\frac{1}{\alpha},1}^{\alpha-1,\alpha} f(z). \end{aligned} \tag{1.1.g}$$

Here the integral representation is an analytical extension of the series to the larger space $\mathfrak{H}_{-1}(\Omega)$ of functions $f(z) = z^p \tilde{f}(z)$, $p > -1$ with $\tilde{f}(z)$ analytic in a starlike domain $\Omega \ni z$. The integer powers of L_α can also be represented in terms of (1.1.7):

$$L_\alpha^{(k)} f(z) = z^k I_{\frac{1}{\alpha},1}^{\alpha-1,k\alpha} f(z), \quad k = 1, 2, \dots .$$

h) An operator of the form (1.1.7), more general than (1.1.g), is the so-called *generalized Gelfond-Leontiev integration with respect to the Mittag-Leffler function* $E_\rho(z;\mu)$ (Džrbashjan-Gelfond-Leontiev integration, see Section 2.2.). For $\rho > 0$, $\mu \in \mathbb{C}$, $\Re\mu > 0$ and for power series $f(z) = \sum_{k=0}^{\infty} a_k z^k$ this operator acts in the following way:

$$l_{\rho,\mu} f(z) = \sum_{k=0}^{\infty} a_k \frac{\Gamma\left(\mu + \frac{k}{\rho}\right)}{\Gamma\left(\mu + \frac{k+1}{\rho}\right)} z^{k+1} \tag{1.1.h$'$}$$

and for functions analytic in a domain Ω starlike with respect to the origin, its extension is the generalized fractional integral

$$l_{\rho,\mu} f(z) = \frac{z}{\Gamma\left(\frac{1}{\rho}\right)} \int_0^1 (1-\sigma)^{\frac{1}{\rho}-1} \sigma^{\mu-1} f\left(z\sigma^{\frac{1}{\rho}}\right) d\sigma = z I_{\rho,1}^{\mu-1,\frac{1}{\rho}} f(z). \tag{1.1.h}$$

For $\mu = 1$ some aspects of these operators are investigated by Tkachenko [488], [491] and by Dimovski [72], [73] who proposes convolutional representations of the operators commuting with $l_{\rho,1}$ or with its integer powers $l_{\rho,1}^k$. For real $\mu > 1$ results for operators $l_{\rho,\mu}$, like convolutions, transmutation operators, corresponding integral transforms (Borel-Džrbashjan transforms) etc. are discussed by Dimovski and Kiryakova in [77], [78] and Kiryakova [196], [206]. Some extensions for complex values of μ are made by Linchouk [258]. More details on this topic are given in Section 2.2.

Other points of view and variants of single (classical) fractional integrals and derivatives, related results and applications can be seen also in: Al-Bassam [7]-[21], Al-Bassam and Kala [22], Bora and Saxena [35], Love [264]-[266], Martic [285], Nahoushev [312]-[314], Nahoushev and Salathidinov [315], Ross and Sachdeva [410], Shulev [449], etc.

ii) m=2. In this case the kernel function of (1.1.6) can be represented in terms of the *Gauss hypergeometric function* ${}_2F_1$ (see [286, p. 64] and [369, p.718]):

$$G_{2,2}^{2,0}\left[\sigma \,\middle|\, \begin{matrix}\gamma_1+\delta_+1, \gamma_2+\delta_2\\ \gamma_1, \gamma_2\end{matrix}\right] = \begin{cases} \frac{\sigma^{\gamma_2}(1-\sigma)^{\delta_1+\delta_2-1}}{\Gamma(\delta_1+\delta_2)}\ {}_2F_1(\gamma_2+\delta_2-\gamma_1, \delta_1; \delta_1+\delta_2; 1-\sigma) & \text{for } \sigma < 1, \\ 0 & \text{for } \sigma > 1. \end{cases} \tag{1.1.18}$$

Therefore, the operators $I_{\beta,2}^{(\gamma_k),(\delta_k)}$ are the so-called *"hypergeometric fractional integrals"*:

$$\begin{aligned} Hf(x) &= I_{\beta,2}^{(\gamma_1,\gamma_2),(\delta_1,\delta_2)} f(x) \\ &= \int_0^1 \frac{\sigma^{\gamma_2}(1-\sigma)^{\delta_1+\delta_2-1}}{\Gamma(\delta_1+\delta_2)}\ {}_2F_1(\gamma_2+\delta_2-\gamma_1, \delta_1; \delta_1+\delta_2; 1-\sigma) f\left(x\sigma^{\frac{1}{\beta}}\right) d\sigma, \end{aligned} \tag{1.1.19}$$

introduced first by Love [262] and considered in different modifications by Tricomi [493], Kalla and Saxena [177]-[178], Sprinkhuizen-Kuiper [457], Koornwinder [233], Saigo [415]-[418], [421], McBride [289], Saigo and Glaeske [422], Saigo and Raina [424] and other authors. For more particular choices of the parameters (γ_1, γ_2), (δ_1, δ_2) fractional integration operators involving confluent hypergeometric functions and Bessel functions (see Lowndes [268]-[269]) as kernels can be derived, among them operators involving orthogonal polynomials and even elementary functions (see for instance the extensive table of examples of $G_{2,2}^{2,0}$-functions in [369, p. 630-727]).

Here we mention only a few examples.

j) The simplest "two-tuple fractional integrals" are: the *two-dimensional fractional Riemann-Liouville integral*

$$R^{\alpha,\beta} f(x) = \int_0^1\int_0^1 \frac{(1-\sigma_1)^{\alpha-1}(1-\sigma_2)^{\beta-1}}{\Gamma(\alpha)\Gamma(\beta)} f(x\sigma_1\sigma_2) d\sigma_1 d\sigma_2 = I_{1,2}^{(0,0),(\alpha,\beta)} f(x) \tag{1.1.j}$$

and the *two-dimensional Weyl integral* (cf. [387]-[390]):

$$W^{\alpha,\beta}f(x)=\int_1^\infty\int_1^\infty\frac{(\sigma_1-1)^{\alpha-1}(\sigma_2-1)^{\beta-1}}{\Gamma(\alpha)\Gamma(\beta)}f(x\sigma_1\sigma_2)d\sigma_1d\sigma_2=W_{1,2}^{(-\alpha,-\beta),(\alpha,\beta)}f(x). \tag{1.1.j'}$$

The reason why (1.1.j), (1.1.j') are referred to as fractional integrals in the sense of Definitions 1.1.1, 1.1.2 *will become clear after obtaining the decomposition theorem in Section 1.2.* (the same holds for the other examples of repeated integrals referred to below, e.g. in Case iii).

k) For the *second-order Bessel type differential operator* $B=x^{\alpha_0}\frac{d}{dx}x^{\alpha_1}\frac{d}{dx}x^{\alpha_2}$ with $\beta=2-\sum_0^2\alpha_k>0$ McBride [289], [291] proposed integral representations of its *fractional powers* $B^{-\alpha},\alpha>0$ (more exactly, they are positive powers of the right inverse integral operator L, see also example o)):

$$\begin{aligned}B^{-\alpha}f(x)=&\frac{\beta^{-2\alpha}x^{-\beta\gamma_1}}{\Gamma(2\alpha)}\int_0^x\left(x^\beta-\tau^\beta\right)^{2\alpha-1}\\&\times{}_2F_1\left(\gamma_2-\gamma_1+\alpha,\alpha;2\alpha;1-\frac{x^\beta}{\tau^\beta}\right)f(\tau)d\left(\tau^\beta\right),\end{aligned}\tag{1.1.k}$$

where $\gamma_1=\frac{1}{\beta}(\alpha_1+\alpha_2-1)$, $\gamma_2=\frac{1}{\beta}\alpha_2$. This means that the operators $B^{-\alpha}$, $\alpha>0$ are generalized fractional integrals of the form (1.1.19):

$$B^{-\alpha}=L^\alpha=\left(\frac{x^\beta}{\beta^2}\right)^\alpha I_{\beta,2}^{\left(\frac{\alpha_1+\alpha_2-1}{\beta},\frac{\alpha_2}{\beta}\right),(\alpha,\alpha)}f(x),\alpha>0.$$

It will be seen further (see Sections 1.6 and 3.1) that the positive powers of B themselves are generalized fractional derivatives. For the more general Bessel type operators of arbitrary order $m\geq 1$ see example o) below.

l) In a series of papers [415]-[418] Saigo introduced the *following pair of operators of fractional integration* :

$$I^{\alpha,\beta,\eta}f(x)=\frac{x^{-(\alpha+\beta)}}{\Gamma(\alpha)}\int_0^x(x-\tau)^{\alpha-1}\,{}_2F_1\left(\alpha+\beta,-\eta;\alpha;1-\frac{\tau}{x}\right)f(\tau)d\tau \tag{1.1.l}$$

and

$$J^{\alpha,\beta,\eta}f(x)=\frac{1}{\Gamma(\alpha)}\int_0^x(\tau-x)^{\alpha-1}\tau^{-(\alpha+\beta)}\,{}_2F_1\left(\alpha+\beta,-\eta;\alpha;1-\frac{\tau}{x}\right)f(\tau)d\tau \tag{1.1.l*}$$

for $\alpha>0$, real β,η and for functions of the spaces (1.1.1), (1.1.2) respectively Written in terms of the notations $I_{\beta,m}^{(\gamma_k)(\delta_k)}$, $W_{\beta,m}^{(\gamma_k)(\delta_k)}$ they look as follows:

$$I^{\alpha,\beta,\eta} = x^{-\beta} I^{(\eta-\beta,0),(-\eta,\alpha+\eta)};$$
$$J^{\alpha,\beta,\eta} = x^{-\beta} W_{1,2}^{(\eta,-2\alpha-\beta-2\eta),(-\eta,\alpha+\eta)}. \qquad (1.1.1')$$

Saigo used these operators ([419]-[421]) in solving *boundary value problems for the Euler-Darboux equation.* A modification of operator (1.1.1), *defined by means of the Hadamard product of series* ($\circ$):

$$F(a,b,c)f(x) = \{x\,{}_2F_1(a,b;c;x)\} \circ f(x) = \frac{\Gamma(c)}{\Gamma(a)\Gamma(b)} I_{1,2}^{(a-2,b-2),(1-a,c-b)} \qquad (1.1.1'')$$

is considered by Hohlov [135]-[136] for the purposes of the theory of univalent functions. *Saigo's operators* (1.1.1-1′) are used from the same point of view in [418].

iii) m > 2. In this case operators of the form (1.1.7), (1.1.8) are considered less often and are used mainly in their alternative representations (1.2.27) without using the G-function. Nevertheless, let us list here some examples:

m) The simplest example is the operator l^n of *n-fold integration* (1.1.a), this time *considered as an n-tuple fractional integral.* To this end, it is necessary to use the following corollary

$$\frac{(1-t)^{n-1}}{(n-1)!} = G_{1,1}^{1,0}\left[t\,\middle|\,\begin{matrix}n\\0\end{matrix}\right] = n^{1-n} G_{n,n}^{n,0}\left[t^n\,\middle|\,\begin{matrix}\left(\frac{k-1}{n}+1\right)_1^n\\ \left(\frac{k-1}{n}\right)_1^n\end{matrix}\right]$$
$$= \left(\frac{t}{n}\right)^{n-1} G_{n,n}^{n,0}\left[t^n\,\middle|\,\begin{matrix}\left(\frac{k}{n}\right)_1^n\\ \left(\frac{k}{n}-1\right)_1^n\end{matrix}\right]$$

of a more general formula (see [272, I, p. 150, (5)] or [286, p. 6, (1.2.5)]). So, l^n *can be written also in the form*

$$l^n f(x) = x^n \int_0^1 \frac{(1-t)^{n-1}}{(n-1)!} f(xt)dt = \frac{x^n}{n^{n-1}} \int_0^1 t^{n-1} G_{n,n}^{n,0}\left[t^n\,\middle|\,\begin{matrix}\left(\frac{k}{n}\right)_1^n\\ \left(\frac{k}{n}-1\right)_1^n\end{matrix}\right] f(xt)dt$$
$$= \left(\frac{x}{n}\right)^n \int_0^1 G_{n,n}^{n,0}\left[\sigma\,\middle|\,\begin{matrix}\left(\frac{k}{n}\right)_1^n\\ \left(\frac{k}{n}-1\right)_1^n\end{matrix}\right] f\left(x\sigma^{\frac{1}{n}}\right) d\sigma = \left(\frac{x}{n}\right)^n I_{n,n}^{\left(\frac{k}{n}-1\right)_1^n,(1)_1^n} f(x). \qquad (1.1.\mathrm{m})$$

n) In [274], [434] Marichev considers an integral operator with the *Horn F_3-function* as kernel ([106], I, p. 223) and $\Re c > 0$:

$$Ff(x) = \int_0^x \frac{(x-\tau)^{c-1}}{\Gamma(c)} F_3(a,a',b,b',1-\frac{x}{\tau},1-\frac{\tau}{x}) f(\tau)d\tau \qquad (1.1.\mathrm{n})$$

for Lebesgue integrable functions of the form (1.1.4). Using the representation [369,

p. 727, (2)]:

$$\frac{(1-\sigma)^{c-1}}{\Gamma(c)} F_3\left(a,a',b,b',c,1-\frac{1}{\sigma},1-\sigma\right) = \frac{(1-\sigma)^{c-1}}{\Gamma(c)} F_3\left(a,a',b,b',c,1-\sigma,1-\frac{1}{\sigma}\right)$$
$$= G_{3,3}^{3,0}\left[\sigma \left| \begin{matrix} a+b, c-a', c-b' \\ a, b, c-a'-b' \end{matrix} \right.\right], \quad \Re c > 0,$$

we can rewrite $Ff(x)$ in the form of a "3-multiple" generalized fractional integral

$$\begin{aligned} Ff(x) &= x^c \int_0^1 \frac{(1-\sigma)^{c-1}}{\Gamma(c)} F_3\left(a,a',b,b',c,1-\frac{1}{\sigma},1-\sigma\right) f(x) d\sigma \\ &= x^c \int_0^1 G_{3,3}^{3,0}\left[\sigma \left| \begin{matrix} a+b, c-a', c-b' \\ a, b, c-a'-b' \end{matrix} \right.\right] f(x\sigma) d\sigma \\ &= x^c I_{1,3}^{(a,b,c-a'-b'),(b,c-a'-b,a')} f(x). \end{aligned} \tag{1.1.n'}$$

Further (see Section 1.2), we establish that this generalized fractional integral can be represented *as a composition of three commuting Erdélyi-Kober operators*, for instance

$$Ff(x) = x^c I_1^{1,b} I_1^{b,c-a'-b} I_1^{c-a'-b',a'} = \ldots \quad , \tag{1.1.n''}$$

or in terms of Riemann-Liouville operators (see also [434], p. 158, (10.46)):

$$Ff(x) = x^{b'} R^{a'} x^{-b'} R^{c-a'-b} x^{-a} R^b x^a. \tag{1.1.n'''}$$

o) In a series of papers [64]-[71] Dimovski considered the *hyper-Bessel integral operator*

$$Lf(x) = \frac{x^\beta}{\beta m} \int_0^1 \ldots \int_0^1 \left[\prod_{k=1}^m t_k^{\gamma_k}\right] f\left[x(t_1,\ldots,t_m)^{\frac{1}{\beta}}\right] dt_1 \ldots dt_m \tag{1.1.o}$$

as a linear right inverse operator ($BL = I$) of the *most general differential operator* B *of Bessel type* and arbitrary order $m \geq 1$:

$$B = x^{-\beta} \prod_{k=1}^m \left(x\frac{d}{dx} + \beta\gamma_k\right) = x^{\alpha_0} \frac{d}{dx} x^{\alpha_1} \ldots \frac{d}{dx} x^{\alpha_m}, \quad \beta > 0.$$

The operator L can be rewritten in the form (see Chapter 3)

$$Lf(x) = \frac{x^\beta}{\beta m} \int_0^1 G_{m,m}^{m,0}\left[\sigma \left| \begin{matrix} (\gamma_k+1)_1^m \\ (\gamma_k)_1^m \end{matrix} \right.\right] f\left(x\sigma^{\frac{1}{\beta}}\right) d\sigma = \frac{x^\beta}{\beta m} I_{\beta,m}^{(\gamma_k),(1)} f(x), \tag{1.1.o'}$$

and therefore it is a generalized fractional integral of arbitrary *multiplicity* $m \geq 1$ with arbitrary *multiweight* $\gamma = (\gamma_1,\ldots,\gamma_m)$ and *multiorder* $\delta = (1,1,\ldots,1)$. Chapter 3 is devoted entirely to the generalized m-tuple fractional derivatives and integrals B and L, called *hyper-Bessel operators*. It is shown that the *transmutation operators* related to them (the so-called *Poisson-Sonine-Dimovski transformations*) are also generalized fractional differintegrals, representable by means of the Meijer's G-function.

p) As will be seen in the next Section 1.2, every composition of m arbitrary Erdélyi-Kober operators with the same $\beta > 0$:

$$I_\beta^{\gamma_1,\delta_1}\left\{I_\beta^{\gamma_2,\delta_2}\ \dots\ \left(I_\beta^{\gamma_m,\delta_m} f(x)\right)\right\}$$
$$= \int_0^1 \dots \int_0^1 \left[\prod_{k=1}^m \frac{(1-t_k)^{\delta_k-1} t_k^{\gamma_k}}{\Gamma(\delta_k)}\right] f\left[x(t_1\dots t_m)^{\frac{1}{\beta}}\right] dt_1 \dots dt_m \tag{1.1.p}$$

has also the $I_{\beta,m}^{(\gamma_k),(\delta_k)}$-representation (1.1.6):

$$I_\beta^{\gamma_1,\delta_1}\left\{I_\beta^{\gamma_2,\delta_2}\ \dots\ \left(I_\beta^{\gamma_m,\delta_m} f(x)\right)\right\} = I_{\beta,m}^{(\gamma_k),(\delta_k)} f(x)$$
$$= \int_0^1 G_{m,m}^{m,0}\left[\sigma \left| \begin{matrix} (\gamma_k+\delta_k)_1^m \\ (\gamma_k)_1^m \end{matrix} \right.\right] f\left(x\sigma^{\frac{1}{\beta}}\right) d\sigma, \quad \forall f \in C_\alpha,\ \alpha \geq \max[-\beta(\gamma_k+1)] \tag{1.1.p$'$}$$

and *this fact is the key for many useful applications of our generalized fractional calculus.* Due to the simple properties of the G-functions and of the $I_{\beta,m}^{(\gamma_k),(\delta_k)}$-operators, we are able to use more effectively the operators

$$I_{\beta,m}^{(\gamma_k),(\delta_k)} \stackrel{\text{def}}{=} \prod_{k=1}^m I_\beta^{\gamma_k,\delta_k}.$$

Most of the results in Chapters 3, 4 and 5 are corollaries of this property which can be considered either as a *composition theorem* for $I_{\beta,m}^{(\gamma_k),(\delta_k)}$-operators (1.1.p), or as a *decomposition theorem* with respect to $I_{\beta,m}^{(\gamma_k),(\delta_k)}$-operators (1.1.p$'$).

1.2. The $\mathbf{I}_{\beta,m}^{(\gamma_k),(\delta_k)}$–operators as integral transforms of convolutional type and as compositions of Erdélyi-Kober operators

Lemma 1.2.1. *The multiple Erdélyi-Kober operators (1.1.6) are well defined in the space* C_α *with* $\alpha \geq \max_k[-\beta(\gamma_k+1)]$ *and preserve there (up to a constant multiplier) the power functions :*

$$I_{\beta,m}^{(\gamma_k),(\delta_k)}\{x^p\} = c_p x^p, \ \textit{where}\ c_p = \prod_{k=1}^m \frac{\Gamma(\gamma_k+\frac{p}{\beta}+1)}{\Gamma(\gamma_k+\delta_k+\frac{p}{\beta}+1)},\ p > \alpha. \tag{1.2.1}$$

Proof. First we verify the correctness of Definition 1.1.1. Let $f(x) = x^p \tilde{f}(x)$ be a function of C_α, i.e. $p > \alpha$, $\tilde{f} \in C[0, \infty)$ which yields the boundedness of $\tilde{f}$ in every finite interval $[0, X]$, $X > 0$. Due to property (A.14) of the G-function, we obtain

$$\begin{aligned} I_{\beta,m}^{(\gamma_k),(\delta_k)} f(x) &= \int_0^1 G_{m,m}^{m,0} \left[\sigma \left| \begin{matrix} (\gamma_k + \delta_k)_1^m \\ (\gamma_k)_1^m \end{matrix} \right. \right] x^p \sigma^{\frac{p}{\beta}} \tilde{f}(x\sigma^{\frac{1}{\beta}})\, d\sigma \\ &= x^p \int_0^1 G_{m,m}^{m,0} \left[\sigma \left| \begin{matrix} (\gamma_k + \delta_k + \frac{p}{\beta})_1^m \\ (\gamma_k + \frac{p}{\beta})_1^m \end{matrix} \right. \right] \tilde{f}(x\sigma^{\frac{1}{\beta}})\, d\sigma. \end{aligned} \tag{1.2.2}$$

According to (1.1.13), $G_{m,m}^{m,0}(\sigma) = \mathcal{O}(\sigma^\mu)$ as $\sigma \to 0$, where $\mu = \min_k (\gamma_k + p/\beta)$. Since $p/\beta > \alpha/\beta$ and $\alpha \geq \max_k [-\beta(\gamma_k + 1)]$ yield $\gamma_k \geq -1 - \frac{\alpha}{\beta}$ for $k = 1, \ldots, m$, we obtain that $\mu > -1$. On the other hand, $\tilde{f}(x\sigma^{1/\beta}) = \mathcal{O}(1)$ as $\sigma \to 0$. Therefore, the integrand of (1.2.2) is $\mathcal{O}(\sigma^\mu)$, $\mu > -1$ as $\sigma \to 0$.

It remains to investigate the asymptotic behaviour of the term $G_{m,m}^{m,0}[\sigma]\tilde{f}(x\sigma^{1/\beta})$ near the other singular point $\sigma = 1$. This is determined by the behaviour of the G-function there. Let us make use of Marichev's results [280] (see also (1.1.14)):

$$G_{m,m}^{m,0}(\sigma) \sim \frac{(1-\sigma)^{\nu_m^*}}{\Gamma(\nu_m^* + 1)}, \qquad \text{as } \sigma \to 1,\ \sigma < 1,$$

for $\nu_m^* = \left(\sum_{k=1}^m \delta_k\right) - 1 \neq 0, \pm 1, \pm 2, \ldots$ Here $\sum_{k=1}^m \delta_k > 0$ and therefore $\nu_m^* > -1$, i.e. $\nu_m^* \neq -1, -2, \ldots$ is fulfilled. In the case that $\sum_{k=1}^m \delta_k = 1, 2, 3, \ldots$ and $\nu_m^* = 0, 1, 2, \ldots$, in the above formula a logarithmic term appears (see [280, p. 390]) which does not weaken the convergence of the integral. So, the conditions

$$\gamma_k \geq -1 - \frac{\alpha}{\beta}, \qquad \delta_k > 0,\ k = 1, \ldots, m, \tag{1.2.3}$$

ensuring that $\mu > -1$, $\nu_m^* > -1$ turn out to be sufficient for the absolute convergence of the improper integral (1.2.2) in C_α and for the correctness of Definition 1.1.1.

It is interesting to evaluate the image of the function $f(x) = x^p$, $p > \alpha$. Equality (1.2.2) implies that

$$I_{\beta,m}^{(\gamma_k),(\delta_k)} x^p = x^p \int_0^1 G_{m,m}^{m,0} \left[\sigma \left| \begin{matrix} (\gamma_k + \delta_k + p/\beta)_1^m \\ (\gamma_k + p/\beta)_1^m \end{matrix} \right. \right] d\sigma = x^p c_p.$$

The value c_p of the integral above can be easily found according to Lemma B.2 (see Appendix, (B.4)) and gives (1.2.1).

Note. For $p > \alpha$ of the form

$$p = \beta[-(\gamma_k + \delta_k + 1) - j] \text{ with some integers } j = 0, 1, 2, \ \ldots\ , \qquad k = 1, \ldots, m,$$

the coefficient $c_p = 0$ and therefore,

$$I_{\beta,m}^{(\gamma_k),(\delta_k)}\{x^p\} = 0. \tag{1.2.1'}$$

This is a corollary of the fact that the corresponding factor $\Gamma\left(\gamma_k + \delta_k + \frac{p}{\beta} + 1\right)$ in the denominator of the constant c_p has a pole $-j = \gamma_k + \delta_k + \frac{p}{\beta} + 1$ $(j = 0, 1, 2, \ldots)$.

In the same way the images of the other basic functions of C_α can be found. For the sake of completeness, we adduce here the following comparatively general result.

Lemma 1.2.2. *The* $I_{\beta,m}^{(\gamma_k),(\delta_k)}$*-image of a G-function of* C_α *is also a G-function whose last three components of the order* $(\mu,\nu;\sigma,\tau)$ *are increased by* m*, namely*

$$I_{\beta,m}^{(\gamma_k),(\delta_k)}\left\{G_{\sigma,\tau}^{\mu,\nu}\left[\omega x^\beta \left| \begin{matrix} (c_i)_1^\sigma \\ (d_j)_1^\tau \end{matrix} \right.\right]\right\} = G_{\sigma+m,\tau+m}^{\mu,\nu+m}\left[\omega x^\beta \left| \begin{matrix} (-\gamma_k)_1^m, (c_i)_1^\sigma \\ (d_j)_1^\tau, (-\gamma_k - \delta_k)_1^m \end{matrix} \right.\right]. \tag{1.2.4}$$

Proof. Due to property (A.14), it is sufficient to consider the case $\sigma \le \tau$. Then

$$G_{\sigma,\tau}^{\mu,\nu}\left[\omega x^\beta\right] = \mathcal{O}\left(x^{d^*}\right) \quad \text{as } x \to 0,$$

where $d^* = \beta \min\limits_j d_j$ and the condition $d^* > \alpha \ge \max\limits_k\left[-\beta\left(\gamma_k + 1\right)\right]$ can be written in the form

$$\min_{1\le j\le \mu} d_j + \min_{1\le k\le m} \gamma_k > -1. \tag{1.2.5}$$

Let $\sigma \ne \tau$, for instance $\sigma < \tau$. Then,

$$f(x) = G_{\sigma,\tau}^{\mu,\nu}\left[\omega x^\beta\right] = x^{d^*}\tilde{f}(x)$$

with $d^* > \alpha$ and a function

$$\tilde{f}(x) = G_{\sigma,\tau}^{\mu,\nu}\left[\omega x^\beta \left| \begin{matrix} \left(c_i - \frac{d^*}{\beta}\right)_1^\sigma \\ \left(d_j - \frac{d^*}{\beta}\right)_1^\tau \end{matrix} \right.\right],$$

continuous in $[0,\infty)$, provided that (1.2.5) is satisfied.

To evaluate the integral (1.2.4) we use the known result (A.29) (see Appendix) for integrals involving products of two G-functions and also the fact that $G_{m,m}^{m,0}(\sigma) \equiv 0$ for $\sigma > 1$:

$$\begin{aligned} If(x) &= \int_0^1 G_{m,m}^{m,0}\left[\sigma \left| \begin{matrix} (\gamma_k + \delta_k)_1^m \\ (\gamma_k)_1^m \end{matrix} \right.\right] G_{\sigma,\tau}^{\mu,\nu}\left[\omega x^\beta \sigma \left| \begin{matrix} (c_i)_1^\sigma \\ (d_j)_1^\tau \end{matrix} \right.\right] d\sigma \\ &= \int_0^\infty G_{m,m}^{m,0}\left[\sigma \left| \begin{matrix} (\gamma_k + \delta_k)_1^m \\ (\gamma_k)_1^m \end{matrix} \right.\right] G_{\sigma,\tau}^{\mu,\nu}\left[\omega x^\beta \sigma \left| \begin{matrix} (c_i)_1^\sigma \\ (d_j)_1^\tau \end{matrix} \right.\right] d\sigma \\ &= G_{\sigma+m,\tau+m}^{\mu,\nu+m}\left[\omega x^\beta \left| \begin{matrix} (-\gamma_k)_1^m, (c_i)_1^\sigma \\ (d_j)_1^\tau, (-\gamma_k - \delta_k)_1^m \end{matrix} \right.\right]. \end{aligned}$$

The major part of the conditions usually required for the validity of the formula (A.29) (see Luke [272, I, p. 159-164]) concern the behaviour of the integrand to infinity and here they are superfluous. So, we shall require the conditions (1.2.5), $\sigma \neq \tau$, $\omega \neq 0$, $|\arg\omega| < \rho\pi$ with $\rho = \mu + \nu - \frac{\sigma+\tau}{2} > 0$, $c_i - d_j \neq 1, 2, 3, \ldots$, $i = 1, 2, \ldots, \nu$, $j = 1, 2, \ldots, \mu$ to be fulfilled. The same result, under some specifications, is true for $\sigma = \tau$, $\rho > 0$ too.

This lemma allows us to find the images of almost all the elementary and special functions, being special cases of the G-functions.

1.2.i. Mellin transform of fractional integrals and corresponding convolutional representations of them

One of the most commonly used tools in studing fractional integration operators and other convolutional type integral transforms (see e.g. Kalla [157]-[161], [163]-[164], Marichev [276], McBride [288]-[290], Tuan [495], [496], Tuan, Marichev and Yakubovich [497], etc.) is the well-known *Mellin transform*

$$\mathfrak{M}(s) = \mathfrak{M}\{f(x); s\} = \int_0^\infty x^{s-1} f(x) dx \tag{1.2.6}$$

with its *complex inversion formula*

$$f(x) = \frac{1}{2\pi i} \int_{\mathcal{C}} \mathfrak{M}(s) x^{-s} ds. \tag{1.2.7}$$

For more details about the contour $\mathcal{C} = \mathcal{L}_{\gamma - i\infty, \gamma + i\infty}$, conditions for existence and invertibility, see for instance [451], [107], [90], [276, p. 31-35], etc. The operation

$$f \circ g(x) = \int_0^\infty f\Big(\frac{x}{\tau}\Big) g(\tau) \frac{d\tau}{\tau} \tag{1.2.8}$$

is a *convolution of the Mellin transform (Theorem for the Mellin convolution)*:

$$\mathfrak{M}\{f \circ g(x); s\} = \mathfrak{M}\{f(x); s\} . \mathfrak{M}\{g(x); s\} . \tag{1.2.9}$$

As is done for instance by Marichev [286, p. 32], here we confine ourselves to the *subspace*

$$C_{\alpha,\alpha^*} = \Big\{ f(x) \in C[0,1] \cap C[1,\infty);\ |f| \leq Ax^{\alpha}, 0 < x < 1;\ |f| \leq Ax^{\alpha^*}, x > 1 \Big\} \tag{1.2.10}$$

of the basic spaces C_α, $C^*_{\alpha^*}$. For $\alpha^* \leq \alpha$ the Mellin transform of $f \in C_{\alpha,\alpha^*}$ exists in the vertical strip $\{\Re s = \gamma, -\alpha \leq \gamma \leq -\alpha^*\}$ and defines there an analytic function $\mathfrak{M}(s)$.

Theorem 1.2.3. *For function of the subspace $C_{\alpha,\alpha^*} \subset C_\alpha$ the operator $I^{(\gamma_k),(\delta_k)}_{\beta,m}$ has the following convolutional representation*

$$I^{(\gamma_k),(\delta_k)}_{\beta,m} f\left(x^{\frac{1}{\beta}}\right) = G^{m,0}_{m,m}\left[\frac{1}{x}\,\middle|\, \begin{matrix} (\gamma_k + \delta_k + 1)_1^m \\ (\gamma_k + 1)_1^m \end{matrix}\right] \circ f\left(x^{\frac{1}{\beta}}\right) \tag{1.2.11}$$

by means of the Mellin convolution (1.2.8) and the Mellin images of $I_{\beta,m}^{(\gamma_k),(\delta_k)} f$ *and* f *are related through the equality*

$$\mathfrak{M}\left\{I_{\beta,m}^{(\gamma_k),(\delta_k)} f(x); s\right\} = \prod_{k=1}^{m} \frac{\Gamma\left(\gamma_k - \frac{s}{\beta} + 1\right)}{\Gamma\left(\gamma_k + \delta_k - \frac{s}{\beta} + 1\right)} \mathfrak{M}\left\{f(x); s\right\}. \tag{1.2.12}$$

Proof. Since $G_{m,m}^{m,0}\left(\frac{\tau}{x}\right) \equiv 0$ for $\tau > x$, then using also (A.15) we obtain

$$\begin{aligned} If\left(x^{\frac{1}{\beta}}\right) &= \int_0^x G_{m,m}^{m,0}\left[\frac{\tau}{x}\left|\begin{matrix}(\gamma_k+\delta_k+1)\\(\gamma_k+1)\end{matrix}\right.\right] f\left(\tau^{\frac{1}{\beta}}\right)\frac{d\tau}{\tau} \\ &= \int_0^\infty G_{m,m}^{0,m}\left[\frac{x}{\tau}\left|\begin{matrix}(-\gamma_k)_1^m\\(-\gamma_k-\delta_k)_1^m\end{matrix}\right.\right] f\left(\tau^{\frac{1}{\beta}}\right)\frac{d\tau}{\tau} \\ &= G_{m,m}^{0,m}\left[x\left|\begin{matrix}(-\gamma_k)_1^m\\(-\gamma_k-\delta_k)_1^m\end{matrix}\right.\right] \circ f\left(x^{\frac{1}{\beta}}\right) = G_{m,m}^{m,0}\left[\frac{1}{x}\left|\begin{matrix}(\gamma_k+\delta_k+1)_1^m\\(\gamma_k+1)_1^m\end{matrix}\right.\right] \circ f\left(\tau^{\frac{1}{\beta}}\right) \end{aligned}$$

which is (1.2.11). According to property (1.2.9) it follows that

$$\mathfrak{M}\left\{I_{\beta,m}^{(\gamma_k),(\delta_k)} f\left(x^{\frac{1}{\beta}}\right); s\right\} = \mathfrak{M}\left\{G_{m,m}^{0,m}\left[\frac{x}{\tau}\left|\begin{matrix}(-\gamma_k)_1^m\\(-\gamma_k-\delta_k)_1^m\end{matrix}\right.\right]; s\right\} \mathfrak{M}\left\{f\left(x^{\frac{1}{\beta}}\right); s\right\}.$$

After routine calculations using the following property of the Mellin transform ([107, I, p. 307, (7)]):

$$\mathfrak{M}\left\{x^\beta F\left(ax^h\right); s\right\} = \frac{a^{-\frac{s+\beta}{h}}}{h} \mathfrak{M}\left\{F(x); \frac{s+\beta}{h}\right\} \tag{1.2.13}$$

for $a > 0$, $h > 0$, one can find for $s_1 = \beta s$:

$$\mathfrak{M}\left\{If(x); s_1\right\} = \mathfrak{M}\left\{G_{m,m}^{0,m}\left[x\left|\begin{matrix}(-\gamma_k)_1^m\\(-\gamma_k-\delta_k)_1^m\end{matrix}\right.\right]; \frac{s_1}{\beta}\right\} \mathfrak{M}\left\{f(x); s_1\right\}.$$

The known formula for the Mellin transform of the G-function (A.25) gives

$$\mathfrak{M}\left\{I_{\beta,m}^{(\gamma_k),(\delta_k)} f(x); s\right\} = \prod_{k=1}^{m} \frac{\Gamma\left(\gamma_k - \frac{s_1}{\beta} + 1\right)}{\Gamma\left(\gamma_k + \delta_k - \frac{s_1}{\beta} + 1\right)} \mathfrak{M}\left\{f(x); s_1\right\},$$

which completes the proof of the theorem.

Theorem 1.2.3′. *Using property (1.2.13) we can find also the Mellin transform of the generalized fractional integrals of form the (1.1.7):* $Rf(x) = x^{\beta\delta_0} I_{\beta,m}^{(\gamma_k),(\delta_k)} f(x)$, $\delta_0 \geq 0$, *namely:*

$$\mathfrak{M}\left\{Rf(x); s\right\} = \prod_{k=1}^{m} \frac{\Gamma\left(\gamma_k - \frac{s}{\beta} - \delta_0 + 1\right)}{\Gamma\left(\gamma_k + \delta_k - \frac{s}{\beta} - \delta_0 + 1\right)} \mathfrak{M}\left\{f(x); s\right\}. \tag{1.2.14}$$

Let us discuss some *special cases*.

Corollary 1.2.4. *The convolutional representation of the Riemann-Liouville fractional integral (1.1.b) is:*

$$\begin{aligned} R^{\delta} &= \int_0^x \frac{(x-\tau)^{\delta-1}}{\Gamma(\delta)} f(\tau)d\tau = x^{\delta-1}\int_0^x G_{1,1}^{1,0}\left[\frac{\tau}{x}\middle|\begin{matrix}\delta\\0\end{matrix}\right] f(\tau)d\tau \\ &= \left\{x^{\delta}G_{1,1}^{1,0}\left[\frac{1}{x}\middle|\begin{matrix}\delta+1\\1\end{matrix}\right]\right\}\circ f(x) = \left\{\frac{x^{\delta-1}}{\Gamma(\delta)}\right\} * f(x), \end{aligned} \tag{1.2.15}$$

where $(*)$ *is the Duhamel convolution (2.1.1).*

Corollary 1.2.5. *For the Erdélyi-Kober operators,* $I_{\beta}^{\gamma,\delta}$ *Theorem 1.2.3 yields the convolutional representation*

$$\begin{aligned} I_{\beta}^{\gamma,\delta} f\left(x^{\frac{1}{\beta}}\right) &= G_{1,1}^{1,0}\left[\frac{1}{x}\middle|\begin{matrix}\gamma+\delta+1\\\gamma+1\end{matrix}\right]\circ f\left(x^{\frac{1}{\beta}}\right) \\ &= \left\{x^{-(\gamma+\delta)}\frac{(x-1)^{\delta+1}}{\Gamma(\delta)}\right\}\circ f\left(x^{\frac{1}{\beta}}\right) \end{aligned} \tag{1.2.16}$$

and the well-known result (cf. McBride [289, Theorem 5.4.]):

$$\mathfrak{M}\left\{I_{\beta}^{\gamma,\delta} f(x);s\right\} = \frac{\Gamma\left(\gamma-\frac{s}{\beta}+1\right)}{\Gamma\left(\gamma+\delta-\frac{s}{\beta}+1\right)}\mathfrak{M}\left\{f(x);s\right\}. \tag{1.2.17}$$

Corollary 1.2.6. *In the case* $m=2$ *the statement of Theorem 1.2.3 takes the form: tbe hypergeometric fractional integrals(1.1.19) have convolutional representations*

$$Hf\left(x^{\frac{1}{\beta}}\right) = G_{2,2}^{2,0}\left[\frac{1}{x}\middle|\begin{matrix}\gamma_1+\delta_1+1,\gamma_2+\delta_2+1\\\gamma_1+1,\gamma_2+1\end{matrix}\right]\circ f\left(x^{\frac{1}{\beta}}\right) \tag{1.2.18}$$

(the $G_{2,2}^{2,0}$*-function can be expressed also in terms of the* ${}_2F_1$*-function). The Mellin transform is:*

$$\mathfrak{M}\left\{Hf(x);s\right\} = \frac{\Gamma\left(\gamma_1-\frac{s}{\beta}+1\right)\Gamma\left(\gamma_2-\frac{s}{\beta}+1\right)}{\Gamma\left(\gamma_1+\delta_1-\frac{s}{\beta}+1\right)\Gamma\left(\gamma_2+\delta_2-\frac{s}{\beta}+1\right)}\mathfrak{M}\left\{f(x);s\right\}. \tag{1.2.19}$$

The above results, specialized for the operators (1.1.l) give:

Corollary 1.2.7. *The Mellin transform acts on Saigo's fractional integrals* $I^{\alpha,\beta,\eta}$ *(1.1.l-l') in the following way:*

$$\mathfrak{M}\left\{I^{\alpha,\beta,\eta}f(x);s\right\} = \frac{\Gamma(\eta-s+1)\Gamma(\beta-s+1)}{\Gamma(1-s)\Gamma\left(\alpha+\beta+\eta-s+1\right)}\mathfrak{M}\left\{x^{-\beta}f(x);s\right\}. \tag{1.2.20}$$

For arbitrary $m > 1$ we have such a particular result.

Corollary 1.2.8. *Let us consider the hyper-Bessel integral operator L (1.1.o-o′). The Mellin convolutional representation*

$$Lf\left(x^{\frac{1}{\beta}}\right) = \frac{x}{\beta^{m-1}}\left\{G^{0,m}_{m,m}\left[x\left|\begin{matrix}(-\gamma_k)_1^m\\(-\gamma_k-1)_1^m\end{matrix}\right.\right]\circ f\left(x^{\frac{1}{\beta}}\right)\right\} \tag{1.2.21}$$

and its generalization for the fractional powers L^λ, $\lambda > 0$ (see McBride [289], Dimovski and Kiryakova [79]) following from Theorem 1.2.3 and are used in Chapter 3 for developing a new approach to the hyper-Bessel operators. Their Mellin images can be derived immediately from (1.2.12′) by taking into acount that $\delta_0 = \delta_1 = \cdots = \delta_m = 1$ and $\frac{\Gamma(\lambda)}{\Gamma(\lambda+1)} = \frac{1}{\lambda}$, for example:

$$\begin{aligned}\mathfrak{M}\left\{Lf(x);s\right\} &= \frac{1}{\beta^m}\left[\prod_{k=1}^{m}\frac{\Gamma\left(\gamma_k-\frac{s}{\beta}\right)}{\Gamma\left(\gamma_k-\frac{s}{\beta}+1\right)}\right]\mathfrak{M}\left\{x^\beta f(x);s\right\}\\ &= \frac{1}{\prod\limits_{k=1}^{m}(\beta\gamma_k-s)}\mathfrak{M}\left\{x^\beta f(x);s\right\}.\end{aligned} \tag{1.2.22}$$

This list of corollaries can be easily extended for any of the special cases of $I^{(\gamma_k),(\delta_k)}_{\beta,m}$, operators mentioned in Section 1.1.iii.

1.2.ii. Decomposition of the multi-Erdélyi-Kober operators

It is known that most of the integral transformations can be represented as compositions of other, usually simpler integral transformations. Decompositions of the $I^{(\gamma_k),(\delta_k)}_{\beta,m}$-operators in some particular cases, especialy for $m = 2$, were proposed by Love [262], Saigo [415] etc. In a more general aspect this problem is considered for convolutional type integral transforms by McBride [289], Brychkov, Glaeske and Marichev [44], Marichev [281], Marichev and Tuan [282]-[284], Tuan [495]-[496], Tuan, Marichev and Yakubov̆ich [497], Nahoushev and Salathidinov [315], Samko, Kilbas and Marichev [434] etc. These papers propose a new approach to integral transformations, developed in details for the G- and H-transforms of convolutional type by Tuan in [496], Nguen Hai and Yakubovich [317], Saigo and Yakubovich [426], etc.

Here we propose a particular result describing the composition structure of the generalized fractional integration operators considered in this book.

Combining the results (1.2.17), (1.2.12), we find that the Mellin image of the operator $I^{(\gamma_k),(\delta_k)}_{\beta,m}$ is a product of the Mellin images of the Erdélyi-Kober operators $I^{\gamma_k,\delta_k}_{\beta}$, $k = 1,\ldots,m$:

$$\mathfrak{M}\left\{I^{(\gamma_k),(\delta_k)}_{\beta,m}f(x);s\right\} = \prod_{k=1}^{m}\mathfrak{M}\left\{I^{\gamma_k,\delta_k}_{\beta}f(x);s\right\}. \tag{1.2.23}$$

This is quite naturaly, since the same is true for their kernel-functions taking part in the convolutional representations (1.2.11), (1.2.16):

$$\mathfrak{M}\left\{G_{m,m}^{m,0}\left[\sigma\left|\begin{matrix}(\gamma_k+\delta_k)\\(\gamma_k)\end{matrix}\right.\right];s\right\}=\prod_{k=1}^{m}\frac{\Gamma(\gamma_k+s)}{\Gamma(\gamma_k+\delta_k+s)}=\prod_{k=1}^{m}\mathfrak{M}\left\{G_{1,1}^{1,0}\left[\sigma\left|\begin{matrix}\gamma_k+\delta_k\\\gamma_k\end{matrix}\right.\right];s\right\}. \tag{1.2.24}$$

This fact suggests the conclusion that the *operator* $I_{\beta,m}^{(\gamma_k),(\delta_k)}$ itself *is a composition of the commuting Erdélyi-Kober operators* $I_\beta^{\gamma_k,\delta_k}$, $k=1,\dots,m$. Before formulating the result, **let us consider the operator generated by such a composition**. We shall denote it *by the same symbol*: $I_{\beta,m}^{(\gamma_k),(\delta_k)}$.

Lemma 1.2.9. *The operator*

$$\begin{aligned}I_{\beta,m}^{(\gamma_k),(\delta_k)}f(x)&\overset{\text{def}}{=}I_\beta^{\gamma_m,\delta_m}\left\{I_\beta^{\gamma_{m-1},\delta_{m-1}}\dots\left(I_\beta^{\gamma_1,\delta_1}f(x)\right)\right\}\\&=\int_0^1\dots\int_0^1\prod_{k=1}^{m}\left[\frac{(1-\sigma_k)^{\delta_k-1}\sigma_k^{\gamma_k}}{\Gamma(\delta_k)}\right]f\left[x(\sigma_1\dots\sigma_m)^{\frac{1}{\beta}}\right]d\sigma_1\dots d\sigma_m,\end{aligned} \tag{1.2.25}$$

where $I_\beta^{\gamma_k,\delta_k}$, $k=1,\dots,m$, *are Erdélyi-Kober fractional integration operators of the form (1.1.17), is well defined in the space* C_α, $\alpha\ge\max_k[-\beta(\gamma_k+1)]$, *does not depend on the arrangement of the multipliers* $I_\beta^{\gamma_k,\delta_k}$, $k=1,\dots,m$ *and the* $I_{\beta,m}^{(\gamma_k),(\delta_k)}$*-image (1.2.25) of any power function* $f(x)=x^p\in C_\alpha$ *coincides with its* $I_{\beta,m}^{(\gamma_k),(\delta_k)}$*-image (1.1.6), i.e. formula (1.2.1) is again true:*

$$\dot{I}_{\beta,m}^{(\gamma_k),(\delta_k)}\{x^p\}=c_px^p\ \textit{with}\ c_p=\prod_{k=1}^{m}\frac{\Gamma\left(\gamma_k+\frac{p}{\beta}+1\right)}{\Gamma\left(\gamma_k+\delta_k+\frac{p}{\beta}+1\right)}\ \textit{for}\ p>\alpha. \tag{1.2.26}$$

Proof. Let $f(x)=x^p\tilde{f}(x)$, $p\ge\max_k[-\beta(\gamma_k+1)]$ and hence, $\frac{p}{\beta}+\gamma_k>-1$, $k=1,\dots,m$. Then the integral (1.2.25):

$$I_{\beta,m}^{(\gamma_k),(\delta_k)}f(x)=x^p\int_0^1\dots\int_0^1\prod_{k=1}^{m}\left[\frac{(1-\sigma_k)^{\delta_k-1}\sigma_k^{\gamma_k+\frac{p}{\beta}}}{\Gamma(\delta_k)}\right]\tilde{f}\left[x(\sigma_1\dots\sigma_m)^{\frac{1}{\beta}}\right]d\sigma_1\dots d\sigma_m$$

is absolutely convergent, since

$$x^p \int_0^1 \ldots \int_0^1 \prod_{k=1}^{m} \left[\frac{(1-\sigma_k)^{\delta_k - 1} \sigma_k^{\gamma_k + \frac{p}{\beta}}}{\Gamma(\delta_k)} \right] \left| \tilde{f}\left[x(\sigma_1 \ldots \sigma_m)^{\frac{1}{\beta}} \right] \right| d\sigma_1 \ldots d\sigma_m$$

$$\leq C x^p \int_0^1 \ldots \int_0^1 \prod_{k=1}^{m} \left[\frac{(1-\sigma_k)^{\delta_k - 1} \sigma_k^{\gamma_k + \frac{p}{\beta}}}{\Gamma(\delta_k)} \right] d\sigma_1 \ldots d\sigma_m$$

$$= C x^p \prod_{k=1}^{m} \frac{\Gamma\left(\gamma_k + \frac{p}{\beta} + 1\right)}{\Gamma\left(\gamma_k + \frac{p}{\beta} + \delta_k + 1\right)}.$$

For $\tilde{f}(x) = 1$, i.e. $f(x) = x^p$ with $p > \alpha$, we obtain formula (1.2.26).

Due to the Weierstrass theorem, a continuous function $\tilde{f} \in C[0, \infty)$ can be uniformly approximated on every finite interval $[0, X]$, $X > 0$ by means of the sequence of polynomials $P_n(x) = \sum_{k=0}^{n} a_{n,k} x^k$, $n = 1, 2, \ldots$. Then the function $f(x) = x^p \tilde{f}(x)$ itself can be approximated by the sequence of functions

$$f_n(x) = x^p P_n(x) = \sum_{k=0}^{n} a_{n,k} x^{p+k}, \quad p > \alpha, \quad n = 1, 2, \ldots$$

By virtue of Lemmas 1.2.1 and 1.2.9, the operators , $I_{\beta,m}^{(\gamma_k),(\delta_k)}$ defined by (1.1.6) and (1.2.25) coincide for such functions, namely:

$$I_{\beta,m}^{(\gamma_k),(\delta_k)} f_n(x) = \sum_{k=0}^{n} a_{n,k} I_{\beta,m}^{(\gamma_k),(\delta_k)} \left\{ x^{p+k} \right\} = \sum_{k=0}^{n} a_{n,k} c_{p+k} x^{p+k}.$$

Hence letting $n \to \infty$ we find that these operators coincide for each function $f(x)$ of space the C_α, being defined correctly there. This proves the following basic proposition for which we also give another direct proof (by means of mathematical induction).

Theorem 1.2.10. *In C_α, $\alpha \geq \max_k [-\beta(\gamma_k + 1)]$ each multiple Erdélyi-Kober operator (1.1.6) with $\delta_k \geq 0$, $k = 1, \ldots, m$*

$$I_{\beta,m}^{(\gamma_k),(\delta_k)} f(x) = \int_0^1 G_{m,m}^{m,0} \left[\sigma \left| \begin{matrix} (\gamma_k + \delta_k)_1^m \\ (\gamma_k)_1^m \end{matrix} \right. \right] f\left(x\sigma^{\frac{1}{\beta}} \right) d\sigma$$

can be represented also by an m-tuple composition of commuting classical Erdélyi-Kober *fractional integrals (1.1.17):*

$$I_{\beta,m}^{(\gamma_k),(\delta_k)} f(x) = \left(\prod_{k=1}^{m} I_\beta^{\gamma_k,\delta_k}\right) f(x)$$
$$= \int_0^1 \cdots \int_0^1 \left[\prod_{k=1}^{m} \frac{(1-\sigma_k)^{\delta_k-1}\sigma_k^{\gamma_k}}{\Gamma(\delta_k)}\right] f\left[x(\sigma_1\ldots\sigma_m)^{\frac{1}{\beta}}\right] d\sigma_1\ldots d\sigma_m, \tag{1.2.27}$$

and conversely.

Proof. To prove (1.2.27) we use *the method of mathematical induction.* Without loss of generality, we can assume for brevity that $\beta = 1$ (see Lemma 1.3.3). For $m = 1$ we have already seen that

$$I_\beta^{\gamma,\delta} f(x) = \int_0^1 \frac{(1-\sigma)^{\delta-1}\sigma^\gamma}{\Gamma(\delta)} f(x\sigma)\, d\sigma$$
$$= \int_0^1 G_{1,1}^{1,0}\left[\sigma \,\middle|\, \begin{matrix}\gamma+\delta\\ \gamma\end{matrix}\right] f(x\sigma)\, d\sigma = I_{\beta,1}^{\gamma,\delta} f(x).$$

Consider the case $m = 2$. Writing $I_{1,2}^{(\gamma_k)_1^2,(\delta_k)_1^2} = \prod\limits_{k=1}^{2} I_1^{\gamma_k,\delta_k}$ in the form:

$$x^{-(\gamma_2+\delta_2)} \int_0^x \frac{(x-t_2)^{\delta_2-1}}{\Gamma(\delta_2)} t_2^{\gamma_2} \left\{ t_2^{-(\gamma_1+\delta_1)} \int_0^{t_2} \frac{(t_2-t_1)^{\delta_1-1}}{\Gamma(\delta_1)} t_1^{\gamma_1} f(t_1)\, dt_1 \right\} dt_2$$
$$= x^{-(\gamma_2+\delta_2)} \int_0^x f(t_1)\, t_1^{\gamma_1}\, dt_1 \int_{t_1}^x \frac{(x-t_2)^{\delta_2-1}}{(\delta_2)} t_2^{\gamma_2-\gamma_1-\delta_1} \frac{(t_2-t_1)^{\delta_1-1}}{\Gamma(\delta_1)}\, dt_2,$$

we can evaluate the inner integral, after changing the order of integrations by virtue of Fubini's theorem. According to [368, p. 301, (1)], its value is

$$\frac{(x-t_1)^{\delta_1+\delta_2-1}}{\Gamma(\delta_1+\delta_2)} t_1^{\gamma_2-\gamma_1-\delta_1} \left(\frac{t_1}{x}\right)^{\delta_1} {}_2F_1(\gamma_2+\delta_2-\gamma_1;\delta_1;\delta_1+\delta_2;1-\frac{t_1}{x}).$$

It remains to change the variable and to use (1.1.18), whence

$$I_{1,2}^{(\gamma_1,\gamma_2),(\delta_1,\delta_2)} f(x) = \int_0^1 G_{2,2}^{2,0}\left[\sigma \,\middle|\, \begin{matrix}\gamma_1+\delta_1,\gamma_2+\delta_2\\ \gamma_1,\gamma_2\end{matrix}\right] f(x\sigma)\, d\sigma = I_1^{\gamma_1,\delta_1}\left\{I_1^{\gamma_2,\delta_2} f(x)\right\}$$

Suppose now that assertion (1.2.27) *is true for the all* $(m-1)$*-tuple operators,* that is:

$$I_{1,m-1}^{(\gamma_k)_1^{m-1}(\delta_k)_1^{m-1}} = \left(\prod_{k=1}^{m-1} I_1^{\gamma_k,\delta_k}\right).$$

Then, *for arbitrary* $m > 1$ we have:

$$I_{1,m}^{(\gamma_k)_1^m,(\delta_k)_1^m} f(x) = I_1^{\gamma_m,\delta_m}\left(\prod_{k=1}^{m-1} I_1^{\gamma_k,\delta_k} f(x)\right) = I_1^{\gamma_m,\delta_m}\left(I_{1,m-1}^{(\gamma_k)_1^{m-1}(\delta_k)_1^{m-1}} f(x)\right)$$

$$= x^{-(\gamma_m+\delta_m)}\int\limits_0^x \frac{(x-t_m)^{\delta_m-1}}{\Gamma(\delta_m)} t_m^{\gamma_m}$$

$$\times\left\{t_m^{-1}\int\limits_0^{t_m} G_{m-1,m-1}^{m-1,0}\left[\frac{t_{m-1}}{t_m}\middle|\begin{matrix}(\gamma_k+\delta_k)_1^{m-1}\\ (\gamma_k)_1^{m-1}\end{matrix}\right] f(t_{m-1})dt_{m-1}\right\} dt_m.$$

Due to Fubini's theorem, we change the order of integrations:

$$I_{1,m}^{(\gamma_k),(\delta_k)} f(x) = x^{-(\gamma_m+\delta_m)}\int\limits_0^x f(t_{m-1})\, t_{m-1}^{\gamma_m-1} dt_{m-1}\int\limits_{t_{m-1}}^x \frac{(x-t_m)^{\delta_m-1}}{\Gamma(\delta_m)}$$

$$\times\, G_{m-1,m-1}^{0,m-1}\left[\frac{t_m}{t_{m-1}}\middle|\begin{matrix}(\gamma_m-\gamma_k)_1^{m-1}\\ (\gamma_m-\gamma_k-\delta_k)_1^{m-1}\end{matrix}\right] dt_m.$$

Since $G_{m-1,m-1}^{0,m-1}\left[\frac{t_m}{t_{m-1}}\right] \equiv 0$ for $t_m < t_{m-1}$ (see (A.12)), the inner integral written as an integral from 0 to x can be evaluated as a Riemann-Liouville fractional integral of a G-function (A.23). Its value is

$$t_m^{\delta_m} G_{m,m}^{m,0}\left[\frac{t_{m-1}}{x}\middle|\begin{matrix}(1-\gamma_m-\delta_m+\gamma_k+\delta_k)_1^{m-1}, 1\\ (1-\gamma_m-\delta_m+\gamma_k)_1^{m-1}, 1-\delta_m\end{matrix}\right].$$

In this manner after routine transformations using (A.14) we obtain :

$$I_{1,m}^{(\gamma_k),(\delta_k)} f(x) = x^{-1}\int\limits_0^x G_{m,m}^{m,0}\left[\frac{t_{m-1}}{x}\middle|\begin{matrix}(\gamma_k+\delta_k)_1^{m-1}, (\gamma_m+\delta_m)\\ (\gamma_k)_1^{m-1}, \gamma_m\end{matrix}\right] f(t_{m-1})\, dt_{m-1}$$

$$= \int\limits_0^1 G_{m,m}^{m,0}\left[\sigma\middle|\begin{matrix}(\gamma_k+\delta_k)_1^m\\ (\gamma_k)_1^m\end{matrix}\right] f(x\sigma)\, d\sigma,$$

i.e. the operators $I_{1,m}^{(\gamma_k),(\delta_k)}$ defined first by (1.1.6) and then by (1.2.27) coincide for arbitrary $m > 1$ in the space C_α, $\alpha \geq \max\limits_k[-\beta(\gamma_k+1)]$, where both of them are well defined. This completes the proof.

Note. The commutability of operators $I_\beta^{\gamma_k,\delta_k}$, $k = 1,\dots,m$ in (1.2.27) is also a simple corollary of the fact that the kernel-function of the operator $I_{\beta,m}^{(\gamma_k),(\delta_k)} = \prod\limits_{k=1}^m I_\beta^{\gamma_k,\delta_k}$, the function $G_{m,m}^{m,0}\left[\sigma\middle|\begin{matrix}(\gamma_k+\delta_k)_1^m\\ (\gamma_k)_1^m\end{matrix}\right]$ in (1.1.6), *is symmetric* in the parameters $(\gamma_k+\delta_k)_{k=1}^m$, $(\gamma_k)_{k=1}^m$.

Consider now some *corollaries.*

Corollary 1.2.11. *The hypergeometric fractional integrals (1.1.19) are compositions of two commuting Erdélyi-Kober operators and can be written also by means of repeated integrals:*

$$\begin{aligned} H\,f(x) &= I_{\beta,2}^{(\gamma_1,\gamma_2),(\delta_1,\delta_2)} f(x) = I_\beta^{\gamma_2,\delta_2} I_\beta^{\gamma_1,\delta_1} f(x) \\ &= I_\beta^{\gamma_1,\delta_1} I_\beta^{\gamma_2,\delta_2} f(x) = \int_0^1\int_0^1 \frac{(1-\sigma_1)^{\delta_1-1}(1-\sigma_2)^{\delta_2-1}}{\Gamma(\delta_1)\Gamma(\delta_2)} \sigma_1^{\gamma_1}\sigma_2^{\gamma_2} f\left[x(\sigma_1\sigma_2)^{\frac{1}{\beta}}\right] d\sigma_1 d\sigma_2. \end{aligned} \tag{1.2.28}$$

Corollary 1.2.12. *In particular, for Saigo's fractional integrals (1.1.l) the following decompositions hold:*

$$\begin{aligned} I^{\alpha,\beta,\eta} f(x) &= x^{-\beta} I_1^{\eta-\beta,-\eta} I_1^{0,\alpha+\eta} f(x) = x^{-\beta} I_1^{0,\alpha+\eta} I_1^{\eta-\beta,-\eta} f(x) \\ &= x^{-\beta}\int_0^1\int_0^1 \frac{(1-\sigma_1)^{-\eta-1}(1-\sigma_2)^{\alpha+\eta-1}}{\Gamma(-\eta)\Gamma(\alpha+\eta)} \sigma_1^{\eta-\beta} f(x\sigma_1\sigma_2) d\sigma_1 d\sigma_2 \end{aligned} \tag{1.2.29}$$

for functions of C_λ, $\lambda \geq \max(-\eta+\beta-1,-1)$ *provided* $\alpha > -\eta > 0$. *The same decompositions are obtained by Saigo [415, p. 138], written in terms of Riemann-Liouville operators as follows:*

$$I^{\alpha,\beta,\eta} f(x) = x^{-\alpha-\beta-\eta} R^{\alpha+\eta} x^{\beta} R^{-\eta} x^{\eta-\beta} f(x) = R^{-\eta} x^{-\alpha-\beta} R^{\alpha+\eta} f(x). \tag{1.2.29'}$$

Corollary 1.2.13. *For* $m=3$ *from Theorem 1.2.10 the decomposition (1.1.n'')*

$$F f(x) = x^c I_1^{a,b} I_1^{b,c-a'-b} I_1^{c-a'-b',a'} f(x)$$

of Marichev's F_3*-fractional integral (1.1.n) can be obtained.*

Corollary 1.2.14. *The representations (1.1.o) and (1.1.o') of the hyper-Bessel integral operator* L *are equivalent, since*

$$L = \frac{x^\beta}{\beta^m} I_{\beta,m}^{(\gamma_k),(1)} = \frac{x^\beta}{\beta^m}\prod_{k=1}^m I_\beta^{\gamma_k,1} = \frac{x^\beta}{\beta^m}\prod_{k=1}^m \left(x^{-\gamma_k-1} R^1 x^{\gamma_k}\right).$$

This result of McBride [289] and Dimovski and Kiryakova [79] is used repeatedly in Chapter 3. Another example of an $(m-1)$*-tuple fractional integration operator is the generalized Sonine transformation*

$$\begin{aligned} \varphi\, f(x) = x^{m(\gamma_m+1)-1} \int_0^1 \ldots \int_0^1 & \left[\prod_{k=1}^{m-1} \frac{(1-\sigma_k)^{\gamma_m-\gamma_k+\frac{k}{m}-1}}{\Gamma\left(\gamma_m-\gamma_k+\frac{k}{m}\right)} \sigma_k^{\gamma_k}\right] \\ & \times f\left[x^{\frac{m}{\beta}}(t_1\ldots t_{m-1})^{\frac{1}{\beta}}\right] dt_1\ldots dt_{m-1}, \end{aligned} \tag{1.2.30}$$

proposed by Dimovski [68]-[71] and playing the role of a transmutation operator from L to the m-fold integration l^m, viz.

$$\varphi L = \left(\frac{m}{\beta}\right)^m l^m \varphi.$$

According to Theorem 1.2.10, the Sonine-Dimovski transformation (1.2.30) can be written also in the form

$$\begin{aligned}\varphi f\left(x^{\frac{\beta}{m}}\right) &= x^{\beta\left(\gamma_m+\frac{m-1}{m}\right)} I_{\beta,m-1}^{(\gamma_k),\left(\gamma_m-\gamma_k+\frac{k}{m}\right)} f(x)\\ &= x^{\beta\left(\gamma_m+\frac{m-1}{m}\right)} \int\limits_0^1 G_{m-1,m-1}^{m-1,0}\left[\sigma \left| \begin{matrix}\left(\gamma_m+\frac{k}{m}\right)_1^m \\ (\gamma_k)_1^m\end{matrix}\right.\right] f\left(x\sigma^{\frac{1}{\beta}}\right) d\sigma.\end{aligned} \tag{1.2.30'}$$

This representation is more useful in various applications; see for instance Sections 3.5, 3.6, 3.7, 3.8. On the base of Theorem 1.2.10 new simpler representations using G-function can be found also for the generalized Poisson-Dimovski transformations (Chapter 3). They are used in Chapter 4 for obtaining new Poisson type integral representations of the hyper-Bessel functions of Delerue and more generally, of the ${}_pF_q$-functions with $p < q$.

These and other examples reveal the important and two-fold role which Theorem 1.2.10 has in the theory of generalized fractional integrals. Firstly, every operator $I_{\beta,m}^{(\gamma_k),(\delta_k)}$ of the form (1.1.6) with $G_{m,m}^{m,0}$-function as kernel can be written by means of repeated integrals without a G-function . Conversely, every composition $I_{\beta,m}^{(\gamma_k),(\delta_k)}$ of an arbitrary number of Erdélyi-Kober operators can be represented by a simple integral involving Meijer's G-function. Such compositions arise very often in various problems of analysis and its applications. Their properties, inversion formulas etc., can be easily found if one uses the simple integral representations and the usefull properties of the G-function . This is the key to the most of the applications considered in Chapters 3, 4 and 5.

1.2.iii The operators $I_{\beta,m}^{(\gamma_k),(\delta_k)}$ as isomorphisms in C_α

Theorem 1.2.15. *The multi-Erdélyi-Kober operator*

$$\begin{aligned}I_{\beta,m}^{(\gamma_k),(\delta_k)} f(x) &= \int\limits_0^1 G_{m,m}^{m,0}\left[\sigma \left| \begin{matrix}(\gamma_k+\delta_k)_1^m \\ (\gamma_k)_1^m\end{matrix}\right.\right] f\left(x\sigma^{\frac{1}{\beta}}\right) d\sigma\\ &= \int\limits_0^1 \cdots \int\limits_0^1 \prod_{k=1}^m \left[\frac{(1-\sigma_k)^{\delta_k-1}}{\Gamma(\delta_k)}\sigma_k^{\gamma_k}\right] f\left[x(\sigma_1\ldots\sigma_m)^{\frac{1}{\beta}}\right] d\sigma_1\ldots d\sigma_m\end{aligned} \tag{1.2.31}$$

is an invertible linear mapping of the space C_α, $\alpha \geq \max\limits_k[-\beta(\gamma_k+1)]$ into itself. If

$$\eta_k = \begin{cases} [\delta_k]+1 & \text{for non integer } \delta_k \\ \delta_k & \text{for integer } \delta_k \end{cases}, \quad k = 1,\ldots,m, \tag{1.2.32}$$

then in the image-space

$$I_{\beta,m}^{(\gamma_k),(\delta_k)}(C_\alpha) = C_\alpha^{(\eta_1+\cdots+\eta_m)} \subset C_\alpha \tag{1.2.33}$$

the following inversion formula holds:

$$f\left((x_1 \dots x_m)^{\frac{1}{\beta}}\right) = \left(\prod_{k=1}^{m} x_k^{-\gamma_k}\right) \frac{\partial^{\delta_1+\cdots+\delta_m}}{\partial x_1^{\delta_1} \dots \partial x_m^{\delta_m}} \left\{ g\left((x_1 \dots x_m)^{\frac{1}{\beta}}\right) \prod_{k=1}^{m} x_k^{\gamma_k+\delta_k} \right\}, \tag{1.2.34}$$

where $g(x) = I_{\beta,m}^{(\gamma_k),(\delta_k)} f(x)$.

Proof. If $f \in C_\alpha$, then $I_{\beta,m}^{(\gamma_k),(\delta_k)} f \in C_\alpha$ too as is seen from the proof of Lemma 1.2.1. This can also be seen as a corollary of the "decomposition" Theorem 1.2.10. Indeed, the well-known result for the Erdélyi-Kober operators:

$$I_\beta^{\gamma_k,\delta_k} : C_\alpha \longrightarrow C_\alpha^{(\eta_k)} \subset C_\alpha$$

yields that

$$I = \prod_{k=1}^{m} I_\beta^{\gamma_k,\delta_k} : C_\alpha \longrightarrow C_\alpha^{(\eta_1+\cdots+\eta_m)} \subset C_\alpha.$$

It remains to prove that $I_{\beta,m}^{(\gamma_k),(\delta_k)}$ is an injective and surjective mapping. We shall prove this directly by establishing that the unique solution of the integral equation

$$I_{\beta,m}^{(\gamma_k),(\delta_k)} f(x) = \int_0^1 G_{m,m}^{m,0}\left[\sigma \,\middle|\, \begin{matrix} (\gamma_k+\delta_k)_1^m \\ (\gamma_k)_1^m \end{matrix}\right] f\left(x\sigma^{\frac{1}{\beta}}\right) d\sigma = 0 \tag{1.2.35}$$

is the function $f(x) \equiv 0$.

Let us put $x^\beta = x_1 \dots x_m$, $\sigma_k x_k = \tau_k$, $k = 1, \dots, m$ in (1.2.31) and multiply by $x_1^{\gamma_1+\delta_1} \dots x_m^{\gamma_m+\delta_m} \neq 0$. Denote by

$$b(x_1, \dots, x_m) = \left[\prod_{k=1}^{m} x_k^{\gamma_k}\right] f\left[(x_1 \dots x_m)^{\frac{1}{\beta}}\right]$$

and

$$c(x_1, \dots, x_m) = \left[\prod_{k=1}^{m} x_k^{\gamma_k+\delta_k}\right] g\left[(x_1 \dots x_m)^{\frac{1}{\beta}}\right].$$

Then the equation $g(x) = I_{\beta,m}^{(\gamma_k),(\delta_k)} f(x)$ takes the form

$$c(x_1 \dots x_m) = \int_0^{x_1} \dots \int_0^{x_m} \left(\prod_{k=1}^{m} \frac{(x_k-\tau_k)^{\delta_k-1}}{\Gamma(\delta_k)}\right) b(\tau_1, \dots, \tau_m)\, d\tau_1 \dots d\tau_m. \tag{1.2.36}$$

When $g(x) = 0$, we obtain equation (1.2.35) written in the more convenient form

$$\int_0^{x_1} \cdots \int_0^{x_1} \left[\prod_{k=1}^{m} \frac{(x_k - \tau_k)^{\delta_k - 1}}{\Gamma(\delta_k)} \right] b(\tau_1, \ldots, \tau_m) \, d\tau_1 \ldots d\tau_m = 0. \tag{1.2.35'}$$

Now the following theorem is applicable.

Theorem of Mikusinski and Ryll-Nardzewski ([307]). *Let the functions $a(x_1, \ldots, x_m)$ and $b(x_1, \ldots, x_m)$ be defined for $x_k \geq 0$, $k = 1, \ldots, m$ and have the form*

$$a(x_1, \ldots, x_m) = x_1^{\lambda_1} \ldots x_m^{\lambda_m} \tilde{a}(x_1, \ldots, x_m); \; b(x_1, \ldots, x_m) = x_1^{\mu_1} \ldots x_m^{\mu_m} \tilde{b}(x_1, \ldots, x_m)$$

with $\lambda_k > -1$, $\mu_k > -1$, $k = 1, \ldots, m$ and continuous functions $\tilde{a}(x_1, \ldots, x_m)$, $\tilde{b}(x_1, \ldots, x_m)$ for $x_k \geq 0$, $k = 1, \ldots, m$. Then the equality

$$\int_0^{x_1} \cdots \int_0^{x_1} a(x_1 - \tau_1, \ldots, x_m - \tau_m) \, b(\tau_1, \ldots, \tau_m) \, d\tau_1 \ldots d\tau_m = 0$$

yields that $a \equiv 0$ or $b \equiv 0$ at least.

Let us choose now $a(x_1, \ldots, x_m) = \prod_k \frac{x_k^{\delta_k - 1}}{\Gamma(\delta_k)}$. The conditions $\delta_k > 0$, $k = 1, \ldots, m$ yield $\delta_k - 1 > -1$, $k = 1, \ldots, m$. The function $b(x_1, \ldots, x_m)$ has the form

$$\prod_{k=1}^{m} x_k^{\gamma_k + \frac{p}{\beta}} \tilde{f}\left[(x_1, \ldots, x_m)^{\frac{1}{\beta}}\right] = \prod_{k=1}^{m} x_k^{\mu_k} \tilde{b}(x_1, \ldots, x_m) \text{ with } \mu_k > -1, \; k = 1, \ldots, m.$$

since $p > \max_k [-\beta(\gamma_k + 1)]$ and $\mu_k = \gamma_k + \frac{p}{\beta}$. Then due to the above cited theorem, it follows that $b(x_1, \ldots, x_m) \equiv 0$, that is $f(x) \equiv 0$.

Equation (1.2.36) can be written in the form

$$R_{x_1}^{\delta_1} R_{x_2}^{\delta_2} \ldots R_{x_m}^{\delta_m} \{b(x_1, \ldots, x_m)\} = c(x_1, \ldots, x_m), \tag{1.2.36'}$$

with Riemann-Liouville fractional integrations $R_{x_1}^{\delta_1}, \ldots, R_{x_m}^{\delta_m}$ with respect to the variables $x_1, \ldots, x_m$. In this manner we find the inversion formula

$$D_{x_1}^{\delta_1} D_{x_2}^{\delta_2} \ldots D_{x_m}^{\delta_m} \{c(x_1, \ldots, x_m)\} = b(x_1, \ldots, x_m),$$

or in terms of the functions $f(x)$, $I_{\beta,m}^{(\gamma_k),(\delta_k)} f(x) = g(x) \in C_\alpha^{(\eta_1 + \cdots + \eta_m)}$:

$$f\left((x_1, \ldots, x_m)^{\frac{1}{\beta}}\right) = \left(\prod_{k=1}^{m} x_k^{-\gamma_k}\right) D_{x_1}^{\delta_1} D_{x_2}^{\delta_2} \ldots D_{x_m}^{\delta_m} \left\{ g\left((x_1 \ldots x_m)^{\frac{1}{\beta}}\right) \prod_{k=1}^{m} x_k^{\gamma_k + \delta_k} \right\},$$

which is (1.2.34). The theorem is proved.

Let us note that inversion formula (1.2.34) is a *modification of Dimovski's formula* [68], [69] *for the generalized Sonine transformation* (1.2.30). *Another type of formula* for inversion of the $I_{\beta,m}^{(\gamma_k),(\delta_k)}$-operators , without the use of functions of many variables, will be proposed in Section 1.5. It is related to the G-function representation of the generalized fractional integrals.

Corollary 1.2.16. *The generalized fractional integration operators*

$$Rf(x) = x^{\beta\delta_0} I_{\beta,m}^{(\gamma_k),(\delta_k)} f(x), \;\; \delta_0 \geq 0$$

map C_α *into* $C_{\alpha+\beta\delta_0}^{(\eta_1+\cdots+\eta_m)} \subset C_{\alpha+\beta\delta_0}$.

Results analogous to Theorems 1.2.10 and 1.2.15 can be proved also for other kinds of functional spaces, for example in $L_{\mu,p}$, or $\mathfrak{H}_\alpha(\Omega)$ (see definitions (1.1.4), (1.1.5)). The corresponding result for Lebesgue integrable functions can be found in [434], namely as a suitable specialization of the more general result [434, Theorem 10.7]. This gives the following *analogue of Theorems 1.2.10 and 1.2.15 for the case* $\beta = 1$.

Theorem 1.2.17. *Let us consider* $L^p(0,\infty)$, $p \geq 1$ *and let the conditions*

$$\beta(1+\gamma_k) > \frac{1}{p}, \qquad \delta_k > 0, \quad k = 1,\ldots,m \tag{1.2.37}$$

be satisfied. Then the operators

$$I_\beta^{\gamma_k,\delta_k} = \left[x^{-(\gamma_k+\delta_k)} R^{\delta_k} x^{\gamma_k}\right]_{x\to x^\beta}, \qquad k = 1,\ldots,m \tag{1.2.38}$$

commute, their composition can be represented also in the more concise form

$$I_{\beta,m}^{(\gamma_k),(\delta_k)} f(x) = \int_0^1 G_{m,m}^{m,0}\left[\sigma \left| \begin{matrix} (\gamma_k+\delta_k)_1^m \\ (\gamma_k)_1^m \end{matrix} \right.\right] f(x\sigma^{\frac{1}{\beta}})d\sigma \tag{1.2.39}$$

and this operator maps boundedly the space L^p *into itself:*

$$I_{\beta,m}^{(\gamma_k),(\delta_k)} : L^p(0,\infty) \longrightarrow L^p(0,\infty),$$

namely:

$$\left\| I_{\beta,m}^{(\gamma_k),(\delta_k)} \right\|_{L^p} \leq \prod_{k=1}^m \frac{\Gamma\left(\gamma_k + 1 - \frac{1}{p\beta}\right)}{\Gamma\left(\gamma_k + \delta_k + 1 - \frac{1}{p\beta}\right)} < \infty.$$

In particular, for $p = 1$ we obtain that the conditions under which the operators $I_1^{\gamma_k,\delta_k}$, $k = 1,\ldots,m$, commute and their product $I_{\beta,m}^{(\gamma_k),(\delta_k)}$ acts from a space into itself, are *equivalent* for the spaces C_{-1} ($\alpha = -1$) and $L^1(0,\infty)$ ($p = 1$), namely:

$$\gamma_k > 0, \quad \delta_k \geq 0, \quad k = 1,\ldots,m. \tag{1.2.40}$$

They are corollaries of conditions (1.2.3) and (1.2.37) for $p = 1$.

More general theorems for operators written by means of Fox's $H^{m,0}_{m,m}$-function and in the space $L_{\mu,p}(0,\infty)$ are proved in Chapter 5 (see Theorems 5.1.3 and 5.2.1). Results concerning *analytic functions in starlike domains* have been mentioned in Examples (1.1.c), (1.1.e), (1.1.g), (1.1.h), (1.1.l). Here, we only meniton the corresponding result, analogous to Theorems 1.2.10, 1.2.15, 1.2.17, namely:

Theorem 1.2.18. *Consider the classes of functions (1.1.5):*

$$\mathfrak{H}_\mu(\Omega) = \left\{ f(z) = z^\mu \tilde{f}(z); \tilde{f}(z) \in \mathfrak{H}(\Omega) \right\}, \quad \mathfrak{H}_0(\Omega) := \mathfrak{H}(\Omega),$$

where $\mu \geq 0$ and $\mathfrak{H}(\Omega)$ denotes the space of analytic functions in a domain $\Omega \subset \mathbb{C}$, starlike with respect to the origin $z = 0$. Suppose additionally that the conditions

$$\gamma_k > -\frac{\mu}{\beta} - 1 \ (i.e. \ \ \mu > \max_k [-\beta(\gamma_k + 1)]), \quad \delta_k > 0, \ k = 1, \ldots, m \tag{1.2.41}$$

are satisfied. Then, the Erdélyi-Kober fractional integrals (1.2.38) commute and their composition is represented as a generalized fractional integral (1.2.39):

$$\left[\prod_{k=1}^m I_\beta^{\gamma_k,\delta_k} \right] f(z) = I_{\beta,m}^{(\gamma_k),(\delta_k)} f(z) = \int_0^1 G^{m,0}_{m,m} \left[\sigma \left| \begin{matrix} (\gamma_k + \delta_k)_1^m \\ (\gamma_k)_1^m \end{matrix} \right. \right] f\left(z\sigma^{\frac{1}{\beta}} \right) d\sigma, \ f \in \mathfrak{H}_\mu(\Omega). \tag{1.2.42}$$

This operator maps $\mathfrak{H}_\mu(\Omega)$ into itself, preserving the power functions up to a constant multiplier (as in formula (1.2.1)) and the image of a power series $f \in \mathfrak{H}_\mu(|z| < R) \subseteq \mathfrak{H}_\mu(\Omega)$:

$$I_{\beta,m}^{(\gamma_k),(\delta_k)} \left\{ z^\mu \sum_{n=0}^\infty a_n z^n \right\} = z^\mu \sum_{n=0}^\infty a_n \left[\prod_{k=1}^m \frac{\Gamma\left(\gamma_k + \frac{n+\mu}{\beta} + 1\right)}{\Gamma\left(\gamma_k + \delta_k + \frac{n+\mu}{\beta} + 1\right)} \right] z^n \tag{1.2.43}$$

has the same radius of convergence $R > 0$.

Proof. See Kiryakova [199].

For example, in [413]-[414] Rusev proved that Uspensky's integral transform $P^{(\alpha)}$, (1.1.e) is an isomorphism of a subspace $\mathfrak{H}(\tau_0)$ of the analytic functions in a strip $S(\tau_0)$ and proposed an inversion formula for it. Now, one can obtain the corresponding result by a suitable specialization of the above theorem for $P^{(\alpha)} = \frac{1}{2} I_{2,1}^{-\frac{1}{2},\alpha+\frac{1}{2}}$. By means of this transformation Rusev solved the problem of the representation of analytic functions by series in Laguerre polynomials.

Erdélyi-Kober fractional integrals and derivatives of analytic functions are widely used in Chapter 2, in studying the so-called Džrbashjan-Gelfond-Leontiev operators (1.1.h) (see Sections 2.2-2.4).

A proof of Theorem 1.2.18, in a more general case of fractional integrals involving Fox's H-functions will be given in details in Section 5.5.

1.3. Basic properties of Riemann-Liouville type generalized fractional integrals

Now using the G-function representations (1.1.6) of the linear and homogeneous operators $I_{\beta,m}^{(\gamma_k),(\delta_k)}$:

$$I_{\beta,m}^{(\gamma_k),(\delta_k)}(f+g) = I_{\beta,m}^{(\gamma_k),(\delta_k)}f + I_{\beta,m}^{(\gamma_k),(\delta_k)}g, \tag{1.3.1}$$

$$I_{\beta,m}^{(\gamma_k),(\delta_k)}(af(cx)) = a\left(I_{\beta,m}^{(\gamma_k),(\delta_k)}f\right)(cx), \tag{1.3.2}$$

we derive a *chain of rules* analogous to the well-known results of the theory of classical fractional integrals of Riemann-Liouville and Erdélyi-Kober: R^{δ}, $I_{\beta}^{\gamma,\delta}$.

Lemma 1.3.1. *For $f \in C_{\alpha'}$, $\alpha' = \alpha - \beta\lambda = \max\limits_k[-\beta(\gamma_k+\lambda+1)]$ the operator $I_{\beta,m}^{(\gamma_k),(\delta_k)}$ can be displaced (shifted) with the power functions $x^{\beta\lambda}$ by the following rule:*

$$I_{\beta,m}^{(\gamma_k),(\delta_k)}x^{\beta\lambda}f(x) = x^{\beta\lambda}I_{\beta,m}^{(\gamma_k+\lambda),(\delta_k)}f(x), \quad \lambda\in\mathbb{R}. \tag{1.3.3}$$

Proof. Property (A.14) of the G-function in representation (1.1.6) be used, namely:

$$\begin{aligned}
I_{\beta,m}^{(\gamma_k),(\delta_k)}\left\{x^{\beta\lambda}f(x)\right\} &= \int_0^1 G_{m,m}^{m,0}\left[\sigma\left|\begin{matrix}(\gamma_k+\delta_k)_1^m\\(\gamma_k)_1^m\end{matrix}\right.\right]x^{\beta\lambda}\sigma^{\lambda}f\left(x\sigma^{\frac{1}{\beta}}\right)d\sigma\\
&= x^{\beta\lambda}\int_0^1 G_{m,m}^{m,0}\left[\sigma\left|\begin{matrix}(\gamma_k+\delta_k)_1^m\\(\gamma_k)_1^m\end{matrix}\right.\right]\sigma^{\lambda}f\left(x\sigma^{\frac{1}{\beta}}\right)d\sigma\\
&= x^{\beta\lambda}\int_0^1 G_{m,m}^{m,0}\left[\sigma\left|\begin{matrix}(\gamma_k+\lambda+\delta_k)_1^m\\(\gamma_k+\lambda)_1^m\end{matrix}\right.\right]f\left(x\sigma^{\frac{1}{\beta}}\right)d\sigma\\
&= x^{\beta\lambda}I_{\beta,m}^{(\gamma_k+\lambda),(\delta_k)}f(x).
\end{aligned}$$

Lemma 1.3.2. *For $f \in C_{\alpha}^{(l)}$, $\alpha \geq \max\limits_k[-\beta(\gamma_k+1)]$, $l \geq 0$ the following relationship between the j-th derivatives ($j = 0,1,2,\ldots,l$) of $I_{\beta,m}^{(\gamma_k),(\delta_k)}f$ and f at the point zero holds:*

$$\left\{I_{\beta,m}^{(\gamma_k),(\delta_k)}f(x)\right\}^{(j)}(0) = c_j f^{(j)}(0), \;\; \textit{where } j = 0,1,2,\ldots,l \tag{1.3.4}$$

and $c_j = \prod\limits_{k=1}^{m}\frac{\Gamma\left(\gamma_k+\frac{j}{\beta}+1\right)}{\Gamma\left(\gamma_k+\delta_k+\frac{j}{\beta}+1\right)}$.

Proof. Since $\left(\frac{d}{dx}\right)^j f\left(x\sigma^{\frac{1}{\beta}}\right) = \sigma^{\frac{j}{\beta}}\left(\frac{d}{du}\right)^j f(u)$, where $u = x\sigma^{\frac{1}{\beta}}$ is substituted, then differentiating under the integral sign in (1.1.6) and using (A.14) again, we obtain:

$$\begin{aligned}
\{If(x)\}_{x=0}^{(j)} &= \left\{\left(\frac{d}{dx}\right)^j \int_0^1 G_{m,m}^{m,0}(\sigma) f\left(x\sigma^{\frac{1}{\beta}}\right) d\sigma\right\}_{x=0} \\
&= \left\{\int_0^1 G_{m,m}^{m,0}(\sigma)\left(\frac{d}{dx}\right)^{(j)} f\left(x\sigma^{\frac{1}{\beta}}\right) d\sigma\right\}_{x=0} \\
&= \left\{\int_0^1 G_{m,m}^{m,0}\left[\sigma \left| \begin{matrix} \left(\gamma_k + \delta_k + \frac{j}{\beta}\right)_1^m \\ \left(\gamma_k + \frac{j}{\beta}\right)_1^m \end{matrix}\right.\right] f^{(j)}\left(x\sigma^{\frac{1}{\beta}}\right) d\sigma\right\}_{x=0} \\
&= f^{(j)}(0) \int_0^1 G_{m,m}^{m,0}\left[\sigma \left| \begin{matrix} \left(\gamma_k + \delta_k + \frac{j}{\beta}\right)_1^m \\ \left(\gamma_k + \frac{j}{\beta}\right)_1^m \end{matrix}\right.\right] d\sigma.
\end{aligned}$$

Then, the above integral is evaluated according to (B.4) (see Appendix, Lemma B.2) and this gives

$$\{If(x)\}_{x=0}^{(j)} = c_j f^{(j)}(0)$$

with c_j as in the statement of Lemma 1.3.2.

Further, for brevity we denote by $\mathfrak{R}_\beta$ the class of all the operators of the form (1.1.6) with the same $\beta > 0$, and analogously, by $\mathfrak{W}_\beta$ the class of operators (1.1.9) with $\beta > 0$.

Lemma 1.3.3. *The mapping*

$$\Xi : f(x) \longrightarrow \hat{f}(x) = f\left(x^\omega\right), \ \omega > 0, \tag{1.3.5}$$

is a similarity between the operators $I_{\beta,m}^{(\gamma_k),(\delta_k)}$ *and* $I_{\beta_1,m}^{(\gamma_k),(\delta_k)}$ *with different* $\beta > 0$, $\beta_1 = \beta\omega > 0$, *that is*

$$I_{\beta,m}^{(\gamma_k),(\delta_k)} f(x) = \Xi^{-1}\left(I_{\beta\omega,m}^{(\gamma_k),(\delta_k)} \Xi f(x)\right), \tag{1.3.6}$$

or briefly:

$$\Xi\mathfrak{R}_\beta = \mathfrak{R}_{\beta\omega}\Xi.$$

This result allows us to consider, without loss of generality, *only the case* $\beta = 1$, *or to transfer the results of the class* $\mathfrak{R}_\beta$ *into the corresponding results for* $\mathfrak{R}_{\beta_1}$.

Lemma 1.3.4. *If* $I = I_{\beta,m}^{(\gamma_k),(\delta_k)} \in \mathfrak{R}_\beta$, $W = W_{\beta,m}^{(\gamma_k),(\delta_k)} \in \mathfrak{W}_\beta$ *are the corresponding generalized fractional integrals of Riemann-Liouville type and of Weyl type respectively, then*

$$x^{-2\beta} I_{\frac{1}{x}}\left\{x^{-2\beta} f\left(\frac{1}{x}\right)\right\} = W_x\{f(x)\}, \tag{1.3.7}$$

where the notation

$$I_{\varphi(x)}f(x) = (If)_{x\to\varphi(x)}\,, W_{\varphi(x)}f(x) = (Wf)_{x\to\varphi(x)}$$

is used. Furthermore, the formula for generalized fractional integration by parts holds:

$$\int_0^\infty xf(x)I_{x^{\frac{1}{\beta}}}\left\{g\left(x^\beta\right)\right\}dx = \int_0^\infty xg(x)W_{x^{\frac{1}{\beta}}}\left\{f\left(x^\beta\right)\right\}dx. \tag{1.3.8}$$

The above results include as *particular cases* some known properties of Riemann-Liouville and Erdélyi-Kober fractional integrals, as for instance:

Corollary 1.3.5.

$$x^\lambda I_\beta^{\gamma,\delta} = I_\beta^{\gamma-\frac{\lambda}{\beta},\delta}x^\lambda f(x) \tag{1.3.3'}$$

$$I_\beta^{\gamma,\delta}f(x) = \left(I_1^{\gamma,\delta}f\left(x^{\frac{1}{\beta}}\right)\right)_{x\to x_\beta} = \left(x^{-(\gamma+\delta)}R^\delta x^\gamma f\left(x^{\frac{1}{\beta}}\right)\right)_{x\to x_\beta}, \tag{1.3.6'}$$

$$\int_0^\infty xf(x)I_1^{\gamma,\delta}g(x)dx = \int_0^\infty xg(x)K_1^{\gamma,\delta}f(x)dx,\ \beta=1 \tag{1.3.8'}$$

(cf. McBride [289, (2.24)], Sneddon [454, p. 41, (2.14)], [455] etc.).

Corollary 1.3.6. *For operators (1.1.l), (1.1.l*) the following results of Saigo [415] can be obtained from (1.3.3) and (1.3.8):*

$$I^{\alpha,\beta,\eta}x^{\beta-\eta}f(x) = I^{\alpha,\eta,\beta}f(x),$$
$$I^{\alpha,\beta,\eta}f(x) = x^{-\alpha-\beta-\eta}I^{\alpha,-\alpha-\eta,-\alpha-\beta}f(x)$$

and

$$\int_0^\infty f(x)I^{\alpha,\beta,\eta}g(x)dx = \int_0^\infty g(x)J^{\alpha,\beta,\eta}f(x)dx.$$

Corollary 1.3.7. *For Uspensky's transformation (1.1.e) Lemma 1.3.2 yields the formulas (see Rusev [414, p. 53]):*

$$\left(P^{(\alpha)}f\right)^{(j)}(0) = \frac{\Gamma\left(\frac{j}{2}+\frac{1}{2}\right)}{2\Gamma\left(\alpha+\frac{j}{2}+\frac{1}{2}\right)}f^{(j)}(0), \quad j=0,1,2,\dots \tag{1.3.4'}$$

The well-known *first index law (law of exponents, semigroup property)* of the classical fractional calculus

$$R^\sigma R^\delta = R^\delta R^\sigma = R^{\sigma+\delta}, \qquad \sigma>0, \delta>0 \tag{1.3.9}$$

has the following *generalization* in our theory.

Theorem 1.3.8. *For fixed* $\beta > 0$, $m \geq 1$,

a) *the operators (1.1.6) of the class* $\mathfrak{R}_\beta$ *commute:*

$$I_{\beta,m}^{(\tau_k),(\sigma_k)} I_{\beta,m}^{(\gamma_k),(\delta_k)} = I_{\beta,m}^{(\gamma_k),(\delta_k)} I_{\beta,m}^{(\tau_k),(\sigma_k)} \tag{1.3.10}$$

for $\sigma_k > 0$, $\delta_k > 0$, $k = 1, \ldots, m$ in the space C_α with

$$\alpha \geq \max_k \{-\beta(\tau_k + 1);\ -\beta(\gamma_k + 1)\}.$$

b) The following *index law*

$$I_{\beta,m}^{(\gamma_k+\delta_k),(\sigma_k)} I_{\beta,m}^{(\gamma_k),(\delta_k)} = I_{\beta,m}^{(\gamma_k),(\delta_k+\sigma_k)}, \quad \sigma_k > 0, \delta_k > 0; k = 1, \ldots, m \tag{1.3.11}$$

holds in C_α, $\alpha \geq \max_k [-\beta(\gamma_k + 1)]$.

c) The composition of two operators $I_{\beta,m_1}^{(\gamma'_k),(\delta'_k)}$, $I_{\beta,m_2}^{(\gamma''_k),(\delta''_k)}$ with the same $\beta > 0$ and different multiplicities $m_1 \geq 1$, $m_2 \geq 1$ is an operator of the same form (1.1.6) but of multiplicity $m = m_1 + m_2$, namely:

$$\begin{aligned} I_{\beta,m_1}^{(\gamma'_k),(\delta'_k)} I_{\beta,m_2}^{(\gamma''_k),(\delta''_k)} f(x) &= I_{\beta,m_1+m_2}^{(\gamma_k),(\delta_k)} f(x) \\ &= \int_0^1 G_{m_1+m_2,m_1+m_2}^{m_1+m_2,0}\left[\sigma \,\middle|\, \begin{matrix}(\gamma_k+\delta_k)_1^{m_1+m_2} \\ (\gamma_k)_1^{m_1+m_2}\end{matrix}\right] f\left(x\sigma^{\frac{1}{\beta}}\right) d\sigma, \end{aligned} \tag{1.3.12}$$

where the set of parameters $(\gamma_k)_1^{m_1+m_2}$, $(\delta_k)_1^{m_1+m_2}$ is taken as follows:

$$\gamma_k = \begin{cases} \gamma'_k, & k = 1, \ldots, m_1 \\ \gamma''_{k-m_1}, & k = m_1+1, \ldots, m_1+m_2 \end{cases}, \quad \delta_k = \begin{cases} \delta'_k, & k = 1, \ldots, m_1 \\ \delta''_{k-m_1}, & k = m_1+1, \ldots, m_1+m_2 \end{cases}.$$

Proof.

a) For functions $f \in C_\alpha$, $\alpha > -\beta(\gamma_k + 1)$, $k = 1, \ldots, m$, the image $f^*(x) = I_{\beta,m}^{(\gamma_k),(\delta_k)} f(x) \in C_\alpha$ and therefore $f^{**}(x) = I_{\beta,m}^{(\tau_k),(\sigma_k)} f^*(x)$ is also defined in C_α, $\alpha > -\beta(\tau_k + 1)$, $k = 1, \ldots, m$, and $f^{**}(x) \in C_\alpha$ too. After a substitution $y = \sigma\tau$ in the inner integral for $f^{**}(x)$ we obtain:

$$\begin{aligned} f^{**}(x) &= \int_0^1 G_{m,m}^{m,0}\left[\sigma \,\middle|\, \begin{matrix}(\tau_k+\sigma_k)_1^m \\ (\tau_k)_1^m\end{matrix}\right] d\sigma \int_0^1 G_{m,m}^{m,0}\left[\tau \,\middle|\, \begin{matrix}(\gamma_k+\delta_k)_1^m \\ (\gamma_k)_1^m\end{matrix}\right] f\left(x\sigma^{\frac{1}{\beta}}\tau^{\frac{1}{\beta}}\right) d\tau \\ &= \int_0^1 G_{m,m}^{m,0}\left[\sigma \,\middle|\, \begin{matrix}(\tau_k+\sigma_k-1)_1^m \\ (\tau_k-1)_1^m\end{matrix}\right] d\sigma \int_0^\sigma G_{m,m}^{m,0}\left[\frac{y}{\sigma} \,\middle|\, \begin{matrix}(\gamma_k+\delta_k)_1^m \\ (\gamma_k)_1^m\end{matrix}\right] f\left(xy^{\frac{1}{\beta}}\right) dy. \end{aligned}$$

Changing the order of integrations, using (A.15) and taking into account (A.12), (A.12′), viz.

$$G_{m,m}^{m,0}(\sigma) \equiv 0 \text{ for } \sigma > 1; \quad G_{m,m}^{m,0}\left(\frac{\sigma}{y}\right) \equiv 0 \text{ for } 0 < \sigma < y,$$

one gets:

$$f^{**}(x)=\int_0^1 f\left(xy^{\frac{1}{\beta}}\right)dy\int_y^1 G_{m,m}^{m,0}\left[\sigma\left|\begin{matrix}(\tau_k+\sigma_k-1)_1^m\\(\tau_k-1)_1^m\end{matrix}\right.\right]G_{m,m}^{0,m}\left[\frac{1}{y}\sigma\left|\begin{matrix}(1-\gamma_k)_1^m\\(1-\gamma_k-\delta_k)_1^m\end{matrix}\right.\right]d\sigma$$

$$=\int_0^1 f\left(xy^{\frac{1}{\beta}}\right)dy\int_0^\infty G_{m,m}^{m,0}\left[\sigma\left|\begin{matrix}(\tau_k+\sigma_k-1)_1^m\\(\tau_k-1)_1^m\end{matrix}\right.\right]G_{m,m}^{0,m}\left[\frac{1}{y}\sigma\left|\begin{matrix}(1-\gamma_k)_1^m\\(1-\gamma_k-\delta_k)_1^m\end{matrix}\right.\right]d\sigma.$$

Then, the new inner integral can be evaluated by formula (A.29) and its value is:

$$yG_{2m,2m}^{2m,0}\left[y\left|\begin{matrix}(\gamma_k+\delta_k-1)_1^m,(\tau_k+\sigma_k-1)_1^m\\(\gamma_k-1)_1^m\,(\tau_k-1)_1^m\end{matrix}\right.\right]=G_{2m,2m}^{2m,0}\left[y\left|\begin{matrix}(\gamma_k+\delta_k)_1^m,(\tau_k+\sigma_k)_1^m\\(\gamma_k)_1^m\,(\tau_k)_1^m\end{matrix}\right.\right]$$

whence:

$$f^{**}(x)=\int_0^1 G_{2m,2m}^{2m,0}\left[y\left|\begin{matrix}(\gamma_k+\delta_k)_1^m,(\tau_k+\sigma_k)_1^m\\(\gamma_k)_1^m,(\tau_k)_1^m\end{matrix}\right.\right]dy. \tag{1.3.12'}$$

For $\alpha>-\beta(\gamma_k+1)$, $\alpha>-\beta(\tau_k+1)$, $\delta_k>0$, $k=1,\ldots,m$, the latter integral is absolutely convergent (which is to justify the correct change in the order of integrations) and defines a $2m$-tuple operator of generalized fractional integration:

$$f^{**}(x)=I_{\beta,2m}^{((\tau_k),(\gamma_k)),((\sigma_k),(\delta_k))}f(x).$$

Its kernel-function $G_{2m,2m}^{2m,0}$ is symmetric in the groups parameters in the first row as well as in the second row. Therefore, the representation obtained for $f^{**}(x)$ is symmetric with respect of the parameters of the m-tuple generalized integrals whose product it is. This shows that they commute:

$$\begin{aligned}I_{\beta,m}^{(\tau_k),(\sigma_k)}I_{\beta,m}^{(\gamma_k),(\delta_k)}&=I_{\beta,2m}^{((\tau_k),(\gamma_k)),((\sigma_k),(\delta_k))}\\&=I_{\beta,m}^{(\gamma_k),(\delta_k)}I_{\beta,m}^{(\tau_k),(\sigma_k)}.\end{aligned}$$

b) The condition $\alpha>-\beta(\gamma_k+1)$, yields also $\alpha>-\beta(\gamma_k+\delta_k+1)$, since $\delta_k>0$, $k=1,\ldots,m$. From the representation of the form (1.3.12′) for composition $I_{\beta,m}^{(\gamma_k+\delta_k),(\sigma_k)}I_{\beta,m}^{(\gamma_k),(\delta_k)}$, we find

$$f^{**}(x)=\int_0^1 G_{2m,2m}^{2m,0}\left[y\left|\begin{matrix}(\gamma_k+\delta_k+\sigma_k)_1^m,(\gamma_k+\delta_k)_1^m\\(\gamma_k+\delta_k)_1^m,(\gamma_k)_1^m\end{matrix}\right.\right]f\left(x\sigma^{\frac{1}{\beta}}\right)d\sigma.$$

Due to property (A.13′), the equal parameters $(\gamma_k+\delta_k)_1^m$ in the first and second rows of the $G^{2m,0}_{2m,2m}$-function cancel each other and it turns into a $G^{m,0}_{m,m}$-function, i.e.

$$f^{**}(x) = \int_0^1 G^{m,0}_{m,m}\left[y \left| \begin{matrix} (\gamma_k + \delta_k + \sigma_k)_1^m \\ (\gamma_k)_1^m \end{matrix} \right.\right] f\left(xy^{\frac{1}{\beta}}\right) dy$$
$$= I^{(\gamma_k+\delta_k),(\sigma_k)}_{\beta,m} f(x),$$

which proves b).

c) The proof is analogous to that of a), leading to the representation of the form (1.3.12′).

The following *index laws* (*product rules*) are the simplest corollaries from Theorem 1.3.8.

Corollary 1.3.9. *For the Erdélyi-Kober operators we obtain the well-known formulas (Sneddon [454, (2.11)], McBride [289; (2.26), (2.29)]):*

$$I^{\tau,\sigma}_{\beta} I^{\gamma,\delta}_{\beta} = I^{\gamma,\delta}_{\beta} I^{\tau,\sigma}_{\beta}, \quad \tau > 0, \sigma > 0; \tag{1.3.10′}$$
$$I^{\gamma+\delta,\sigma}_{\beta} I^{\gamma,\delta}_{\beta} = I^{\gamma,\sigma+\delta}_{\beta}, \quad \sigma > 0, \delta > 0. \tag{1.3.11′}$$

Corollary 1.3.10. *The following product rules for the hypergeometric fractional integrals (see Saigo [415]-[420]):*

$$\begin{aligned} I^{\alpha,\beta,\eta} I^{\gamma,\delta,\alpha+\eta} &= I^{\alpha+\gamma,\beta+\delta,\eta} \\ I^{\alpha,\beta,\eta} I^{\gamma,\delta,\eta-\beta-\gamma-\delta} &= I^{\alpha+\gamma,\beta+\delta,\eta-\gamma-\delta} \end{aligned} \tag{1.3.13}$$

follow from Theorem 1.3.8, b) for the two-tuple operators ($m = 2$)

$$I^{\alpha,\beta,\eta} = x^{-\beta} I^{(\eta-\beta,0),(-\eta,\alpha+\eta)}_{1,2}.$$

Then, the product of two operators of this kind (see [415, (3.12)]) can be represented in the form

$$\begin{aligned} I^{\alpha,\beta,\eta} I^{\gamma,\delta,\zeta} f(x) &= x^{-\beta} I^{(\eta-\beta,0),(-\eta,\alpha+\eta)}_{1,2} x^{-\delta} I^{(\zeta-\delta,0),(-\zeta,\gamma+\zeta)}_{1,2} \\ &= x^{-(\beta+\delta)} I^{(\eta-\beta-\delta,-\delta,\zeta-\delta,0),(-\eta,\alpha+\eta,-\zeta,\gamma+\zeta)}_{1,4} f(x) \\ &= x^{-(\beta+\delta)} \int_0^1 G^{4,0}_{4,4}\left[\sigma \left| \begin{matrix} -(\beta+\delta), \alpha+\eta-\delta, -\delta, \gamma+\zeta \\ \eta-\beta-\delta, -\delta, \zeta-\delta, 0 \end{matrix} \right.\right] f(x\sigma) d\sigma \end{aligned}$$

(a corollary of (1.3.12)). Taking into account property (A.13′):

$$G^{4,0}_{4,4}\left[\sigma \left| \begin{matrix} -(\beta+\delta), \alpha+\eta-\delta, -\delta, \gamma+\zeta \\ \eta-\beta-\delta, -\delta, \zeta-\delta, 0 \end{matrix} \right.\right] = G^{3,0}_{3,3}\left[\sigma \left| \begin{matrix} -(\beta+\delta), \alpha+\eta-\delta, \gamma+\zeta \\ \eta-\beta-\delta, \zeta-\delta, 0 \end{matrix} \right.\right],$$

we obtain the following new result: The product of two Saigo's operators of the form (1.1.l-l') can be considered also as a 3-tuple generalized fractional integral having the single integral representation

$$I^{\alpha,\beta,\eta} I^{\gamma,\delta,\zeta} f(x) = x^{-(\beta+\delta)} \int_0^1 G_{3,3}^{3,0} \left[\sigma \left| \begin{matrix} -(\beta+\delta), \alpha+\eta-\delta, \gamma+\zeta \\ \eta-\beta-\delta, \zeta-\delta, 0 \end{matrix} \right. \right] f(x\sigma) d\sigma, \quad (1.3.14)$$

or equivalently, as a composition of three Erdélyi-Kober operators

$$\begin{aligned} I^{\alpha,\beta,\eta} I^{\gamma,\delta,\zeta} f(x) &= x^{-(\beta+\delta)} I_{1,3}^{(\eta-\beta-\delta,\zeta-\delta,0),(-\eta,\alpha+\eta-\zeta,\gamma+\zeta)} f(x) \\ &= x^{-(\beta+\delta)} I_1^{\eta-\beta-\delta,-\eta} I_1^{\zeta-\delta,\alpha+\eta-\zeta} I_1^{0,\gamma+\zeta} f(x). \end{aligned} \quad (1.3.15)$$

Compositions of the form $I^{\alpha,\beta,\eta} J^{\gamma,\delta,\zeta}$, $J^{\alpha,\beta,\eta} I^{\gamma,\delta,\zeta}$ involving both Riemann-Liouville and the Weyl type hypergeometric fractional integrals are considered by Saigo [420] and Srivastava and Saigo [480].

For *arbitrary* $m \geq 1$ the above properties will be used in Chapters 3, 4 and 5.

1.4. Some properties of Weyl type generalized fractional integrals $W_{\beta,m}^{(\gamma_k),(\delta_k)}$

First, let us note that referring to the operator

$$W^\delta f(x) = \int_x^\infty \frac{(t-x)^{\delta-1}}{\Gamma(\delta)} f(t) dt = x^\delta \int_0^1 \frac{(\sigma-1)^{\delta-1}}{\Gamma(\delta)} f(x\sigma) d\sigma \quad (1.4.1)$$

as the *Weyl operator* is a historical misunderstanding since the fractional integration over an infinite integral had been introduced by Liouville (1834) (see [434, p.13]). That is why there is a trend in some papers to call the Riemann-Liouville type operators R^δ (1.1.b), $I_\beta^{\gamma,\delta}$ (1.1.17), $I_{\beta,m}^{(\gamma_k),(\delta_k)}$ (1.1.6) *"right-hand sided" fractional integrals* and respectively, the operators W^δ (1.1.b*)-(1.4.1), $K_\beta^{\gamma,\delta}$ (1.1.17*), $W_{\beta,m}^{(\tau_k),(\alpha_k)}$ (1.1.9) *"left-hand sided" fractional integrals.* Nevertheless, to keep the established tradition, we refer to operators (1.1.9):

$$W_{\beta,m}^{(\tau_k),(\alpha_k)} f(x) = \int_1^\infty G_{m,m}^{m,0} \left[\frac{1}{\sigma} \left| \begin{matrix} (\tau_k+\alpha_k+1)_1^m \\ (\tau_k+1)_1^m \end{matrix} \right. \right] f\left(x\sigma^{\frac{1}{\beta}}\right) d\sigma, \quad (1.4.2)$$

as *Weyl type generalized (m-tuple) fractional integrals.*

The Weyl fractional operator (1.4.1) has been defined e.g. by Miller [309], in the so-called *class $\mathfrak{A}$ of good functions* introduced by Lighthill [257, p. 15]. One says that a function $f(x)$ is a good function, if it is everywhere differentiable any number of times and if it and all its derivatives are $\mathcal{O}\left(x^{-N}\right)$ as x increases without limit, for all N. Examples of good functions are $f(x) = P(x)\exp(-\gamma x)$, where $P(x)$ is a polynomial, $\gamma > 0$.

Similarly, two-dimensional analogues of (1.4.1), the so-called *two-dimensional Erdélyi-Kober operators of fractional integration* of orders $\alpha > 0$, $\beta > 0$:

$$K^{\eta,\alpha} K^{\delta,\beta} f(x,y) = \frac{x^{\eta} y^{\delta}}{\Gamma(\alpha)\Gamma(\beta)} \int_x^{\infty} \int_y^{\infty} t^{-\eta-\alpha} \tau^{-\delta-\beta} (t-x)^{\alpha-1} (\tau - y)^{\beta-1} f(t,\tau) dt d\tau,$$

have been considered by different authors in the *modified class $\mathfrak{A}$ of good functions*: $f(x,y)$ – differentiable any number of times and having partial derivatives which are $\mathcal{O}\left(|x|^{-\varepsilon_1}, |y|^{-\varepsilon_2}\right)$ as $x \to \infty$, $y \to \infty$, for all $\varepsilon_1 > 0$, $\varepsilon_2 > 0$. See e.g. Raina [384]-[388], Raina and Koul [393], Raina and Kiryakova [390], Saxena and Ram [441], R. Srivastava [483], Saxena and Kiryakova [438], Saxena, Kiryakova and Davie [439], etc.

Here, we consider operators (1.4.2) in the spaces

$$C^*_{\alpha^*} = \left\{ f(x) = x^q \hat{f}(x);\ q < \alpha^*, \hat{f} \in C[0,\infty),\ \left|\hat{f}\right| \le A_{\hat{f}} \right\} \tag{1.4.3}$$

and in the more general ones (1.1.2), since it is natural that the integration over an infinite interval requires additional conditions on the growth of $f(x)$ near $x = \infty$.

These properties can be proved either directly, following the method of proving their Riemann-Liouville type counterparts (see Sections 1.2, 1.3), or by using the relationship (1.3.7) (Lemma 1.3.4) between the Riemann-Liouville and Weyl type integrals.

Lemma 1.4.1. *The multiple Erdélyi-Kober fractional integrals (1.4.2) are well defined in the spaces $C^{\alpha}_{\alpha^*}$ with $\alpha^* \le \min_k (\beta\tau_k)$ and preserve the power functions up to constant multipliers:*

$$W^{(\tau_k),(\alpha_k)}_{\beta,m} \{x^q\} = d_q x^q, \qquad \textit{where} \tag{1.4.4}$$

$$d_q = \prod_{k=1}^{m} \frac{\Gamma\left(\tau_k - \frac{q}{\beta}\right)}{\Gamma\left(\tau_k + \alpha_k - \frac{q}{\beta}\right)}, \qquad q < \alpha^*.$$

Lemma 1.4.2. *The $W^{(\tau_k),(\alpha_k)}_{\beta,m}$-image of a G-function of $C^*_{\alpha^*}$ is another G-function, for example:*

$$W^{(\tau_k),(\alpha_k)}_{\beta,m} \left\{ G^{\mu,\nu}_{\sigma,\tau} \left[\omega x^{\beta} \middle| \begin{matrix} (c_i)_1^{\sigma} \\ (d_j)_1^{\tau} \end{matrix} \right] \right\} = G^{\mu+m,\nu}_{\sigma+\mu,\tau+\mu} \left[\omega x^{\beta} \middle| \begin{matrix} (c_i)_1^{\sigma}, (\tau_k+\alpha_k)_1^m \\ (\tau_k)_1^m, (d_j)_1^{\tau} \end{matrix} \right]. \tag{1.4.5}$$

Then, as particular cases the generalized Weyl images of various elementary and special functions can be obtained.

Theorem 1.4.3. *In the subspace (1.2.10): $C_{\alpha,\alpha^*} \subset C^*_{\alpha^*}$, the Mellin transform of the generalized Weyl fractional integrals is represented by the relation*

$$\mathfrak{M}\left\{ W_{\beta,m}^{(\tau_k),(\alpha_k)} f(x); s \right\} = \prod_{k=1}^{m} \frac{\Gamma\left(\tau_k + \frac{s}{\beta}\right)}{\Gamma\left(\tau_k + \alpha_k + \frac{s}{\beta}\right)} \mathfrak{M}\left\{f(x); s\right\}. \tag{1.4.6}$$

Theorem 1.4.4. *The generalized (m-tuple) Weyl type fractional integrals (1.4.2) can be represnted in $C^*_{\alpha^*}$, $\alpha^* \le \min\limits_k (\beta\tau_k)$ as m-tuple compositions of commuting Erdélyi-Kober operators of Weyl type of the form (1.1.17*):*

$$K_\beta^{\tau,\alpha} f(x) = \int_1^\infty \frac{(\sigma-1)^{\alpha-1}\sigma^{-(\tau+\alpha)}}{\Gamma(\alpha)} f(x\sigma) d\sigma := W_{\beta,1}^{\tau,\alpha} f(x), \tag{1.4.7}$$

namely:

$$\begin{aligned} W_{\beta,m}^{(\tau_k),(\alpha_k)} f(x) &= \left(\prod_{k=1}^{m} K_\beta^{\tau_k,\alpha_k} \right) f(x) \\ &= \int_1^\infty \cdots \int_1^\infty \left[\prod_{k=1}^{m} \frac{(\sigma-1)^{\alpha_k-1}\sigma^{-(\tau_k+\alpha_k)}}{\Gamma(\alpha_k)} \right] f\left[x(\sigma_1 \ldots \sigma_m)^{\frac{1}{\beta}} \right] d\sigma_1 \ldots d\sigma_m. \end{aligned} \tag{1.4.8}$$

Conversely, the operators of the form (1.4.8) can be put in the single integral form (1.4.2) involving Meijer's $G^{m,0}_{m,m}$-function.

Lemma 1.4.5. *For $f \in C^*_{\alpha^{**}}$, $\alpha^{**} = \alpha^* - \beta\lambda \le \min\limits_k [\beta(\tau_k - \lambda)]$, the following shift property holds:*

$$W_{\beta,m}^{(\tau_k),(\alpha_k)} x^{\beta\lambda} f(x) = x^{\beta\lambda} W_{\beta,m}^{(\tau_k-\lambda),(\alpha_k)} f(x). \tag{1.4.9}$$

Corollary 1.4.6. *For the Weyl type Erdélyi-Kober operator (1.1.17*) and Saigo's operator (1.1.l*) the shift properties take the forms:*

$$\begin{gathered} K_\beta^{\tau+\frac{\lambda}{\beta},\alpha} x^\lambda = x^\lambda K_\beta^{\tau,\alpha}, \\ J^{\alpha,\beta,\eta} x^{\alpha+\beta+\eta} = J^{\alpha,-\alpha-\eta,-\alpha-\beta} \end{gathered} \tag{1.4.10}$$

(cf. McBride [289, (2.25)], Saigo [415, (1.12)]).

Theorem 1.4.7. *For fixed $\beta > 0$ and integer $m \ge 1$, if $\alpha_k > 0$, $\delta_k > 0$, $k = 1, \ldots, m$, then the Weyl type multiple fractional integrals (1.4.2) have the following properties:*

a) *They commute in* $C^*_{\alpha^*}$, $\alpha^* \le \min_k (\beta\tau_k, \beta\gamma_k)$:

$$\begin{aligned} W^{(\tau_k),(\alpha_k)}_{\beta,m} W^{(\gamma_k),(\delta_k)}_{\beta,m} &= W^{(\gamma_k),(\delta_k)}_{\beta,m} W^{(\tau_k),(\alpha_k)}_{\beta,m} \\ &= W^{((\gamma_k),(\delta_k)),((\tau_k),(\alpha_k))}_{\beta,2m} \end{aligned} \tag{1.4.11}$$

(in general, the product of m_1-tuple and m_2-tuple fractional integrals is a (m_1+m_2)-tuple one);

b) *The index law has the form:*

$$W^{(\tau_k+\alpha_k),(\delta_k)}_{\beta,m} W^{(\tau_k),(\alpha_k)}_{\beta,m} = W^{(\tau_k),(\delta_k+\alpha_k)}_{\beta,m}; \tag{1.4.12}$$

c) *The formal inversion formula holds:*

$$\left\{ W^{(\tau_k),(\alpha_k)}_{\beta,m} \right\}^{-1} = W^{(\tau_k+\alpha_k),(-\alpha_k)}_{\beta,m}, \tag{1.4.13}$$

since $W^{(\tau_k),(0,\dots,0)}_{\beta,m} = I$ *is adopted to be the identity operator in* $C^*_{\alpha^*}$, $\alpha^* \le \min_k (\beta\tau_k)$.

The explicit inversion formula corresponding to (1.4.13) can be written by means of a differintegral expression in the way used in Section 1.5 for the Riemann-Liouville type operators. This is to generalize the *fractional derivatives of Weyl type* (Weyl fractional integrals (1.4.1), (1.4.7) of negative order):

$$W^{-\delta} f(x) := D^{\delta}_{*} f(x) = (-1)^n \left(\frac{d}{dx}\right)^n W^{n-\delta} f(x) \tag{1.4.14}$$

with $\delta > 0$ and integer n so that $0 \le n - \delta < 1$;

$$\begin{aligned} \left\{K_2^{\tau,\alpha}\right\}^{-1} f(x) &= K_2^{\tau+\alpha,-\alpha} f(x) \\ &= (-1)^{\eta} x^{2\tau+2\alpha-1} \left(\frac{1}{2}\frac{d}{dx} x^{-1}\right)^{\eta} x^{2\eta+2\tau+2\alpha+1} \left\{ K_2^{\tau-\eta+\alpha,\eta-\alpha} f(x) \right\} \end{aligned} \tag{1.4.15}$$

with $\alpha > 0$, $\beta = 2$ and integer η so that $0 \le \eta - \alpha < 1$ (see Sneddon [452], [455, p. 9]).

Another type of inversion formula for operators $W^{(\tau_k),(\alpha_k)}_{\beta,m}$ in terms of functions of many variables and similar to (1.2.34) holds too. A special case of it is found and used in Kiryakova [191], see Chapter 3 (Section 3.7.iii). From (1.4.4) and (1.4.13) one can easily deduce the following theorem.

Theorem 1.4.8. *The Weyl type generalized fractional integral (multiple Erdélyi-Kober operator)* $W^{(\tau_k),(\alpha_k)}_{\beta,m}$, $\alpha_k > 0$, $k = 1,\dots,m$, *is an invertable linear mapping of* $C^*_{\alpha^*}$, $\alpha^* \le \min_k (\beta\tau_k)$ *into itself. (If* α^* *is fixed, then we require* $\tau_k > \frac{\alpha^*}{\beta}$, $\alpha_k > 0$, $k = 1,\dots,m$.*)*

The corresponding result for the functions of the spaces $L_{\mu,p}(0,\infty)$ and for more general fractional integrals, involving Fox's H-function instead of Meijer's G-function, is discussed in Chapter 5. For the simpler case of operators (1.4.2) with $\beta = 1$ and spaces $L^p(0,\infty)$, it can be found for example in Samko, Kilbas and Marichev [434, p. 159, Theorem 10.7]. Here we state the following theorema.

Theorem 1.4.9. *Let us consider $L^p(0,\infty)$, $p \geq 1$, and let the conditions*

$$\beta\tau_k > \frac{1}{p}, \quad \alpha_k > 0, \quad k = 1,\dots,m \tag{1.4.16}$$

be satisfied. Then, the operators $K_\beta^{\tau_k,\alpha_k}$, $k = 1,\dots,m$, of the form (1.4.7) commute, their composition can be represented in the single integral form (1.4.2) and the operator $W_{\beta,m}^{(\tau_k),(\alpha_k)}$, obtained in this way, maps boundedly the space $L^p(o,\infty)$ into itself, namely:

$$\left\| K_{\beta,m}^{(\tau_k),(\alpha_k)} \right\|_{L_p} \leq \prod_{k=1}^{m} \frac{\Gamma\left(\tau_k + \frac{1}{p\beta}\right)}{\Gamma\left(\tau_k + \alpha_k + \frac{1}{p\beta}\right)} < \infty. \tag{1.4.17}$$

1.5. Generalized operators of fractional differentiation. Inversion formula for the operators $I_{\beta,m}^{(\gamma_k),(\delta_k)}$

The generalized fractional integrals $I_{\beta,m}^{(\gamma_k),(\delta_k)}$ were defined by (1.1.6) only for a strongly positive multiorder of integration , that is for $\delta_k > 0$, $k = 1,\dots,m$. Now we are going to extend this definition to the case when some (or all) of the components δ_k of the multiorder can be zero or negative numbers. This definition has to coincide with the original Definition 1.1.1, when all the δ_k are positive. Furthermore, in the case of zero multiorder of integration $\delta = (0,\dots,0)$ it must generalize in a suitable manner the identities from the classical fractional calculus:

$$R^0 = I, \quad I_\beta^{\gamma,0} = I, \tag{1.5.1}$$

where I denotes the identity operator ,γ is arbitrary real and $\delta = 0$.

Lemma 1.5.1. *For arbitrary multiweight $\gamma = (\gamma_1,\dots,\gamma_m)$ and zero multiorder of integration $\delta_1 = \delta_2 = \dots = \delta_m = 0$ the multi-Erdélyi-Kober operator $I_{\beta,m}^{(\gamma_k),(\delta_k)}$ coincides with the identity operator I in the space C_α, $\alpha \geq \max_k[-\beta(\gamma_k+1)]$:*

$$I_{\beta,m}^{(\gamma_1,\dots,\gamma_m),(0,\dots,0)} f(x) = f(x), \quad f \in C_\alpha. \tag{1.5.2}$$

More generally, $R = x^{\beta\delta_0} I_{\beta,m}^{(\gamma_k),(\delta_k)} = I$ for $\delta_0 = \delta_1 = \dots = \delta_m = 0$.

Proof. As $I_{\beta,m}^{(\gamma_k),(\delta_k)} = \prod_{k=1}^{m} I_{\beta}^{\gamma_k,\delta_k}$, the proof is evident in view of (1.5.1). It is seen also by the Mellin transform approach: formula (1.2.12) with $\delta_k = 0$, $k = 1,\ldots,m$ yields

$$\mathfrak{M}\left\{ I_{\beta,m}^{(\gamma_k),(0,\ldots,0)} f(x); s \right\} = \prod_{k=1}^{m} \frac{\Gamma\left(\gamma_k - \frac{s}{\beta} + 1\right)}{\Gamma\left(\gamma_k + 0 - \frac{s}{\beta} + 1\right)} \mathfrak{M}\left\{ f(x); s \right\},$$

i.e. $I_{\beta,m}^{(\gamma_k),(0,\ldots,0)} f = f$ almost everythere in $(0,\infty)$. In particular, if $f \in C_{\alpha,\alpha^*} \subset C_\alpha$, then (1.5.2) follows.

Nevertheless, a direct proof using the G-function property (A.13$'$) and leading to the next more general statement (Corollary 1.5.2) seems to be useful from a *propaedentic point of view.*

First, let us assume that $0 < \delta_1 < 1$ and $\delta_2 = \cdots = \delta_m = 0$. Since in this case $\nu_m^* = \delta_1 - 1 > -1$, $\nu_m^* \neq 0, \pm 1, \pm 2, \ldots$, the asymptotic formula (1.1.14) yields that the integrand in $I_{\beta,m}^{(\gamma_k),(\delta_1,0,\ldots,0)} f(x)$ is $\mathcal{O}\left((1-\sigma)^{\nu_m^*}\right)$ with $\nu_m^* > -1$ as $\sigma \to 1$ and, therefore integral (1.1.6) is still convergent.

Property (A.13$'$) gives for the kernel function:

$$G_{m,m}^{m,0}\left[\sigma \left| \begin{matrix} \gamma_1 + \delta_1, \gamma_2, \ldots, \gamma_m \\ \gamma_1, \ldots, \gamma_m \end{matrix} \right.\right] = G_{1,1}^{1,0}\left[\sigma \left| \begin{matrix} \gamma_1 + \delta_1 \\ \gamma_1 \end{matrix} \right.\right] = \frac{(1-\sigma)^{\delta_1 - 1}\sigma^{\gamma_1}}{\Gamma(\delta_1)},$$

i.e. $I_{\beta,m}^{(\gamma_k),(\delta_1,0,\ldots,0)} f(x) = I_\beta^{\gamma_1,\delta_1} f(x)$.

If we assume additionally that $\delta_1 = 0$ and take into account (1.5.1), then the multiple Erdélyi-Kober fractional integral (1.1.6) of multiorder $(0,\ldots,0)$ turns into the identity operator in C_α.

For a kernel-function with $\delta_1 = \delta_2 = \cdots = \delta_s = 0$, $1 \leq s \leq m-1$, property (A.13$'$) takes the form

$$\begin{aligned} &G_{m,m}^{m,0}\left[\sigma \left| \begin{matrix} \gamma_1, \ldots, \gamma_s, \gamma_{s+1} + \delta_{s+1}, \ldots, \gamma_m + \delta_m \\ \gamma_1, \ldots, \gamma_s, \gamma_{s+1}, \ldots, \gamma_m \end{matrix} \right.\right] \\ &\quad = G_{m-s,m-s}^{m-s,0}\left[\sigma \left| \begin{matrix} \gamma_{s+1} + \delta_{s+1}, \ldots, \gamma_m + \delta_m \\ \gamma_{s+1}, \ldots, \gamma_m \end{matrix} \right.\right] \end{aligned}$$

and yields the following corollary.

Corollary 1.5.2. *Let s $(1 \leq s \leq m-1)$ of the components γ_k, $k = 1,\ldots,m$ be zero and the rest be positive, e.g.*

$$0 = \delta_1 = \ \ldots \ = \delta_s \ < \ \delta_{s+1} \leq \ \ldots \ \leq \delta_m.$$

Then, the operator $I_{\beta,m}^{(\gamma_k),(\delta_k)}$ *reduces itself to an* $(m-s)$*-dimensional one:*

$$\begin{aligned} I_{\beta,m}^{(\gamma_k)_1^m,(0,\dots,0,\delta_{s+1},\dots,\delta_m)} f(x) &= I_{\beta,m-s}^{(\gamma_k)_{s+1}^m,(\delta_k)_{s+1}^m} f(x) \\ &= \int_0^1 G_{m-s,m-s}^{m-s,0}\left[\sigma \,\middle|\, \begin{matrix} \gamma_{s+1}+\delta_{s+1},\dots,\gamma_m+\delta_m \\ \gamma_{s+1},\dots,\gamma_m \end{matrix}\right] f\left(x\sigma^{\frac{1}{\beta}}\right) d\sigma. \end{aligned} \tag{1.5.3}$$

The following *inversion formulas* for the classical Riemann-Liouville and Erdélyi-Kober operators of fractional integration hold:

$$\left(R^{\delta}\right)^{-1} = R^{-\delta}, \quad \left(I_{\beta}^{\gamma,\delta}\right)^{-1} = I_{\beta}^{\gamma+\delta,-\delta}, \tag{1.5.4}$$

where the *"integrations"* $R^{-\delta}$, $I_{\beta}^{\gamma+\delta,-\delta}$ *of negative order* $\delta' = -\delta < 0$ are defined as differintegral operators usually referred to as *fractional derivatives.*

In a similar way, let us formally put $\sigma_k = -\delta_k$, $k = 1,\dots,m$, in the index law (1.3.11). Using in addition Lemma 1.5.1 we obtain

$$I_{\beta,m}^{(\gamma_k+\delta_k),(-\delta_k)} I_{\beta,m}^{(\gamma_k),(\delta_k)} = I_{\beta,m}^{(\gamma_k),(0)} = I.$$

From here the following *formal inversion formula* is suggested (it will be justified below):

$$\left\{I_{\beta,m}^{(\gamma_k),(\delta_k)}\right\}^{-1} = I_{\beta,m}^{(\gamma_k+\delta_k),(-\delta_k)} \tag{1.5.5}$$

as a generalization of the known result (1.5.4) for Erdélyi-Kober operators . It remains to give a meaning to the symbol $I_{\beta,m}^{(\gamma_k+\delta_k),(-\delta_k)}$ for negative components $\delta_k' = -\delta_k < 0$, $k = 1,\dots,m$, of the multiorder.

Before giving this extended definition, we offer some suggestive reasons for it. Let us go back again to the classical example of the Riemann-Liouville integral of order $\delta > 0$:

$$R^{\delta} f(x) = \int_0^x \frac{(x-t)^{\delta-1}}{\Gamma(\delta)} f(t)dt = \int_0^x \Phi_{\delta}(x,t) f(t)dt. \tag{1.5.6}$$

For $\delta < 0$ denote by η the smallest positive integer number so that $\delta + \eta > 0$ (i.e. $\eta = [-\delta]+1$ if δ is non integer and $\eta = -\delta$ for integer $\delta < 0$). Then, the Riemann-Liouville operator is analytically extended to a differintegral operator, the so-called *fractional derivative* of order $\delta' = -\delta > 0$):

$$D^{\delta'} f(x) = R^{\delta} f(x) = \left(\frac{d}{dx}\right)^{\eta} R^{\delta+\eta} f(x) = \left(\frac{d}{dx}\right)^{\eta} \int_0^x \frac{(x-t)^{\delta+\eta-1}}{\Gamma(\delta+\eta)} f(t)dt \tag{1.5.7}$$

with kernel-function (if $\left(\frac{d}{dx}\right)^{\eta}$ is formally put under the integral sign):

$$\left(\frac{d}{dx}\right)^{\eta} \Phi_{\delta+\eta}(x,t) = \left(\frac{d}{dx}\right)^{\eta} \frac{(x-t)^{\delta+\eta-1}}{\Gamma(\delta+\eta)}.$$

As (1.5.7) is an analytical continuation of (1.5.6), their kernel functions can be equated, i.e.

$$\left(\frac{d}{dx}\right)^{\eta} \Phi_{\delta+\eta}(x,t) = \Phi_{\delta}(x,t), \tag{1.5.8}$$

due to the differential relation

$$\left(\frac{d}{dx}\right)^{\eta} \frac{(x-t)^{\delta+\eta-1}}{\Gamma(\delta+\eta)} = \frac{(x-t)^{\delta-1}}{\Gamma(\delta)}. \tag{1.5.9}$$

Our operators $I_{\beta,m}^{(\gamma_k),(\delta_k)}$, defined for $\delta_k > 0$, $k = 1, \dots, m$, by means of the integral (1.1.6′), have as a kernel the G-function

$$\Phi_{\delta_1,\dots,\delta_m}(x,t) = G_{m,m}^{m,0}\left[\left(\frac{t}{x}\right)^{\beta} \middle| \begin{matrix} (\gamma_k+\delta_k)_1^m \\ (\gamma_k)_1^m \end{matrix}\right]. \tag{1.5.10}$$

If we try to find an analytical continuation of $I_{\beta,m}^{(\gamma_k),(\delta_k)}$ for $\delta_k < 0$, $k = 1, \dots, m$, then it is quite natural to look for some differential relations satisfied by $\Phi_{\delta_1,\dots,\delta_m}(x,t)$ and analogous to (1.5.8)-(1.5.9). Fortunately, such relations have been found by the author in [202]. In the general case of a $G_{p,q}^{m,n}$-function there are formulated and proved in Appendix, Lemma B.3. As a particular case, from Corollary B.6, (B.12) we obtain:

$$\begin{aligned} G_{m,m}^{m,0}\left[\left(\frac{t}{x}\right)^{\beta} \middle| \begin{matrix} (\gamma_i+\delta_i)_1^m \\ (\gamma_i)_1^m \end{matrix}\right] &= \left[\prod_{k=1}^{m}\prod_{j=1}^{\eta_k}\left(\frac{1}{\beta}x\frac{d}{dx} + \gamma_k + \delta_k + j - 1\right)\right] \\ &\times G_{m,m}^{m,0}\left[\left(\frac{t}{x}\right)^{\beta} \middle| \begin{matrix} (\gamma_i+\delta_i+\eta_i)_1^m \\ (\gamma_i)_1^m \end{matrix}\right]. \end{aligned} \tag{1.5.11}$$

This differential relation can be written also in the form

$$D'_{\eta}\Phi_{\delta_1+\eta_1,\dots,\delta_m+\eta_m}(x,t) = \Phi_{\delta_1,\dots,\delta_m}(x,t), \tag{1.5.12}$$

where D'_{η} denotes the differential operator

$$D'_{\eta} = D'_{\eta_1,\dots,\eta_m} = \prod_{k=1}^{m}\prod_{j=1}^{\eta_k}\left(\frac{1}{\beta}x\frac{d}{dx} + \gamma_k + \delta_k + j - 1\right), \tag{1.5.13}$$

a polynomial of the Euler differential operator $\delta_x = x\frac{d}{dx}$ of degree $\eta = \eta_1 + \cdots + \eta_m$. Now we are ready to propose the following definition.

Definition 1.5.3. Let us consider an arbitrary multiweight $\gamma' = (\gamma_1', \ldots, \gamma_m')$ and a *multiorder of "integration" $\delta' = (\delta_1', \ldots, \delta_m')$ whose components δ_k' are also arbitrary real numbers.* We introduce the integers

$$\eta_k' = \begin{cases} [-\delta_k'] + 1 & \text{for non integer } \delta_k' < 0, \\ -\delta_k' & \text{for integer } \delta_k' < 0, \\ 0 & \text{for } \delta_k' \geq 0, \end{cases} \qquad k = 1, \ldots, m. \tag{1.5.14}$$

Then, by the symbol $I_{\beta,m}^{(\gamma_k'),(\delta_k')}$ we mean the following differintegral operator defined in the space $C_{\alpha'}^{(\eta_1' + \cdots + \eta_m')}$ by

$$I_{\beta,m}^{(\gamma_k'),(\delta_k')} = D_{\eta'}' I_{\beta,m}^{(\gamma_k'),(\delta_k' + \eta_k')}, \tag{1.5.15}$$

where the differential operator $D_{\eta'}'$ of order $\eta' = (\eta_1' + \cdots + \eta_m')$ has the form

$$D_{\eta'}' = \prod_{k=1}^{m} \Delta_{\eta_k'}' = \prod_{k=1}^{m} \prod_{j=1}^{\eta_k'} \left(\frac{1}{\beta} x \frac{d}{dx} + \gamma_k' + \delta_k' + j \right). \tag{1.5.16}$$

It is evident that if some δ_k' are non negative, then the corresponding $\eta_k' = 0$ and therefore, the factor $\Delta_{\eta_k'}'$ in $D_{\eta'}'$ is lacking. When all the δ_k' are positive, that is, all the η_k' are equal to zero, then $D_{\eta'}'$ being the empty product is interpreted as the identity operator I $I_{\beta,m}^{(\gamma_k'),(\delta_k' + \eta_k')} = I_{\beta,m}^{(\gamma_k'),(\delta_k')}$. Therefore, *in this case the extended Definition 1.5.3 coincides with the initial Definition 1.1.1.*

Now we are able to interpret the symbol $I_{\beta,m}^{(\gamma_k + \delta_k),(-\delta_k)}$ from the "formal" inversion formula (1.5.5). Let us put $\gamma_k' = \gamma_k + \delta_k$, $\delta_k' = -\delta_k$, $k = 1, \ldots, m$ in (1.5.15). Thus we obtain the following meaning of the symbol inverting the operator $I_{\beta,m}^{(\gamma_k),(\delta_k)}$:

$$\left(I_{\beta,m}^{(\gamma_k),(\delta_k)} \right)^{-1} = I_{\beta,m}^{(\gamma_k + \delta_k),(-\delta_k)} = D_\eta' I_{\beta,m}^{(\gamma_k + \delta_k),(\eta_k - \delta_k)}. \tag{1.5.17}$$

Usually, the differintegral operators inverting fractional integrals are called fractional derivatives, or operators of fractional differentiation. Then, it is natural to give the following definition.

Definition 1.5.4. Let γ_k, $\delta_k \geq 0$, $k = 1, \ldots, m$, be arbitrary real numbers and

$$\eta_k = \begin{cases} [\delta_k] + 1 & \text{for non integer } \delta_k, \\ \delta_k & \text{for integer } \delta_k, \end{cases} \qquad k = 1, \ldots, m. \tag{1.5.18}$$

The differintegral operator

$$D_{\beta,m}^{(\gamma_k),(\delta_k)} \stackrel{\text{def}}{=} D_\eta I_{\beta,m}^{(\gamma_k+\delta_k),(\eta_k-\delta_k)}$$
$$= \left[\prod_{k=1}^{m}\prod_{j=1}^{\eta_k}\left(\frac{1}{\beta}x\frac{d}{dx}+\gamma_k+j\right)\right] I_{\beta,m}^{(\gamma_k+\delta_k),(\eta_k-\delta_k)}, \tag{1.5.19}$$

defined for functions of $C_\alpha^{(\eta_1+\cdots+\eta_m)}$, is said to be a *generalized (m-tuple) Erdélyi-Kober fractional derivative* . More generally, the *generalized operators of fractional differentiation*, corresponding to the generalized fractional integrals $R = x^{\beta\delta_0} I_{\beta,m}^{(\gamma_k),(\delta_k)}$, $\delta_0 \geq 0$, are defined as

$$Df(x) = x^{-\beta\delta_0} D_{\beta,m}^{(\gamma_k-\delta_0),(\delta_k)} f(x) = D_{\beta,m}^{(\gamma_k),(\delta_k)} x^{-\beta\delta_0} f(x),\ \ \delta_0 \geq 0. \tag{1.5.20}$$

This definition is justified by:

Theorem 1.5.5. *Let $\delta_k \geq 0$, $k = 1, \ldots, m$. The generalized fractional integral $I_{\beta,m}^{(\gamma_k),(\delta_k)}$ is a linear right inverse operator of the generalized fractional derivative $D_{\beta,m}^{(\gamma_k),(\delta_k)}$, that is:*

$$D_{\beta,m}^{(\gamma_k),(\delta_k)} I_{\beta,m}^{(\gamma_k),(\delta_k)} f(x) = f(x) \text{ for every } f \in C_\alpha, \alpha \geq \max_k\left[-\beta(\gamma_k+1)\right]. \tag{1.5.21}$$

In terms of notations (1.1.7) and (1.5.20) this means that

$$DRf(x) = D_{\beta,m}^{(\gamma_k),(\delta_k)} x^{-\beta\delta_0} x^{\beta\delta_0} I_{\beta,m}^{(\gamma_k),(\delta_k)} f(x) = f(x), \tag{1.5.21'}$$

i.e. $DR = I$.

Proof. For a *direct and simple proof*, using the differential relation (1.5.11), even in a more general case of operators, see the proof of Theorem 5.1.9. Here we illustrate the use of decomposition Theorem 1.2.10. For brevity, denote by

$$\Delta\eta_k = \prod_{j=1}^{\eta_k}\Delta_{\eta_k,j} = \prod_{j=1}^{\eta_k}\left(\frac{1}{\beta}x\frac{d}{dx}+\gamma_k+j\right)$$

the multipliers in the differential operator $D_\eta = \prod_{k=1}^{m}\Delta_{\eta_k}$, being a polynomial of the Euler differential operator $\delta_x = x\frac{d}{dx}$ of order $\eta = (\eta_1 + \cdots + \eta_m)$ with zeros $\mu_{k,j} = -\beta(\gamma_k + j)$, $k = 1, \ldots, m$, $j = 1, \ldots, \eta_k$.

i) First consider the case when *all the $\delta_k = \eta_k$, $k = 1, \ldots, m$ are integers.* Then, by Definition 1.5.4 and Lemma 1.5.1 it follows that

$$\begin{aligned} D_{\beta,m}^{(\gamma_k),(\delta_k)} &= D_\eta I_{\beta,m}^{(\gamma_k+\eta_k),(0)} = D_\eta = \prod_{k=1}^{m} \Delta\eta_k \\ &= \prod_{k=1}^{m}\prod_{j=1}^{\eta_k}\left(\frac{1}{\beta}x\frac{d}{dx}+\gamma_k+j\right) = \left[\prod_{k=1}^{m}\prod_{j=1}^{\eta_k}\left(z\frac{d}{dz}+\gamma_k+j\right)\right]_{z\to x^\beta}. \end{aligned} \tag{1.5.22}$$

Since

$$\Delta_{\eta_k,j} = \left(z\frac{d}{dz}+\nu_{k,j}\right) = z^{-\nu_{k,j}+1}\frac{d}{dz}z^{\nu_{k,j}},$$

substituting $\nu_{k,j}=\gamma_k+j$, $z=x^\beta$ we receive

$$\Delta\eta_k = \prod_{j=1}^{\eta_k}\left(z^{-\gamma_k-j+1}\frac{d}{dz}z^{\gamma_k+j}\right) = z^{-\gamma_k}\left(\frac{d}{dz}\right)^{\eta_k} z^{\gamma_k+\eta_k}. \tag{1.5.23}$$

On the other hand, $I_{\beta,m}^{(\gamma_k),(\delta_k)}$ is a composition of the Erdélyi-Kober operators

$$\left(I_\beta^{\gamma_k,\delta_k}\right)_x = \left(I_1^{\gamma_k,\delta_k}\right)_z = z^{-(\gamma_k+\delta_k)}R^{\delta_k}z^{\gamma_k},\ k=1,\dots,m,$$

namely:

$$I_{\beta,m}^{(\gamma_k),(\delta_k)} = \prod_{k=1}^{m} I_\beta^{\gamma_k,\delta_k},\ \text{in this case with } \delta_k=\eta_k,\ k=1,\dots,m. \tag{1.5.24}$$

Since

$$\Delta_{\eta_k}I_1^{\gamma_k,\eta_k} = \left(z^{-\gamma_k}\left(\frac{d}{dz}\right)^{\eta_k}z^{\gamma_k+\eta_k}\right)\left(z^{-(\gamma_k+\varepsilon_k)}R^{\varepsilon_k}z^{\gamma_k}\right) = z^{-\gamma_k}\left(\frac{d}{dz}\right)^{\eta_k}R^{\eta_k}z^{\gamma_k} = I,$$

for $k=1,\dots,m$, combining (1.5.22), (1.5.24) we obtain identity (1.5.21):

$$\left[D_{\beta,m}^{(\gamma_k),(\delta_k)}I_{\beta,m}^{(\gamma_k),(\delta_k)}\right]_{z=x^\beta} = (\Delta_{\eta_1}\dots\Delta_{\eta_m})\left(I_1^{\gamma_m,\eta_m}\dots I_1^{\gamma_1,\eta_1}\right) = \dots = \Delta_{\eta_1}I_1^{\gamma_1,\eta_1} = I.$$

ii) It remains to consider the case when *some or all the δ_k are non integers.* By definition, $D_{\beta,m}^{(\gamma_k),(\delta_k)} = D_\eta I_{\beta,m}^{(\gamma_k+\delta_k),(\eta_k-\delta_k)}$. By the law of indices (1.3.11):

$$I_{\beta,m}^{(\gamma_k+\delta_k),(\eta_k-\delta_k)}I_{\beta,m}^{(\gamma_k),(\delta_k)} = I_{\beta,m}^{(\gamma_k),(\eta_k)}$$

is an operator of integration of integer multiorder $\eta=(\eta_1,\dots,\eta_m)$ and we obtain that

$$D_{\beta,m}^{(\gamma_k),(\delta_k)}I_{\beta,m}^{(\gamma_k),(\delta_k)} = D_\eta I_{\beta,m}^{(\gamma_k),(\eta_k)} = I$$

by virtue of i). The proof of the theorem is over.

This theorem shows that the generalized fractional integrals and derivatives act by the scheme:

$$C_\alpha \xrightarrow{I} \left(C_\alpha^{(\eta_1+\cdots+\eta_m)} \subset C_\alpha \right) \xrightarrow{D} C_\alpha. \tag{1.5.25}$$

It also gives the meaning of the formal inversion (1.5.5), realized by means of the generalized fractional derivatives, namely:

Inversion formula *for the generalized fractional integrals* $I_{\beta,m}^{(\gamma_k),(\delta_k)}$, $\delta_k \geq 0$, $k = 1,\ldots,m$:

$$\begin{aligned}\left(I_{\beta,m}^{(\gamma_k),(\delta_k)}\right)^{-1} &= D_{\beta,m}^{(\gamma_k),(\delta_k)} g(x) = I_{\beta,m}^{(\gamma_k+\delta_k),(-\delta_k)} g(x) \\ &= \left[\prod_{k=1}^{m}\prod_{j=1}^{\eta_k}\left(\frac{1}{\beta}x\frac{d}{dx}+\gamma_k+j-1\right)\right]\int\limits_0^1 G_{m,m}^{m,0}\left[\sigma \left| \begin{matrix}(\gamma_k+\eta_k)_1^m \\ (\gamma_k+\delta_k)_1^m\end{matrix}\right.\right] g\left(x\sigma^{\frac{1}{\beta}}\right) d\sigma,\end{aligned} \tag{1.5.26}$$

defined in the functional space $C_\alpha^{(\eta_1+\cdots+\eta_m)}$.

Corollary 1.5.6. *For* $g \in C_\alpha^{(\eta_1+\cdots+\eta_m)}$ *the integral equation*

$$I_{\beta,m}^{(\gamma_k),(\delta_k)} f(x) = g(x) \tag{1.5.27}$$

has the solution $f = \left(I_{\beta,m}^{(\gamma_k),(\delta_k)}\right)^{-1} g \in C_\alpha$, *defined by (1.5.26).*

Special cases of inversion formula (1.5.26) are:

Corollary 1.5.7. *(for Erdélyi-Kober operators, cf. formula (1.5.4)):*

$$\left(I_\beta^{\gamma,\delta}\right)^{-1} = I_\beta^{\gamma+\delta,-\delta}$$

(Sneddon, [453, p. 50], [454, p. 40], [455]; Kalla [164], McBride [289, (2.18)], etc.)

Corollary 1.5.8. *(for* $m = 2$, *Saigo's operators (1.1.l)):*

$$\left(I^{\alpha,\beta,\eta}\right)^{-1} = I^{-\alpha,-\beta,\alpha+\eta}$$

(cf. Saigo [415], [420], [421]).

In both these cases, the symbols $I_\beta^{\gamma+\delta,-\delta}$, $I^{-\alpha,-\beta,\alpha+\eta}$ are interpreted by means of differintegral expressions, being special cases of (1.5.15)-(1.5.16), (1.5.19). However, we point out that such differintegral inversion formulas have not been considered as new objects, namely, as generalized fractional derivatives. This interpretation, for special cases, will be done in the next section.

1.6. Properties and examples of generalized (multiple) Erdélyi-Kober derivatives.

Most of the properties of the generalized fractional integrals $I_{\beta,m}^{(\gamma_k),(\delta_k)}$ have their counterparts for the generalized derivatives $D_{\beta,m}^{(\gamma_k),(\delta_k)}$, defined by (1.5.19).

Lemma 1.6.1. *The multi-Erdélyi-Kober derivatives (1.5.19)*

$$D_{\beta,m}^{(\gamma_k),(\delta_k)}:C_\alpha^{(\eta_1+\cdots+\eta_m)}\to C_\alpha \tag{1.6.1}$$

preserve the power functions of C_α *(they are simultaneosly of the subspace* $C_\alpha^{(\eta_1+\cdots+\eta_m)}$*):*

$$D_{\beta,m}^{(\gamma_k),(\delta_k)}\{x^p\}=x^p\prod_{k=1}^{m}\frac{\Gamma\left(\gamma_k+\delta_k+\frac{p}{\beta}+1\right)}{\Gamma\left(\gamma_k+\frac{p}{\beta}+1\right)},\quad p>\alpha. \tag{1.6.2}$$

Proof. Formally, this follows from (1.5.21), (1.2.1). The regular proof exploits definition (1.5.19), (1.2.1):

$$\begin{aligned}D_{\beta,m}^{(\gamma_k),(\delta_k)}\{x^p\}&=D_\eta I_{\beta,m}^{(\gamma_k+\delta_k),(\eta_k-\delta_k)}\{x^p\}\\&=\prod_{k=1}^{m}\frac{\Gamma\left(\gamma_k+\delta_k+\frac{p}{\beta}+1\right)}{\Gamma\left(\gamma_k+\delta_k+\eta_k-\delta_k+\frac{p}{\beta}+1\right)}D_\eta\{x^p\}.\end{aligned}$$

The differential formula ([272, I, p. 24, (3)]):

$$\left[\prod_{i=1}^{n}\left(z\frac{d}{dz}+a_i\right)\right]\left\{z^\lambda\right\}=z^\lambda\prod_{i=1}^{n}(\lambda+a_i)$$

yields (with $z=x^\beta$):

$$\begin{aligned}D_\eta\{x^p\}=D_\eta\left\{z^{\frac{p}{\beta}}\right\}&=\left[\prod_{k=1}^{m}\prod_{j=1}^{\eta_k}\left(z\frac{d}{dz}+\gamma_k+j\right)\right]\left\{z^{\frac{p}{\beta}}\right\}\\&=z^{\frac{p}{\beta}}\left[\prod_{k=1}^{m}\prod_{j=1}^{\eta_k}\left(\frac{p}{\beta}+\gamma_k+j\right)\right]=x^p\prod_{k=1}^{m}\left(\prod_{j=1}^{\eta_k}\left(\gamma_k+\frac{p}{\beta}+j\right)\right).\end{aligned}$$

Since

$$\prod_{j=1}^{\eta_k}\left(\gamma_k+\frac{p}{\beta}+j\right)=\prod_{k=1}^{m}\frac{\Gamma\left(\gamma_k+\eta_k+\frac{p}{\beta}+1\right)}{\Gamma\left(\gamma_k+\frac{p}{\beta}+1\right)},$$

we obtain

$$D\{x^p\}=x^p\prod_{k=1}^{m}\left[\frac{\Gamma\left(\gamma_k+\delta_k+\frac{p}{\beta}+1\right)}{\Gamma\left(\gamma_k+\eta_k+\frac{p}{\beta}+1\right)}\cdot\frac{\Gamma\left(\gamma_k+\eta_k+\frac{p}{\beta}+1\right)}{\Gamma\left(\gamma_k+\frac{p}{\beta}+1\right)}\right].$$

Lemma 1.6.2. *The shift property for the generalized fractional derivatives in* $C_{\alpha-\beta\lambda}^{(\eta_1+\ldots\eta_m)}$

has the form

$$D_{\beta,m}^{(\gamma_k),(\delta_k)} x^{\beta\lambda} f(x) = x^{\beta\lambda} D_{\beta,m}^{(\gamma_k+\lambda),(\delta_k)} f(x). \tag{1.6.3}$$

Proof. For integers $\delta_k = \eta_k \geq 0$, $k = 1, \ldots, m$, property (1.6.3) in $C_{\alpha-\beta\lambda}^{(\eta_1+\ldots\eta_m)}$ follows from the identity

$$\left(\frac{1}{\beta} x \frac{d}{dx} + \gamma_k + j\right) x^{\beta\lambda} f(x) = x^{\beta\lambda} \left(\frac{1}{\beta} x \frac{d}{dx} + (\gamma_k + \lambda) + j\right) f(x),$$

i.e. denoting

$$D_{\beta,m}^{(\gamma_k),(\delta_k)} := D_\eta^{(\gamma)}; \quad D_{\beta,m}^{(\gamma_k+\lambda),(\eta_k)} = D_\eta^{(\gamma+\lambda)},$$

we obtain

$$D_\eta^{(\gamma)} x^{\beta\lambda} f(x) = x^{\beta\lambda} D_\eta^{(\gamma+\lambda)} f(x). \tag{1.6.4}$$

Then, for arbitrary $\delta_k > 0$, $k = 1, \ldots, m$, we use (1.6.4) and the shift property (1.3.3) for the multi-Erdélyi-Kober integrals, namely:

$$\begin{aligned} D_{\beta,m}^{(\gamma_k),(\delta_k)} x^{\beta\lambda} f(x) &= D_\eta^{(\gamma)} I_{\beta,m}^{(\gamma_k+\delta_k),(\eta_k-\delta_k)} x^{\beta\lambda} f(x) = D_\eta^{(\gamma)} x^{\beta\lambda} I_{\beta,m}^{(\gamma_k+\delta_k+\lambda),(\eta_k-\delta_k)} f(x) \\ &= x^{\beta\lambda} D_\eta^{(\gamma+\lambda)} I_{\beta,m}^{(\gamma_k+\lambda+\delta_k),(\eta_k-\delta_k)} f(x) = x^{\beta\lambda} D_{\beta,m}^{(\gamma_k+\lambda),(\delta_k)} f(x). \end{aligned}$$

An analogue of Lemma 1.3.3 is the following lemma.

Lemma 1.6.3.

$$D_{\beta,m}^{(\gamma_k),(\delta_k)} f(x) = \left[D_{1,m}^{(\gamma_k),(\delta_k)} f\left(x^{\frac{1}{\beta}}\right)\right]_{x\to x^\beta}. \tag{1.6.5}$$

Examples of generalized fractional derivatives of the form (1.5.19), (1.5.20) are all the differintegral operators whose linear right-inverse operators are the generalized fractional integrals listed in Section 1.1.iii. Some authors, whose names are associated with the examples there, introduce as their starting point the generalized differential operators and use them exclusively (e.g. Iliev [146], [147], Gelfond and Leontiev [120], Ruscheweyh [412], Dimovski [64]-[69], McBride [289], etc.).

Other authors consider operators of fractional and generalized integration only, inverting them by formal (divergent) or contour integrals (Kalla [156]-[164], Parashar [355]). In other series of papers, along with the generalized integrals, differintegral (differential, integro-differential) operators are introduced naturally as their inverse operators but are not considered as separate objects with their own meaning like generalized derivatives. Often, these differintegrals are seen just as analytical continuations of the generalized fractional integrals for negative (multi)orders of integration (McBride [289], [291], Saigo [415]-[421], etc.).

Our aim is to consider the operators of generalized fractional integration and differentiation simultaneously and in parallel, as a *united object: generalized fractional differintegrals*. Sometimes we study and use mainly the "integrals", otherwise we prefer dealing with the "derivatives" (being differential or integro-differential operators) and this depends on the multiorder of integration (differentiation), particular conditions and purposes.

First, let us consider the form of the generalized fractional derivatives (1.5.19), (1.5.20) when $m = 1$. As special cases (some of them listed below) many differential (differintegral) operators, introduced by various authors can be obtained.

Due to Lemma 1.6.3,

$$D_{\beta,1}^{\gamma,\delta} f(x) = \left[D_{1,1}^{\gamma,\delta} f\left(x^{\frac{1}{\beta}}\right)\right]_{x\to x^\beta}$$

and so, without loss of generality one can consider the 1-tuple Erdélyi-Kober fractional derivative

$$D_1^{\gamma,\delta} f(x) := D_{1,1}^{\gamma,\delta} f(x).$$

By Definition 1.5.4,

$$D_1^{\gamma,\delta} = D_\eta I_1^{\gamma+\delta,\eta-\delta}. \tag{1.6.6}$$

Since (see the proof of Theorem 1.5.5)

$$D_\eta = \Delta_\eta = x^{-\gamma}\left(\frac{d}{dx}\right)^\eta x^{\gamma+\eta}, \quad I_1^{\gamma+\delta,\eta-\delta} = x^{-(\gamma+\eta)} R^{\eta-\delta} x^{\gamma+\delta},$$

taking into account that (see e.g. [101], [404])

$$\left(\frac{d}{dx}\right)^\eta R^{\eta-\delta} = \left(\frac{d}{dx}\right)^\delta \quad \text{for } \eta > \eta-\delta \geq 0,\ \delta > 0,$$

we obtain the following *formal representation* of (1.6.6):

$$D_1^{\gamma,\delta} f(x) = \left[x^{-\gamma}\left(\frac{d}{dx}\right)^\delta x^{\gamma+\delta}\right] f(x). \tag{1.6.7'}$$

For the general case with $\beta > 0$:

$$D_\beta^{\gamma,\delta} f(x) := D_{\beta,1}^{\gamma,\delta} f(x) = \left[x^{-\gamma}\left(\frac{d}{dx}\right)^\delta x^{\gamma+\delta} f\left(x^{\frac{1}{\beta}}\right)\right]_{x\to x^\beta}. \tag{1.6.7}$$

One should have in mind however, that behind the Riemann-Liouville fractional derivative $D^\delta = \left(\frac{d}{dx}\right)^\delta$, $\delta > 0$, a differintegral expression is hidden.

This operator (1.6.7) has an important role and applications in our considerations and in fractional calculus generally, but until now *it has not been referred to by a special name*. Its *linear right inverse operator* is the *Erdélyi-Kober fractional integral* $I_\beta^{\gamma,\delta}$ (1.1.17), symbolically written as:

$$I_\beta^{\gamma,\delta} f(x) = \left[x^{-(\gamma+\delta)} r^\delta x^\gamma f\left(x^{\frac{1}{\beta}}\right)\right]_{x\to x^\beta}. \tag{1.6.8}$$

That is why, we introduce the following definition.

Definition 1.6.4. By *Erdélyi-Kober fractional derivatives* (1-tuple operators of fractional differentiation) we mean the differintegral operators

$$D_{\beta}^{\gamma,\delta} f(x) = D_{\eta} I_{\beta}^{\gamma+\delta,\eta-\delta} f(x) = \left[\prod_{j=1}^{\eta} \left(\frac{1}{\beta} x \frac{d}{dx} + \gamma + j\right)\right] I_{\beta}^{\gamma+\delta,\eta-\delta} f(x), \tag{1.6.9}$$

defined in spaces

$$C_{\alpha}^{(\eta)} \text{ with } \alpha \geq -\beta(\gamma+1) \text{ and } \eta = \begin{cases} \delta, & \text{if } \delta \text{ is integer}, \\ [\delta]+1 & \text{if } \delta \text{ is non integer}, \end{cases} \tag{1.6.10}$$

and having the symbolical form (1.6.7).

In particular, for $\beta = 1$ operators (1.6.6), (1.6.7′) have the explicit form

$$D_{\beta}^{\gamma,\delta} f(x) = \left[\prod_{j=1}^{\eta} \left(x \frac{d}{dx} + \gamma + j\right)\right] \int_0^1 \frac{(1-\sigma)^{\eta-\delta-1} \sigma^{\gamma+\delta}}{\Gamma(\eta-\delta)} f(x\sigma) d\sigma. \tag{1.6.11}$$

Now, we give a list of several *examples of Erdélyi-Kober fractional derivatives.*

i) $m = 1$.

a) the classical *Riemann-Liouville fractional derivatives*, including the m-fold differentiation:

$$\begin{aligned} D^{\delta} &= \left(\frac{d}{dx}\right)^{\delta} = x^{-\delta} D_1^{-\delta,\delta} = D_1^{0,\delta} x^{-\delta}, \delta > 0 \\ D^m &= \left(\frac{d}{dx}\right)^m, \quad m = \delta \geq 1 \quad \textit{integer}; \end{aligned} \tag{1.6.a}$$

b) *Hardy-Littlewood differentiation* (cf. (1.1.c))

$$D_{m,n} = x^{-n} \frac{d}{dx} x^m = D_1^{n,1} x^{m-n-1}, \quad n > m-1, \tag{1.6.b}$$

where n and m are integers;

c) *Gelfond-Leontiev generalized differentiations* with respect to Mittag-Leffler functions $E_{\rho}(x;\mu)$ ([120], cf. (1.1.h-h′)), defined for power series:

$$D_{\rho}\left\{\sum_{k=0}^{\infty} a_k x^k\right\} = \sum_{k=1}^{\infty} a_k \frac{\Gamma\left(\frac{k}{\rho}+1\right)}{\Gamma\left(\frac{k-1}{\rho}+1\right)} x^{k-1}, \; p > 0, \tag{1.6.c}$$

and analytically extended by the differintegral expressions

$$D_{\rho} f(x) = \frac{1}{\rho}\left(x \frac{d}{dx} + \rho\right) x^{-1} \int_0^1 \frac{(1-\sigma)^{\frac{1}{\rho}}}{\Gamma\left(1-\frac{1}{\rho}\right)} f\left(x\sigma^{\frac{1}{\rho}}\right) d\sigma - \frac{x^{-1}}{\Gamma\left(1-\frac{1}{\rho}\right)} f(0), \; \rho > 1,$$

or

$$D_\rho f(x) = \frac{d}{dx} f(x), \qquad \text{if } \rho = 1, \tag{1.6.c$'$}$$

in the space of functions analytic in a starlike domain. Section 2.2 is devoted to operators (1.6.c-c$'$) and the more general operators $D_{\rho,\mu}$ (the so-called *Džrbashjan-Gelfond-Leontiev operators*) and their inverse integral operators $l_{\rho,\mu}$. It is seen that

$$D_\rho = D_{\rho,1} = x^{-1} D_\rho^{-\frac{1}{\rho},\frac{1}{\rho}}; \quad D_{\rho,\mu} = x^{-1} D_\rho^{\mu-\frac{1}{\rho}-1,\frac{1}{\rho}}.$$

d) The so-called *Rusheweyh derivatives*:

$$D_\alpha f(x) = \left\{ \frac{x}{(1-x)^{1+\alpha}} \right\} \circ f(x), \quad \alpha \geq 0 \tag{1.6.d}$$

are defined by means of the Hadamard product ($\circ$) of power series:

$$\left(\sum_{n=0}^{\infty} a_n x^n \right) \circ \left(\sum_{n=0}^{\infty} b_n x^n \right) = \sum_{n=0}^{\infty} a_n b_n x^n.$$

However, D_α can be represented also as an Erdélyi-Kober fractional derivative of order α, namely,

$$D_\alpha f(x) = \frac{1}{\Gamma(\alpha+1)} x \left(\frac{d}{dx} \right)^\alpha x^{\alpha-1} f(x) = \frac{1}{\Gamma(\alpha+1)} D_1^{-1,\alpha} f(x). \tag{1.6.d$'$}$$

More details about the Rusheweyh derivatives, their multiple analogues and other (multiple) Erdélyi-Kober fractional derivatives used in the *theory of univalent functions* can be found in Section 5.4.

ii) $m = 2$. By Definition 1.5.4, the 2-tuple Erdélyi-Kober fractional derivatives $D_{\beta,2}^{(\gamma_k),(\delta_k)}$ have the form

$$\begin{aligned} D_{\beta,2}^{(\gamma_1,\gamma_2),(\delta_1,\delta_2)} &= \Delta_{\eta_1} \Delta_{\eta_2} I_{\beta,2}^{(\gamma_k+\delta_k),(\eta_k-\delta_k)} \\ &= \left[\prod_{i=1}^{\eta_1} \left(\frac{1}{\beta} x \frac{d}{dx} + \gamma_1 + i \right) \prod_{j=1}^{\eta_2} \left(\frac{1}{\beta} x \frac{d}{dx} + \gamma_2 + j \right) \right] I_\beta^{\gamma_1+\delta_1,\eta_1-\delta_1} I_\beta^{\gamma_2+\delta_2,\eta_2-\delta_2}, \end{aligned} \tag{1.6.12}$$

but as we show in Chapters 3 and 5 (even for arbitrary $m \geq 2$), they can be put in the symbolical form

$$D_{\beta,2}^{(\gamma_1,\gamma_2),(\delta_1,\delta_2)} = x^{\alpha_0} \left(\frac{d}{dx} \right)^{\delta_1} x^{\alpha_1} \left(\frac{d}{dx} \right)^{\delta_2} x^{\alpha_2}, \tag{1.6.13}$$

$\delta_1 > 0$, $\delta_2 > 0$; α_0, α_1, α_2 are determined by β, γ_1, γ_2.

Such differential operators are related to the differential equations satisfied by the Gauss, Bessel and Wright functions and the polynomials of mathematical physics.

EXAMPLES:

e) The 2-tuple *("2-dimensional") fractional differentiations* ($\alpha > 0$, $\beta > 0$):

$$d^{\alpha,\beta} = \left(\frac{d}{dx}\right)^{\alpha} x^{\alpha} \left(\frac{d}{dx}\right)^{\beta} x^{\beta} = D_{1,2}^{(0,0),(\alpha,\beta)}, \tag{1.6.e}$$

extensions of the "two-dimensional Riemann-Liouville and Weyl fractional integrals" (1.1.j-j$'$) (see Raina [387], Raina and Kiryakova [390]).

f) Special cases of operators (1.6.13), when $\delta_1 = \delta_2 = 1$, are the *Bessel type differential operators of second order*:

$$B = x^{\alpha_0}\frac{d}{dx}x^{\alpha_1}\frac{d}{dx}x^{\alpha_2} = D_{\beta,2}^{(\gamma_1,\gamma_2),(1,1)}\frac{x^{-\beta}}{\beta^2} = \frac{x^{\beta}}{\beta^2}D_{\beta,2}^{(\gamma_1-1,\gamma_2-1),(1,1)}, \tag{1.6.f}$$

where $\beta = 2-(\alpha_0+\alpha_1+\alpha_2) > 0$, $\gamma = \frac{\alpha_1+\alpha_2-1}{\beta}$, $\gamma_2 = \frac{\alpha_2}{\beta}$. The best-known examples of (1.6.f) are the Bessel operators

$$B_\nu = \frac{d^2}{dx^2} - \frac{1}{x}\frac{d}{dx} + \frac{\nu^2}{x^2}, \quad \ddot{B}_\nu = \frac{d^2}{dx^2} - \frac{\nu}{x}\frac{d}{dx}, \tag{1.6.f$'$}$$

related to the Bessel functions $J_\nu(x)$. One can see e.g. McBride [289], Sprinkhnizen-Kuiper [457], and also Chapter 3 (for $m := 2$).

g) Saigo [415]-[421] extends his operators of fractional integration $I^{\alpha,\beta,\eta}$, $J^{\alpha,\beta,\eta}$ (1.1.l-l$'$) to the case of negative order $\alpha' = -\alpha < 0$ by means of differintegral operators, e.g.

$$I^{\alpha',\beta',\eta'} := \left(\frac{d}{dx}\right)^{n} I^{\alpha'+n,\beta'-n,\eta'-n},$$

where n is the smallest integer such that $0 < \alpha'+n \leq 1$. Then, in our sense his inversion formula is realized by means of the 2-tuple Erdélyi-Kober fractional derivative:

$$\left(I^{\alpha,\beta,\eta}\right)^{-1} = I^{-\alpha,-\beta,\alpha+\eta} = \left(\frac{d}{dx}\right)^{n} I^{n-\alpha,n-\beta,\alpha+\eta-n} \tag{1.6.g$'$}$$

$$= x^{\beta}D_{1,2}^{(\eta,\beta),(-\eta,\alpha+\eta)} = x^{\beta}D_1^{\beta,\alpha+\eta}D_1^{\eta_1-\eta} = \left(\frac{d}{dx}\right)^{\alpha+\eta} x^{\alpha+\beta}\left(\frac{d}{dx}\right)^{-\eta}.$$

We denote this differintegral operator by

$$D^{\alpha,\beta,\eta} := \left(\frac{d}{dx}\right)^{\alpha+\eta} x^{\alpha+\beta}\left(\frac{d}{dx}\right)^{-\eta} = \left(I^{\alpha,\beta,\eta}\right)^{-1} \tag{1.6.g}$$

and call it, naturally, *Saigo's fractional derivative.*

h) It is of special interest to consider a differential operator, related to *Wright's functions* (D.2), (E.36) (also called Bessel-Maitland functions, see Sections D and E of Appendix and Wright [511]-[512], Stankovic [484]-[485], Gajic and Stankovic [117], Marichev [276]):

$$J_\nu^{(\mu)}(x) = \sum_{k=0}^{\infty} \frac{(-x)^k}{k!\,\Gamma(1+\nu+\mu k)}.$$

When $\mu = \frac{p}{q}$ is rational, Pathak [360, p. 50] shows that $J_\nu^{(\mu)}(x)$ satisfies a differential equation of order $(p+q)$ (see also [511, p. 74], [485, p. 120]):

$$D_\nu^{(\mu)} J_\nu^{(\mu)}(x) = \frac{(-1)^q}{\mu} J_\nu^{(\mu)}(x),$$

where $D_\nu^{(\mu)}$ can be put in the form of 2-tuple Erdélyi-Kober derivatives (1.6.13) of integer multiorder (p,q):

$$D_\nu^{(\mu)} = x^{1-\frac{\nu+1}{\mu}} \left(\frac{d}{dx}\right)^p x^{\frac{\nu}{\mu}+q} \left(\frac{d}{dx}\right)^q = x^{1-p-\frac{1}{\mu}} D_{1,2}^{\left(\frac{\nu}{\mu}-p,-q\right),(p,q)}. \tag{1.6.h}$$

A similar differential operator, introduced by Krätzel [239] leads to an open problem considered in Section 5.4, see (5.4.3).

iii) for arbitrary $m \geq 2$ we establish the analogue of decomposition Theorem 1.2.10, namely:

Lemma 1.6.5. *The generalized operators of fractional differentiation (m-tuple Erdélyi-Kober fractional derivatives) $D_{\beta,m}^{(\gamma_k),(\delta_k)}$ of the form (1.5.19) can be represented by compositions of m commuting Erdélyi-Kober fractional derivatives $D_{\beta,m}^{(\gamma_k),(\delta_k)}$ (1.6.9), (1.6.7), namely:*

$$D_{\beta,m}^{(\gamma_k),(\delta_k)} = D_\beta^{\gamma_1,\delta_1} D_\beta^{\gamma_2,\delta_2} \dots D_\beta^{\gamma_m,\delta_m}, \tag{1.6.14}$$

and therefore, they have the symbolical form:

$$D_{\beta,m}^{(\gamma_k),(\delta_k)} = \left[x^{\alpha_0} \left(\frac{d}{dx}\right)^{\delta_1} x^{\alpha_1} \left(\frac{d}{dx}\right)^{\delta_2} \dots \left(\frac{d}{dx}\right)^{\delta_m} x^{\alpha_m} f\left(x^{\frac{1}{\beta}}\right) \right]_{x \to x^\beta} \tag{1.6.15}$$

with $\delta_1 > 0, \dots, \delta_m > 0$ and real $\gamma_1, \dots, \gamma_m$, respectively $\alpha_1, \dots, \alpha_m$.

Proof. In view of Lemma 1.6.3 we can consider the simple case $\beta = 1$. For functions $f(x) = x^p$, the Erdélyi-Kober derivatives (1.6.9) have values

$$D_1^{\gamma,\delta}\{x^p\} = x^p \frac{\Gamma(\gamma+\delta+p+1)}{\Gamma(\gamma+p+1)}, \quad p > -(\gamma+1).$$

Then, operators $D_1^{\gamma_k,\delta_k}$, $k = 1, \dots, m$, applied subsequently to x^p give:

$$D'\{x^p\} = D_1^{\gamma_1,\delta_1} \dots D_1^{\gamma_m,\delta_m}\{x^p\} = x^p \prod_{k=1}^m \frac{\Gamma(\gamma_k+\delta_k+p+1)}{\Gamma(\gamma_k+p+1)},$$

the same as given by Lemma 1.6.1, (1.6.2) for $D_{1,m}^{(\gamma_k),(\delta_k)}\{x^p\}$.

As like in the proof of Lemma 1.2.9, on each finite interval $[0,x]$, $x > 0$, one can approximate uniformly a function $f(x) \in C_\alpha^{(\eta_1+\dots+\eta_m)}$ i.e. $f(x) = x^p \tilde{f}(x)$; $p > \alpha; \tilde{f} \in$

$C_\alpha^{(\eta_1+\cdots+\eta_m)}[0,\infty)$ by means of the sequence of functions

$$f_n(x) = \sum_{k=0}^{n} a_{n,k} x^{p+k}, \quad p > \alpha,$$

for which operators D' and $D_{\beta,m}^{(\gamma_k),(\delta_k)}$ coincide. By letting $X \to \infty$ and $n \to \infty$ we obtain that operators $D' = \prod_{k=1}^{m} D_1^{\gamma_k,\delta_k}$ and $D_{1,m}^{(\gamma_k),(\delta_k)}$ coincide for each $f \in C_\alpha^{(\eta_1+\cdots+\eta_m)}$, $\alpha \geq \max_k [-(\gamma_k+1)]$, where η_k, $k = 1,\ldots,m$, are the smallest integers greater than δ_k, $k = 1,\ldots,m$. For arbitrary $\beta > 0$, Lemma 1.6.5 holds in $C_\alpha^{(\eta_1+\cdots+\eta_m)}$, $\alpha \geq \max_k [-\beta(\gamma_k+1)]$.

From (1.6.7) we have also

$$D_{1,m}^{(\gamma_k),(\delta_k)} = \left[x^{-\gamma_1}\left(\frac{d}{dx}\right)^{\delta_1} x^{\gamma_1+\delta_1}\right]\left[x^{-\gamma_2}\left(\frac{d}{dx}\right)^{\delta_2} x^{\gamma_2+\delta_2}\right]\ldots\left[x^{-\gamma_m}\left(\frac{d}{dx}\right)^{\delta_m} x^{\gamma_m+\delta_m}\right]$$

and putting

$$\alpha_0 = -\gamma_1; \quad \alpha_k = \gamma_k + \delta_k - \gamma_{k+1}, \; k = 1,\ldots,m-1; \quad \alpha_m = \gamma_m + \delta_m, \tag{1.6.16}$$

we find the following *general (but formal) representation of the multiple fractional derivatives* of order $\delta = \delta_1 + \cdots + \delta_m > 0$:

$$D_{1,m}^{(\gamma_k),(\delta_k)} = x^{\alpha_0}\left(\frac{d}{dx}\right)^{\delta_1} x^{\alpha_1}\left(\frac{d}{dx}\right)^{\delta_2}\ldots x^{\alpha_{m-1}}\left(\frac{d}{dx}\right)^{\delta_m} x^{\alpha_m}. \tag{1.6.17}$$

The most typical operators of this form are the *hyper-Bessel differential operators* ($\delta_1 = \cdots = \delta_m = 1$) considered in Chapter 3.

— || —

THE MAIN RESULTS OF CHAPTER 1 HAVE BEEN PUBLISHED IN: Kiryakova [194], [196], [201]-[202], [208] and Dimovski and Kiryakova [79].

2 Recent aspects of classical Erdelyi-Kober operators

In a series of papers [104], [109], [220] Erdélyi and Kober investigated the properties of the fractional integrals

$$\frac{x^{-\eta-\alpha+1}}{\Gamma(\alpha)}\int\limits_0^x (x-t)^{\alpha-1}t^{\eta-1}f(t)dt, \qquad \frac{x^{\eta}}{\Gamma(\alpha)}\int\limits_x^\infty (t-x)^{\alpha-1}t^{-\alpha-\eta}f(t)dt,$$

($\alpha > 0,\ \eta > 0$) which are obvious generalizations of the Riemann-Liouville and Weyl fractional integrals (1.1.b), (1.1.b*). Sneddon [452], [454]-[455] considered the following modifications of the Erdélyi-Kober operators:

$$\left.\begin{aligned} I_{\eta,\alpha}f(x) &= \tfrac{2x^{-2\alpha-2\eta}}{\Gamma(\alpha)}\int\limits_0^x (x^2-t^2)^{\alpha-1}\, t^{2\eta+1}f(t)dt \\ K_{\eta,\alpha}f(x) &= \tfrac{2x^{2\eta}}{\Gamma(\alpha)}\int\limits_x^\infty (t^2-x^2)^{\alpha-1}\, t^{-2\alpha-2\eta+1}f(t)dt \end{aligned}\right\} , \quad \alpha > 0,\ \eta \geq -\frac{1}{2}$$

and illustrated their various applications in applied mathematics, in particular, in solving mixed boundary value problems of potential theory, solutions of dual and triple integral equations, problems of GASPT and diffraction theory, in electrostatics and elastostatics, etc. Sneddon revealed also the close relationship of these operators with the modified operators $S_{\eta,\alpha}$ of the Hankel transforms, their Mellin transforms, and further generalizations involving the Bessel functions (studied also by Lowndes [268]-[269]). As shown in Rooney [399]-[400], Kalla [161], [164], McBride [289], etc, the same considerations are easily transferred to a more general case with an additional parameter $\beta > 0$ (one should put $\beta = 1$, $\beta = 2$ to obtain the above operators). In our notation, these are operators (1.1.17), (1.1.17*):

$$\begin{aligned} I_\beta^{\gamma,\delta}f(x) &= \int\limits_0^1 \frac{(1-\sigma)^{\delta-1}\sigma^\gamma}{\Gamma(\delta)} f(x\sigma^{\frac{1}{\beta}})\, d\sigma \\ &= x^{-\beta(\gamma+\delta)}\int\limits_0^x \frac{(x^\beta-\tau^\beta)^{\delta-1}}{\Gamma(\delta)}\tau^{\beta\gamma}f(\tau)\, d(\tau^\beta) = I_{\beta,1}^{\gamma,\delta} \end{aligned}$$

and

$$K_\beta^{\gamma,\delta} f(x) = \int_1^\infty \frac{(\sigma-1)^{\delta-1}\sigma^{-(\gamma+\delta)}}{\Gamma(\delta)} f(x\sigma^{\frac{1}{\beta}})\, d\sigma$$

$$= x^{\beta\gamma} \int_x^\infty \frac{(\tau^\beta - x^\beta)^{\delta-1}}{\Gamma(\delta)} \tau^{-\beta(\gamma+\delta)} f(\tau)\, d(\tau^\beta) = W_{\beta,1}^{\gamma,\delta} f(x),$$

referred to simply in this work as *Erdélyi-Kober fractional integration operators (integrals).* Since an overall exposition of their basic properties and applications has been given in the above-cited literature (as well as in other works from the References), in this chapter we are aiming only to add some more recent aspects and results on the Erdélyi-Kober operators which are closely related to the ideas of the generalized fractional calculus developed here.

2.1. Convolutions of the Erdélyi-Kober fractional integration operators

The notion of convolution is important in many problems of analysis and its applications. The best known examples of convolutions: are the convolution of the Fourier transform

$$(f\#g)(x) = \int_{-\infty}^\infty f(x-t)g(t)dt$$

and the *Duhamel convolution* (corresponding to the Laplace transform)

$$(f*g)(x) = \int_0^x f(x-t)g(t)dt. \tag{2.1.1}$$

The latter operation has been used by *Mikusinski* [306] as a basis for his direct algebraic approach to the *Heaviside operational calculus.*

The more general notion of convolution of a linear operator mapping a linear space into itself, introduced by Dimovski in 1966 (see e.g. [64], [69]-[73]), is conceived as a "multiplication" in this space such that the space becomes a commutative and associative algebra. By means of this generalized concept for a convolution, it is possible to find non trivial analytical results using simple algebraic considerations only.

Definition 2.1.1. (Dimovski [64], [70], [73]) Let $\mathcal{X}$ be a linear space and let $L : \mathcal{X} \longrightarrow \mathcal{X}$ be a linear operator . A bilinear, commutative and associative operation $* : \mathcal{X}\times\mathcal{X} \longrightarrow \mathcal{X}$ is said to be a *convolution of the linear operator* L in $\mathcal{X}$ if the relation

$$L(f*g) = (Lf)*g \tag{2.1.2}$$

is fulfilled for all $f,\ g \in \mathcal{X}$.

In the sense of this definition the Duhamel convolution (2.1.1) is a convolution of the *Volterra integration operator*

$$lf(x) = \int_0^x f(t)dt \tag{2.1.3}$$

in the space $\mathcal{X} = C([0,\infty))$, of the continuous functions in $[0,\infty)$, into itself. Definition 2.1.1 is the basis of Dimovski's *convolutional calculus* (see [73]). This definition can be considered also as an inversion of the definition of a multiplier of a commutative and associative algebra, the basis of Larsen's general theory of multipliers. One of the applications of such convolutions is in developing operational calculi for the operators L. In [70] and [73] a general scheme of such operational calculi is proposed together with its particular realizations for the most commonly used and sufficiently general cases of operators. Convolutions are found for the linear right inverse operators of the general non singular differential operators of first and second order, singular differential operators of second order related to Bessel, Legendre and Laguerre functions, and singular differential operators of Bessel type of arbitrary order $m \geq 1$ etc.

Here we propose a family of convolutions of the Erdélyi-Kober fractional integration operators of the form

$$\begin{aligned} Lf(x) = L^{(\delta)}f(x) &= x^{\beta\delta} I_{\beta}^{\gamma,\delta} f(x) \\ &= \begin{cases} x^{\beta\delta} \int\limits_0^1 \frac{(1-\sigma)^{\delta-1}}{\Gamma(\delta)} \sigma^{\gamma} f\left(x\sigma^{\frac{1}{\beta}}\right) d\sigma, & \delta > 0 \\ f(x), & \delta = 0, \end{cases} \end{aligned} \tag{2.1.4}$$

acting from the space C_α, $\alpha \geq -\beta(\gamma+1)$ into itself, namely, $L^{(\delta)} : C_\alpha \longrightarrow C_{\alpha+\beta\delta} \subset C_\alpha$. The *following proposition is inspired by a series of papers of Dimovski* (e.g. [64], [68]-[69]) *containing a family of convolutions of the hyper-Bessel operators of order* $m \geq 1$.

Theorem 2.1.2. (Kiryakova [206]) *Denote by* $(\circ)$ *the auxiliary operation*

$$(f \circ g)(x) = x^{\beta} \int_0^1 f\left[x(1-\sigma)^{\frac{1}{\beta}}\right] g\left(x\sigma^{\frac{1}{\beta}}\right) [\sigma(1-\sigma)]^{\gamma} d\sigma \tag{2.1.5}$$

and by T_λ *the Erdélyi-Kober fractional integration operators*

$$T_\lambda = x^{\beta\lambda} I_{\beta}^{2\gamma,\lambda-\gamma} \text{ with arbitrary real } \lambda \geq \gamma. \tag{2.1.6}$$

For arbitrary real γ, $\delta \geq 0$, $\beta > 0$ *the operations*

$$f \overset{\lambda}{*} g = T_\lambda(f \circ g), \quad \lambda \geq \gamma \tag{2.1.7}$$

are convolutions without divisors of zero of every Erdélyi-Kober operator (2.1.4) of fractional integration of order $\delta \geq 0$, *i.e. of the operator*

$$L^{(\delta)} = x^{\beta\delta} I_\beta^{\gamma,\delta} \text{ in the space } C_\alpha,\ \alpha \geq -\beta(\gamma+1).$$

Proof. For $\lambda > \gamma$ the linear operator T_λ has the explicit integral representation

$$T_\lambda h(x) = x^{\beta\lambda} \int_0^1 \frac{(1-\sigma)^{\lambda-\gamma-1}}{\Gamma(\lambda-\gamma)} \sigma^{2\gamma} h\left(x\sigma^{\frac{1}{\beta}}\right) d\sigma. \tag{2.1.6'}$$

For $\lambda = \gamma$, since $I_\beta^{2\gamma,0} = I$ is the identity operator, T_λ reduces to a multiplication by $x^{\beta\lambda}$. The case $\lambda < \gamma$ is possible too but the functional space should be suitably confined and T_λ should be considered as a fractional derivative of Erdélyi-Kober type.

The commutability and bilinearity of $(\circ)$ and hence, of the operations $\left(\overset{\lambda}{*}\right)$ are evident. It remains to check if the other two properties in Definition 2.1.1 are also fulfilled. First, we consider functions in C_α of the form

$$f(x) = x^p, \quad g(x) = x^q, \quad h(x) = x^r \quad \text{with } p, q, r > \alpha. \tag{2.1.8}$$

After routine calculations one can find

$$x^p \circ x^q = x^{(p+q)+\beta} \frac{\Gamma\left(\gamma + \frac{p}{\beta} + 1\right)\Gamma\left(\gamma + \frac{q}{\beta} + 1\right)}{\Gamma\left(\gamma + \frac{p+q}{\beta} + 2\right)}$$

and

$$T_\lambda\left(x^r\right) = x^{r+\beta\lambda} \frac{\Gamma\left(2\gamma + \frac{r}{\beta} + 1\right)}{\Gamma\left(\gamma + \lambda + \frac{r}{\beta} + 2\right)}$$

(see Lemma 1.2.1, (1.2.1)), whence

$$x^p \overset{\lambda}{*} x^q = x^{\beta(\gamma+1)+(p+q)} \frac{\Gamma\left(\gamma + \frac{p}{\beta} + 1\right)\Gamma\left(\gamma + \frac{q}{\beta} + 1\right)}{\Gamma\left(\gamma + \lambda + \frac{p+q}{\beta} + 1\right)}. \tag{2.1.9}$$

So,

$$C_\alpha \times C_\alpha \xrightarrow{\circ} C_{2\alpha+\beta} \xrightarrow{T_\lambda} C_{(2\alpha+\beta)+\beta\lambda},$$

or:

$$\left(\overset{\lambda}{*}\right) : C_\alpha \times C_\alpha \longrightarrow C_{2\alpha+\beta(\lambda+1)} \subset C_\alpha.$$

Further we obtain that the expression

$$\left(x^p \overset{\lambda}{*} x^q\right) \overset{\lambda}{*} x^r = x^{2\beta(\lambda+1)+(p+q+r)} \frac{\Gamma\left(\gamma+\frac{p}{\beta}+1\right)\Gamma\left(\gamma+\frac{q}{\beta}+1\right)\Gamma\left(\gamma+\frac{r}{\beta}+1\right)}{\Gamma\left(\gamma+\frac{p+q+r}{\beta}+2\lambda+3\right)}$$

is symmetric with respect to x^p, x^q and x^r and this fact yields the associativity of $\left(\overset{\lambda}{*}\right)$ for functions of the form (2.1.8). To prove that

$$\left(f \overset{\lambda}{*} g\right) \overset{\lambda}{*} h = f \overset{\lambda}{*} \left(g \overset{\lambda}{*} h\right)$$

holds for arbitrary $f, g, h \in C_\alpha$, i.e. for functions

$$f(x) = x^p \tilde{f}(x), \quad g(h) = x^q \tilde{g}(x), \quad h(x) = x^r \tilde{h}(x) \quad \text{with } p, q, r > \alpha,$$

one can approximate the continuous functions $\tilde{f}, \tilde{g}, \tilde{h} \in C[0, \infty)$ on every finite interval $[0, X]$, $X > 0$, by means of sequences of polynomials, respectively:

$$P_m(x) = \sum_{k=0}^{m} a_{m,k} x^k, \quad Q_n(x) = \sum_{l=0}^{n} b_{n,l} x^l, \quad R_s(x) = \sum_{i=0}^{s} c_{s,i} x^i,$$

according to the Weierstrass theorem. Then, $f, g, h \in C_\alpha$ are approximated by the sequences of functions

$$f_m(x) = x^p P_m(x), \quad g_n = x^q Q_n(x), \quad h_s(x) = x^r R_s(x).$$

Using the bilinearity of operation $\left(\overset{\lambda}{*}\right)$ and its associativity for power functions, it is easily seen that

$$\left(f_m \overset{\lambda}{*} g_n\right) \overset{\lambda}{*} h_s = f_m \overset{\lambda}{*} \left(g_n \overset{\lambda}{*} h_s\right).$$

By letting $m, n, s \longrightarrow \infty$ and then, $X \longrightarrow \infty$, from this equality we obtain the associativity of $\left(\overset{\lambda}{*}\right)$.

The propositions that $T_\lambda h(x) = 0$ yields $h \equiv 0$ and that $(f \circ g)(x) = 0$ iff $f \equiv 0$ or $g \equiv 0$, are corollaries of a known theorem of Mikusinski and Ryll-Nardzewski (see the proof of Theorem 1.2.3). This proves the absence of divisors of zero of the operations $\left(\overset{\lambda}{*}\right)$ and thus, Theorem 2.1.2.

Remark 1. Let us note that the operations $\left(\overset{\lambda}{*}\right)$ defined by (2.1.7) do not depend on the fractional power of integration $\delta \geq 0$. Actually, they are convolutions of the basic operator $L^{(1)} = x^\beta I_\beta^{\gamma,1}$ of generalized integration of order $\delta = 1$, and therefore of each of its fractional powers: $L = L^{(\delta)} = \left(L^{(1)}\right)^\delta$, since $\left(x^\beta I_\beta^{\gamma,1}\right)^\delta = x^{\beta\delta} I_\beta^{\gamma,\delta}$.

Remark 2. If we look for a convolution of $L^{(1)}$, $L^{(\delta)}$ having the constant function $\{1\}$ as unit element:

$$\left(1 \overset{\lambda}{*} f\right)(x) = f(x), \ f \in C_\alpha, \tag{2.1.10}$$

then we have to choose $\lambda = -1$ in (2.1.6), (2.1.7). Normalizing by means of a suitable constant multiplier, we get then the convolution

$$\begin{aligned}(f\tilde{*}g)(x) &= \frac{1}{\Gamma(\gamma+1)}\left(f \overset{-1}{*} g\right)(x) \qquad (2.1.11)\\ &= \frac{x^{-\beta}}{\Gamma(\gamma+1)} I_\beta^{2\gamma,-1-\gamma}\left(x^\beta \int_0^1 f\left[x(1-\tau)^{\frac{1}{\beta}}\right] g\left(x\tau^{\frac{1}{\beta}}\right) d\tau\right), \gamma \neq -1, -2, \ldots,\end{aligned}$$

satisfying conditions (2.1.10).

Some applications of Theorem 2.1.2 are given in Sections 2.2 and 2.3.

Corollary 2.1.3. *For $\gamma = 0$, $\beta = 1$ the operators T_λ (2.1.6) have the form*

$$T_\lambda = x^\lambda I_1^{0,\lambda} = R^\lambda, \quad \lambda \geq 0$$

of Riemann-Liouville operators (1.1.b). Thus we obtain the operations

$$\begin{aligned}\left(f \overset{\lambda}{*} g\right)(x) &= R^\lambda \left\{ x \int_0^1 f\left[x(1-\tau)\right] g(x\tau) d\tau \right\}\\ &= R^\lambda \left\{ \int_0^x f(x-t) g(t) dt \right\}, \quad \lambda \geq 0\end{aligned} \tag{2.1.12}$$

as a one-parameter family of convolutions of Volterra integration operator (2.1.3) $l = R^1$ and of each fractional power: $l^\delta = R^\delta$, $\delta > 0$. In particular, *for $\lambda = 0$ the Duhamel convolution (2.1.1) is obtained ($R^0 = I$):*

$$\left(f \overset{\circ}{*} g\right)(x) = (f * g)(x) = \int_0^x f(x-t) g(t) dt.$$

For $g(x) = \{1\}$ the above equality gives the well-known Duhamel convolutional representation of the Volterra integration operator:

$$lf(x) = \int_0^x f(t) dt = \left(\{1\} \overset{\circ}{*} f\right)(x) = (\{1\} * f)(x). \tag{2.1.13}$$

On the other hand, the unique convolution of the form (2.1.12), for which condition (2.1.10) is fulfilled, is the operation

$$(f\tilde{*}g)(x) = \frac{d}{dx}\left(\int_0^x f(x-t)g(t)dt\right) = \frac{d}{dx}\left(f \overset{\circ}{*} g\right)(x), \qquad (2.1.14)$$

obtained from (2.1.11) by taking $\gamma = 0$, $\beta = 1$ *and interpreting the symbol* $x^{-1}I_1^{0,-1} = R^{-1}$ *as the usual derivative* $D^1 = \frac{d}{dx}$ *of first order.*

It is reasonable to state the more general problem for convolutions of the multiple Erdélyi-Kober fractional integration operators of form:

$$L = x^{\beta\delta_0} I_{\beta,m}^{(\gamma_k),(\delta_k)}, \quad \delta_0 > 0. \qquad (2.1.15)$$

Till recently, this has been an open problem, except for the case of the hyper-Bessel operators with $\delta_0 = \delta_1 = \cdots = \delta_m = 1$ (see Chapter 3) with convolutions found by Dimovski. For quite general fractional integrals, close to (2.1.15) but involving Fox's H-function as a kernel (instead of the $G_{m,m}^{m,0}$-function), convolutions have been found by Luchko and Yakubovich [270]-[271] and Nguyen Hai and Yakubovich [317]. On the base of their results, we find some new convolutions in Section 5.4.

2.2. Džrbashjan-Gelfond-Leontiev operators of generalized differentiation and integration. Convolutions

2.2.i. Generalized Gelfond-Leontiev differentiation operators

Let the function

$$f(z) = \sum_{k=0}^{\infty} a_k z^k \qquad (2.2.1)$$

be analytic in the unit disk $\{|z| < 1\}$. Then, its Riemann-Liouville fractional derivative of order $\alpha > 0$ has the form

$$D^\alpha f(z) = z^{-\alpha} \sum_{k=0}^{\infty} a_k \frac{\Gamma(k+1)}{\Gamma(k-\alpha+1)} z^k. \qquad (2.2.2)$$

A natural way to generalize differintegration (2.2.2) is to change the multiplier $\frac{\Gamma(k+1)}{\Gamma(k-\alpha+1)}$ by means of a more general expression. In 1951 Gelfond and Leontiev [120] introduced such an operation, more general than the usual differntiation $\frac{d}{dz}$, using as a starting point the fact that the multiplier corresponding to $\frac{d}{dz}$ is $\frac{\Gamma(k+1)}{\Gamma(k)}$.

Definition 2.2.1. Let the function

$$\varphi(\lambda) = \sum_{k=0}^{\infty} \varphi_k \lambda^k \qquad (2.2.3)$$

be an entire function of order $\rho > 0$ and $\sigma \neq 0$ and such that

$$\lim_{k \to \infty} k^{\frac{1}{\rho}} \sqrt[k]{|\varphi_k|} = (\sigma e \rho)^{\frac{1}{\rho}}. \tag{2.2.4}$$

Then, the operation

$$D_{\varphi}^{(n)} f(z) = \sum_{k=0}^{\infty} a_k \frac{\varphi_{k-n}}{\varphi_k} z^{k-n} \tag{2.2.5}$$

is said to be a *Gelfond-Leontiev operator of generalized differentiation* (of order $n = 1, 2, \ldots$) *with respect to the function* $\varphi(\lambda)$.

In particular ($n = 1$),

$$D_{\varphi} f(z) = \sum_{k=0}^{\infty} a_k \frac{\varphi_{k-1}}{\varphi_k} z^{k-1}. \tag{2.2.6}$$

For modified conditions (2.2.4) see Tkachenko, [490, p. 662-663]. From the theory of entire functions, it is known that (2.2.4) always holds for $\overline{\lim_{k \to \infty}}$. However, condition (2.2.4) yields that there exists $\lim_{k \to \infty} \sqrt[k-n]{\left| \frac{\varphi_{k-n}}{\varphi_k} \right|} = 1$ and therefore, by the Cauchy-Hadamard formula, series (2.2.5), (2.2.6) have the same radius of convergence as (2.2.1). It is evident that if $\varphi(\lambda) = \exp \lambda$, i.e. $\varphi_k = \frac{1}{\Gamma(k+1)}$, $k = 0, 1, \ldots$, then (2.2.5) and (2.2.6) give $\left(\frac{d}{dz}\right)^n$, respectively $\frac{d}{dz}$.

Definition 2.2.2. The right-inverse operators of (2.2.5), (2.2.6), respectively

$$I_{\varphi}^{(n)} f(z) = \sum_{k=0}^{\infty} a_k \frac{\varphi_{k+n}}{\varphi_k} z^{k+n}, \tag{2.2.7}$$

$$I_{\varphi} f(z) = \sum_{k=0}^{\infty} a_k \frac{\varphi_{k+1}}{\varphi_k} z^{k+1} \quad (n = 1), \tag{2.2.8}$$

are called *generalized Gelfond-Leontiev integration operators with respect to the function* $\varphi(\lambda)$.

One can go further, generalizing Gelfond-Leontiev operators by replacing the multipliers $\frac{\Gamma(k+1)}{\Gamma(k-\alpha+1)}$ in (2.2.2) by means of an arbitrary sequence $\{b_k\}_{k=0}^{\infty}$ satisfying suitable conditions. In this way, the so-called *Hadamard product (convolution)* is introduced, namely: if

$$b(z) = \sum_{k=0}^{\infty} b_k z^k$$

and $f(z)$ given by (2.2.1) are analytic functions in the unit disk, then the Hadamard product is defined as:

$$D\{b; f\} = (b \circ f)(z) := \sum_{k=0}^{\infty} a_k b_k z^k. \tag{2.2.9}$$

If $b_k \to \infty$ as $k \to \infty$, then operation (2.2.9) can be considered as a *generalized differentiation*. For $b_k \neq 0$, $k = 1, 2, \ldots$, its inverse operation

$$I\{b; f\} = \sum_{k=0}^{\infty} \frac{a_k}{b_k} z^k = D\{b_*; f\} \tag{2.2.10}$$

should be a *generalized integration*, or Hadamard product with the "reciprocal" function

$$b_*(z) = \sum_{k=0}^{\infty} \frac{z^k}{b_k}.$$

More details and examples of the generalized differintegration (2.2.9), (2.2.10) can be seen, for example, in Samko, Kilbas and Marichev [434, §22, 3°] and Section 5.5.

Next we consider Gelfond-Leontiev generalized integration and differentiation with respect to a particular entire function $\varphi(\lambda)$.

2.2.ii. Džrbashjan-Gelfond-Leontiev operators, related to the Mittag-Leffler functions

The Mittag-Leffler function (E.22):

$$\varphi(\lambda) = E_\alpha(\lambda) = \sum_{k=0}^{\infty} \frac{\lambda^k}{\Gamma(1 + k\alpha)}, \quad \alpha > 0 \tag{2.2.11}$$

is one of the most characteristic examples of an entire function of order $\rho = \frac{1}{\alpha} > 0$ and of type $\sigma = 1$.

In a number of papers [98]-[103], summarized in the monograph [101], Džrbashjan introduced and investigated the properties of the more general functions of the same kind (E.24):

$$\varphi(\lambda) = E_\rho(\lambda; \mu) = \sum_{k=0}^{\infty} \frac{\lambda^k}{\Gamma\left(\mu + \frac{k}{\rho}\right)}, \tag{2.2.12}$$
$$\rho > 0, \ \mu \in \mathbb{C}.$$

Due to the free choice of the arbitrary complex parameter μ, these functions have many more applications. Usually, they are referred to as *functions of Mittag-Leffler type*. For the sake of brevity we call them *Mittag-Leffler (M.-L.) functions*. The function (2.2.11), written in the form (2.2.12), is

$$E_\alpha(\lambda) = E_\rho(\lambda; 1) \text{ with } \rho = \frac{1}{\alpha}.$$

More about the Mittag-Leffler functions and their multiple counterpart ([103]) can be seen in Appendix, Section E.ii. Obviously, since their coefficients

$$e_k = \left[\Gamma\left(\mu + \frac{k}{\rho}\right)\right]^{-1}, \quad k = 0, 1, 2, \dots \tag{2.2.13}$$

satisfy conditions (2.2.4), therefore generalized Gelfond-Leontiev differentiation and integration operators with respect to Mittag-Leffler function $E_\rho(\lambda;\mu)$ can be considered. These were introduced and investigated by Dimovski and Kiryakova [77]-[78], Kiryakova [196] for real values of the parameter μ. These considerations were extended to complex parameter μ in 1985 by Linchouk [258].

Definition 2.2.3. Let $\rho > 0$, let μ be a complex parameter with $\Re\mu > 0$ and let $\mathfrak{H}(\Delta_R)$ denote the space of functions analytic in the circle $\Delta_R : |z| < R$. For a function

$$f(z) = \sum_{k=0}^{\infty} a_k z^k \in \mathfrak{H}(\Delta_R) \tag{2.2.14}$$

we define the *Džrbashjan-Gelfond-Leontiev (D.-G.-L.) operator of differentiation* by means of the series

$$D_{\rho,\mu} f(z) = \sum_{k=1}^{\infty} a_k \frac{\Gamma\left(\mu + \frac{k}{\rho}\right)}{\Gamma\left(\mu + \frac{k-1}{\rho}\right)} z^{k-1}. \tag{2.2.15}$$

Its linear right inverse operator of the form

$$l_{\rho,\mu} f(z) = \sum_{k=0}^{\infty} a_k \frac{\Gamma\left(\mu + \frac{k}{\rho}\right)}{\Gamma\left(\mu + \frac{k+1}{\rho}\right)} z^{k+1} \tag{2.2.16}$$

is said to be a *Džrbashjan-Gelfond-Leontiev (D.-G.-L.) integration operator.*

Strictly speaking, any linear right inverse operator of $D_{\rho,\mu}$ $(D_{\rho,\mu} L_{\rho,\mu} = I)$ has the form

$$L_{\rho,\mu} f = l_{\rho,\mu} f + \chi(f)$$

with arbitrary linear functional χ in $\mathfrak{H}(\Delta_R)$. The *defining projector of* $l_{\rho,\mu}$ (or the *initial value operator* for this notion see [370], [73] and Section 3.2, is

$$F_{\rho,\mu} f(z) = (I - l_{\rho,\mu} D_{\rho,\mu}) f(z) = f(0).$$

Lemma 2.2.4. *In $\mathfrak{H}(\Delta_R)$ the Džrbashjan-Gelfond-Leontiev integration operator (2.2.16) has also the following integral representation*

$$l_{\rho,\mu} f(z) = \frac{z}{\Gamma\left(\frac{1}{\rho}\right)} \int_0^1 (1-\sigma)^{\frac{1}{\rho}-1} \sigma^{\mu-1} f\left(z\sigma^{\frac{1}{\rho}}\right) d\sigma. \tag{2.2.16'}$$

Proof. It is enough to verify the validity of (2.2.16′) for an arbitrary $f(z) = z^k$, $k \geq 0$. We have:

$$l_{\rho,\mu}\left\{z^k\right\} = \frac{z^{k+1}}{\Gamma\left(\frac{1}{\rho}\right)}\int_0^1 (1-\sigma)^{\frac{1}{\rho}-1}\sigma^{\mu+\frac{k}{\rho}-1}d\sigma = \frac{\Gamma\left(\mu+\frac{k}{\rho}\right)}{\Gamma\left(\mu+\frac{k+1}{\rho}\right)}.$$

Then, if $f(z) = \sum_{k=0}^{\infty} a_k z^k$ is an arbitrary function of $\mathfrak{H}(\Delta_R)$,

$$l_{\rho,\mu}f(z) = \sum_{k=0}^{\infty} a_k l_{\rho,\mu}\left\{z^k\right\} = \sum_{k=0}^{\infty} a_k \frac{\Gamma\left(\mu+\frac{k}{\rho}\right)}{\Gamma\left(\mu+\frac{k+1}{\rho}\right)} z^{k+1},$$

which proves the lemma.

The operator $l_{\rho,\mu}$ is defined by means of (2.2.16′) also in a more general space $\mathfrak{H}(\Omega)$, so it can be considered as an extension of (2.2.16). Here $\mathfrak{H}(\Omega)$ denotes the *space of functions $f(z)$ analytic in a domain Ω starlike with respect to the origin $z = 0$*. Furthermore, (2.2.16′) defines $l_{\rho,\mu}$ as a special case of the 1-tuple generalized fractional integrals, that is, as an Erdélyi-Kober fractional integration operator (1.1.17) with $\gamma = \mu - 1$, $\delta = \frac{1}{\rho}$, $\beta = \rho$ (see example (1.1.h-h′), Chapter 1):

$$l_{\rho,\mu}f(z) = zI_{\rho,1}^{\mu-1,\frac{1}{\rho}}f(z) = z^{\rho\frac{1}{\rho}}I_{\rho}^{\mu-1,\frac{1}{\rho}}f(z). \tag{2.2.17}$$

As we have mentioned in Chapter 1 (Theorem 1.2.18), the generalized fractional integration operators $I_{\beta,m}^{(\gamma_k),(\delta_k)}$ can be considered also in the space

$$\mathfrak{H}_\alpha(\Omega) = \left\{f(z) = z^p\tilde{f}(z);\ p > \alpha, \tilde{f} \in \mathfrak{H}(\Omega)\right\}, \tag{2.2.18}$$

when Ω denotes a domain starlike with respect to the point $z = 0$. The more general results with proofs are given in Section 5.5. In the case of operator (2.2.17), the corresponding parameter α is to be taken greater or equal to $-\beta(\gamma+1) = -\mu\rho$. So, *further we consider $l_{\rho,\mu} = zI_{\rho,1}^{\mu-1,\frac{1}{\rho}}$ in the space $\mathfrak{H}_{-\mu\rho}(\Omega) \supset \mathfrak{H}(\Omega)$.*

The "differentiation" operator $D_{\rho,\mu}$ has its analytic continuation in $\mathfrak{H}(\Omega)$ and in $\mathfrak{H}_{-\mu\rho}(\Omega)$ too. In the case $\rho \geq 1$ this proposition was established by Dimovski and Kiryakova [77]-[78], namely:

Lemma 2.2.5. *The generalized fractional differentiation operator, defined in $\mathfrak{H}_{-\rho\mu}(\Omega)$ by*

$$D_{\rho,\mu}f(z) = D_{\rho}^{\mu-1,\frac{1}{\rho}}z^{-1}f(z) - \frac{f(0)\Gamma(\mu)}{\Gamma\left(\mu-\frac{1}{\rho}\right)}z^{-1}, \tag{2.2.19}$$

that is,

$$D_{\rho,\mu}f(z)=\begin{cases}\left(\frac{1}{\rho}z\frac{d}{dz}+\mu\right)z^{-1}\int\limits_0^1\frac{(1-\sigma)^{-\frac{1}{\rho}}}{\Gamma\left(1-\frac{1}{\rho}\right)}f\left(z\sigma^{\frac{1}{\rho}}\right)d\sigma-\frac{f(0)\Gamma(\mu)}{\Gamma\left(\mu-\frac{1}{\rho}\right)}z^{-1} & \text{for } \rho>1,\\ \left(z\frac{d}{dz}+\mu\right)z^{-1}f(z)-f(0)(\mu-1)z^{-1} & \text{for } \rho=1,\end{cases}\qquad (2.2.19')$$

coincides with the Džrbashjan-Gelfond-Leontiev differentiation operator (2.2.15) in the space $\mathfrak{H}(\Delta_R)$, *when* $\Delta_R\subseteq\Omega$.

The case $\rho\geq 1$ is chosen only for the sake of simplicity of the explicit representation (2.2.19′). In the general case $\rho>0$, the same representation (2.2.19) holds and this is a corollary of the results of Section 1.5. According to Theorem 1.5.1, the operator $l_{\rho,\mu}$ (2.2.17) is a linear right inverse of the generalized fractional derivative

$$d_{\rho,\mu}=D_\rho^{\mu-1,\frac{1}{\rho}}z^{-1}=z^{-1}D_\rho^{\mu-\frac{1}{\rho}-1,\frac{1}{\rho}},$$

and also, of each operator of generalized differentiation

$$D_{\rho,\mu}f(z)=D_\rho^{\mu-1,\frac{1}{\rho}}z^{-1}+f(0)\frac{c}{z}.\qquad (2.2.20)$$

For $c=\frac{\Gamma(\mu)}{\Gamma\left(\mu-\frac{1}{\rho}\right)}$ this operator (2.2.19) coincides with (2.2.15) in $\mathfrak{H}(\Delta_R)$, since:

$$D_{\rho,\mu}\left\{z^k\right\}=D_\rho^{\mu-1,\frac{1}{\rho}}\left\{z^{k-1}\right\}-0.\frac{\Gamma(\mu)}{\Gamma\left(\mu\frac{1}{\rho}\right)}z^{-1}=\frac{\Gamma\left(\mu-1+\frac{1}{\rho}+\frac{k-1}{\rho}+1\right)}{\Gamma\left(\mu-1+\frac{k-1}{\rho}+1\right)}$$
$$=\frac{\Gamma\left(\mu+\frac{k}{\rho}\right)}{\Gamma\left(\mu+\frac{k-1}{\rho}\right)}z^{k-1},\quad k\geq 1,$$

due to Lemma 1.6.1 (see Section 1.6), and

$$D_{\rho,\mu}\left\{z^0\right\}=\frac{\Gamma(\mu)}{\Gamma\left(\mu-\frac{1}{\rho}\right)}z^{-1}-\frac{\Gamma(\mu)}{\Gamma\left(\mu-\frac{1}{\rho}\right)}z^{-1}=0,\quad k=0.$$

In the general case $\rho>0$, define

$$\eta=\begin{cases}\left[\frac{1}{\rho}\right]+1 & \text{for non integer } \frac{1}{\rho},\\ \frac{1}{\rho} & \text{for integer } \frac{1}{\rho}.\end{cases}\qquad (2.2.21)$$

Then, the explicit representation of (2.2.19) follows by Definition 1.5.4, namely:

$$\begin{aligned} D_{\rho,\mu}f(z) &= \left[\prod_{j=1}^{\eta}\left(\frac{1}{\rho}z\frac{d}{dz}+\mu-1+j\right)\right] I_{\rho}^{\mu-1+\frac{1}{\rho},\eta-\frac{1}{\rho}}\left(z^{-1}f(z)\right) \\ &= \left[\prod_{j=1}^{\eta}\left(\frac{1}{\rho}z\frac{d}{dz}+\mu-1+j\right)\right] I_{\rho}^{\mu-1+\frac{1}{\rho},\eta-\frac{1}{\rho}} z^{-1}f(z). \end{aligned} \tag{2.2.19''}$$

For $\rho \geq 1$, i.e. for $\eta = 1$, we get the simpler form (2.2.19′).

All these considerations lead to the following extended definition.

Definition 2.2.3′. The generalized fractional integration operator

$$\begin{aligned} l_{\rho,\mu}f(z) &= zI_{\rho}^{\mu-1,\frac{1}{\rho}}f(z) \\ &= z\int_0^1 \frac{(1-\sigma)^{\frac{1}{\rho}-1}}{\Gamma\left(\frac{1}{\rho}\right)}\sigma^{\mu-1}f\left(z\sigma^{\frac{1}{\rho}}\right)d\sigma, \quad \mu>0,\ \rho>0, \end{aligned} \tag{2.2.22}$$

defined in $\mathfrak{H}_{-\mu\rho}(\Omega)$ is said to be a *Džrbashjan-Gelfond-Leontiev integration operator*. Analogously, the generalized fractional differentiation operator (2.2.19) is called a *Džrbashjan-Gelfond-Leontiev differentiation operator* in $\mathfrak{H}_{-\mu\rho}(\Omega)$.

From the proofs of Lemma 1.2.1 and Theorem 1.2.15, it is evident that the Džrbashjan-Gelfond-Leontiev integration operator $l_{\rho,\mu}$, $\rho > 0$, $\mu > 0$ is a linear operator mapping the linear space $\mathfrak{H}_{-\mu\rho}(\Omega)$ into itself:

$$l_{\rho,\mu} : \mathfrak{H}_{-\mu\rho}(\Omega) \longrightarrow \mathfrak{H}_{-\mu\rho+1}(\Omega) \subset \mathfrak{H}_{-\mu\rho}(\Omega).$$

So, *it is reasonable to look for a convolution of* $l_{\rho,\mu}$ *in* $\mathfrak{H}_{-\mu\rho}(G)$ in the sense of Definition 2.1.1. Since this operator is quite a special case of the Erdélyi-Kober fractional integration operator $x^{\rho\delta}I_{\rho}^{\gamma,\delta}$, its convolutions can be found by specialization of the general Theorem 2.1.2. It is enough to put $\gamma = \mu - 1$, $\delta = \frac{1}{\rho} > 0$, $\beta = \rho > 0$ in (2.1.6) and to modify the statement for spaces $\mathfrak{H}_{\alpha}(\Omega)$. Thus, we obtain the following

Theorem 2.2.6. *Let* $\left(\overset{\rho,\mu}{\circ}\right)$ *be the operation*

$$\left(f \overset{\rho,\mu}{\circ} g\right)(z) = z^{\rho}\int_0^1 f\left[z(1-\zeta)^{\frac{1}{\rho}}\right] g\left(z\zeta^{\frac{1}{\rho}}\right)[\zeta(1-\zeta)]^{\mu-1}\,d\zeta, \tag{2.2.23}$$

and let the operators T_{λ} *have the form*

$$T_{\lambda} = \begin{cases} z^{\rho\lambda}I_{\rho}^{2(\mu-1),\lambda-\mu+1} & \text{if } \lambda \geq \mu-1, \\ z^{\rho\lambda}D_{\rho}^{\mu-1+\lambda,\mu-1-\lambda} & \text{if } \lambda < \mu-1. \end{cases} \tag{2.2.24}$$

(Here $D_\beta^{\gamma,\delta}$ denotes the "Erdélyi-Kober fractional derivative" (1.6.6)). Then, each operation of the form

$$\left(f \overset{\lambda}{\underset{\rho,\mu}{*}} g\right)(z) = T_\lambda \left(f \overset{\rho,\mu}{\circ} g\right)(z) \tag{2.2.25}$$

is a convolution of $l_{\rho,\mu}$ in $\mathfrak{H}_{-\mu\rho}(\Omega)$ if $\lambda \geq \mu-1$, or in the subspace $\mathfrak{H}(\Omega)$ only, if $\lambda < \mu-1$.

If we look for a convolution of $l_{\rho,\mu}$, *uniquely determined* by the condition $\{1\} \overset{\lambda}{\underset{\rho,\mu}{*}} f = f$, $f \in \mathfrak{H}(\Omega)$, then as noted in Section 2.1, we have to choose $\lambda = -1$. Indeed, $\{1\} \overset{-1}{\underset{\rho,\mu}{*}} z^p = z^p$ for every $p \geq 0$. For arbitrary $f(z) \in \mathfrak{H}(\Omega)$ the same condition can be verified with the help of *Runge's approximation theorem*, valid in a starlike domain Ω being also a simply-connected domain. The operation $\left(\overset{-1}{\underset{\rho,\mu}{*}}\right)$ is generated by means of the differintegral operator

$$T_{-1} = z^{-\rho} D_\rho^{\mu-2,\mu} = D_\rho^{\mu-1,\mu} z^{-\rho}, \tag{2.2.26}$$

which does not preserve the basic space $\mathfrak{H}_{-\rho\mu}(\Omega)$. So, we have to confine ourselves to the *subspace* $\mathfrak{H}(\Omega) \subset \mathfrak{H}_{-\rho\mu}(\Omega)$ *of functions analytic in* Ω. Then,

$$\left(\overset{-1}{\underset{\rho,\mu}{*}}\right) = T_{-1}\left(\overset{\rho,\mu}{\circ}\right) \tag{2.2.27}$$

is an operation in $\mathfrak{H}(\Omega)$, since

$$\mathfrak{H}(\Omega) \times \mathfrak{H}(\Omega) \overset{\overset{\rho,\mu}{\circ}}{\longrightarrow} z^\rho \mathfrak{H}(\Omega) \subset \mathfrak{H}_\rho(\Omega) \overset{T_{-1}}{\longrightarrow} \mathfrak{H}_{\rho-\rho}(\Omega) = \mathfrak{H}_0(\Omega) \subseteq \mathfrak{H}(\Omega).$$

According to (1.6.9) for

$$\eta = \begin{cases} [\mu]+1 & \text{if } \mu \text{ is non integer,} \\ \mu & \text{if } \mu \text{ is integer,} \end{cases}$$

the operator (2.2.26) can be written explicitly in the form

$$\begin{aligned} T_{-1} = D_\rho^{\mu-1,\mu} z^{-\rho} &= \prod_{j=0}^{\eta-1} \left(\frac{1}{\rho} z \frac{d}{dz} + \mu + j\right) I_\rho^{2\mu-1,\eta-\mu} \\ &= \begin{cases} \prod_{j=0}^{[\mu]} \left(\frac{1}{\rho} z \frac{d}{dz} + \mu + j\right) I_\rho^{2\mu-1,1-\{\mu\}} & \text{for non integer } \mu, \\ \prod_{j=0}^{\mu-1} \left(\frac{1}{\rho} z \frac{d}{dz} + \mu + j\right) & \text{for integer } \mu. \end{cases} \end{aligned}$$

In this manner we obtain the following result of Dimovski and Kiryakova [77]-[78], Kiryakova [196].

Corollary 2.2.7. *For $\rho > 0$, $\mu > 0$ the operation*

$$\left(f \overset{-1}{\underset{\rho,\mu}{*}} g\right)(z) = \begin{cases} B_{\rho,\mu}^{[\mu]+1} \int\limits_0^1 \frac{(1-\sigma)^{-\{\mu\}}}{\Gamma(1-\{\mu\})} \sigma^{2\mu-1} d\sigma \int\limits_0^1 \frac{(1-t)^{\mu-1}}{\Gamma(\mu)} t^{\mu-1} & \\ \quad \times f\left[z\sigma^{\frac{1}{\rho}}(1-t)^{\frac{1}{\rho}}\right] g\left(z\sigma^{\frac{1}{\rho}} t^{\frac{1}{\rho}}\right) dt, & \text{when } \mu \text{ is non integer,} \\ B_{\rho,\mu}^{\mu} \int\limits_0^1 \frac{(1-t)^{\mu-1}}{\Gamma(\mu)} t^{\mu-1} f\left[z(1-t)^{\frac{1}{\rho}}\right] g\left(z t^{\frac{1}{\rho}}\right) dt, & \text{when } \mu \text{ is an integer,} \end{cases} \tag{2.2.28}$$

where $B_{\rho,\mu}^k$ denotes the differential operator

$$B_{\rho,\mu}^k = \prod_{j=0}^{k-1} \left(\frac{1}{\rho} z \frac{d}{dz} + \mu + j\right), \quad k = 1, 2, \ldots, \tag{2.2.29}$$

is a convolution of the Džrbashjan-Gelfond-Leontiev integration operator $l_{\rho,\mu}$ in $\mathfrak{H}(\Omega)$. This convolution has an unit element in $\mathfrak{H}(\Omega)$, namely the constant function $\{1\}$:

$$\{1\} \overset{-1}{\underset{\rho,\mu}{*}} f(z) = f(z). \tag{2.2.30}$$

The operator $l_{\rho,\mu}$ itself has the following $\left(\overset{-1}{\underset{\rho,\mu}{*}}\right)$-convolutional representation

$$l_{\rho,\mu} f(z) = \left(r \overset{-1}{\underset{\rho,\mu}{*}} f\right)(z) \tag{2.2.31}$$

with

$$r(z) = \frac{\Gamma(\mu)}{\Gamma\left(\mu + \frac{1}{\rho}\right)} z \in \mathfrak{H}(\Omega). \tag{2.2.32}$$

Proof. Since

$$l_{\rho,\mu} = z I_\rho^{\mu-1,\frac{1}{\rho}} = z^{\rho\frac{1}{\rho}} I_\rho^{\mu-1,\frac{1}{\rho}}$$

is an Erdélyi-Kober fractional integral of the form

$$z^{\beta\delta} I_\beta^{\gamma,\delta} \text{ with } \beta = \rho > 0, \quad \gamma = \mu - 1, \quad \delta = \frac{1}{\rho} > 0,$$

then according to Theorem 2.1.2, for arbitrary $\lambda \in \mathfrak{R}$ the operator

$$f \overset{\lambda}{\underset{\rho,\mu}{*}} g = T_\lambda(f \circ g) = \begin{cases} z^{\rho\lambda} I_\rho^{2(\mu-1),\lambda-(\mu-1)}(f \circ g), & \text{if } \lambda \geq \mu - 1, \\ z^{\rho\lambda} D_\rho^{(\mu-1)+\lambda,(\mu-1)-\lambda}(f \circ g), & \text{if } \lambda < \mu - 1, \end{cases}$$

where $(\circ)$ is the auxiliary operation

$$(f \circ g)(z) = z^\rho \int\limits_0^1 \frac{(1-t)^{\mu-1} t^{\mu-1}}{\Gamma(\mu)} f\left[z(1-t)^{\frac{1}{\rho}}\right] g\left[z t^{\frac{1}{\rho}}\right] dt,$$

is a convolution of $l_{\rho,\mu}$ in $\mathfrak{H}(\Omega)$.

If we choose $\lambda = -1$, then $\mu \geq 1$ yields $\lambda < \mu - 1$ and the "correcting" operator T_λ is:

$$T_{-1} = z^{-\rho} D_\rho^{\mu-2,\mu} = D_\rho^{\mu-1,\mu} z^{-\rho}.$$

Then, the operation $\left(\underset{\rho,\mu}{\overset{\lambda}{*}}\right)$ takes the form

$$\left(\underset{\rho,\mu}{\overset{-1}{*}}\right)(z) = D_\rho^{\mu-1,\mu} \left\{ \int_0^1 \frac{(1-t)^{\mu-1} t^{\mu-1}}{\Gamma(\mu)} f\left[z(1-t)^{\frac{1}{\rho}}\right] g\left[zt^{\frac{1}{\rho}}\right] dt \right\}.$$

According to (1.6.9), for

$$\eta = \begin{cases} [\mu] + 1, & \text{if } \mu \text{ is non integer,} \\ \mu, & \text{if } \mu \text{ is integer,} \end{cases}$$

the Erdélyi-Kober fractional derivative $D_\rho^{\mu-1,\mu}$ has the form:

$$\begin{aligned} D_\rho^{\mu-1,\mu} &= \left[\prod_{j'=1}^{\eta} \left(\frac{1}{\rho} z \frac{d}{dz} + \mu - 1 + j' \right) \right] I_\rho^{\mu-1+\mu,\eta-\mu} \\ &= \begin{cases} \left[\prod_{j=0}^{[\mu]} \left(\frac{1}{\rho} z \frac{d}{dz} + \mu + j \right) \right] I_\rho^{2\mu-1,1-\{\mu\}}, & \text{if } \mu \text{ is non integer,} \\ \prod_{j=0}^{\mu-1} \left(\frac{1}{\rho} z \frac{d}{dz} + \mu + j \right), & \text{if } \mu \text{ is integer,.} \end{cases} \end{aligned}$$

Hence, the operation $\left(\underset{\rho,\mu}{\overset{-1}{*}}\right)$ takes the form (2.2.28).

On the other hand, for each $f \in \mathfrak{H}(\Omega)$:

$$\{1\} \underset{\rho,\mu}{\overset{-1}{*}} f(z) = D_\rho^{\mu-1,\mu} z^{-\rho} (\{1\} \circ f(z)) = D_\rho^{\mu-1,\mu} z^{-\rho} \left(z^\rho I_\rho^{\mu-1,\mu} f(z) \right) = f(z),$$

which is (2.2.30). Since

$$l_{\rho,\mu}\{1\} = z I_\rho^{\mu-1,\frac{1}{\rho}} \left\{ z^0 \right\} = z \frac{\Gamma(\mu)}{\Gamma\left(\mu + \frac{1}{\rho}\right)} := r(z),$$

by the convolutional property

$$l_{\rho,\mu} \left(\{1\} \underset{\rho,\mu}{\overset{-1}{*}} f \right) = (l_{\rho,\mu}\{1\}) \underset{\rho,\mu}{\overset{-1}{*}} f,$$

we find (2.2.31): $l_{\rho,\mu} f = r \underset{\rho,\mu}{\overset{-1}{*}} f$. The proof is over.

Remark 1. A *more concise* (but only symbolic) *form of convolution* (2.2.28) is proposed by Dimovski and Kiryakova [77], viz:.

$$\left(f \overset{-1}{\underset{\rho,\mu}{*}} g\right)(z) = \left\{t^{1-\mu}\left(\frac{d}{dt}\right)^{\mu} \int\limits_0^t \frac{(t-x)^{\mu-1}}{\Gamma(\mu)} f\left[(t-x)^{\frac{1}{\rho}}\right] x^{\mu-1} g\left(x^{\frac{1}{\rho}}\right) dx \right\}_{t\to z^\rho} . \quad (2.2.33)$$

Remark 2. In a *modified form*, the same convolution of $l_{\rho,\mu}$ was found by Linchouk [258] in the case of a complex parameter μ with $\Re\mu > 0$. Let $\mu_1 = \mu - [\Re\mu]$ and consider the operators

$$R_{\rho,\mu} f(z) = \int\limits_0^1 \frac{(1-\tau)^{-\mu_1}}{\Gamma(1-\mu_1)} \tau^{\mu_1 - \operatorname{sgn}(\Re\mu_1)} f\left(z\tau^{\frac{1}{\rho}}\right) d\tau$$

$$S_{\rho,\mu} f(z) = \left[(\mu_1 - 1)\operatorname{sgn}(\Re\mu_1) + \frac{z}{\rho}\frac{d}{dz}\right] \prod_{k=2}^{[\Re\mu]} \left(\mu_1 + k - 1\frac{z}{\rho}\frac{d}{dz}\right) f(z)$$

and

$$T_{\rho,\mu} = S_{\rho,\mu} R_{\rho,\mu}.$$

Then, the operation

$$(f * g)(z) = f(0)g(z) + \frac{z}{\rho\Gamma(\mu)} \int\limits_0^1 (1-t)^{\frac{1}{\rho}-1} t^{\mu-1} \left(T_{\rho,\mu} f\right)' \left[z(1-t)^{\frac{1}{\rho}}\right] g\left(zt^{\frac{1}{\rho}}\right) dt \quad (2.2.34)$$

is a continuous convolution of the Džrbashjan-Gelfond-Leontiev integration operator $l_{\rho,\mu}$ in $\mathfrak{H}(\Omega)$.

It is evident that Linchouk's result [258] is another form of the convolution $\left(\overset{-1}{\underset{\rho,\mu}{*}}\right)$ (2.2.28) obtained by putting the differential operator $B^k_{\rho,\mu}$ under the sign of the integral ($\mu_1 = 0$ for real integer μ and $\mu_1 = \mu - [\mu]$ for non integer μ).

As an application, it is interesting to find the $\left(\overset{-1}{\underset{\rho,\mu}{*}}\right)$*-convolutional products of some basic functions in* $\mathfrak{H}(\Omega)$.

Lemma 2.2.8. *For* $p \geq 0$, $q \geq 0$ *we have*

$$\{z^p\} \overset{-1}{\underset{\rho,\mu}{*}} \{z^q\} = z^{p+q} \frac{\Gamma\left(\mu + \frac{p}{\rho}\right)\Gamma\left(\mu + \frac{q}{\rho}\right)}{\Gamma\left(\mu + \frac{p+q}{\rho}\right)}. \quad (2.2.35)$$

Lemma 2.2.9. *Let α, β be two different complex numbers. Then,*

$$\{E_\rho(\alpha z;\mu)\} \overset{-1}{\underset{\rho,\mu}{*}} \{E_\rho(\beta z;\mu)\} = \frac{\alpha E_\rho(\alpha z;\mu) - \beta E_\rho(\beta z;\mu)}{\Gamma(\mu)(\alpha-\beta)}. \tag{2.2.36}$$

Proof. From (2.2.11) and (2.2.16) it follows easily that

$$l_{\rho,\mu} E_\rho(\alpha z;\mu) = \frac{1}{\alpha} E_\rho(\alpha z;\mu) - \frac{1}{\alpha\Gamma(\mu)} \tag{2.2.37}$$

and

$$l_{\rho,\mu} E_\rho(\beta z;\mu) = \frac{1}{\beta} E_\rho(\beta z;\mu) - \frac{1}{\beta\Gamma(\mu)}, \tag{2.2.38}$$

respectively. Multiplying convolutionally (2.2.37) by $E_\rho(\beta z;\mu)$ and (2.2.38) by $E_\rho(\alpha z;\mu)$, we use the property (2.1.2) to obtain one and the same result on the left-hand sides. It remains to apply (2.2.30) to the right-hand sides which leads to (2.2.16).

Let us mention the following analogue of (2.2.37) concerning $D_{\rho,\mu}$ (see (E.27)):

$$D_{\rho,\mu} E_\rho(\alpha z;\mu) = \alpha E_\rho(\alpha z;\mu), \quad \alpha \neq 0. \tag{2.2.39}$$

For $\mu = \rho = 1$, since $E_\rho(\alpha z;\mu) = E_1(\alpha z;1) = e^{\alpha z}$ and $D_{\rho,\mu} = \frac{d}{dz}$ is the usual differentiation, we obtain the well-known differential relation

$$\frac{d}{dz} e^{\alpha z} = \alpha e^{\alpha z}, \quad \alpha \neq 0.$$

In the same case, (2.2.35) and (2.2.36) turn into the relations

$$\{z^p\} \overset{-1}{\underset{1,1}{*}} \{z^p\} = z^{p+q}\frac{p!q!}{(p+q)!} \text{ for integers } p \geq 0,\ q \geq 0,$$

$$\{e^{\alpha z}\} \overset{-1}{\underset{1,1}{*}} \left\{e^{\beta z}\right\} = \frac{\alpha e^{\alpha z} - \beta e^{\beta z}}{\alpha - \beta}, \quad \alpha \neq \beta,$$

where $\left(\overset{-1}{\underset{1,1}{*}}\right)$ is the "differentiated" Duhamel convolution

$$\left(f \overset{-1}{\underset{1,1}{*}} g\right)(z) = \frac{d}{dz}\int_0^z f(z-t)g(t)dt = \frac{d}{dz}(f * g)(z).$$

This scope of results for the generalized integrations $l_{\rho,\mu}$ and their applications in describing the commutants of $l_{\rho,\mu}$ (see next Section 2.3), was proposed by *Dimovski* [72],

[73], first in the case $\mu = 1$. The convolution in $\mathfrak{H}(\Omega)$ of the Gelfond-Leontiev integration operator with respect to the function $E_\rho(\lambda; 1)$:

$$l_\rho f(z) = l_{\rho,1} f(z) = z \int_0^1 \frac{(1-\sigma)^{\frac{1}{\rho}-1}}{\Gamma\left(\frac{1}{\rho}\right)} f\left(z\sigma^{\frac{1}{\rho}}\right) d\sigma, \tag{2.2.40}$$

found by Dimovski has the simpler form

$$\left(f \overset{-1}{\underset{\rho,1}{*}} g\right)(z) = \left(\frac{1}{\rho} z \frac{d}{dz} + 1\right) \int_0^1 f\left[z(1-\tau)^{\frac{1}{\rho}}\right] g\left(z\tau^{\frac{1}{\rho}}\right) d\tau. \tag{2.2.41}$$

It seems interesting to discuss *the relationship between the Gelfond-Leontiev integration operators $l_{\rho,\mu}$, $l_{\rho,1}$ and Riemann-Liouville operator $R^{\frac{1}{\rho}}$*. It turns out that these operators are similar in the sense that *there exist transmutation operators* which are isomorphisms in the corresponding spaces and transform each of these operators into another such operator. Furthermore, these transmutation operators are also fractional integration operators . For instance, the following proposition holds.

Theorem 2.2.10. *The fractional integration operator*

$$\Phi f(z) = I_\rho^{0,\mu-1} f(z) = \int_0^1 \frac{(1-\sigma)^{\mu-2}}{\Gamma(\mu-1)} f\left(z\sigma^{\frac{1}{\rho}}\right) d\sigma, \quad \mu \geq 1 \tag{2.2.42}$$

is a similarity from the Gelfond-Leontiev integration operator $l_{\rho,1}$ (2.2.40) to the more general Džrbashjan-Gelfond-Leontiev operator $l_{\rho,\mu}$ (2.2.22) with $\mu \geq 1$:

$$\Phi : l_{\rho,1} \longrightarrow l_{\rho,\mu}, \quad i.e. \quad \Phi l_{\rho,1} = l_{\rho,\mu}\Phi \quad in\ \mathfrak{H}(\Omega). \tag{2.2.43}$$

On the other hand, the mapping $\Xi^{-1} : f(z) \longrightarrow f(z^\rho)$ $(\rho > 0)$ is similarity from the Riemann-Liouville operator $R^{\frac{1}{\rho}}$ to the operator $l_{\rho,1}$:

$$\Xi^{-1} : R^{\frac{1}{\rho}} \longrightarrow l_{\rho,1}, \quad i.e. \quad \Xi^{-1} R^{\frac{1}{\rho}} = l_{\rho,1}\Xi^{-1} \quad in\ \mathfrak{H}(\Omega). \tag{2.2.44}$$

Then, the composition $\Phi\Xi^{-1}$, that is, the integration operator

$$\Psi f(z) = I_\rho^{0,\mu-1}\Xi^{-1} f(z) = \int_0^1 \frac{(1-\sigma)^{\mu-2}}{\Gamma(\mu-1)} f(z^\rho \sigma)\, d\sigma \tag{2.2.45}$$

is a similarity (transmutation operator) from $R^{\frac{1}{\rho}}$ to $l_{\rho,\mu}$, namely:

$$\Psi : R^{\frac{1}{\rho}} \longrightarrow l_{\rho,\mu} \quad or \quad \Psi R^{\frac{1}{\rho}} = l_{\rho,\mu}\Psi \quad in\ \mathfrak{H}(\Omega). \tag{2.2.46}$$

Proof. Lemma 1.3.1, (1.3.3) and Theorem 1.3.8 (the index law (1.3.11)) yield:

$$\Phi l_{\rho,1} = I_\rho^{0,\mu-1}\left(zI_\rho^{0,\frac{1}{\rho}}\right) = zI_\rho^{\frac{1}{\rho},\mu-1}I_\rho^{0,\frac{1}{\rho}} = zI_\rho^{0,\mu-1+\frac{1}{\rho}}$$

and

$$l_{\rho,\mu}\Phi = zI_\rho^{\mu-1,\frac{1}{\rho}}I_\rho^{0,\mu-1} = zI_\rho^{0,\mu-1+\frac{1}{\rho}},$$

therefore (2.2.43) is fulfilled in $\mathfrak{H}(\Omega)$. Relation (2.2.44) is obvious. Then,

$$\Psi R^{\frac{1}{\rho}} = \Phi\Xi^{-1}R^{\frac{1}{\rho}} = \Phi l_{\rho,1}\Xi^{-1} = l_{\rho,\mu}\Phi\Xi^{-1} = l_{\rho,\mu}\Psi,$$

which is (2.2.46). From Lemma 1.3.3, relation (1.3.6), for the product of $I_\rho^{0,\mu-1}$ and Ξ^{-1} we find the representation (2.2.45):

$$\begin{aligned}\Psi f(z) &= I_\rho^{0,\mu-1}\Xi^{-1}f(z) = \Xi^{-1}I_1^{0,\mu-1}f(z)\\ &= \left[I_1^{0,\mu-1}f(z)\right]_{z\to z^\rho} = \int_0^1 \frac{(1-\sigma)^{\mu-2}}{\Gamma(\mu-1)}f(z^\rho\sigma)\,d\sigma.\end{aligned}$$

This ends the proof.

Note. The operators Φ, Ψ have also the representations

$$\Phi f(z) = \rho\frac{z^{\rho(\mu-1)}}{\Gamma(\mu-1)}\int_0^z (z^\rho-\zeta^\rho)^{\mu-2}\zeta^{\rho-1}f(z)d\zeta \tag{2.2.42'}$$

and

$$\Psi f(z) = z^{\rho(\mu-1)}\int_0^{z^\rho}\frac{(z^\rho-\zeta)^{\mu-2}}{\Gamma(\mu-1)}f(\zeta)d\zeta \tag{2.2.45'}$$

Knowledge of the transmutation operators between two given operators , or more generally, between two given problems, allows us to transfer the known results.for one of them (usually for the simpler) into results for the other one (usually the more complicated). This is the key idea of the *transmutation method* in general (for more details see Section 3.5). In this sense, Theorem 2.2.10 can be useful in finding a convolution $\left(\underset{\rho,\mu}{\overset{-1}{*}}\right)$ of $l_{\rho,\mu}$ on the basis of the known convolution $\left(\underset{\rho,1}{\overset{-1}{*}}\right)$ (Dimovski [72], [73]) of $l_{\rho,1}$. We have to use the following general theorem of Dimovski proposing one of the most effective approaches in finding new convolutions.

Theorem (Th. 1.3.6, [72, p.36]). *If $T : X \longrightarrow \tilde{\mathcal{X}}$ is an isomorphism of a linear space $\mathcal{X}$ onto a linear space $\tilde{\mathcal{X}}$ and $\tilde{L} : \tilde{\mathcal{X}} \longrightarrow \tilde{\mathcal{X}}$ is a linear operator in $\tilde{\mathcal{X}}$ with a convolution $\tilde{*} : \tilde{\mathcal{X}} \times \tilde{\mathcal{X}} \longrightarrow \tilde{\mathcal{X}}$, then the operation*

$$f * g = T^{-1}(Tf\tilde{*}Tg) \tag{2.2.47}$$

is a convolution of the similar operator $L = T^{-1}\tilde{L}T$ in $\mathcal{X}$.

Now let us choose $\tilde{L} = l_{\rho,1}$, $L = l_{\rho,\mu}$, $T = \Phi^{-1}$ and $\mathcal{X} = \tilde{\mathcal{X}} = \mathfrak{H}(\Omega)$. Then, the above theorem yields that the operation

$$\left(f \overset{\rho,\mu}{*} g\right)(z) = \Phi\left\{(\Phi^{-1}f) \underset{\rho,1}{\overset{-1}{*}} (\Phi^{-1}g)\right\} \tag{2.2.48}$$

is a convolution of $l_{\rho,\mu}$ in $\mathfrak{H}(\Omega)$, where $\left(\underset{\rho,1}{\overset{-1}{*}}\right)$ denotes Dimovski's convolution (2.2.41). It is easy to calculate that (2.2.36) is fulfilled for the convolution (2.2.48) too. Since the system of Mittag-Leffler functions $\{E_\rho(\alpha z;\mu), \alpha \in \mathbb{C}\}$ is complete in $\mathfrak{H}(\Omega)$, it is easy to verify the coincidence of the convolutions $\left(\underset{\rho,\mu}{\overset{-1}{*}}\right)$, $\left(\overset{\rho,\mu}{*}\right)$ in $\mathfrak{H}(\Omega)$.

One of the most frequently encountered cases of the Džrbashjan-Gelfond-Leontiev operators is the case $\mu = \frac{1}{\rho} > 0$. An example of such an operator is the generalized differentiation operator D_α, $\alpha > 0$ of *Iliev* [146], [147] (mentioned in Chapter 1, Example (1.1.g′)). Its linear right inverse operator (1.1.g) is the Džrbashjan-Gelfond-Leontiev integration operator with $\mu = \frac{1}{\rho} = \alpha > 0$:

$$L_\alpha = zI_{\frac{1}{\alpha}}^{\alpha-1,\alpha} = l_{\frac{1}{\alpha},\alpha}.$$

Its convolution $\left(\underset{\frac{1}{\alpha},\alpha}{\overset{-1}{*}}\right)$, denoted by $\left(\overset{\alpha}{*}\right)$, has the form

$$\left(f \overset{\alpha}{*} g\right)(z) = \left\{t^{1-\alpha}\left(\frac{d}{dt}\right)^\alpha \int_0^t \frac{[(t-x)x]^{\alpha-1}}{\Gamma(\alpha)} f\left[(t-x)^\alpha\right] g\left(x^\alpha\right) dx\right\}_{t\to z^{\frac{1}{\alpha}}}. \tag{2.2.49}$$

Then, the $\left(\overset{\alpha}{*}\right)$-convolutional product of two power functions is

$$\{z^p\} \overset{\alpha}{*} \{z^q\} = \left[\binom{p+q}{p}_\alpha\right]^{-1} z^{p+q}, \tag{2.2.50}$$

where the *"generalized binomial coefficients"* introduced by Iliev [147] are used:

$$\binom{p+q}{p}_\alpha = \frac{\Gamma(\alpha(p+q+1))}{\Gamma(\alpha(p+1))\,\Gamma(\alpha(q+1))}. \tag{2.2.51}$$

For other results related to the Gelfond-Leontiev derivatives, see also Kapoor and Patel [183].

2.3. Representations of the commutants of Džrbashjan-Gelfond-Leontiev integration operators

Once a convolution for a linear operator L in a space $\mathcal{X}$ is found, the problem of finding the multipliers of this convolution can be considered and this problem turns out to be closely related to that for finding the commutant of L in $\mathcal{X}$.

Definition 2.3.1. Let $L : \mathcal{X} \longrightarrow \mathcal{X}$ be a linear operator mapping the linear space $\mathcal{X}$ into itself. The set of all continuous linear operators $M : \mathcal{X} \longrightarrow \mathcal{X}$ commuting with L:

$$MLf = LMf, \quad f \in \mathcal{X}$$

is said to be a *commutant of L in $\mathcal{X}$*.

Definition 2.3.2. An element k is said to be a *cyclic element of L* if the set of all the linear combinations of $\{L^n k\}$, $n = 0, 1, 2, \dots$ denoted by *span* $\{L^n k\}_{n=0}^{\infty}$, is dense in $\mathcal{X}$.

If we have a continuous convolution of L in $\mathcal{X}$, then the following assertion allows finding the explicit form of the operators commuting with L, under some additional conditions.

Theorem 2.3.3. (Dimovski [73, p. 43], Th. 1.3.11) *If $(*)$ is a continuous and annihilator-free convolution of the linear continuous operator $L : \mathcal{X} \longrightarrow \mathcal{X}$ in a Fréchet space $\mathcal{X}$ with a cyclic element, then the multiplier ring of the convolutional algebra $(\mathcal{X}, *)$ coincides with the commutant of the operator L in $\mathcal{X}$.*

2.3.i. Commutant of the Džrbashjan-Gelfond-Leontiev operator $l_{\rho,\mu}$ in $\mathfrak{H}(\Omega)$

In [72], [73] Dimovski proved that the linear operators $M : \mathfrak{H}(\Omega) \longrightarrow \mathfrak{H}(\Omega)$ commuting with the Gelfond-Leontiev integration operator $l_\rho = l_{\rho,1}$ in $\mathfrak{H}(\Omega)$ have the form

$$Mf(z) = \left(m \overset{-1}{\underset{\rho,1}{*}} f \right)(z) \tag{2.3.1}$$

with $m(z) = M\{1\} \in \mathfrak{H}(\Omega)$, where $\left(\overset{-1}{\underset{\rho,1}{*}} \right)$ is the convolution (2.2.41).

The more general result concerning convolutional representation of the commutant of the Džrbashjan-Gelfond-Leontiev integration operator $l_{\rho,\mu}$ is given by the following assertion, proved by Dimovski and Kiryakova [77], [78].

Theorem 2.3.4. *A linear operator $M : \mathfrak{H}(\Omega) \longrightarrow \mathfrak{H}(\Omega)$ commutes with the Džrbashjan-Gelfond-Leontiev integration operator $l_{\rho,\mu}$, iff it admits a representation of the form*

$$Mf(z) = \left(m \overset{-1}{\underset{\rho,\mu}{*}} f \right)(z) \tag{2.3.2}$$

with $m(z) = M\{1\} \in \mathfrak{H}(\Omega)$, where operation (2.2.28) is denoted by $\left(\overset{-1}{\underset{\rho,\mu}{*}} \right)$.

Proof. To use the general theorem of Dimovski mentioned above, one should establish that the operator $l_{\rho,\mu}$ has a cyclic element in the Fréchet space $\mathfrak{H}(\Omega)$. Indeed, the constant function $\{1\}$ is a cyclic element of $l_{\rho,\mu}$, since the span of $\left(l^n_{\rho,\mu}\{1\}\right)_{n=0}^{\infty}$ coincides with the space of the polynomials in $\mathfrak{H}(\Omega)$. But, according to the Runge approximation theorem in the simply-connected starlike domain $\Omega \subset \mathbb{C}$, the polynomials are dense in $\mathfrak{H}(\Omega)$. Further, if $M : \mathfrak{H}(\Omega) \longrightarrow \mathfrak{H}(\Omega)$ is a linear operator commuting with $l_{\rho,\mu}$ in $\mathfrak{H}(\Omega)$, then M is a multiplier of the convolution $\left(\overset{-1}{\underset{\rho,\mu}{*}}\right)$. From (2.2.30): $f = \{1\} \overset{-1}{\underset{\rho,\mu}{*}} f$, we obtain

$$Mf(z) = (M\{1\}) \overset{-1}{\underset{\rho,\mu}{*}} f(z) = \left(m \overset{-1}{\underset{\rho,\mu}{*}} f\right)(z),$$

thus proving representation (2.3.2). Conversely, each convolutional operator of the form (2.3.2) obviously commutes with $l_{\rho,\mu}$. For more details see [73], [77].

An extension of this result to complex values of the parameter μ is made by Linchouk. Using the same convolution but written in the form (2.2.34) and slightly different arguments, he proves the following.

Theorem 2.3.5. (Linchouk [258], Theorems 1, 2) *A linear continuous operator* $M : \mathfrak{H}(\Omega) \longrightarrow \mathfrak{H}(\Omega)$ *commutes with the Džrbashjan-Gelfond-Leontiev operator* $l_{\rho,\mu}$ *in* $\mathfrak{H}(\Omega)$ *iff it has a representation of the form*

$$Mf(z) = m(z)f(0) + \frac{z}{\rho\Gamma(\mu)} \int_0^1 (1-t)^{\frac{1}{\rho}-1} t^{\mu-1} \left(T_{\rho,\mu} f\right)' \left[z(1-t)^{\frac{1}{\rho}}\right] m\left(zt^{\frac{1}{\rho}}\right) dt \qquad (2.3.3)$$

with $m = M\{1\}$, *or letting* $\varphi(z) = [\Gamma(\mu)]^{-1} T_{\rho,\mu} m(z) \in \mathfrak{H}(\Omega)$*:*

$$Mf(z) = \varphi(0)f(z) + \frac{z}{\rho} \int_0^1 (1-t)^{\frac{1}{\rho}-1} t^{\mu-1} \varphi' \left[z(1-t)^{\frac{1}{\rho}}\right] f\left(zt^{\frac{1}{\rho}}\right) dt. \qquad (2.3.3')$$

For $\mu = 1$, the latter expression coincides with the following result of Dimovski [72]-[73]:

$$Mf(z) = m(0)f(z) + \frac{z}{\rho} \int_0^1 (1-t)^{\frac{1}{\rho}-1} m' \left[z(1-t)^{\frac{1}{\rho}}\right] f\left(zt^{\frac{1}{\rho}}\right) dt, \qquad (2.3.4)$$

following immediately from (2.3.1), if one uses the representation (see [73, p. 107])

$$\left(f \overset{-1}{\underset{\rho,1}{*}} g\right)(z) = f(0)g(z) + \frac{z}{\rho} \int_0^1 (1-t)^{\frac{1}{\rho}-1} f' \left[z(1-t)^{\frac{1}{\rho}}\right] g\left(zt^{\frac{1}{\rho}}\right) dt \qquad (2.3.5)$$

of the convolution $\left(\overset{-1}{\underset{\rho,1}{*}}\right)$.

For arbitrary μ and $\rho = \mu$, representations (2.3.2) and (2.3.3′) coincide with the results of *Kiryutenko* [215].

Alternative representations of the commutant of $l_{\rho,1}$ in $\mathfrak{H}(\Omega)$ were proposed also by *Tkachenko* [488]-[492].

As a special case of these considerations, when $\mu = \rho = 1$ we obtain the previous results of Raichinov [373]-[374], namely, the representation

$$Mf(z) = m(0)f(z) + \int_0^z m'(z-\zeta)f(\zeta)d\zeta \quad \text{with} \quad m(z) \in \mathfrak{H}(\Omega) \tag{2.3.6}$$

of the operators M commuting in $\mathfrak{H}(\Omega)$ with the *Volterra integration operator*

$$lf(z) = l_{1,1}f(z) = \int_0^z f(\zeta)d\zeta.$$

2.3.ii. Convolutional representation of the commutant of a fixed integer power of the Gelfond-Leontiev integration operator

Let m be a fixed positive integer. If $m > 1$, then the operators $M : \mathfrak{H}(\Omega) \longrightarrow \mathfrak{H}(\Omega)$ commuting with $l_{\rho,\mu}^m$ in $\mathfrak{H}(\Omega)$ are more than those commuting with $l_{\rho,\mu}$. In particular, Raichinov [378] has found an explicit representation of the commutant of $l^m = l_{1,1}^m$ in a *m-symmetric* domain Ω. A domain Ω is said to be *m-symmetric*, if $z \in \Omega$ implies $\omega z \in \Omega$, where ω is an arbitrary m-th root of unity, i.e. a complex number with $\omega^m = 1$. His result can be summarized in the following theorem.

Theorem 2.3.6. (Raichinov [378]) *A linear operator M in $\mathfrak{H}(\Omega)$ commutes with the mth power l^m of the Volterra integration operator l (1.1.a) in $\mathfrak{H}(\Omega)$ iff it admits a representation of the form*

$$Mf(z) = \sum_{k=0}^{m-1} \frac{d^{k+1}}{dz^{k+1}} \int_0^z m_k(z-\zeta)f_k(\zeta)d\zeta \tag{2.3.7}$$

with arbitrary $m_k(z) \in \mathfrak{H}(\Omega)$, $k = 0, 1, 2, \ldots, m-1$, and $f_k(z)$ given by

$$f_k(z) = \frac{1}{m}\sum_{j=0}^{m-1} \omega^{-kj} f\left(\omega^j k\right), \quad k = 0, 1, 2, \ldots, m-1, \tag{2.3.8}$$

where ω is a primitive m-th root of unity.

A more general result concerning the commutant of the operators $l_{\rho,1}^m$, $m = 1, 2, \ldots$, $(\mu = 1)$ is proposed by Dimovski [73], namely:

Theorem 2.3.7. (Dimovski [73, Th. 2.5.3]) *A linear operator* $M : \mathfrak{H}(\Omega) \longrightarrow \mathfrak{H}(\Omega)$, $\mathfrak{H}(\Omega)$ *being the space of the analytic functions in a* m*-symmteric and starlike domain* Ω *commutes with the* m^{th} *power* $l_{\rho,1}^m$ *of the Gelfond-Leontiev integration operator* $l_{\rho,1}$ *(2.2.40) if and only if it admits a representation of the form*

$$Mf = \sum_{k=0}^{m-1} D_{\rho,1}^k \left(m_k \overset{-1}{\underset{\rho,1}{*}} f_k \right), \tag{2.3.9}$$

where $D_{\rho,1}^k$ *is the Gelfond-Leontiev differentiation operator ((2.2.19-19') with* $\mu = 1$*):*

$$D_{\rho,1} f(z) = \frac{z^{\frac{1}{\rho}-1}}{\rho^{\frac{1}{\rho}}} \left(\frac{d}{dz} \right)^{\frac{1}{\rho}} f(z) - \frac{f(0)}{z\Gamma\left(1 - \frac{1}{\rho}\right)},$$

$m_k \in \mathfrak{H}(\Omega)$ *are arbitrary,*

$$f_k(z) = \frac{1}{m} \sum_{k=0}^{m-1} \omega^{-kj} f\left(\omega^j z\right), \quad k = 0, 1, \ldots, m-1,$$

ω *is a primitive* m*-th root of unity and* $\left(\overset{-1}{\underset{\rho,1}{*}} \right)$ *is the convolution (2.3.5) of* $l_{\rho,1}$.

Proof. The proof is based on a series of lemmas (see [73, p. 108-112]). The basic steps are as follows. The space $\mathfrak{H}(\Omega)$ is represented as a direct sum of subspaces

$$\mathfrak{H}_k(\Omega) = \left\{ f \in \mathfrak{H}(\Omega);\ f(\omega z) = \omega^k f(z),\ z \in \Omega, \omega^m = 1 \right\}, \quad k = 0, 1, \ldots, m-1,$$

namely:

$$\mathfrak{H}(\Omega) = \mathfrak{H}_0(\Omega) \oplus \mathfrak{H}_1(\Omega) \oplus \ldots \oplus \mathfrak{H}_{m-1}(\Omega).$$

Each of these subspaces is shown to be invariant for $l_{\rho,1}^m$. The functions

$$e_k(z) = \frac{z^k}{\Gamma\left(\frac{k}{\rho} + 1\right)} \in \mathfrak{H}_k(\Omega), \quad k = 0, 1, \ldots, m-1$$

are cyclic elements of $l_{\rho,1}^m$ in $\mathfrak{H}_k(\Omega)$. Then, in each subspace $\mathfrak{H}_k(\Omega)$ the linear operators M, commuting with $l_{\rho,1}^m$, have the form

$$Mf = (Me_k) \overset{k}{*} f = m_k \overset{k}{*} f,$$

where

$$m_k \overset{k}{*} f = D_{\rho,1}^k \left(m_k \overset{-1}{\underset{\rho,1}{*}} f \right)$$

is a convolution of $l_{\rho,1}^m$ in its invariant subspace $\mathfrak{H}_k(\Omega)$.

Similar constructions for the commutants of the powers of the differentiation operator, generalized backward shift operators, Euler differentiation operator, etc. in spaces of analytic functions can be found in Dimovski and Vasilev [83], Vasilev [500-501], Raichinov [372]-[377], [379]-[381] and Raichinov and Raichinov [382]-[383]; see also the book of Fage and Nagnibida [112].

Operators commuting with the differentiation (integration) in spaces of functions of several variables are considered ,for example, by Raichinov [378] and Napalkov [316].

In particular, for the Hardy-Littlewood integration operators (1.1.c):

$$L_{1,0}f(z) = \frac{1}{z}\int_0^z f(\zeta)d\zeta; \quad L_{m,n}f(z) = z^{-m}\int_0^z \zeta^n f(\zeta)d\zeta,$$

as special cases of the Erdélyi-Kober fractional integral:

$$L_{m,n}f(z) = z^{n-m+1} I_1^{n,1} f(z); \qquad \text{integers } m, n; \quad n > m-1,$$

the problem for the commutants of $L_{m,n}$ and of their fixed integer powers $L^p_{m,n}$, $p = 1, 2, \ldots$ is considered by Hristova [139]-[141]. Her approach is rather different, including recurrent formulas and matrix methods. There, the functional spaces are the algebra of polynomials with complex coefficients and the space $\mathfrak{H}(\Delta_R)$ of analytic functions in the disk $|z| < R$. An analytic description of the operators commuting with $L_{m,n}$ and $L^p_{m,n}$ is found in terms of series as well as in integral form. The isomorphic properties of the mappings $M : \mathfrak{H}(\Delta_R) \longrightarrow \mathfrak{H}(\Delta_R)$, generated by the linear operators M of the commutants, are investigated. It is proved also that $L^p_{m,n}$ is *finitely minimally commutative* iff $p = 1$ and $m = n$.

2.4. Borel-Džrbashjan integral transform: operational properties and convolution

Considering the *Gelfond-Leontiev differentiation and integration operators* $D_{\rho,\mu}$, $l_{\rho,\mu}$ and their convolutions, it is natural to state *the problem of the existence of a corresponding Laplace type integral transform*. Such an integral transformation should have the same convolution and should transform differintegrals $D_{\rho,\mu}$, $l_{\rho,\mu}$ into algebraical operations. For a more general aspect on this problem, see Section 5.6.

In Dimovski and Kiryakova [78], Kiryakova [196] we have shown that the role of such a transformation can be played by a generalization of the *Borel transform*

$$\mathfrak{B}\left\{\sum_{k=0}^{\infty} a_k z^k\right\} = \sum_{k=0}^{\infty} \frac{a_k}{k! z^{k+1}}. \tag{2.4.1}$$

This generalization has been introduced for other purposes by Džrbashjan [101, p.323], [98]-[100], considering entire functions of order $\rho > 0$ and finite type and more generally, functions analytic in angular domains and satisfying some conditions on their growth.

The so-called *generalized Borel transformation with respect to the Mittag-Leffler function* $E_\rho(z;\mu)$ is defined for entire functions of order $\rho > 0$:

$$f(z) = \sum_{k=0}^{\infty} a_k z^k \tag{2.4.2}$$

by means of the series

$$\mathfrak{B}_{\rho,\mu}\{f(z);\zeta\} = \sum_{k=0}^{\infty} \Gamma\left(\mu + \frac{k}{\rho}\right) \frac{a_k}{\zeta^{k+1}}, \tag{2.4.3}$$

but it has also the integral representation

$$\mathfrak{B}_{\rho,\mu}\{f;\zeta\} = \rho\zeta^{\mu\rho-1} \int_0^{\infty} \exp\left(-\zeta^\rho t^\rho\right) t^{\mu\rho-1} f(t)dt. \tag{2.4.4}$$

Usually the generalized Borel transforms are considerd in *ρ-convex domains* of $\mathbb{C}$. Thus Džrbashjan [98],[101] and later on, Tkachenko [488]-[489], [492], considered the transform

$$\mathfrak{B}_{\rho,\mu}\{f;\zeta\} = \rho\left(e^{-i\theta}\zeta\right)^{\mu\rho} \zeta^{-1} \int_0^{\infty} f\left(e^{-i\theta}t\right) \exp\left[-t^\rho \left(e^{-i\theta}\zeta\right)^\rho\right] t^{\mu\rho-1}dt,$$

with an arbitrary $\theta \in (-\pi, \pi]$, in the domain

$$\mathfrak{D}_\rho(\theta;\nu) = \left\{\zeta : \Re\left(e^{-i\theta}\zeta\right)^\rho > \nu, |\arg\zeta - \theta| < \frac{\pi}{2\rho}\right\}, \tag{2.4.5}$$

whose contour $\mathfrak{L}_\rho(\theta,\nu)$ is defined by the equation $\left(e^{-i\theta}\zeta\right)^\rho = \nu + i\tau$, $0 \le \nu < \infty$, $-\infty < \tau < \infty$.

Using the above integral representation of $\mathfrak{B}_{\rho,\mu}$, Džrbashjan ([99]-[100], [101, p. 383-404]) extends its definition to functions, continuous on a system of rays in $\mathbb{C}$ or analytic in angular domains. In accordance with his considerations, we give the following definition.

Definition 2.4.1. Let us denote by $A^{(\alpha)}[\rho_1,\sigma_1]$ the class of functions $f(z)$, analytic in the angular domain

$$\Delta(\alpha;0) = \left\{z : |\arg z| < \frac{\pi}{2\alpha}, \quad 0 < |z| < \infty, \quad \frac{1}{2} < \alpha < \infty\right\}$$

and satisfying growth conditions of the form

$$|f(z)| \le M_f \exp\left(\sigma_1 |z|^{\rho_1}\right).$$

For $\mu > 0$, $\rho \geq \max\left\{\rho_1, \frac{\alpha}{2\alpha-1}\right\}$ and $f \in A^{(\alpha)}[\rho_1, \sigma_1]$, the functions

$$\mathfrak{B}_{\rho,\mu}\{f;\zeta\} = \rho\left(e^{-i\theta}\zeta\right)^{\mu\rho}\zeta^{-1}\int_0^\infty f\left(e^{-i\theta}t\right)\exp\left[-t^\rho\left(e^{-i\theta}\zeta\right)^\rho\right]t^{\mu\rho-1}dt, \qquad (2.4.6)$$

depending on the parameter $\theta \in \left[-\frac{\pi}{2\alpha}, \frac{\pi}{2\alpha}\right]$ are said to be *Borel-Džrbashjan (B.-D.) transforms*. For brevity, we have omitted the additional superscript (θ) in the exact denotation $\mathfrak{B}^{(\theta)}_{\rho,\mu}$ for (2.4.6).

It is seen that for $\theta = 0$, (2.4.6) gives the simpler representation (2.4.4).

Further, denote

$$\mathfrak{D}^{(\alpha)}_\rho(\nu) = \bigcup_{-\frac{\pi}{2\alpha}\leq\theta\leq\frac{\pi}{2\alpha}} \mathfrak{D}_\rho(\theta;\nu)$$

and let $\mathfrak{L}^{(\alpha)}_\rho(\nu)$ be the contour encircling $\mathfrak{D}^{(\alpha)}_\rho(\nu)$ in a negative (clock-wise) direction. Then, the function $\mathfrak{B}_{\rho,\mu}\{f;\zeta\}$ is an analytic function of ζ in $\mathfrak{D}^{(\alpha)}_\rho(\nu_0)$, where

$$\nu_0 = \begin{cases} \sigma_1 & \text{if } \rho = \rho_1 \\ 0 & \text{if } \rho > \rho_1. \end{cases}$$

For $\frac{1}{2} < \mu < \frac{1}{2} + \frac{1}{\rho}$, the following *inversion formula* of the Borel-Džrbashjan transform (2.4.6) holds (see Džrbashjan [101, p. 397]):

$$f(z) = \frac{1}{2\pi i}\int_{\mathfrak{L}} E_\rho(z\zeta;\mu)\,\mathfrak{B}_{\rho,\mu}\{f,\zeta\}d\zeta, \quad z \in \Delta(\alpha;0), \qquad (2.4.7)$$

where $\nu > \nu_0$, $\mathfrak{L} = \mathfrak{L}^{(\alpha)}_\rho(\nu)$ and $E_\rho(z;\mu)$ is the Mittag-Leffler function (2.2.12).

In [78], [196] we have proved a series of properties of the Borel-Džrbashjan transform that show its *relationship with the Gelfond-Leontiev operators and their convolution* $\left(\overset{-1}{\underset{\rho,\mu}{*}}\right)$. Here we state some of them.

Theorem 2.4.2. *If $f(z) \in A^\alpha[\rho_1, \sigma_1]$, then the Gelfond-Leontiev integration operator (2.2.16-16′) is algebrized by the Borel-Džrbashjan transform in the following way:*

$$\mathfrak{B}_{\rho,\mu}\{l_{\rho,\mu}f;\zeta\} = \frac{1}{\zeta}\mathfrak{B}_{\rho,\mu}\{f;\zeta\}. \qquad (2.4.8)$$

Proof. There is no loss of generality in taking $\theta = 0$ in the expression (2.4.6) for $\mathfrak{B}_{\rho,\mu}$. This means that z varies on the real half-half $\mathbb{R}^+$ only and we exhibit this by putting $z = x$. Since $l_{\rho,\mu}$ (2.2.16′) can be written in the form

$$l_{\rho,\mu}f(x) = \frac{\rho(1-\mu)}{\Gamma\left(1+\frac{1}{\rho}\right)}\int_0^\infty (x^\rho - \tau^\rho)^{\frac{1}{\rho}-1}\tau^{\mu\rho-1}f(\tau)d\tau,$$

the left-hand side of (2.4.8) is:

$$\mathfrak{B}_{\rho,\mu}\left\{l_{\rho,\mu}f;\zeta\right\} = \frac{\rho\zeta^{\mu\rho-1}}{\Gamma\left(1+\frac{1}{\rho}\right)}\int\limits_0^\infty \exp\left(-\zeta^\rho t^\rho\right) t^{\rho-1} dt \times \int\limits_0^t \left(t^\rho - \tau^\rho\right)^{\frac{1}{\rho}-1} \tau^{\mu\rho-1} f(\tau) d\tau.$$

After the substitution $t^\rho = \sigma$ and interchanging the order of integrations, we get

$$\mathfrak{B}_{\rho,\mu}\left\{l_{\rho,\mu}f;\zeta\right\} = \rho\zeta^{\mu\rho-1}\int\limits_0^\infty \tau^{\mu\rho-1} f(\tau) \frac{d\tau}{\Gamma\left(\frac{1}{\rho}\right)} \int\limits_{\tau^\beta}^\infty \left(\sigma - \tau^\rho\right)^{\frac{1}{\rho}-1} \exp\left(-\zeta^\rho \sigma\right) d\sigma.$$

The Weyl fractional integral ([107, II, p. 202, (11)]):

$$\frac{1}{\Gamma\left(\frac{1}{\rho}\right)}\int\limits_y^\infty (\sigma - y)^{\frac{1}{\rho}-1} \exp(-\alpha\sigma) d\sigma = \alpha^{-\frac{1}{\rho}} \exp(-\alpha y)$$

with $y = \tau^\rho$, $\alpha = \zeta^\rho$, gives

$$\begin{aligned}\mathfrak{B}_{\rho,\mu}\left\{l_{\rho,\mu}f;\zeta\right\} &= \rho\zeta^{\mu\rho-1}\frac{1}{\zeta}\int\limits_0^\infty \exp\left(-\zeta^\rho \tau^\rho\right) \tau^{\mu\rho-1} f(\tau) d\tau.\\ &= \frac{1}{\zeta}\mathfrak{B}_{\rho,\mu}\left\{f;\zeta\right\}.\end{aligned}$$

Thus, relation (2.4.8) is proved for real variables and by the principle of analytic continuation, it holds for complex variables too.

Theorem 2.4.3. *For functions $f \in A^{(\alpha)}\left[\rho_1, \sigma_1\right]$ continuously differentiable at $z = 0$, the following differential property of the Borel-Džrbashjan transform holds:*

$$\mathfrak{B}_{\rho,\mu}\left\{\mathfrak{D}_{\rho,\mu}f;\zeta\right\} = \zeta\mathfrak{B}_{\rho,\mu}\{f;\zeta\} - \Gamma(\mu)f(0). \tag{2.4.9}$$

Proof. For simplicity we consider the case $\rho \geq 1$ only, when $\mathfrak{D}_{\rho,\mu}$ has the explicit representation (2.2.19′). It is easy to verify that $\mathfrak{D}_{\rho,\mu} l_{\rho,\mu} f(z) = f(z)$ but

$$l_{\rho,\mu}\mathfrak{D}_{\rho,\mu}f(z) = f(z) - f(0). \tag{2.4.10}$$

Applying (2.4.8) to (2.4.10) we obtain

$$\begin{aligned}\mathfrak{B}_{\rho,\mu}\{f;\zeta\} - f(0)\mathfrak{B}_{\rho,\mu}\{1;\zeta\} &= \mathfrak{B}_{\rho,\mu}\left\{l_{\rho,\mu}\left[\mathfrak{D}_{\rho,\mu}f\right];\zeta\right\}\\ &= \frac{1}{\zeta}\mathfrak{B}_{\rho,\mu}\left\{\mathfrak{D}_{\rho,\mu}f;\zeta\right\}\end{aligned}$$

and then taking into account that

$$\mathfrak{B}_{\rho,\mu}\{1;\zeta\} = \zeta^{\mu\rho-1}\mathfrak{L}\left\{(t^\rho)^{\mu-1};\zeta^\rho\right\} = \frac{\Gamma(\mu)}{\zeta^{\mu\rho}},$$

where $\mathfrak{L}\{f(t);\zeta\}$ stands for the Laplace transform, we get relation (2.4.9).

Theorem 2.4.4. (Convolution of the B.-D. transform) *The convolution of the Gelfond-Leontiev integration operator* $l_{\rho,\mu}$*:*

$$f \overset{\rho,\mu}{*} g := f \underset{\rho,\mu}{\overset{-1}{*}} g, \tag{2.4.11}$$

defined by (2.2.27), (2.2.28), (2.2.33), is a convolution also of the Borel-Džrbashjan transform (2.4.6) in $A^{(\alpha)}[\rho_1,\sigma_1]$*, that is, the identity*

$$\mathfrak{B}_{\rho,\mu}\left\{f \overset{\rho,\mu}{*} g;\zeta\right\} = \frac{\zeta}{\Gamma(\mu)}\mathfrak{B}_{\rho,\mu}\{f;\zeta\}\mathfrak{B}_{\rho,\mu}\{g;\zeta\} \tag{2.4.12}$$

holds.

Proof. Using inversion formula (2.4.7), one can write:

$$f(z) = \frac{1}{2\pi i}\int\limits_{\mathfrak{L}_1} E_\rho(z\zeta;\mu)\mathfrak{B}_{\rho,\mu}\{f;\zeta\}d\zeta \tag{2.4.13}$$

and

$$g(z) = \frac{1}{2\pi i}\int\limits_{\mathfrak{L}_2} E_\rho(z\eta;\mu)\mathfrak{B}_{\rho,\mu}\{g;\eta\}d\eta \tag{2.4.14}$$

with

$$\mathfrak{L}_1 = \mathfrak{L}_p^{(\alpha)}(\nu_1), \quad \mathfrak{L}_2 = \mathfrak{L}_p^{(\alpha)}(\nu_2),$$

$\nu_2 > \nu_1 > \nu_0$, provided z is inside $\mathfrak{D}_\rho^{(\alpha)}(\nu_2) \subset \mathfrak{D}_\rho^{(\alpha)}(\nu_1)$. Then, multiplying convolutionally (2.4.13) and (2.4.14) we have

$$\begin{aligned}\left(f \overset{\rho,\mu}{*} g\right)(z) &= \frac{1}{(2\pi i)^2}\iint\limits_{\mathfrak{L}_1\times\mathfrak{L}_2}\left[E_\rho(z\zeta;\mu) \overset{\rho,\mu}{*} E_\rho(z\eta;\mu)\right] \\ &\times \mathfrak{B}_{\rho,\mu}\{f;\zeta\}\mathfrak{B}_{\rho,\mu}\{g;\eta\}d\zeta d\eta \\ &= \frac{1}{(2\pi i)^2}\iint\limits_{\mathfrak{L}_1\times\mathfrak{L}_2}\left[\frac{\zeta E_\rho(z\zeta;\mu) - \eta E_\rho(z\eta;\mu)}{\zeta-\eta}\right] \\ &\times \mathfrak{B}_{\rho,\mu}\{f;\zeta\}\mathfrak{B}_{\rho,\mu}\{g;\eta\}d\zeta d\eta,\end{aligned}$$

due to formula (2.2.36) for the convolutional product of Mittag-Leffler functions.

Now, let us take into account that $\mathfrak{L}_2$ is contained inside $\mathfrak{L}_1$ and both contours are traced in the negative direction. Then, the above expression can be rewritten in the form:

$$\left(f \overset{\rho,\mu}{*} g\right)(z) = \frac{1}{2\pi i}\int\limits_{\mathfrak{L}_1} \frac{\zeta}{\Gamma(\mu)} E_\rho(z\zeta;\mu)\mathfrak{B}_{\rho,\mu}\{f;\zeta\}\left[\frac{1}{2\pi i}\int\limits_{\mathfrak{L}_2}\frac{\mathfrak{B}_{\rho,\mu}\{g;\eta\}}{\zeta-\eta}\right]d\zeta$$

$$-\frac{1}{2\pi i}\int\limits_{\mathfrak{L}_2} \frac{\eta}{\Gamma(\mu)} E_\rho(z\eta;\mu)\,\mathfrak{B}_{\rho,\mu}\{g;\eta\}\left[\frac{1}{2\pi i}\int\limits_{\mathfrak{L}_1}\frac{\mathfrak{B}_{\rho,\mu}\{f;\zeta\}}{\zeta-\eta}\right]d\eta$$

$$=\frac{1}{2\pi i}\int\limits_{\mathfrak{L}_2} E_\rho(z\eta;\mu)\,\frac{\eta}{\Gamma(\mu)}\mathfrak{B}_{\rho,\mu}\{f;\eta\}\mathfrak{B}_{\rho,\mu}\{g;\eta\}d\eta.$$

Hence, by using inversion formula (2.4.7) once again, we obtain convolutional property (2.4.12).

The above properties show that the Borel-Džrbashjan transform is an analogue of the Laplace transform, if the Gelfond-Leontiev integration and differentiation operators are considered instead of the usual integration and differentiation and their integer powers.

Obviously, for $\mu = \rho = 1$ the Borel-Džrbashjan integral transform (2.4.4) turns into the *Laplace transform*, the integral analogue of the Borel transform (2.4.1):

$$\mathfrak{B}_{1,1}\{f;\zeta\} = \mathfrak{L}\{f;\zeta\} = \int\limits_0^\infty \exp(-\zeta t)\, f(t)dt \tag{2.4.15}$$

and Theorems 2.4.2, 2.4.3, 2.4.4 turn into its well-known operational properties:

$$\mathfrak{L}\{lf;\zeta\} = \frac{1}{\zeta}\mathfrak{L}\{f;\zeta\}, \tag{2.4.8'}$$

$$\mathfrak{L}\left\{\frac{d}{dt}f;\zeta\right\} = \zeta\mathfrak{L}\{f;\zeta\} - f(0), \tag{2.4.9'}$$

$$\mathfrak{L}\left\{f \underset{1,1}{\overset{-1}{*}} g;\zeta\right\} = \zeta\mathfrak{L}\{f;\zeta\}\,\mathfrak{L}\{g;\zeta\}, \tag{2.4.12'}$$

where

$$lf(t) = \int\limits_0^t f(\tau)d\tau$$

is the Volterra integration operator and (cf. (2.2.33)):

$$\left(f \underset{1,1}{\overset{-1}{*}} g\right)(t) = \frac{d}{dt}\int\limits_0^t f(t-x)g(x)dx = \frac{d}{dt}(f*g)(t)$$

is the *"differentiated" Duhamel convolution*

$$(f * g)(t) = \int_0^t f(t-x)g(x)dx. \tag{2.4.16}$$

It is seen that (2.4.12′) is a modification of the *classical convolutional property* of the Laplace transform:

$$\mathfrak{L}\{f * g; \zeta\} = \mathfrak{L}\{f; \zeta\} \, \mathfrak{L}\{g; \zeta\} \tag{2.4.17}$$

with $(*)$ as in (2.1.1), (2.4.16).

For the Gelfond-Leontiev operators $l_{\rho,\mu}$, $D_{\rho,\mu}$ with $\mu = 1$ (considered by Tkachenko [488]-[489] and Dimovski [72]-[73]), the *"generalized Borel transform"* follows from (2.4.4), (2.4.6), has the form

$$\mathfrak{B}_{\rho,1}\{f; \zeta\} = \rho \left(e^{-i\theta}\zeta\right)^{\rho} \zeta^{-1} \int_0^\infty \exp\left[-t^\rho \left(e^{-i\theta}\zeta\right)^{\rho}\right] t^{\rho-1} dt \tag{2.4.18}$$

and has properties which are corollaries of Theorems 2.4.2, 2.4.3, 2.4.4 for $\mu = 1$.

Note. Let us mention that the Borel type integral transform, strictly corresponding to Erdélyi-Kober fractional integration operator (2.1.4) $L = x^{\beta\delta} I_\beta^{\gamma,\delta}$, instead of to the Džrbashjan-Gelfond-Leontiev operator $l_{\rho,\mu}$, has the form

$$\mathfrak{B}_{\beta,\gamma}\{f(x);\ z\} = \beta \int_0^\infty \exp\left(-zx^\beta\right) x^{\beta(\gamma+1)-1} f(x) dx. \tag{2.4.19}$$

Its convolutions are given by operations (2.1.7) and its relation (5.6.18) with the Laplace transform can be seen in Section 5.6.

— || —

THE MAIN RESULTS OF CHAPTER 2 HAVE BEEN PUBLISHED IN: Dimovski and Kiryakova [77]-[78] and Kiryakova [196], [206].

3 Hyper-Bessel differential and integral operators and equations

The theory of the one-tuple ($m = 1$) operators of fractional integration and differentiation, i.e. of *the classical Riemann-Liouville and Erdélyi-Kober fractional integrals and derivatives*, seems to be comparatively complete. Their study and applications traced back to the pioneers of the fractional calculus: Leibnitz, Liouville, Riemann, Euler, Abel and continued in our century by Kober, Erdélyi, Sneddon and many other contemporary mathematicians. The basic facts can be found in a large number of papers, including the proceedings of the international conferences of fractional calculus [404], [291], [329] as well as in the Samko, Kilbas, Marichev encyclopaedia on this subject [434]. In Chapters 1 and 2 we have simply added a few results and new interpretations: a unified approach to the Erdélyi-Kober fractional integrals and derivatives, convolutions related to them, interpretations of the Gelfond-Leontiev operators and corresponding generalized Borel transforms, etc.

The two-tuple ($m = 2$) fractional integrals (1.1.19), the so-called *hypergeometric fractional integrals* have been also investigated in detail and intensively used in analysis and mathematical physics; see Love [262], Kalla and Saxena [177]-[178], Saigo [415]-[421], McBride [289]-[291], Sprinkhuizen-Kuiper [457], etc. In Chapter 1 we have just mentioned some of their applications and stressed the interpretation of the corresponding *hypergeometric fractional derivatives.*

However, multiple ($m > 2$) operators of fractional integration like those introduced in Chapter 1 (Definition 1.1.1):

$$I_{\beta,m}^{(\gamma_k),(\delta_k)} f(x) = \int\limits_0^1 G_{m,m}^{m,0}\left[\sigma \left| \begin{matrix} (\gamma_k+\delta_k)_1^m \\ (\gamma_k)_1^m \end{matrix} \right.\right] f\left(x\sigma^{\frac{1}{\beta}}\right) d\sigma$$

have not been investigated to any great extent and have not been used effectively. Some authors, e.g. Kalla [157]-[161], [164], Parashar [355] have studied fractional integration operators involving the arbitrary $G_{p,q}^{m,n}$-functions of Meijer but usually only a chain of rules and formal inversion formulas are derived. In particular, no relations to corresponding differential (differintegral) operators and Laplace-type integral transforms are mentioned. Together with the lack of a decomposition representation like

$$I_{\beta,m}^{(\gamma_k),(\delta_k)} f(x) = \left[\prod_{k=1}^{m} I_{\beta}^{\gamma_k,\delta_k} \right] f(x)$$

$$= \int_0^1 \cdots \int_0^1 \left[\prod_{k=1}^{m} \frac{(1-\sigma_k)^{\delta_k - 1} \sigma_k^{\gamma_k}}{\Gamma(\delta_k)} \right] f\left[x (\sigma_1 \ldots \sigma_m)^{\frac{1}{\beta}} \right] d\sigma_1 d\sigma_2 \ldots d\sigma_m,$$

this fact has prevented their wide applications.

A good example of multiple (m-tuple, $m \geq 2$) Erdélyi-Kober fractional integrals $I_{\beta,m}^{(\gamma_k),(\delta_k)}$ and derivatives $D_{\beta,m}^{(\gamma_k),(\delta_k)}$ with various interpretations and application, is provided by the hyper-Bessel integral and differential operators, considered in this chapter.

The hyper-Bessel differential operators, called *Bessel type differential operators of arbitrary order* $m > 1$ are introduced by Dimovski (1966) in the form

$$B = x^{\alpha_0} \frac{d}{dx} x^{\alpha_1} \frac{d}{dx} x^{\alpha_2} \cdots \frac{d}{dx} x^{\alpha_m}, \quad 0 < x < \infty.$$

In a series of papers [64]-[71] he proposes several approaches in building *operational calculi for the hyper-Bessel operators.* One of them is the direct algebraic method of Mikusinski based on the notion of convolution of a linear operator mapping a linear space into itself (see Definition 2.1.1, Chapter 2). Another approach is to use transmutation operators (here we call them *Poisson-Sonine-Dimovski transformations*) which allow us to transfer the results of the classical (Mikusinski's) operational calculus to the case of hyper-Bessel operators. Last but not least, we mention the use of the *Obrechkoff integral transform* [339] proposed by Dimovski as an analogue of the Laplace transform. For the same operators, McBride [289]-[291] finds a decomposition representation in terms of the Erdélyi-Kober operators and fractional powers, represented by integral operators involving Meijer's $G_{m,m}^{m,0}$-function. Special cases ($m = 2$, or $m > 2$ but with particular parameters) have been considered in a large number of papers, by different authors and methods (see Examples in Section 3.3) and have found various applications.

Here we consider the hyper-Bessel operators *as generalized "fractional" integrals and derivatives of multiorder* $\delta = (1, 1, \ldots, 1)$. These considerations have a two-fold role. First, the general theory of Chapter 1 is illustrated particularly by useful results which allow us to deal easily with the hyper-Bessel operators and hyper-Bessel differential equations

$$By(x) = \lambda y(x) + f(x).$$

On the other hand, the generalized fractional calculus clarifies the important role of Meijer's G-functions in the theory of hyper-Bessel operators and equations. They can serve either as kernel-functions of the generalized "fractional" (hyper-Bessel) integrals, or as solutions of initial value problems for the corresponding O.D.E. Finally, the Obrechkoff integral transform (Sections 3.9 and 3.10) turns out to be a G-transformation and what is more, a composition of the Laplace transform with a suitable transmutation (Sonine-Dimovski) operator (which in turn is a multiple fractional integral too!).

Hence, in terms of the generalized fractional calculus and Meijer's G-functions, we give a unified exposition and a new insight on this topic, combined with new results and applications.

3.1. The hyper-Bessel operators as generalized fractional differintegrals

3.1.i. Definitions and notations

Let $m > 1$ be an arbitrary integer. We consider special kinds of m-th order differential operators on the half-line $0 \leq x < \infty$ which are far-reaching generalization of the second order Bessel differential operator. As in Chapter 1, the basic functional space suitable for practical applications is the space of real-valued functions which are continuous (or sufficiently smooth) on $[0, \infty)$ and may have power singularities at $x = 0$, namely:

$$C_{\alpha}^{(k)} := \left\{ f(x) = x^{p} \tilde{f}(x); \quad p > \alpha, \ \tilde{f} \in C^{(k)}[0, \infty) \right\}, \tag{3.1.1}$$
$$C_{\alpha}^{(0)} := C_{\alpha},$$

with arbitrary real α and integer $k \geq 0$ (see Section 1.1.i). It is easily seen that $\alpha_1 \leq \alpha_2$ and $k_1 \leq k_2$ yield $C_{\alpha_1}^{(k_1)} \supseteq C_{\alpha_2}^{(k_2)}$ and $f \in C_{\alpha}^{(k)}$, $\alpha \geq 0$ yields $\lim\limits_{x \to +0} f(x) = 0$.

Definition 3.1.1. By a *hyper-Bessel differential operator*, or a *Bessel type differential operator* of order $m > 1$ we mean each linear differential operator of the form

$$B = x^{\alpha_0} \frac{d}{dx} x^{\alpha_1} \frac{d}{dx} \ldots x^{\alpha_{m-1}} \frac{d}{dx} x^{\alpha_m} \tag{3.1.2}$$

with arbitrary real $\alpha_0, \alpha_1, \ldots, \alpha_m$ such that

$$\beta := m - (\alpha_0 + \alpha_1 + \cdots + \alpha_m) > 0;$$

or of the alternative forms:

$$B = x^{-\beta} Q_m \left(x \frac{d}{dx} \right) = x^{-\beta} \prod_{k=1}^{m} \left(x \frac{d}{dx} + \beta \gamma_k \right), \tag{3.1.3}$$

where $\beta > 0$, Q_m is an m-th degree polynomial of the Euler differential operator $\delta = x \frac{d}{dx}$ with zeros $\mu_k = -\beta \gamma_k$, $k = 1, \ldots, m$; and

$$B = x^{-\beta} \left(x^m \frac{d^m}{dx^m} + a_1 x^{m-1} \frac{d^{m-1}}{dx^{m-1}} + \cdots + a_{m-1} x \frac{d}{dx} + a_m \right), \tag{3.1.4}$$

with $\beta > 0$ and arbitrary real $a_1, \ldots, a_m$.

Originally, the hyper-Bessel diffrential operators were introduced by Dimovski [64]-[71] under the name *Bessel type operators of arbitrary order*.

The assumption that x is a real positive variable and the parameters in (3.1.2)-(3.1.4) are real is not essential since most of the results can be easily transferred to the case of a complex variable in starlike domains (as in Sections 2.3, 2.4 and 5.5) and complex values of parameters, e.g. complex $\alpha_0, \alpha_1, \ldots, \alpha_m$ such that $\beta := \Re\,[m - (\alpha_0 + \alpha_1 + \cdots + \alpha_m)] > 0$.

It is seen that (3.1.2), (3.1.3) and (3.1.4) are equivalent representations for the same differential operators B, defined by means of the $(m+1)$-tuples:

$$\{\alpha_0, \alpha_1, \ldots, \alpha_m\} \text{ with } m - (\alpha_0 + \cdots + \alpha_m) > 0, \tag{3.1.2$'$}$$
$$\{\beta > 0;\ \gamma_1, \ldots, \gamma_m\}, \tag{3.1.3$'$}$$
$$\{\beta > 0;\ a_1, \ldots, a_m\}. \tag{3.1.4$'$}$$

For example, (3.1.2) and (3.1.3) follow each from the other by choosing:

$$\gamma_k = \frac{1}{\beta}(\alpha_k + \alpha_{k+1} + \cdots + \alpha_m - m + k), \quad k = 1, \ldots, m, \tag{3.1.5}$$

or conversely,

$$\alpha_0 = -\beta - \beta\gamma_1 + 1; \quad \alpha_k = \beta\gamma_k - \beta\gamma_{k+1} + 1; \quad k = 1, \ldots, m-1; \quad \alpha_m = \beta\gamma_m. \tag{3.1.6}$$

Taking into account relation (3.1.6) and the formulas

$$\left(x\frac{d}{dx} - \mu_k\right) = x^{\mu_k+1}\frac{d}{dx}x^{-\mu_k}, \quad k = 1, \ldots, m,$$

from (3.1.2) we obtain:

$$\begin{aligned} B &= x^{-\beta}\left(x^{-\beta\gamma_1+1}\frac{d}{dx}x^{\beta\gamma_1}\right)\left(x^{-\beta\gamma_2+1}\frac{d}{dx}x^{\beta\gamma_2}\right)\cdots\left(x^{-\beta\gamma_m+1}\frac{d}{dx}x^{\beta\gamma_m}\right) \\ &= x^{-\beta}\prod_{k=1}^{m}\left(x\frac{d}{dx} + \beta\gamma_k\right) = z^{-\beta}Q_m\left(x\frac{d}{dx}\right), \end{aligned}$$

i.e. representation (3.1.3). Conversely, for fixed $\beta > 0$ and different arrangements of $\gamma_1, \ldots, \gamma_m$ in (3.1.3) (i.e. of the zeros $\mu_1, \ldots, \mu_m$ of Q_m) there are different ways (just $m!$ if $\gamma_i \neq \gamma_j$ for $i \neq j$) of putting the operator B into the form (3.1.2) with α's as in (3.1.6). However, if we assume that

$$\begin{aligned} &\gamma_1 \le \gamma_2 \le \cdots \le \gamma_m, \quad \text{or strictly:} \\ &\gamma_1 \le \gamma_2 \le \cdots \le \gamma_s < \gamma_{s+1} = \cdots = \gamma_m, \quad 0 \le s \le m-1, \end{aligned} \tag{3.1.7}$$

then representation (3.1.2) is uniquely determined. The case $s = 0$ corresponds to a multiple (m-tuple) zero $\mu = -\beta\gamma$ of the polynomial $Q_m(\mu)$, i.e.

$$\gamma_1 = \gamma_2 = \cdots = \gamma_m.$$

Hyper-Bessel operators (3.1.2), (3.1.3) can also be put in the form

$$B = x^{\kappa}\frac{d^m}{dx^m} + a_1 x^{\kappa-1}\frac{d^{m-1}}{dx^{m-1}} + \cdots + a_m x^{\kappa-m}, \quad \kappa = m - \beta, \tag{3.1.8}$$

i.e. (3.1.4), where the coefficients (3.1.4′) are defined by means of (3.1.5) and

$$a_{m-k} = \sum_{j=0}^{k} \left[\frac{(-1)^j}{j!(k-j)!} \prod_{i=1}^{m} (\beta\gamma_i + k - j) \right], \quad k = 0, 1, \ldots, m-1. \tag{3.1.9}$$

This representation, as well as the more general representation of polynomials of hyper-Bessel differential operators $P_n(B) = \sum\limits_{k=0}^{n} c_{n-k} B^k$ as polynomials of the usual differentiation $D = \frac{d}{dx}$, namely:

$$P_n(B) = P_n \left[x^{-\beta} Q_m \left(x \frac{d}{dx} \right) \right] = \sum_{s=0}^{mn} d_{mn-s}(x) \left(\frac{d}{dx} \right)^s, \tag{3.1.10}$$

$$d_{mn-s}(x) = \sum_{j=\left[\frac{s+m-1}{m}\right]}^{n} \left\{ x^{s-\beta j} c_{n-j} \sum_{l=0}^{s} \left[\frac{(-1)^l}{l!(s-l)!} \prod_{i=1}^{j} (\beta\gamma_i + (s-l)(m-\beta-1)) \right] \right\},$$

can be obtained by (3.1.3) and the formula:

$$\delta^n = \left(x \frac{d}{dx} \right)^n = \sum_{k=1}^{n} \binom{n}{n-k} B_{n-k}^{(k)} x^k \frac{d^k}{dx^k}, \tag{3.1.11}$$

where $B_l^{(k)}$ stand for the *generalized Bernoulli numbers* (see [272]).

Representations (3.1.4), (3.1.8), (3.1.10) generalize the results of Osipov [342]-[343] for the Bessel type operator $B_\alpha = x^{-\alpha} \frac{d}{dx} x^{\alpha+1} \frac{d}{dx}$. Klučantčev [216]-[218] considers hyper-Bessel operators only of the form (3.1.4), (3.1.8) with $\kappa = 0$, i.e. $\beta = m$. On the other hand, it is seen that the substitution $y = x^\beta$ reduces the hyper-Bessel operators with an arbitrary $\beta > 0$ into operators with $\beta = 1$, since $\delta_x = x\frac{d}{dx} = \beta y \frac{d}{dy} = \beta \delta_y$.

To examine the hyper-Bessel operators (3.1.2)-(3.1.4) from the point of view of the generalized hypergeometric functions, let us consider the general differential equation of order $\max(p, q)$ satisfied by Meijer's function $y(x) = G_{p,q}^{m,n}(x)$ (see (A.19), also [106, I]):

$$\left[\prod_{k=1}^{q} \left(x \frac{d}{dx} - b_k \right) - (-1)^{p-m-n} x \prod_{k=1}^{p} \left(x \frac{d}{dx} - a_k + 1 \right) \right] y(x) = 0.$$

If we put $n = p = 0$, $q = m$, $b_k = -\gamma_k$, $k = 1, \ldots, m$ and divide by $x \neq 0$, we obtain:

$$\left[x^{-1} \prod_{k=1}^{m} \left(x \frac{d}{dx} + \gamma_k \right) - (-1)^m \right] y(x) = 0, \tag{3.1.12}$$

i.e.

$$By(x) = (-1)^m y(x).$$

This equation characterizes the hyper-Bessel operators B as special cases of the *"generalized hypergeometric operators"*:

$$
\begin{aligned}
H &= (-1)^m x^{-1} \prod_{k=1}^{m} \left(x\frac{d}{dx} - b_k \right) - (-1)^{p-n} \prod_{k=1}^{p} \left(x\frac{d}{dx} - a_k + 1 \right) \\
&= B^{(1)} - B^{(0)},
\end{aligned} \tag{3.1.13}
$$

being *differences of two hyper-Bessel differential operators*: the former $B^{(1)}$ with $\beta = 1 > 0$ and the latter $B^{(0)}$ with $\beta = 0$ (i.e. a differential operator of Euler type). An example of a hypergeometric operator (3.1.13), viz.

$$
x^{-m} \left(\frac{d}{dx} \right)^n + \sum_{i=1}^{n} b_i x^{n-i} \left(\frac{d}{dx} \right)^i = B^{(1)} - B^{(0)},
$$

is related to the differential equation of n-th order, considered by Bainov and Shopolov [26], following the method of Karanikoloff [184] for the equation

$$
x^{-m} \frac{d^n}{dx^n} y(x) = y(x).
$$

Differential equation (3.1.12) and the hyper-Bessel operators B (3.1.2)-(3.1.4) have *two singular points*: $x = 0$ (a regular one) and $x = \infty$ (an essential singularity). Further we study in detail, the fundamental system of solutions of ODE (3.1.12) and its general solution in terms of Meijer's G-function (see Section 3.4).

All these notes characterize the hyper-Bessel differential operators as *singular linear differential operators of arbitrary order and with variable coefficients*, written down in the equivalent forms (3.1.2), (3.1.3), (3.1.4).

3.1.ii. The hyper-Bessel operators as multiple Erdélyi-Kober derivatives and integrals

Lemma 3.1.2. *The hyper-Bessel differential operator (3.1.3) of order $m > 1$:*

$$
B = x^{-\beta} \prod_{k=1}^{m} \left(x\frac{d}{dx} + \beta\gamma_k \right) = \beta^m x^{-\beta} \prod_{k=1}^{m} \left(\frac{1}{\beta} x \frac{d}{dx} + \gamma_k \right)
$$

is a generalized operator of fractional differentiation of the form (1.5.20) and multiorder $\delta = (1, 1, \dots, 1)$, namely:

$$
B = \beta^m x^{-\beta} D_{\beta,m}^{(\gamma_k - 1),(1)} = \beta^m D_{\beta,m}^{(\gamma_k),(1)} x^{-\beta}. \tag{3.1.14}
$$

Proof. Since $\delta_k = 1$ yields $\delta_k = \eta_k = 1$, $k = 1, \ldots, m$ and $I_{\beta,m}^{(\gamma_k+\eta_k),(0,\ldots,0)} = I$ is the identity operator in C_α, then from (1.6.3) and (1.5.19) (see Definition 1.5.4),

$$\begin{aligned} \beta^m D_{\beta,m}^{(\gamma_k),(1)} x^{-\beta} &= \beta^m x^{-\beta} D_{\beta,m}^{(\gamma_k-1),(1)} \\ &= \beta^m x^{-\beta} \left[\prod_{k=1}^{m} \left(\frac{1}{\beta} x \frac{d}{dx} + (\gamma_k - 1) + 1 \right) \right] I_{\beta,m}^{(\gamma_k+\eta_k),(0)} \\ &= x^{-\beta} \prod_{k=1}^{m} \left(x \frac{d}{dx} + \beta\gamma_k \right) = B, \end{aligned}$$

i.e. B is a generalized (m-tuple) fractional derivative (1.5.20) with

$$\delta_0 = 1 \text{ and } \delta_1 = \cdots = \delta_m = 1.$$

By the way, from (3.1.14) and (1.6.2) the following useful corollary follows: *the B-image of the power functions $f(x) = x^p$ is:*

$$B\{x^p\} = \beta^m x^{p-\beta} \prod_{k=1}^{m} \frac{\Gamma\left(\gamma_k + \frac{p}{\beta} + 1\right)}{\Gamma\left(\gamma_k + \frac{p}{\beta}\right)},$$

i.e.

$$B\{x^p\} = \begin{cases} \beta^m x^{p-\beta} \prod\limits_{k=1}^{m} \left(\gamma_k + \frac{p}{\beta}\right) & \text{if } p \neq -\beta\gamma_k, \quad k = 1, \ldots, m \\ 0, & \text{if } p = -\beta\gamma_k \text{ for some } k = 1, \ldots, m. \end{cases} \tag{3.1.15}$$

It is seen that for functions $f(x) \in C_{\alpha'}^{(l)}$, $l \geq m$, $\alpha' \geq \alpha + \beta$, the hyper-Bessel differential operator B reduces the powers by $\beta > 0$ and the order of smoothness l by m. That is why it does not map $C_{\alpha'}^{(l)}$ into itself:

$$B : C_{\alpha'}^{(l)} \longrightarrow C_{\alpha'-\beta}^{(l-m)} \not\subset C_{\alpha'}^{(l)}. \tag{3.1.16}$$

On the other hand, sometimes it is more convenient to deal with linear operators mapping a linear space into itself (the so-called *endomorphisms*) and especially, the general notion of *convolution* of Dimovski [73] is related to these kinds of operators.

Hence, when solving problems related to the hyper-Bessel differential operators B (like Cauchy problems), we naturally deal with *their right inverse (integral) operators* L, defined by suitable initial conditions and *mapping C_α into itself.* Let us recall the following general definition.

Definition 3.1.3. Let $\mathcal{X}$ be a linear space, $\mathcal{X}_B$ its subspace and let $B : \mathcal{X}_B \to \mathcal{X}$ be a linear operator. A linear operator $L : \mathcal{X} \to \mathcal{X}_B$ is said to be a *right inverse* of B if $BLf = f$ for each $f \in \mathcal{X}$.

To look upon the hyper-Bessel differential operators B as *right-invertible operators* (see Przeworska-Rolewicz [365], also Dimovski [73]), we consider them not in the basic space $\mathcal{X} = C_\alpha$ but in the subspace

$$\mathcal{X} = C_{\alpha+\beta}^{(m)} \subset C_\alpha.$$

Then, L is sought so that

$$C_\alpha \xrightarrow{L} \left(C_{\alpha+\beta}^{(m)} \subset C_\alpha \right) \xrightarrow{B} C_\alpha \tag{3.1.17}$$

and $BLf(x) = f(x)$ for each $f \in C_\alpha$, $\alpha = \max_{1 \le k \le m} [-\beta(\gamma_k + 1)]$.

Definition 3.1.4. The linear right inverse operator L of B (3.1.2) defined by the solution $y(x) = Lf(x)$ of the initial value problem

$$\begin{aligned} &By(x) = f(x) \\ &\lim_{x \to +0} B_k y(x) = \lim_{x \to +0} x^{\alpha_k} \frac{d}{dx} x^{\alpha_{k+1}} \dots \frac{d}{dx} x^{\alpha_m} y(x) = 0, k = 1, \dots, m, \end{aligned} \tag{3.1.18}$$

is said to be a *hyper-Bessel integral operator* of order $m > 1$, or Bessel type integral operator. Let us note that initial conditions (3.1.18) can be stated alternatively to correspond with representations (3.1.3), (3.1.4) of B, namely, (3.1.18) *are equivalent to*

$$\lim_{x \to +0} B_k y(x) = \lim_{x \to +0} \left[x^{\beta\gamma_k} \prod_{j=k+1}^{m} \left(x \frac{d}{dx} + \beta\gamma_j \right) \right] y(x) = 0, \quad k = 1, \dots, m. \tag{3.1.18$'$}$$

Also, "classical" initial value conditions

$$\lim_{x \to +0} y^{(k-1)}(x) = \beta_k, \quad k = 1, \dots, m \tag{3.1.19}$$

that correspond to (3.1.4) can be considered, in this case with $\beta_k = 0$, $k = 1, \dots, m$.

The form of the hyper-Bessel integral operator L in C_α with

$$\alpha = \max_{0 \le k \le m-1} (\alpha_0 + \alpha_1 + \dots + \alpha_k - k - 1) = \max_{1 \le k \le m} [-\beta(\gamma_k + 1)], \tag{3.1.20}$$

corresponding to (3.1.2) and (3.1.18), is found by Dimovski [64]-[65] in terms of a multiple integral, namely:

$$\begin{aligned} Lf(x) = &x^{-\alpha_m} \int_0^x x_1^{-\alpha_{m-1}} dx_1 \int_0^{x_1} x_2^{-\alpha_{m-2}} dx_2 \\ &\dots \int_0^{x_{m-1}} x_m^{-\alpha_0} dx_m. \end{aligned}$$

After a substitution, this gives:

$$Lf(x) = \frac{x^{\beta}}{\beta^m} \int_0^1 \cdots \int_0^1 \left[\prod_{k=1}^{m} x_k^{\gamma_k} \right] f\left[x\,(x_1 \dots x_m)^{\frac{1}{\beta}} \right] dx_1 dx_2 \dots dx_m. \tag{3.1.21}$$

However, according to (1.1.7) and Theorem 1.2.10 integral operator (3.1.21) is nothing but a *generalized (Riemann-Liouville) fractional integral* (in the sense of Definition 1.1.1):

$$Lf(x) = \frac{x^{\beta}}{\beta^m} I_{\beta,m}^{(\gamma_k),(1)} f(x)$$

of multiorder $\delta = (1, 1, \dots, 1)$.

The same result (3.1.21) together with an alternative single integral representation of L by means of the Meijer's G-function also follows directly by the general investigations in Chapter 1. Thus, we obtain the following theorem.

Theorem 3.1.5. *The hyper-Bessel integral operator L, defined in C_α as a solution of the initial value problem defined by conditions (3.1.18′):*

$$By(x) = f(x),$$

$$\lim_{x\to+0} B_k y(x) = \lim_{x\to+0} \left[x^{\beta\gamma_k} \prod_{j=k+1}^{m} \left(x\frac{d}{dx} + \beta\gamma_j \right) \right] y(x) = 0, \quad k = 1, \dots, m,$$

is a generalized fractional integral

$$L = \frac{x^{\beta}}{\beta^m} I_{\beta,m}^{(\gamma_k),(1)} \tag{3.1.22}$$

and possesses the single integral representation equivalent to (3.1.21):

$$y(x) = Lf(x) = \frac{x^{\beta}}{\beta^m} \int_0^1 G_{m,m}^{m,0} \left[\sigma \left| \begin{matrix} (\gamma_k + 1)_1^m \\ (\gamma_k)_1^m \end{matrix} \right. \right] f\left(x\sigma^{\frac{1}{\beta}} \right) d\sigma, \tag{3.1.23}$$

whose kernel-function is Meijer's G-function of the form (1.1.11).

Proof. Due to Theorem 1.5.5,

$$D_{\beta,m}^{(\gamma_k-1),(1)} I_{\beta,m}^{(\gamma_k-1),(1)} = I \quad \text{in} \quad C_\alpha, \quad \alpha = \max_k \left[-\beta\,(\gamma_k + 1) \right]$$

and therefore, the linear right inverse operator of $B = \beta^m x^{-\beta} D_{\beta,m}^{(\gamma_k-1),(1)}$ is (see also (1.3.3)):

$$L = \frac{1}{\beta^m} I_{\beta,m}^{(\gamma_k-1),(1)} x^{\beta} = \frac{x^{\beta}}{\beta^m} I_{\beta,m}^{(\gamma_k),(1)}.$$

It remains to establish that (3.1.22) satisfies the zero initial conditions (3.1.18′). To this end, we consider the expressions $B_k L f(x)$, $k = 1, \ldots, m$. Since

$$x^{\beta\gamma_k} \prod_{j=k+1}^{m} \left(x\frac{d}{dx} + \beta\gamma_j \right) = \beta^{m-k} x^{\beta\gamma_k} \prod_{j=k+1}^{m} \left(\frac{1}{\beta} x \frac{d}{dx} + (\gamma_j - 1) + 1 \right)$$

$$= \beta^{m-k} x^{\beta\gamma_k} D_{\beta,m-k}^{(\gamma_j-1)_{k+1}^{m},(1)}$$

and due to (1.3.12):

$$L = I_{\beta,m}^{(\gamma_j-1)_1^m,(1)} \frac{x^\beta}{\beta^m} = I_{\beta,m-k}^{(\gamma_j-1)_{k+1}^m,(1)} I_{\beta,k}^{(\gamma_j-1)_1^k,(1)} \frac{x^\beta}{\beta^m},$$

then according to Theorem 1.5.5 again,

$$B_k L f(x) = \beta^{m-k} x^{\beta\gamma_k} D_{\beta,m-k}^{(\gamma_j-1)_{k+1}^m,(1)} I_{\beta,m-k}^{(\gamma_j-1)_{k+1}^m,(1)} I_{\beta,k}^{(\gamma_j-1)_1^k,(1)} \frac{x^\beta}{\beta^m} f(x)$$

$$= \beta^{m-k} x^{\beta\gamma_k} I_{\beta,k}^{(\gamma_j-1)_1^k,(1)} \frac{x^\beta}{\beta^m} f(x)$$

$$= \frac{x^{\beta(\gamma_k+1)}}{\beta^k} I_{\beta,k}^{(\gamma_j)_1^k,(1)} f(x),$$

by (1.3.3). But if $f \in C_\alpha$, then $I_{\beta,k}^{(\gamma_j)_1^k,(1)} f \in C_\alpha^{(k)}$ and therefore,

$$B_k L f(x) \in C_{\alpha+\beta(\gamma_k+1)}^{(k)} = C_{\beta(\gamma_k-\gamma_1)}^{(k)} \subseteq C_0^{(k)},$$

since $\gamma_1 \leq \gamma_k$, $k = 1, \ldots, m$. This means,

$$\lim_{x \to +0} B_k L f(x) = 0, \quad k = 1, \ldots, m,$$

i.e. (3.1.18′) are fulfilled.

From Corollary 1.2.16,

$$L : C_\alpha \longrightarrow C_{\alpha+\beta}^{(m)} \subset C_\alpha \tag{3.1.24}$$

is an operator mapping C_α, $\alpha = \max_k [\beta(\gamma_k + 1)]$ into itself.

Once we have shown that the hyper-Bessel integral operator L is a generalized fractional integral in the sense of Chapter 1, we can use directly all the results stated there to simplify and shorten the considerations and calculations related in particular with the hyper-Bessel operators and equations. Hence, from Lemmas 1.2.1 and 1.2.2 we obtain immediately the L-images of some elementary and special functions as shown bellow.

Lemma 3.1.6. *For $p > \alpha$, $\alpha = \max_k [\beta(\gamma_k + 1)]$,*

$$L\{x^p\} = c_p x^{p+\beta}, \tag{3.1.25}$$

where

$$c_p = \beta^{-m}\prod_{k=1}^{m}\frac{\Gamma\left(\gamma_k+\frac{p}{\beta}+1\right)}{\Gamma\left(\gamma_k+\frac{p}{\beta}+2\right)} = \left[\beta^m \prod_{k=1}^{m}\left(\gamma_k+\frac{p}{\beta}+1\right)\right]^{-1}.$$

Lemma 3.1.7. *If a Meijer G-function*

$$f(x) = G^{\mu,\nu}_{\sigma,\tau}\left[\omega x^\beta \middle| \begin{matrix}(c_i)_1^\sigma \\ (d_j)_1^\tau\end{matrix}\right]$$

belongs to C_α*, then its L-image is another G-function belonging to* $C^{(m)}_{\alpha+\beta}$*, namely:*

$$Lf(x) = \frac{x^\beta}{\beta^m} I^{(\gamma_k),(1)}_{\beta,m} f(x) = \frac{1}{\omega\beta^m} G^{\mu,\nu+m}_{\sigma+m,,\tau+m}\left[\omega x^\beta \middle| \begin{matrix}(-\gamma_k+1)_1^m, (c_i+1)_1^\sigma \\ (d_j+1)_1^\tau, (-\gamma_k)_1^m\end{matrix}\right], \quad (3.1.26)$$

provided (as in Lemma 1.2.2),

$$\begin{gathered}\min_{1\le j\le\mu} d_j + \min_{1\le k\le m}\gamma_k > -1, \quad \sigma\ne\tau, \quad \omega\ne 0,\\ \rho=\mu+\nu-\frac{\sigma+\tau}{2}, \quad |\arg\omega|<\rho\pi, \quad c_i-d_j\ne 1,2,3,\dots\\ i=1,\dots,\nu; \quad j=1,\dots,\mu.\end{gathered}$$

Lemma 3.1.8. *For* $p>\alpha$, $\lambda\ge 0$,

$$\begin{aligned}L\left\{x^p \exp\left(-\lambda x^{\frac{\beta}{m}}\right)\right\} =&\beta^{-m}\sqrt{m(2\pi)^{1-m}}\left(\frac{m}{\lambda}\right)^{m\left(\frac{p}{\beta}+1\right)}\\ &\times G^{m,m}_{m,2m}\left[\left(\frac{\lambda}{m}\right)^m x^\beta \middle| \begin{matrix}(-\gamma_k+1)_1^m \\ \left(\frac{p}{\beta}+\frac{k}{m}+1\right)_0^{m-1}, (-\gamma_k)_1^m\end{matrix}\right].\end{aligned} \quad (3.1.27)$$

Proof. The function

$$f(x) = x^p\exp\left(-\lambda x^{\frac{\beta}{m}}\right) = \lambda^{-\frac{pm}{\beta}} G^{1,0}_{0,1}\left[\lambda x^{\frac{\beta}{m}} \middle| \frac{pm}{\beta}\right]$$

is a distinguished representative of the subspace $\Omega = C_\alpha^{\exp} \subset C_\alpha$ of the functions which are *Obrechkoff transformable* (see Section 3.9 on the Obrechkoff transform). These are the functions from C_α having an exponential growth near $x=\infty$. The image of this particular function, being a G-function, can be found as in Lemma 3.1.6 but using the more general integral formula (A.28) (see Appendix), namely:

$$
\begin{aligned}
Lf(x) &= \frac{x^{\beta}}{\lambda^{\frac{pm}{\beta}}\beta^m} \int\limits_0^1 G_{m,m}^{m,0}\left[\sigma \,\middle|\, \begin{matrix} (\gamma_k+1)_1^m \\ (\gamma_k)_1^m \end{matrix}\right] G_{0,1}^{1,0}\left[\left(\lambda x^{\frac{\beta}{m}}\right)\sigma^{\frac{1}{m}} \,\middle|\, \frac{pm}{\beta}\right] d\sigma \\
&= \beta^{-m}\sqrt{m(2\pi)^{1-m}}\left(\frac{m}{\lambda}\right)^{\frac{pm}{\beta}} x^{\beta} G_{m,2m}^{m,m}\left[\left(\frac{\lambda}{m}\right)^m x^{\beta} \,\middle|\, \begin{matrix} (-\gamma_k)_1^m \\ \left(\frac{p}{\beta}+\frac{k}{m}\right)_0^{m-1}, (-\gamma_k-1)_1^m \end{matrix}\right] \\
&= \frac{\sqrt{m(2\pi)^{1-m}}\left(\frac{m}{\lambda}\right)^{\frac{pm}{\beta}+m}}{\beta^m} G_{m,2m}^{m,m}\left[\left(\frac{\lambda}{m}\right)^m x^{\beta} \,\middle|\, \begin{matrix} (-\gamma_k+1)_1^m \\ \left(\frac{p}{\beta}+\frac{k}{m}+1\right)_0^{m-1}, (-\gamma_k)_1^m \end{matrix}\right].
\end{aligned}
$$

3.2. Defining projectors (initial operators) of the hyper-Bessel operators

The following notion plays an important role in the algebraical analysis and *theory of right-invertible operators* (see Przeworska-Rolewicz [365]) and in operational calculus (Dimovski [70], [73]).

Definition 3.2.1. ([365]) A linear operator $F : \mathcal{X}_F \longrightarrow \mathcal{X}$ is said to be a *defining projector (initial operator or operator of the initial conditions)* of the operator $L : \mathcal{X} \longrightarrow \mathcal{X}_B \subset \mathcal{X}$, linear right inverse of the a linear operator $B : \mathcal{X}_B \longrightarrow \mathcal{X}$, if:

a) $\mathcal{X}_B \subseteq \mathcal{X}_F \subset \mathcal{X}$,

b) the range of F coincides with the kernel-space of B, i.e.

$$F(\mathcal{X}_F) = \ker B := \{f : Bf = 0\},$$

c) F is a projector, i.e. $F^2 = F$,

d) F is a left annihilator of L, i.e. $FL = 0$.

We recall the following general assertion.

Theorem ([365]). *If L is a linear right inverse operator of the right-invertible operator $B : \mathcal{X}_B \longrightarrow \mathcal{X}$, i.e. $BL = I$, then the operator defined in $\mathcal{X}_F = \mathcal{X}_B$ by*

$$F = I - LB, \tag{3.2.1}$$

where I is the identity operator, is a defining projector of L. Furthermore, the restriction of each defining projector F of L to the subspace $\mathcal{X}_B$ coincides with operator (3.2.1).

In particular, for the hyper-Bessel operators B, L we have:

$$\mathcal{X} = C_{\alpha}, \quad \mathcal{X}_B = C_{\alpha+\beta}^{(m)} \tag{3.2.2}$$

with $\alpha = \max_k [\beta(\gamma_k + 1)]$.

Before specifying the domain $\mathcal{X}_F$ of F, we formulate the following auxiliary result.

Lemma 3.2.2. *Let us suppose the condition*

$$\gamma_1 < \gamma_2 < \cdots < \gamma_m < \gamma_1 + 1 \tag{3.2.3}$$

for the parameters γ_k of operators B. The fundamental system of solutions of the homogeneous hyper-Bessel differential equation

$$By(x) = \left[x^{-\beta} \prod_{k=1}^{m} \left(x\frac{d}{dx} + \beta\gamma_k \right) \right] y(x) = 0 \tag{3.2.4}$$

in a neighbourhood of the origin $x = 0$ consists of the power functions

$$y_k(x) = x^{-\beta\gamma_k} = x^{-\alpha_k - \alpha_{k-1} - \cdots - \alpha_m + m - k}, \quad k = 1, \ldots, m. \tag{3.2.5}$$

Then, the solution of (3.2.4) under initial conditions of the form (3.1.18′):

$$\begin{aligned} \lim_{x \to +0} B_k y(x) &= \lim_{x \to 0} \left[x^{\beta\gamma_k} \prod_{j=k+1}^{m} \left(x\frac{d}{dx} + \beta\gamma_j \right) \right] y(x) \\ &= b_k, \quad k = 1, \ldots, m, \end{aligned} \tag{3.2.6}$$

has the form

$$y(x) = \sum_{i=1}^{m} c_i x^{-\beta\gamma_i} \in C_\alpha^{(m)} \subset C_\alpha \tag{3.2.7}$$

with constants defined as follows:

$$c_i = \left[\prod_{j=k+1}^{m} \beta(\gamma_j - \gamma_k) \right]^{-1} b_i, \quad i = 1, \ldots, m. \tag{3.2.8}$$

Proof. After multiplying by $x^\beta \neq 0$, equation (3.2.4) becomes an m-th order Euler differential equation

$$Q_m\left(x\frac{d}{dx}\right) y(x) = \left[\left(x\frac{d}{dx} + \beta\gamma_1 \right) \cdots \left(x\frac{d}{dx} + \beta\gamma_m \right) \right] y(x) = 0.$$

Conditions (3.2.3) yields that the zeros $\mu_k = -\beta\gamma_k$, $k = 1, \ldots, m$ of $Q_m(\mu) = 0$ are all different and so, the f.s.s. of equation $By = 0$ consists of the functions (3.2.5):

$$y_k(x) = x^{\mu_k} = x^{-\beta\gamma_k} \in C_\alpha^{(\infty)}, \quad k = 1, \ldots, m.$$

Due to (3.2.3) again, they belong to $C_\alpha^{(\infty)} \subset C_\alpha^{(m)} \subset C_\alpha$ since in this case:

$$\begin{aligned} \alpha &= \max_{1 \le i \le m} [-\beta(\gamma_i + 1)] = -\beta(\gamma_1 + 1); \\ -\beta\gamma_k &\ge -\beta(\gamma_1 + 1), \quad k = 1, \ldots, m. \end{aligned}$$

Then, the general solution of $By = 0$ is a linear combination (3.2.7). Suppose it satisfies initial conditions (3.2.6). Let us consider the expressions

$$B_k y(x) = \sum_{i=1}^{m} c_i B_k \{y_i(x)\} = \sum_{i=1}^{m} c_i B_k \left\{x^{-\beta\gamma_i}\right\}.$$

Using the formula ([272, p. 24, (3)]):

$$\left[\prod_{j=k+1}^{m} \left(x\frac{d}{dx} + \beta\gamma_j\right)\right] x^p = x^p \prod_{j=k+1}^{m} (p + \beta\gamma_j),$$

we find

$$B_k \left\{x^{-\beta\gamma_i}\right\} = \begin{cases} x^{\beta(\gamma_k - \gamma_i)} \prod\limits_{j=k+1}^{m} \left[\beta\left(\gamma_j - \gamma_i\right)\right], & i \leq k \\ 0 & i > k. \end{cases}$$

Since $i < k$ yields $\gamma_i < \gamma_k$,

$$\lim_{x \to +0} B_k \left\{x^{-\beta\gamma_i}\right\} = \begin{cases} 0, & i < k \\ \prod\limits_{j=k+1}^{m} \left[\beta\left(\gamma_j - \gamma_k\right)\right], & i = k \\ 0, & i > k \end{cases}$$

and therefore,

$$\lim_{x \to +0} B_k y(x) = \sum_{i=1}^{m} c_i \lim_{x \to +0} B_k \left\{x^{-\beta\gamma_i}\right\} = c_k \prod_{j=k+1}^{m} \left[\beta\left(\gamma_j - \gamma_k\right)\right] = b_k,$$

whence we determine the constants

$$c_k = \left[\prod_{j=k+1}^{m} \beta\left(\gamma_j - \gamma_k\right)\right]^{-1} b_k, \quad k = 1, \dots, m,$$

as in (3.2.8).

It is seen that condition (3.2.3) is to ensure that the characteristic polynomial $Q_m(\mu) = 0$ has m different zeros and the corresponding functions $y_k \in C_\alpha$. In essence, it means that the *parameters* γ_k, $k = 1, \dots, m$, *of the hyper-Bessel operator* B *are different and do not differ by integers*:

$$\gamma_i - \gamma_j \neq l, \quad l = 0, \pm 1, \pm 2, \dots .$$

Conversely, if we admit coincidence of two or more parameters γ_k, $k = 1, \dots, m$, then a multiple zero of $Q_m(\mu) = 0$ arises, leading to the so-called *logarithmic case*. It is well known that if $\mu_i = -\beta\gamma_i$ is an s-tuple zero of $Q_m(\mu) = 0$, $1 \leq s \leq m$, then the functions

$$x^{-\beta\gamma_i}, \quad x^{-\beta\gamma_i} \ln x, \dots, x^{-\beta\gamma_i} (\ln x)^{s-1} \tag{3.2.9}$$

form a part (or the whole) f.s.s. of $By = 0$. More about the *special logarithmic case*

$$B = \frac{d}{dx} x \frac{d}{dx} \dots x \frac{d}{dx} = x^{-1}\left(x\frac{d}{dx}\right)^m; \quad \gamma_1 = \dots = \gamma_m = 0$$

can be found , for example in Ditkin and Prudnikov [87], Botashev [36] and especially, for a f.s.s., in Adamchik and Marichev [5].

Meanwhile, Lemma 3.2.2 shows that the *kernel-space of* B in the case (3.2.3) coincides with the set of all linear combinations of functions (3.2.5), i.e.

$$\ker B = \operatorname{span}\left\{x^{-\beta\gamma_k}\right\}_{k=1}^m \subset C_\alpha^{(m)} \subset C_\alpha, \tag{3.2.10}$$

and so it is not in the domain $\mathcal{X}_B = C_{\alpha+\beta}^{(m)}$ of B (chosen to make B right-invertible) but lies in the wider subspace $C_\alpha^{(m)}$.

That is why it is suitably *to choose the domain* $\mathcal{X}_F$ *of the defining projector* F to be

$$\begin{aligned}\mathcal{X}_F &= \ker B \oplus \mathcal{X}_B \\ &= \left[\operatorname{span}\left\{x^{-\beta\gamma_k}\right\}_{k=1}^m \oplus C_{\alpha+\beta}^{(m)}\right] \subset C_\alpha^{(m)} \subset C_\alpha = \mathcal{X},\end{aligned} \tag{3.2.11}$$

the wider space of functions of the form

$$\begin{aligned} f(x) &= \left(f_1 x^{-\beta\gamma_1} + \dots + f_m x^{-\beta\gamma_m}\right) + \hat{f} \\ &= \mathring{f} + \hat{f}, \quad \text{where } \mathring{f} \in \ker B, \quad \hat{f} \in C_{\alpha+\beta}^{(m)}.\end{aligned} \tag{3.2.11'}$$

Now we can find the *explicit form of the defining projector* of the hyper-Bessel integral operator L.

Theorem 3.2.3. *Let* $\mathcal{X}_F \subset \mathcal{X} = C_\alpha$ *be defined by (3.2.11). Then, the operator* F *defined in* $\mathcal{X}_F$ *as*

$$F = I - LB,$$

is a defining projector of L *and in the case (3.2.3) has the representation*

$$Ff(x) = \sum_{k=1}^m c_k(f) x^{-\beta\gamma_k} \tag{3.2.12}$$

with coefficients

$$c_k(f) = \left[\prod_{j=k+1}^m \beta\left(\gamma_j - \gamma_k\right)\right]^{-1} \lim_{x\to+0} B_k f(x). \tag{3.2.13}$$

Proof. Since

$$\left[\mathcal{X}_B = C_{\alpha+\beta}^{(m)}\right] \subset \left[\mathcal{X}_F = \operatorname{span}\left\{x^{-\beta\gamma_k}\right\}_{k=1}^m \oplus C_{\alpha+\beta}^{(m)}\right] \subset \left[\mathcal{X} = C_\alpha\right],$$

then the condition a) of Definition 3.2.1 is fulfilled.

The operator F is defined in $\mathcal{X}_F$. Indeed, for $f = \overset{\circ}{f} + \hat{f}$ as in (3.2.11′), we have $B\left(\overset{\circ}{f}\right) = 0$, $LB\left(\overset{\circ}{f}\right) = 0$ (due to the regularity of the operator L, i.e. $L\{0\} = 0$) and therefore,

$$\begin{aligned} F(f) &= F\left(\overset{\circ}{f}\right) + F(\hat{f}) = \overset{\circ}{f} - LB\left(\overset{\circ}{f}\right) + F\left(\hat{f}\right), \\ F(f) &= \overset{\circ}{f} + F\left(\hat{f}\right) = f - LB\left(\hat{f}\right), \end{aligned} \tag{3.2.14}$$

the latter expression being well defined for $\hat{f} \in \mathcal{X}_B$. Then,

$$\begin{aligned} B(Ff) = B\left(\overset{\circ}{f} + \hat{f} - LB\left(\hat{f}\right)\right) &= B\overset{\circ}{f} + B\hat{f} - (BL)B\hat{f} \\ &= 0 + B\hat{f} - B\hat{f} = 0, \end{aligned}$$

i.e. condition b) is also satisfied for $F = I - LB$ in $\mathcal{X}_F$.

The other conditions c), d) are easily verified, namely:

$$\begin{aligned} FLf &= (I - LB)Lf = Lf - L(BL)f = Lf - Lf = 0, \quad f \in \mathcal{X} = C_\alpha, \\ F^2 f &= F(I - LB)f = Ff - FL(Bf) = Ff. \end{aligned}$$

This means $F = I - LB$ is a defining projector also in $\mathcal{X}_F \supset \mathcal{X}_B$.

Condition b) and Lemma 3.2.2 yield that the defining projector F has the form (3.2.12) with coefficients as follows:

$$c_k(f) = \left[\prod_{j=k+1}^{m} \beta\left(\gamma_j - \gamma_k\right)\right]^{-1} \lim_{x \to +0} B_k F f(x).$$

But for functions $f = \overset{\circ}{f} + \hat{f} \in \mathcal{X}_F$ (i.e. $\hat{f} \in C^{(m)}_{\alpha+\beta}$):

$$f_1 = Bf = B\overset{\circ}{f} + B\hat{f} = B\hat{f} \in \mathcal{X} = C_\alpha,$$

and therefore, by Definition 3.1.4 of operator L, the function $LBf = Lf_1$ satisfies the initial conditions

$$\lim_{x \to +0} B_k(LBf(x)) = 0, \quad k = 1, \ldots, m.$$

Then,

$$\lim_{x \to +0} B_k F f(x) = \lim_{x \to +0} B_k f(x) = b_k \qquad \text{(as in (3.2.6))},$$

i.e. we obtain (3.2.13).

Representation (3.2.12)-(3.2.13) justifies the alternative name *initial operator, operator of the initial conditions* for the hyper-Bessel defining projector $F = I - LB$. Another proof of the same representation can be seen in Kiryakova [190], Dimovski and Kiryakova [74].

Corollary 3.2.4. *The hyper-Bessel defining projector $F = I - LB$ of the form (3.2.12) vanishes identically on the subspace* $\left[\mathcal{X}_B = C^{(m)}_{\alpha+\beta}\right] \subset \mathcal{X}_F$:

$$F|_{\mathcal{X}_B} \equiv 0, \tag{3.2.15}$$

i.e. the operators B and L are mutually inverse in $C^{(m)}_{\alpha+\beta} \subset C_\alpha : BL = LB = I$.

Proof. We show that if $\hat{f} \in C^{(m)}_{\alpha+\beta}$, then

$$\lim_{x\to+0} B_k\hat{f}(x) = 0, \quad k = 1, \dots, m.$$

Indeed, if $\hat{f} \in C^{(l)}_{\alpha'}$, then $B_k\hat{f} \in C^{(l-(m-k))}_{\alpha'+\beta\gamma_k}$. For $l = m$, $\alpha' = \alpha + \beta = -\beta\gamma_1$ (if (3.2.3) is satisfied), we find subsequently:

$$\begin{aligned}
&B_m\hat{f}(x) = x^{\beta\gamma_m}\hat{f}(x) \in C^{(m)}_{\beta(\gamma_m-\gamma_1)} \subset C^{(m)}_0, && \text{since } \beta(\gamma_m - \gamma_1) > 0;\\
&\quad\vdots && \quad\vdots\\
&B_k\hat{f}(x) \in C^{(m-m+k)}_{\beta(\gamma_k-\gamma_1)} \subset C^{(k)}_0, && \text{since } \beta(\gamma_k - \gamma_1) > 0;\\
&\quad\vdots && \quad\vdots\\
&B_1\hat{f}(x) \in C^{(m-m+1)}_0 = C^{(1)}_0, && \text{since } \beta(\gamma_1 - \gamma_1) = 0.
\end{aligned}$$

But property $B_k\hat{f} \in C^{(k)}_0$ yields immmediately $\lim\limits_{x\to+0} B_k\hat{f}(x) = 0$ and this means that in (3.2.12) all the coefficients are:

$$c_k\left(\hat{f}\right) = 0, \quad k = 1, \dots, m, \quad \text{i.e. } F\hat{f}\Big|_{C^{(m)}_{\alpha+\beta}} \equiv 0,$$

provided (3.2.3) holds: $\gamma_1 < \gamma_2 < \dots < \gamma_m < \gamma_1 + 1$.

Corollary 3.2.5. *In the subspace $\ker B \subset \mathcal{X}_F$ the defining projector F coincides with the identity operator:*

$$F|_{\ker B} = I, \quad \textit{i.e. } F\left\{\sum_{k=1}^m f_k x^{-\beta\gamma_k}\right\} = \sum_{k=1}^m f_k x^{-\beta\gamma_k}. \tag{3.2.16}$$

Combining Corollaries 3.2.4 and 3.2.5 we find that for an arbitrary function $f = \overset{\circ}{f} + \hat{f} \in \mathcal{X}_F$, the coefficients $c_k(f)$ in (3.2.12) coincide with coefficients f_k, $k = 1, \dots, m$, of $\overset{\circ}{f} \in \ker B$, namely:

$$Ff(x) = \sum_{k=1}^m f_k x^{-\beta\gamma_k} \tag{3.2.17}$$

with

$$f_k = \left[\prod_{j=k+1}^m \beta(\gamma_j - \gamma_k)\right]^{-1} \lim_{x\to+0} B_k f(x).$$

3.3. Examples of hyper-Bessel differential operators and their use in mathematical physics

The best-known example giving rise to the name Bessel type operators for differential operators (3.1.2), (3.1.3), (3.1.4), is:

a) The second order *differential operator of Bessel* ($m = \beta = 2$, $\gamma_1 = \frac{\nu}{2}$, $\gamma_2 = -\frac{\nu}{2}$):

$$\begin{aligned} B_\nu &= x^{-2}\left(x\frac{d}{dx}+\nu\right)\left(x\frac{d}{dx}-\nu\right) = x^{\nu-1}\frac{d}{dx}x^{-2\nu+1}\frac{d}{dx}x^{\nu} \\ &= x^{-\nu-1}\frac{d}{dx}x^{2\nu+1}\frac{d}{dx}x^{-\nu} = \frac{d^2}{dx^2}+\frac{1}{x}\frac{d}{dx}-\frac{\nu^2}{x^2}, \end{aligned} \tag{3.3.a}$$

related to the *Bessel function* $y(x) = J_\nu(x)$, satisfying the equation $B_\nu y(x) = -y(x)$, i.e.

$$x^2y''(x) + xy'(x) + \left(x^2 - \nu^2\right) y(x) = 0.$$

If $\nu \geq 0$, then the basic functional space is $\mathcal{X} = C_{\nu-2}$ with $\alpha = \nu - 2$, while $\mathcal{X}_B = C_\nu^{(2)}$. The corresponding Bessel integral operator has the representations

$$\begin{aligned} L_\nu f(x) &= \frac{x^2}{4}\int_0^1\int_0^1 x_1^{-\frac{\nu}{2}} x_2^{\frac{\nu}{2}} f\left(x\sqrt{x_1x_2}\right) dx_1 dx_2 \\ &= \frac{x^2}{4}\int_0^1 G_{2,2}^{2,0}\left[\sigma \,\middle|\, \begin{matrix} \frac{\nu}{2}+1, -\frac{\nu}{2}+1 \\ \frac{\nu}{2}, -\frac{\nu}{2} \end{matrix}\right] f\left(x\sqrt{\sigma}\right) d\sigma \\ &= \frac{x^2}{4}\int_0^1 (1-\sigma)\sigma^{-\frac{\nu}{2}}\, {}_2F_1\left(-\nu+1, 1; 2; 1-\sigma\right) d\sigma. \end{aligned} \tag{3.3.a$'$}$$

In relation to Section 3.2, the defining projector F_ν in

$$\mathcal{X}_{F_\nu} = \left\{ f(x) = f_0 x^\nu + \hat{f}(x), \hat{f} \in C_\nu^{(2)} \right\}$$

has the form following from (3.2.12) :

$$F_\nu f(x) = f_0 x^\nu, \qquad \text{where } f_0 = \lim_{x\to+0} x^{-\nu} f(x) \tag{3.3.a$''$}$$

(see Dimovski [73, p. 155]).

Among the other popular and useful *Bessel type operators of second order* ($m = 2$), it is worth mentioning also:

b) The *Weinstein operator*

$$B_k = \frac{d^2}{dx^2} + \frac{k}{x}\frac{d}{dx} = x^{-k}\frac{d}{dx}x^{k+1}\frac{d}{dx}, \qquad k \geq 1 \tag{3.3.b}$$

from the so-called Darboux-Weinstein relation (see [509]). This differential operator appears very often in the PDEs of mathematical physics, for instance in the *generalized (Bessel) heat equation*

$$y''_{xx}(x,t) + \frac{k}{x} y'_x(x,t) = a y'_t(x,t), \quad 0 < x < \infty, \quad 0 \le t < \infty$$

and other equations of GASP theory (Weinstein [508]). Recently, operator (3.3.b) has been investigated by Dimovski [70] (as an illustrative case of his general theory), Koprinski [234] (operational calculus for the purposes of the generalized heat equation) and Sprinkhuizen-Kuiper [457] (negative powers of B_k).

In this case $\beta = m = 2$, $\gamma_1 = \frac{k-1}{2}$, $\gamma_2 = 0$, $\alpha = -2$ and hence the hyper-Bessel integral operator for which the operational calculi are developed is $L_k : C_{-2} \longrightarrow C_0^{(2)}$,

$$\begin{aligned} L_k f(x) &= \frac{x^2}{4} \int_0^1 \int_0^1 x_1^{\frac{k-1}{2}} f\left(x\sqrt{x_1 x_2}\right) dx_1 dx_2 \\ &= \frac{x^2}{4} \int_0^1 (1-\sigma)\, {}_2F_1\left(\frac{3-k}{2}, 1; 2; 1-\sigma\right) f\left(x\sqrt{\sigma}\right) d\sigma. \end{aligned} \tag{3.3.b$'$}$$

The defining projector (Koprinski [234]) has the form

$$\begin{aligned} F_k f(x) &= (I - L_k B_k) f(x) = c_1 + c_2 x^{1-k} \\ &= \frac{1}{k-1} \lim_{x \to +0} \left(x^k f'(x)\right) + x^{1-k} \lim_{x \to +0} f(x). \end{aligned} \tag{3.3.b$''$}$$

In the subspace $\mathcal{X}_{B_k} = C_0^{(2)} : c_1 = c_2 = 0 \Rightarrow F_k \equiv 0$.

c) The so-called *Bessel-Clifford operators*

$$\begin{aligned} B_\nu &= \frac{d^2}{dx^2} + \frac{\nu+1}{x}\frac{d}{dx} = x^{-\nu} \frac{d}{dx} x^{1+\nu} \frac{d}{dx} \\ &= \frac{d}{dx} x^{1-\nu} \frac{d}{dx} x^{\nu}; \quad m = 2,\ \beta = 1,\ \gamma_1 = \nu, \gamma_2 = 0 \end{aligned} \tag{3.3.c}$$

coincide in essence with operators (3.3.b) but have been considered by many authors under this name and form. They are closely related to the *Bessel-Clifford functions* $C_\nu(x)$ and operational calculi for them have been developed in different ways (both algebraic and transform approaches). Let us note the paper [126] of Hayek, proposing a full investigation of the Bessel-Clifford functions (differential equation, properties, integral representations, zeros, asymptotic expansions and relation to other special functions and integral transforms). Following the algebraic approach of Mikusinski, Meller [298]-[299] developed an operational calculus for the operator B_ν with $-1 < \nu < 1$ and found a convolution for its right inverse operator (i.e. for (3.3.b$'$)) by the *method of similarity*. Koh [221]-[222] and Rodriguez [398] extended these results to the cases of $\nu \in [-1, \infty)$ and arbitrary ν. Osipov [342]-[343] found an expansion of a polynomial $P_n(B_\nu)$ of this operator in powers of the usual differentiation (similar to (3.1.10)). In the special case $\nu = 0$ (respectively $k = 0$) operators (3.3.c), (3.3.b) turn into the operator $B_0 = \frac{d}{dx} x \frac{d}{dx}$, considered by Ditkin [85].

d) The most general Bessel type differential operator of second order:

$$B = x^{\alpha_0} \frac{d}{dx} x^{\alpha_1} \frac{d}{dx} x^{\alpha_2} \tag{3.3.d}$$

has been considered by McBride [289]-[290] (fractional powers of its right inverse operator L), Betancor [32] (generalized Hankel transforms and Cauchy problems of the form

$$\begin{cases} \frac{\partial}{\partial t} u(x,t) = P(B) u(x,t) \\ u(x, t_0) = \phi_0(x), \end{cases}$$

where P is a polynomial, B is acting on variable x), etc. From the point of view of the Operational Calculus and Integral Transforms, various special cases of (3.3.d) like:

$$\begin{aligned} B_\nu &= x^{-\nu-1} \frac{d}{dx} x^{2\nu+1} \frac{d}{dx} x^{-\nu}, \quad B_{\mu,\nu} = x^{-\nu-\mu-1} \frac{d}{dx} x^{2\nu+1} \frac{d}{dx} x^{\mu-\nu}, \\ B_{\alpha,\beta} &= x^{1-\alpha-\beta} \frac{d}{dx} x^{\alpha} \frac{d}{dx} x^{\beta}, \\ B_\mu &= \frac{d^2}{dx^2} - \frac{4\mu^2-1}{4x^2} = x^{-\mu-\frac{1}{2}} \frac{d}{dx} x^{2\mu+1} \frac{d}{dx} x^{-\mu-\frac{1}{2}}, \end{aligned}$$

have been investigated by different authors: see Rodriguez [395]-[398], Zemanian [519] and Gonzalez [121] (integral transforms of Watson type).

e) The *operators of Tricomi and Gelersted*

$$B_1 = \frac{1}{x} \frac{d^2}{dx^2}, \quad B_n = x^{-n} \frac{d^2}{dx^2}, \quad n > 0 \text{ integer};$$

arise in *PDEs of mixed type*, e.g. in the Tricomi and Gelersted equations

$$x u''_{yy} + u''_{xx} = 0; \quad x^n u''_{yy} - u''_{xx} = 0,$$

the latter considered by Nahoushev [312]-[313]. Usually, the boundary conditions imposed on the solution $u(x,y)$ are combinations of its fractional derivatives on both characteristic lines. Now, the corresponding hyper-Bessel integral operator L and the defining projector of L have the form

$$\begin{aligned} Lf(x) &= \frac{x^{n+2}}{(n+2)^2} \int_0^1 \int_0^1 x_1^{-\frac{1}{n+2}} f\left[x (x_1 x_2)^{\frac{1}{n+2}} \right] dx_1 dx_2, \quad f \in C_{-n-1}; \\ Ff(x) &= f(0) + x f'(0). \end{aligned}$$

PDEs of mixed type involving Bessel type operators of the form (3.3.c), (3.3.d) are also considered.

Let us consider now the case of an arbitrary integer $m > 2$.

f) The simplest higher order hyper-Bessel differential operator is the *operator of m-fold differentiation*:

$$B = D^m = \left(\frac{d}{dx}\right)^m, \quad m > 1 \text{ integer}, \tag{3.3.f}$$

with parameters

$$\{\alpha_0 = \alpha_1 = \cdots = \alpha_m = 0,\ \beta = m\} \quad \text{or} \quad \left\{\beta = m,\ \gamma_k = \frac{k}{m} - 1,\ k = 1, \ldots, m\right\},$$

and alternative representation

$$B = x^{-m}\left\{\left(x\frac{d}{dx}\right)\left(x\frac{d}{dx} - 1\right)\cdots\left(x\frac{d}{dx} - m + 1\right)\right\} \tag{3.3.f'}$$

(cf. Luke [272]). Then, the linear right inverse operator of B, i.e. the *operator of m-fold integration* can be represented by the G-functions in two different ways:

$$\begin{aligned} Lf(x) = l^m f(x) &= \int_0^x \frac{(x-t)^{m-1}}{(m-1)!} f(t)dt \\ &= x^m \int_0^1 G_{1,1}^{1,0}\left[\sigma \middle| \begin{matrix} m \\ 0 \end{matrix}\right] f(x\sigma)\, d\sigma = x^m I_1^{0,m} f(x) \end{aligned} \tag{3.3.1}$$

and

$$\begin{aligned} Lf(x) = l^m f(x) &= \left(\frac{x}{m}\right)^m \int_0^1 \cdots \int_0^1 \left[\prod_{k=1}^m \sigma_k^{\frac{k}{m}-1}\right] f\left[x(\sigma_1 \ldots \sigma_m)^{\frac{1}{m}}\right] d\sigma_1 \ldots d\sigma_m \\ &= \left(\frac{x}{m}\right)^m \int_0^1 G_{m,m}^{m,0}\left[\sigma \middle| \begin{matrix} \left(\frac{k}{m}\right)_1^m \\ \left(\frac{k}{m} - 1\right)_1^m \end{matrix}\right] f\left(x\sigma^{\frac{1}{m}}\right) d\sigma = \left(\frac{x}{m}\right)^m I_{m,m}^{\left(\frac{k}{m}-1\right),(1)} f(x), \end{aligned} \tag{3.3.2}$$

the latter corresponding to (3.3.f′).

These two different representations of the m-fold integration $L = l^m = R^m$ are due to the alternative representations (3.3.f) and (3.3.f′) of $B = \left(\frac{d}{dx}\right)^m$ and also to the relation between the kernel G-functions of operators $I_{m,m}^{\left(\frac{k}{m}-1\right),(1)}$ and $I_1^{0,m}$, namely:

$$\begin{aligned} G_{1,1}^{1,0}\left[\sigma \middle| \begin{matrix} m \\ 0 \end{matrix}\right] &= m^{1-m} G_{m,m}^{m,0}\left[\sigma^m \middle| \begin{matrix} \left(\frac{k-1}{m} + 1\right) \\ \left(\frac{k-1}{m}\right) \end{matrix}\right] \\ &= \left(\frac{\sigma}{m}\right)^{m-1} G_{m,m}^{m,0}\left[\sigma^m \middle| \begin{matrix} \left(\frac{k}{m}\right) \\ \left(\frac{k}{m} - 1\right) \end{matrix}\right], \end{aligned}$$

see, for example, [286, p. 6, (1.2.5)].

The defining projector $F_m = I - LB = I - l^m D^m$ in this example has the form of a Taylor polynomial:

$$F_m f(x) = \sum_{k=0}^{m-1} \frac{f^{(k)}(0)}{k!} x^k, \tag{3.3.3}$$

depending on the initial conditions $f(0), f'(0), \dots, f^{(m-1)}(0)$.

g) It seems the first operational calculi combined also with integral transforms for a higher order differential operator with variable coefficients were developed by Ditkin and Prudnikov [86]-[87], [89] and Botashev [36]-[37]. They considered the m-th order differential hyper-Bessel operator

$$B_m = \frac{d}{dx} x \frac{d}{dx} \dots x \frac{d}{dx} = \frac{1}{x}\left(x\frac{d}{dx}\right)^m \tag{3.3.g}$$

with $\beta = 1$, $\gamma_1 = \gamma_2 = \dots = \gamma_m = 0$, an example of a logarithmic case hyper-Bessel differential operator (multiple zeros of the characteristic polynomial). A convolution of the integral operator

$$\begin{aligned} L_m f(x) &= x \int_0^1 \dots \int_0^1 f(x\sigma_1 \dots \sigma_m)\, d\sigma_1 \dots d\sigma_m \\ &= x \int_0^1 G_{m,m}^{m,0}\left[\sigma \left| \begin{matrix} 1,1,\dots,1 \\ 0,0,\dots,0 \end{matrix} \right.\right] f(x\sigma) d\sigma \end{aligned} \tag{3.3.g$'$}$$

in the space C_{-1} was found and corresponding operational calculi as well as Meijer-Obrechkoff type integral transforms were proposed. For the f.s.s. of ODE $B_m u(x) + u(x) = 0$, see Adamchik and Marichev [5].

h) Further generalization of the Bessel operators belongs to Krätzel [235]-[239]. He developed an operational calculus for the differential operator

$$B_{n,\nu} = \frac{d}{dx} x^{\frac{1}{n}-\nu} \left(x^{1-\frac{1}{n}} \frac{d}{dx}\right)^{n-1} x^{\nu+1-\frac{2}{n}} \tag{3.3.h}$$

and investigated the corresponding Laplace type integral transformation $\mathfrak{L}_{n,\nu}$ (convolution, properties, inversion formulae, etc.). In this case,

$$m \to n; \quad \alpha_0 = 0,\ \alpha_1 = 1-\nu,\ \alpha_2 = \dots = \alpha_{n-1} = 1 - \frac{1}{n},\ \alpha_n = \nu + 1 - \frac{2}{n};$$

i.e.

$$\beta = 1, \quad \gamma_1 = 0, \quad \gamma_k = \nu + \frac{k-2}{n},\ k = 2,\dots,n; \quad \alpha = -1 \Rightarrow C_{\alpha+\beta} = C_{-1}.$$

More exactly, the operational calculus and the convolution are related to the corresponding integral operator which, according to our results, can be written in C_{-1} by means of a $G^{n,0}_{n,n}$-function:

$$L_{n,\nu}f(x) = x\int_0^1 G^{n,0}_{n,n}\left[\sigma \left| \begin{matrix} 1, \left(\nu + \frac{k+n-2}{n}\right)^n_{k=2} \\ 0, \left(\nu + \frac{k-2}{n}\right)^n_{k=2} \end{matrix} \right.\right] f(x\sigma)d\sigma. \tag{3.3.h$'$}$$

i) While not having this purpose in mind, in 1958 the Bulgarian mathematician Nikola Obrechkoff [339] introduced and investigated a rather general integral transform of Laplace type, which can be used as a transform for the operational calculus for the most general hyper-Bessel operator (3.1.2) of arbitrary order $m > 1$. This was establised later by Dimovski [66]-[67] who called it the *Obrechkoff transform.* It turned out that the Obrechkoff transform incorporates as special cases all the other integral transforms, related to particular hyper-Bessel operators like (3.3.a), (3.3.c), (3.3.d), (3.3.g), (3.3.h), etc. It is suitable for a hyper-Bessel operator of order $m = p + 1 > 1$:

$$B = x^{\beta_p}\frac{d}{dx}x^{\beta_{p-1}-\beta_p+1}\frac{d}{dx}\cdots x^{\beta_1-\beta_2+1}\frac{d}{dx}x^{\beta_1}\frac{d}{dx}, \tag{3.3.i}$$

which is in essence the general hyper-Bessel operator but with $\beta = 1$ and $\alpha_{p+1} = 0$, namely:

$$\alpha_0 = \beta_p; \quad \alpha_k = \beta_{p-k} - \beta_{p-k+1} + 1,\ k = 1, \ldots, p-1; \quad \alpha_p = -\beta_1; \quad \alpha_{p+1} = 0;$$

i.e.

$$\beta = 1; \quad \gamma_k = -\beta_{p-k+1} - 1,\ k = 1, \ldots, p; \quad \gamma_{p+1} = 0.$$

The condition (3.2.3): $\gamma_1 < \gamma_2 < \cdots < \gamma_{p+1} < \gamma_1 + 1$, was imposed by Obrechkoff too, in his denotation:

$$-1 < \beta_1 < \beta_2 < \cdots < \beta_p < 0.$$

More details about Obrechkoff's results [339] and their extensions, due to Dimovski and Kiryakova [74]-[76], Kiryakova [190], [192]-[193], [196], [201], Nikolic-Despotovic and Kiryakova [320], can be seen in Sections 3.9 and 3.10.

j) A rather general hyper-Bessel differential operator was investigated by Delsarte in 1957 (see [61, II, p. 893-948]):

$$\begin{aligned} B &= x^{-1}\prod_{j=1}^n \left[x^{\frac{q_j}{s_j}+1}\frac{d}{dx}x^{\frac{1}{s_j}+1}\frac{d}{dx}x^{\frac{2}{s_j}+1}\frac{d}{dx}\cdots x^{\frac{-q_j+s_j-1}{s_j}+1}\frac{d}{dx}\right] \\ &= x^{-1}\prod_{j=1}^n\left[\left(x\frac{d}{dx}+\frac{q}{m}-\frac{q_j}{s_j}\right)\left(x\frac{d}{dx}+\frac{q}{m}-\frac{q_j+1}{s_j}\right)\cdots\left(x\frac{d}{dx}+\frac{q}{m}-\frac{q_j+s_j-1}{s_j}\right)\right], \end{aligned} \tag{3.3.j}$$

where s_j and q_j, $j = 1, \ldots, n$, are integers and $s_1 + \cdots + s_n = m$. In a sense, this operator is more particular and involved than (3.3.i). Delsarte stated, for the first time, and solve in part *the problem of the transmutation operator*, related to (3.3.j). These are are related to the generalized Poisson-Sonine-Dimovski transformations (see Section 3.5) found by Dimovski [68]-[71]. Since Delsarte's works [61] were published posthumously in 1972, his result has had little influenced on the development of the Bessel type operational calculi the later authors.

k) For the general hyper-Bessel differential operator of arbitrary order $m > 1$

$$B = x^{\alpha_0}\frac{d}{dx}x^{\alpha_1}\frac{d}{dx}\dots x^{\alpha_{m-1}}\frac{d}{dx}x^{\alpha_m}, \quad \beta = m - (\alpha_0 + \dots + \alpha_m) > 0 \tag{3.3.k}$$

the first fully developed operational calculus, following Mikusinski's scheme, belongs to Dimovski [64]-[65], [68]-[70]. Recently, elements of such operational calculi have been developed by Hayek and Hernandez [129], following Delerue [60] and Dimovski's approach (for $m = 3$ see also Hayek and Hernandez [127]-[128]).

3.4. Solutions to the homogeneous hyper-Bessel differential equations. Hyper-Bessel functions

It is to point out that the history of the Bessel functions and corresponding differential operators can be traced back to Bernoulli (about 1700) and since Euler (1764) and Poisson (1823) they have been associated mainly with the partial differential equations of potential, wave motion or diffusion. Actually, in these kinds of problems as well as in generalized axially/biaxially symmetric potential equations, equations of elasticity and hydro-aerodynamics and other PDE-s of mathematical physics, hyper-Bessel operators arise quite often, at least in one of the variables x, y (Cartesian coordinates) or r (polar, cylindrical or spherical coordinates), for example in the forms:

$$\frac{1}{r}\frac{\partial}{\partial r}r, \quad \frac{1}{r}\frac{\partial}{\partial r}r^2, \quad \frac{1}{r^n}\left(\frac{\partial}{\partial r}\right)^2, \quad \frac{\partial}{\partial r}\frac{1}{r}\frac{\partial}{\partial r}r\frac{\partial}{\partial r}. \tag{3.4.1}$$

Usually, by separating variables or applying a suitable integral transforms one can reduce these problems to initial value problems for ordinary differential equations involving hyper-Bessel operators.

Definition 3.4.1. Let B be an arbitrary Bessel-type operator (3.1.2), (3.1.3) or (3.1.4) of order $m > 1$. An ordinary differential equation of the form

$$By(x) = \lambda y(x) + f(x), \quad \lambda = \text{const}, \quad f(x) \text{ a given function}, \tag{3.4.2}$$

is said to be a *hyper-Bessel differential equation.*

Cauchy (initial value) problems for equations (3.4.2) can be stated either in terms of the classical initial conditions

$$\lim_{x \to +0} y^{(k-1)}(x) = \beta_k, \; k = 1, \dots, m, \tag{3.4.3}$$

or by means of the equivalent set of *Bessel type initial conditions* (of type (3.1.18)-(3.1.18')):

$$\lim_{x\to+0} B_k y(x) = b_k, \ k = 1,\dots,m, \tag{3.4.4}$$

where B_k are the *"truncated"* hyper-Bessel operators

$$B_k = x^{\beta\gamma_k} \prod_{j=k+1}^{m} (xD + \beta\gamma_j) = x^{\alpha_k} D x^{\alpha_{k+1}} \dots D x^{\alpha_m}, \quad k = 1,\dots,m-1,$$

$$B_m = x^{\beta\gamma_m} = x^{\alpha_m}.$$

In view of this, the problem of finding a *fundamental system of solutions of hyper-Bessel equation* (3.4.2) and the solution of initial value problem (3.4.2), (3.4.4) in the general case, as is *our purpose here*, seems to be important. Till now, many authors have tried to find solutions of (3.4.2) in various particular cases (mainly for $m = 2$) as in more as possible general cases. Here we shall mention only several authors whose results are closely related to our topic. First of all, the papers of Kummer [241] and Delerue [60] should be distinguished. In the latter, unfortunately not very popular paper, the so-called hyper-Bessel functions $J^{(n)}_{\nu_1,\dots,\nu_n}(x)$ (see Appendix) are introduced as generalizations of the Bessel functions and as solutions of a hyper-Bessel differential equation of $(n+1)$-th order. Further, Klučantčev [216]-[217] has considered the following three kinds of Bessel-type equations

$$By(x) = 0, \quad By(x) = f(x), \quad By(x) = \lambda y(x) \tag{3.4.5}$$

with B of the form (3.1.4) and $\beta = m$. He proposes an algorithm for solving such equations by means of a Poisson-type integral transformation. We have extended the same transmutation method in [79]-[80], [196]-[197], [210] by representing the more general transformation of Dimovski [68], [71] in a concise form and obtaining the solutions of (3.4.5) in a closed explicit form. Section 3.5 goes on further in this direction. In another paper [218] Klučantčev studies the hyper-Bessel functions and the generalized trigonometric functions of order $m > 1$ (see Appendix), considered as solutions of the equations $By(x) = \lambda y(x)$. For special cases of $2n$-th order self-adjoint hyper-Bessel operators, the same equation has been solved by Exton [111] and Agarwal [6] in terms of the "n-Bessel" (for $n = 1$: "di-Bessel") functions; see also Sarabia [435] and Gonzalez [122]. Recently, investigations of the so-called *Bessel-Clifford functions of order n* have been made by Hayek and Hernandez [127]-[129], considering them as solutions of Bessel type equations. Solutions of other hyper-Bessel differential equations have been found also by Hriptun [138], Bondarenko [34], Paris [356], Paris and Wood [357] and Trimeche [494].

Fundamental systems of solutions for a more general case, in neighbourhoods of the singular points $x = 0$ and $x = \infty$, have been proposed by Adamchik and Marichev [5] and Adamchik [1] in terms of Mellin-Barnes-type integrals, and especially, a more sophisticated investigation of the so-called "logarithmic" cases has been carried out there.

A relation between Meijer's G-functions and the hyper-Bessel operators has been observed in Kiryakova [192]-[193], [196] and Dimovski and Kiryakova [75], [79]. First it has been shown that the kernel-function of the Obrechkoff transform is nothing but a $G^{m,0}_{m,m}$-function. Later, the fractional powers of the integral operators L have been represented by integral operators involving the $G^{m,0}_{0,m}$-function (see also McBride [289]). The transmutation operators between two different hyper-Bessel operators of order $m > 1$ are also representable by integral transforms with G-functions in the kernel. We have also established that the $G^{m,0}_{0,m}$-function is one of the solutions of the equation $By(x) = \lambda y(x)$. On the other hand, the hyper-Bessel and the n-Bessel functions as well as the solutions found by Adamchik [1] and Adamchik and Marichev [5] in terms of Mellin-Barnes type integrals are also particular cases of the G-function. For the non homogeneous equation $By(x) = f(x)$, its solution $y(x) = Lf(x)$, L being the linear right inverse operator of B, has been expressed in [79], [193], [196] as an integral operator of $f(x)$ whose kernel is the $G^{m,0}_{0,m}$-function. It has become clear that Meijer's G-functions are playing an important role in the theory of hyper-Bessel operators. Our aim here is to find the explicit solutions of the initial-value problems for (3.4.2), (3.4.5), all of them represented in terms of the G-functions.

For simplicity, throughout this section we consider only the *non logarithmic case*:

$$\gamma_1 < \gamma_2 < \cdots < \gamma_m < \gamma_1 + 1. \tag{3.4.6}$$

The details concerning the logarithmic case, when some of the parameters $\gamma_1, \gamma_2, \ldots, \gamma_m$ coincide or differ by an integer, could be easily settled by following the pattern of Adamchik and Marichev [1], [5].

The functional space in which we seek the solutions is a weighted version of the space of functions continuous and m-times differentiable on $[0, \infty)$, namely of the form (3.2.11),

$$\mathcal{X} = \ker B \oplus C^{(m)}_{\alpha+\beta} = \left\{ y(x) = y_0(x) + Y(x);\ y_0 \in \operatorname{span}\left\{ x^{-\beta\gamma_k} \right\}_1^m,\ Y \in C^{(m)}_{\alpha+\beta} \right\}. \tag{3.4.7}$$

First of all, let us recall that the initial value problem (3.2.4), (3.2.6) for the simplest Bessel type equation:

$$\begin{cases} By(x) = 0 \\ \lim\limits_{x\to+0} B_k y(x) = b_k, \quad k = 1, \ldots, m \end{cases} \tag{3.4.8}$$

has been solved in Lemma 3.2.2 and its solution is:

$$y(x) = \sum_{i=1}^{m} c_i x^{\beta\gamma_i} \in C^{(m)}_{\alpha} \tag{3.4.9}$$

with constants c_i, defined by (3.2.8).

The next step is to solve explicitly the non homogeneous hyper-Bessel differential equations of the form $By = f$ under arbitrary initial conditions.

Theorem 3.4.2. *Let (3.4.6) be satisfied and $f \in C_\alpha$. Consider the non homogeneous hyper-Bessel differential equation*

$$By(x) = f(x), \ \textit{that is,}\ x^{\alpha_0} D x^{\alpha_1} D \ldots x^{\alpha_{m-1}} D x^{\alpha_m} y(x) = f(x) \tag{3.4.10}$$

and the initial value problem for it, defined by the conditions (3.4.4). Its solution $y(x) \in \mathcal{X}$ *has the representation*

$$y(x) = y_0(x) + Y(x)$$

with

$$y_0(x) = \sum_{k=1}^{m} c_k x^{-\beta\gamma_k}; \; c_k = \left[\prod_{j=k+1}^{m} \beta(\gamma_j - \gamma_k)\right]^{-1} b_k \;\; \textit{(as in (3.2.8))}$$

and

$$Y(x) = Lf(x) = \beta^{-m} x^{\beta} \int_0^1 G^{m,0}_{m,m}\left[\sigma \left| \begin{matrix} (\gamma_i + 1)_1^m \\ (\gamma_i)_1^m \end{matrix} \right.\right] f(x\sigma^{\frac{1}{\beta}})\, d\sigma. \tag{3.4.11}$$

Proof. Denote by $Y(x) = Lf(x)$ the linear right inverse operator of B defined by zero initial conditions. This is the solution (3.1.23) of the initial value problem

$$By(x) = f(x); \quad \lim_{x\to+0} B_k y(x) = 0,\; k = 1, \ldots, m, \tag{3.4.12}$$

given by Theorem 3.1.5 and having the form (3.4.11).

Then, the sought solution can be found as a sum

$$y(x) = y_0(x) + Y(x),$$

where $Y(x) \in C^{(m)}_{\alpha+\beta}$ is the solution (3.4.11) of (3.4.12) and $y_0(x) \in \operatorname{span}\left\{x^{-\beta\gamma_k}\right\} = \ker B$ has the form (3.4.9), according to Lemma 3.2.2. Therefore, $y(x) \in \left[\mathcal{X} = \ker B \oplus C^{(m)}_{\alpha+\beta}\right]$ and this completes the proof.

Further, we consider the last of the hyper-Bessel differential equations (3.4.5), namely: $By(x) = \lambda y(x)$. Solving the initial value problem for it is related to the *important problem of finding the resolvent of the hyper-Bessel operator.* In a particular form, we have solved this problem in [190], generalizing the result of Obrechkoff [339]. In [192] we have shown that the *kernel-function of the Obrechkoff transform*, namely the G-function:

$$y(x) = G^{m,0}_{0,m}\left[(-1)^m \frac{\lambda x^\beta}{\beta^m} \left| (-\gamma_i)_1^m \right.\right] \tag{3.4.13}$$

is a solution of this m-th order differential equation, having in general m lineary independent solutions. Now we prove to complete result.

Theorem 3.4.3. *Let conditions (3.4.6) be satisfied. Then, the fundamental system of solutions of the hyper-Bessel equation:*

$$By(x) = \lambda y(x), \quad \lambda = \text{const}, \tag{3.4.14}$$

in a neighbourhood of $x = 0$ *consists of Meijer's G-functions:*

$$y_k(x) = G^{1,0}_{0,m}\left[-\frac{\lambda x^\beta}{\beta^m} \left| -\gamma_k, -\gamma_1, \ldots, -\gamma_{k-1}, -\gamma_{k+1}, \ldots, -\gamma_m \right.\right], \quad (k = 1, 2, \ldots, m) \tag{3.4.15}$$

or, up to constant multipliers, of the "Bessel-type" generalized hypergeometric functions

$$\tilde{y}_k(x) = \left(\frac{\lambda x^\beta}{\beta^m}\right)^{-\gamma_k} {}_0F_{m-1}\left((1+\gamma_j-\gamma_k)_{j\neq k}; \frac{\lambda x^\beta}{\beta^m}\right), \quad k=1,\ldots,m, \tag{3.4.16}$$

which belong to the space $\mathcal{X} = \mathrm{span}]\left\{x^{-\beta\gamma_k}\right\}_1^m \bigoplus C_{\alpha+\beta}^{(m)}$. *Then the solution of the initial value problem (3.4.14), (3.4.4) is given by the linear combination*

$$y(x) = a_1\tilde{y}_1(x) + \cdots + a_m\tilde{y}_m(x) \tag{3.4.17}$$

with coefficients:

$$a_k = b_k.\lambda^{\gamma_k}\beta^{k-m(\gamma_k+1)}\left[\prod_{j=k+1}^{m}(\gamma_j-\gamma_k)\right]^{-1}, \quad k=1,\ldots,m-1,$$
$$a_m = b_m.\lambda^{\gamma_m}\beta^{-m\gamma_m}.$$

Proof. We consider equation (3.4.14) written in the form:

$$\left[\lambda x^\beta - \prod_{j=1}^{m}\left(x\frac{d}{dx}+\beta\gamma_j\right)\right] y(x) = 0. \tag{3.4.18}$$

After the substitution $t = \frac{\lambda x^\beta}{(-\beta)^m}$, it turns into a special case of the generalized hypergeometric differential equation, namely:

$$\left[(-1)^{-m}t^1 - \prod_{j=1}^{m}(t\frac{d}{dt}+\gamma_j)\right]\hat{y}(t) = 0, \text{ where } \hat{y}(t) := y(x).$$

According to [106, I, p. 206, (2)] the G-functions (3.4.15) form a fundamental system of solutions of equation (3.4.18) near the singular point $x=0$. Moreover, each of the G-functions:

$$G_{0,s}^{m,0}\left[(-1)^s(\frac{\lambda x^\beta}{\beta^m})\middle|(-\gamma_k)_1^m\right], \quad 1\leq s\leq m,$$

is also a solution of (3.4.18), that is, of (3.4.14) and, especially, this is true for the particular $G_{0,m}^{m,0}$-function ($s=m$) in Kiryakova [192]. Up to the constant multipliers

$$d_k = \exp(\pi_i\gamma_k)\prod_{j\neq k}(1+\gamma_j-\gamma_k), k=1,\ldots,m,$$

the G-functions (3.4.15) are the generalized hypergeometric functions (3.4.16):

$$\tilde{y}_k(x) = d_k y_k(x), \quad k=1,\ldots,m$$

(see [106, I, p. 215, (1)]) which satisfy the "modified" initial conditions:

$$\lim_{x\to+0} B_i\tilde{y}_k(x) = c_{ik} = \begin{cases} \frac{(\beta^{m(\gamma_k+1)-k)}}{\lambda^{\gamma_k}} \left[\prod_{j=k+1}^{m} (\gamma_j - \gamma_k)\right], & \text{if } i = k, \\ 0, & \text{if } i \neq k, \quad i = 1, \dots, m. \end{cases}$$

Thus, the theorem is proved.

It is worth mentioning that the generalized hypergeometric functions (3.4.16) are closely related to the so-called *hyper-Bessel functions* (D.3), (D.5) introduced by Delerue [60] as generalizations of the Bessel functions with respect to the number of indices (see Appendix, Section D), namely, the following assertion equivalent to Theorem 3.4.3 holds and *explains the use of the name hyper-Bessel differential operators given to the operators* B, (3.1.2)-(3.1.4).

Corollary 3.4.4. *For* $\gamma_1 < \gamma_2 < \cdots < \gamma_m < \gamma_1 + 1$ *the hyper-Bessel functions of Delerue:*

$$J^{(m-1)}_{1+\gamma_1-\gamma_k,\dots,*,\dots,1+\gamma_m-\gamma_k}\left[(-\lambda)^{\frac{1}{m}}\frac{m}{\beta}x^{\frac{\beta}{m}}\right], \quad k = 1, \dots, m \tag{3.4.19}$$

form a fundamental system of solutions of the hyper-Bessel differential equation $By(x) = \lambda y(x)$ *in a neighbourhood of* $x = 0$.

Till now, we have managed to find explicit solutions to Cauchy (initial) value problems for the hyper-Bessel differential equations of the forms (3.4.5): $By(x) = 0$, $By(x) = f(x)$, $By(x) = \lambda y(x)$, (Lemma 3.2.2, Theorem 3.4.2, Theorem 3.4.3) in terms of Meijer's G-functions. We can state also the following corollary.

Corollary 3.4.5. *The solution of the initial value problem:*

$$\begin{cases} By(x) = \lambda y(x) + f(x), & \lambda \neq 0, \ f \neq 0 \\ \lim_{x\to+0} B_k y(x) = b_k, & k = 1, \dots, m \text{ of form (3.4.4)}, \end{cases} \tag{3.4.20}$$

for the most general non homogeneous hyper-Bessel differential equation (3.4.2) can be found as a sum

$$y(x) = y_0(x) + \tilde{y}(x) \tag{3.4.21}$$

of a particular solution $y_0(x)$ *of the problem*

$$\begin{cases} By(x) = \lambda y(x) + f(x), \\ \lim_{x\to+0} B_k y(x) = 0, \quad k = 1, \dots, m \end{cases} \tag{3.4.20$'$}$$

and the solution $\tilde{y}(x)$ *of* $B\tilde{y}(x) = \lambda\tilde{y}(x)$ *with arbitrary initial conditions (3.4.4), found in Theorem 3.4.3, namely:*

$$\tilde{y}(x) = \sum_{i=1}^{m} \frac{A_i \lambda^{\gamma_i}}{\beta^{m(\gamma_i+1)-i}} \left[\prod_{j=i+1}^{m} \Gamma(\gamma_j - \gamma_i)\right]^{-1} \tilde{y}_i(x), \tag{3.4.22}$$

where $\tilde{y}_i(x)$, $i = 1, \dots, m$ are the generalized hypergeometric functions (3.4.16).

Since the particular solution $y_0(x)$ of (3.4.20′) is easier to find again found (also in terms of the G-functions) by the method of transmutations, considered in Section 3.5, we give it after in Section 3.8. In this way, the general Cauchy problem (3.4.20) for the hyper-Bessel differential equations will be completely solved.

Now, let us consider *some examples* of equation (3.4.14) and show the relation with some special and elementary functions like the $\cos_m$-function, the n- (di-) Bessel functions, the hyper-Bessel functions, etc. These results are corollaries of Theorem 3.4.3.

EXAMPLE 3.4.6. Let B be a hyper Bessel differential operator (3.1.3) with $\beta = m$ and one of the parameters γ_k being zero, for example:

$$\beta = m > 1; \quad \gamma_1 < \gamma_2 < \cdots < \gamma_m = 0 < \gamma_1 + 1.$$

Consider the Cauchy initial value problem for (3.4.14) with $\lambda = -1$:

$$\begin{cases} By(x) = -y(x), \\ y(0) = 1, \ y'(0) = \cdots = y^{(m-1)}(0) = 0. \end{cases} \tag{3.4.23}$$

These initial conditions can be written in the Bessel-type form (3.4.4) as follows:

$$\lim_{x \to +0} B_k y(x) = b_k = 0, \ k = 1, \dots, m-1; \ \lim_{x \to +0} B_m y(x) = b_m = 1.$$

Then, according to Theorem 3.4.3, the solution of (3.4.23) is:

$$\begin{aligned} y(x) &= a_m y_m(x) = \frac{a_m}{d_m} \tilde{y}_m(x) = {}_0F_{m-1}((1+\gamma_j)_1^{m-1}; -\left(\frac{x}{m}\right)^m) \\ &= j_{\gamma_1, \dots, \gamma_{m-1}}^{(m-1)}(x), \end{aligned} \tag{3.4.24}$$

that is, a *normalized hypergeometric function* (D.4) of order $(m-1)$ (also called a *Bessel-Clifford function of order* $(m-1)$, see Appendix). For different initial conditions other modifications of the hyper-Bessel functions (D.3) appear as solutions.

If we take the same problem but with $\lambda = 1$, that is, for the equation $By(x) = y(x)$, then we obtain the solution

$$y(x) = {}_0F_{m-1}((1+\gamma_j)^{m-1}; \left(\frac{x}{m}\right)^m) = i_{\gamma_1, \dots, \gamma_{m-1}}^{(m-1)}(x), \tag{3.4.25}$$

namely, the normalized version (D.6) of the so-called *modified hyper-Bessel function*

$$I_{\gamma_1, \dots, \gamma_{m-1}}^{(m-1)}(x),$$

defined by (D.5) and analogous to the Bessel function $I_\nu(x)$. For details, see Appendix and as well as [60], [218], [80].

EXAMPLE 3.4.7. The simplest case of problem (3.4.23) occurs for the "hyper-Bessel" operator

$$B = D^m = \left(\frac{d}{dx}\right)^m \text{ with } \beta = m; \; \gamma_k = \frac{k}{m} - 1, \; k = 1, \dots, m.$$

Then, the initial value problem

$$\begin{cases} y^{(m)}(x) = \lambda y(x), & \lambda = \pm 1 \\ y(0) = 1, & y'(0) = \cdots = y^{(m-1)}(0) = 0 \end{cases} \tag{3.4.26}$$

is a special case of Example 3.4.6, and its solution is given by the generalized cosine function of m-th order (D.9), respectively by the m-ht order hyperbolic h_m-function (D.14).

In particular, for $m = 2$ the classical trigonometric functions $\cos(x)$, $\cosh(x)$ turn out to be the solutions of (3.4.26) (for $\lambda = \pm 1$), namely of the problems:

$$y''(x) = \pm y(x); \quad y(0) = 1, \quad y'(0) = 0.$$

Now, let us see how Theorem 3.4.3 works in the case $m = 3$.

EXAMPLE 3.4.8. (Hayek and Hernandez [127]) Consider the following problem: find a fundamental system of solutions of the *general 3-rd order Bessel type differential equation* in a neighbourhood of $x = 0$:

$$x^3 y'''(x) + (\lambda_1 + \lambda_2 + 3)x^2 y''(x) + (1 + \lambda_1 + \lambda_2 + \lambda_1\lambda_2)xy'(x) + xy(x) = 0, \tag{3.4.27}$$

if its parameters fulfil the conditions:

$$\lambda_1, \lambda_2 \in \mathfrak{R}; \quad \lambda_1, \lambda_2, \lambda_1 - \lambda_2 \neq \text{integers}.$$

This means that condition (3.4.6) holds and the case is non logarithmic. Equation (3.4.27) can be written in the form $By(x) = -y(x)$, with

$$\begin{aligned} &B = x^{-1}(xD + \lambda_1)(xD + \lambda_2)xD = x^{-\lambda_1} D x^{\lambda_1 - \lambda_2 + 1} D x^{\lambda_2 + 1} D, \\ &\lambda = -1, \; m = 3, \; \beta = 1; \; \gamma_1, \gamma_2, \gamma_3 \longrightarrow 0, \lambda_1, \lambda_2. \end{aligned}$$

According to Theorem 3.4.3, in terms of G-functions, the f.s.s. (3.4.15) is:

$$G^{1,0}_{0,3}\left[x \,|0, -\lambda_1, -\lambda_2\right], \quad G^{1,0}_{0,3}\left[x \,|-\lambda_1, -\lambda_2, 0\right], \quad G^{1,0}_{0,3}\left[x \,|-\lambda_2, -\lambda_1, 0\right],$$

or in the form (3.4.16):

$$\begin{aligned} \tilde{y}_1(x) &= x^{-\lambda_1} {}_0F_2(1 + \lambda_2 - \lambda_1, 1 - \lambda_1; -x), \\ \tilde{y}_2(x) &= x^{-\lambda_2} {}_0F_2(1 + \lambda_1 - \lambda_2, 1 - \lambda_2; -x), \\ \tilde{y}_3(x) &= {}_0F_2(1 + \lambda_1, 1 + \lambda_2; -x), \end{aligned}$$

or by means of the hyper-Bessel functions (3.4.19):

$$x^{\lambda_1-\lambda_2}J^{(2)}_{\lambda_2-\lambda_1,-\lambda_1}(3x), \quad x^{\lambda_2-\lambda_1}J^{(2)}_{\lambda_1-\lambda_2,-\lambda_2}(3x), \quad x^{-\lambda_1-\lambda_2}J^{(2)}_{\lambda_1,\lambda_2}(3x).$$

The latter system can be written also in terms of the equivalent *Bessel-Clifford functions of third order* [127], viz.:

$$C_{\lambda_1,\lambda_2}(x), \quad x^{-\lambda_1}C_{-\lambda_1,\lambda_2-\lambda_1}(x), \quad x^{-\lambda_2}C_{-\lambda_2,\lambda_1-\lambda_2}(x).$$

Then each solution of (3.4.27) has the form:

$$y(x) = c_1\tilde{y}_1(x) + c_2\tilde{y}_2(x) + c_3\tilde{y}_3(x)$$

with c_1, c_2, c_3 defined by the initial value conditions (3.4.3): $y(0)$, $y'(0)$, $y''(0)$, or by (3.4.4):

$$y(0), \quad \left[x^{\lambda_2+1}y'(x)\right]_{x=0}, \quad \left[x^{\lambda_1-\lambda_2+1}\left(x^{\lambda_2+1}y'(x)\right)'\right]_{x=0}.$$

EXAMPLE 3.4.9. Consider the third order hyper-Bessel equation (Hriptun [137]):

$$x^3y'''(x) + (\frac{3}{2})x^2y''(x) - \frac{6p^2+3p+1}{2}xy'(x) + (2p^3+3p^2-(\frac{1}{4})x^3)y(x) = 0, \quad (3.4.28)$$

with a parameter $p \in \Re$, $p \neq \frac{k}{3}$, $p \neq \frac{2k-3}{6}$, $k = 0, \pm1, \pm2, \ldots$ This can be rewritten in the form $By(x) = \lambda y(x)$ with $\lambda = \frac{1}{4}$ and

$$B = D^3 + \frac{3}{2}.\frac{1}{x}D^2 - \frac{6p^2+3p+1}{2}\frac{1}{x^2}D + (2p^3+3p^2)\frac{1}{x^3}$$

of the form $B = x^{-3}Q_3(xD)$. Now $m = 3$, $\beta = 3$ and the zeros of Q_3 are p, $p+\frac{3}{2}$, $-2p$. The conditions on p ensure that none of them coincide or differ by an integer, that is, the case is non logarithmic. From Theorem 3.4.3 we obtain that the generalized hypergeometric functions (being also hyper-Bessel functions $J^{(2)}_{\nu_1,\nu_2}$):

$$y_p(1) = x^p{}_0F_2(\frac{1}{2}, p+1; \frac{1}{4}\frac{x^3}{2}7) = \sum_{n=0}^{\infty}\frac{(\frac{x}{3})^{p+3n}}{(2n)!\Gamma(n+p+1)},$$

$$y_p(2) = x^{p+\frac{3}{2}}{}_0F_2(\frac{3}{2}, p+\frac{3}{2}; \frac{1}{4}\frac{x^3}{2}7) = \sum_{n=0}^{\infty}\frac{(\frac{x}{3})^{p+3n+\frac{3}{2}}}{(2n+1)!\Gamma(n+p+\frac{3}{2})},$$

$$y_p(3) = x^{-2p}{}_0F_2(1-p, -p+\frac{1}{2}; \frac{1}{4}\frac{x^3}{2}7) = \sum_{n=0}^{\infty}\frac{\Gamma(2p+2n)(\frac{x}{3})^{-2p+3n}}{\Gamma(2p)n!},$$

are three linearly independent solutions of (3.4.28). They coincide with the power series solutions found by Hriptun [137] in a different manner.

EXAMPLE 3.4.10. The hyper-Bessel equation

$$\frac{d^n y(x)}{dx^n} \pm x^q y(x) = 0 \tag{3.4.29}$$

has been studied by different authors under some restrictions on the parameters n, q. A fundamental system of solutions has been proposed by Delerue [60, p. 267-268], if $n + q > 0$. Kummer [241] shows that for integer $q = m > 0$, the integral

$$y(x) = \int_0^\infty u^{m-1} \exp\left(-\frac{u^{m+n}}{m+n}\right) \psi(xu) du$$

gives a solution of (3.4.29), provided $\psi(x)$ satisfies the equation

$$\frac{d^{n+1}\psi(x)}{dx^{n+1}} = x^{m-1}\psi(x).$$

For arbitrary values of q, Karanikoloff [184] finds a series solution and later Bainov and Shopolov [26] use the same method for solving a more general hypergeometric equation involving the Bessel type operator $x^{-q}\frac{d^n}{dx^n}$. Equations (3.4.29) incorporate the so-called *Airy equation*

$$\frac{d^2 u(x)}{dx^2} - xu(x) = 0 \qquad (n = 2, \ q = 1).$$

In [191] and Section 3.7 we give a new explanation of the Stokes phenomenon for the Airy functions (see also Heading [132]-[133]).

Here, *from Theorem 3.4.3* we obtain a fundamental system of solutions of (3.4.29) and specify the initial value conditions satisfied by these solutions, namely:

Equation (3.4.29) can be rewritten in the form

$$x^{-q}\left(\frac{d}{dx}\right)^n y(x) = -y(x)$$

with $B = x^{-q}\left(\frac{d}{dx}\right)^n$, $\beta = n + q > 0$, $\gamma_k = \frac{k-n}{n+q}$, $k = 1, \ldots, n$, and the condition $\gamma_1 < \gamma_2 < \cdots < \gamma_n = 0 < \gamma_1 + 1$ is satisfied. Then, the functions

$$y_k(x) = G^{1,0}_{0,n}\left[\frac{x^{n+q}}{(n+q)^n}\middle| \frac{n-k}{n+q}, \left(\frac{n-j}{n+q}\right)_{j\neq k}\right], \quad k = 1, \ldots, n, \tag{3.4.30}$$

or equivalently, the hyper-Bessel functions

$$\begin{aligned} \tilde{y}_k(x) &= \left[\frac{x^{n+q}}{(n+q)^n}\right]^{\frac{n-k}{n+q}} {}_0F_{n-1}\left(\left(1 + \frac{j-k}{n+q}\right)_{j\neq k}; -\frac{x^{n+q}}{(n+q)^n}\right) \\ &= (n+q)^{-\frac{n(n-k)}{n+q}} x^{n-k} j^{(n-1)}_{\left(1+\frac{j-k}{n+q}\right)_{j\neq k}}\left(\frac{n}{n+q} x^{\frac{n}{n+q}}\right), \quad k = 1, \ldots, n, \end{aligned} \tag{3.4.30'}$$

form a f.s.s. of (3.4.29). *This system (3.4.30′) of hyper-Bessel functions can be shown to be equivalent to the system proposed by Delerue* [60] in the form:

$$J_k(x) = x^{\frac{n-1}{2}} J^{(n-1)}_{\frac{1}{n+q},\frac{2}{n+q},\dots,\frac{n-k-1}{n+q},\frac{-k}{n+q},\dots,\frac{-1}{n+q}}\left(\frac{n}{n+q} x^{\frac{n}{n+q}}\right), \quad k = 1, \dots, n. \qquad (3.4.30'')$$

Let us consider also *some examples* of *hyper-Bessel equations* $By(x) = \lambda y(x)$ and the corresponding hyper-Bessel functions *in the logarithmic case*, i.e. provided condition (3.4.6) is not fulfilled. It is seen that by the arguments and approaches used in [1], [5], the results of Theorem 3.4.3 work again.

EXAMPLE 3.4.11. Consider the following initial value problem for the $2n$-*th order self-adjoint ODE*:

$$\begin{cases} By(x) = y(x), & \text{that is }, \ [(xD)^n (xD+\nu)^n - x]\, y(x) = 0, \\ y(0) = \{\Gamma(\nu+1)\}^{-n}, & y'(0) = \dots = y^{(2n-1)}(0) = 0 \end{cases} \qquad (3.4.31)$$

with

$$B = x^{-1}(xD)^n (xD+\nu)^n,$$

that is,

$$m = 2n,\ \beta = 1,\ \gamma_1 = \dots = \gamma_n = 0,\ \gamma_{n+1} = \dots = \gamma_{2n} = \nu.$$

The solution of (3.4.31) is expressible by means of *the n-Bessel functions* (D.20):

$$y(x) = (-x)^{-\frac{\nu}{2}} A_{\nu,n}\left(2i\sqrt{x}\right),$$

as shown in Agarwal [6].

EXAMPLE 3.4.12. The initial value problem (3.4.31) for $n = 2$ has the form

$$\begin{cases} [(xD)^2 (xD+\nu)^2 - x]\, y(x) = (x^{\nu+3} y'')'' + (\nu+1)(x^{\nu+1} y')' - x^\nu y = 0, \\ y(0) = \{\Gamma(\nu+1)\}^{-2},\ y'(0) = y''(0) = y'''(0) = 0. \end{cases} \qquad (3.4.32)$$

The solution of this fourth order self-adjoint ODE, according to Exton [111], is *the di-Bessel function* $y(x) = (-x)^{-\frac{\nu}{2}} A_\nu(2i\sqrt{x})$.

3.5. The transmutation method. The Poisson-Sonine-Dimovski transformations

Along with the algebraic approach of Mikusinski [306] and the method based on a suitable integral transform (like the Laplace transform), there is also another approach for building operational calculi for linear operators in the general sense of Dimovski [73]. This is the widely used *similarity method* (or *transmutation method*), an extention to Meller's approach [299]. It is based on the use of *transmutation operators* (*similarity operators, transmutations, transformation operators*, etc) for isomorphic transfer of some (known) elements from the theory of a right-invertible operator $\tilde{B}$ into corresponding (new) elements for another, more complicated or less familiar operator B.

The essence of this method lies in the natural striving for seeking solutions of new complicated problems by their reduction to well-known or simpler ones, by means of a specific *"translator"*. In a narrow sense, the notion of transmutation operator originates from the works of Delsarte and Lions [61, I, p. 427], [62], [259]-[261]. If $\tilde{B} : \mathcal{X} \longrightarrow \mathcal{X}$ and $B : \mathcal{X} \longrightarrow \mathcal{X}$ are operators acting in a space $\mathcal{X}$, then the isomorphism $T : \mathcal{X} \longrightarrow \mathcal{X}$ is said to transmute $\tilde{B}$ into B, if $T\tilde{B} = BT$ in $\mathcal{X}$. In this sense, the transmutation method has been widely used in mathematical analysis and mainly in differential equations and problems of mathematical physics (see Delsarte [61], Delsarte and Lions [62], Lions [259]-[261], Hearsh [134], Bregg [43], Khachatrjan [185]-[187], Klučantčev [216]-[217]) in operational calculus (see Carroll [53]-[59]), etc. For applications to analysis and operational calculus, one can see Meller [299], Dimovski [68]-[71], [73], Bozhinov [38]-[39], Koprinski [234], etc. Some of these authors use "similarities" in a wider sense, as below.

Definition 3.5.1. (Dimovski [70], [73]). An isomorphism $T : \tilde{\mathcal{X}} \longrightarrow \mathcal{X}$ of the linear space $\mathcal{X}$ into another linear space $\mathcal{X}$, is said to be a *similarity* (*similarity operator*) from a linear operator $\tilde{L} : \tilde{\mathcal{X}} \longrightarrow \tilde{\mathcal{X}}$ to the linear operator $L : \mathcal{X} \longrightarrow \mathcal{X}$, if $T\tilde{L} = LT$ holds in $\mathcal{X}$. We say also that *the operator $\tilde{L}$ is similar to L* under similarity T.

The *similarity relation*

$$T\tilde{L} = LT \quad \text{in} \quad \tilde{\mathcal{X}} \tag{3.5.1}$$

can be represented also by means of the *commutative diagram*

$$\begin{array}{ccc} \tilde{\mathcal{X}} & \xrightarrow{\tilde{L}} & \tilde{\mathcal{X}} \\ T\big\downarrow & & \big\downarrow T \\ \mathcal{X} & \xrightarrow[L]{} & \mathcal{X} \end{array} \tag{3.5.2}$$

or written in the forms

$$\tilde{L} = T^{-1}LT \quad \text{in} \quad \tilde{\mathcal{X}}, \qquad \text{or} \quad L = T\tilde{L}T^{-1} \quad \text{in} \quad \mathcal{X}. \tag{3.5.3}$$

Definition 3.5.2. If the operators $\tilde{L}$ and L are the corresponding linear right inverses of the right-invertible operators $\tilde{B} : \left[\tilde{\mathcal{X}}_B \subset \tilde{\mathcal{X}}\right] \longrightarrow \tilde{\mathcal{X}}$ and $B : [\mathcal{X}_B \subset \mathcal{X}] \longrightarrow \mathcal{X}$, we say *also* that *operator T transmutes (transforms) $\tilde{B}$ into B* (in a wider sense), although the relation $T\tilde{B} = BT$ may not always be fulfilled.

Lemma 3.5.3. *If T transmutes the rigth-invertible linear operator $\tilde{B} : \tilde{\mathcal{X}}_B \longrightarrow \mathcal{X}$ into the right-invertible operator $B : \mathcal{X}_B \longrightarrow \mathcal{X}$ and if $\tilde{F}$, F are the corresponding defining projectors (initial operators) of their right inverse operators $\tilde{L}$, L, then the following relations hold:*

$$\begin{aligned} T\tilde{B} &= BT - BT\tilde{F} && \textit{in } \tilde{\mathcal{X}}_B \subset \tilde{\mathcal{X}}, \\ T^{-1}B &= \tilde{B}T^{-1} - -\tilde{B}T^{-1}F && \textit{in } \mathcal{X}_B \subset \mathcal{X}. \end{aligned} \tag{3.5.4}$$

The *proof* is trivial. If $BT\tilde{F} \equiv 0$ (e.g. if we have zero initial conditions), it follows that $\tilde{B}$ and B are also similar.

Now, the right-invertible operators $\tilde{B}$ and B can be *two arbitrary hyper-Bessel differential operators of the same order* $m > 1$.

First, dealing with functions of many variables $x_1, \ldots, x_m$ in $\{x_k > 0,\ k = 1, \ldots, m\}$ Dimovski [70]-[71] showed that the linear transformation

$$Xf(x_1, \ldots, x_m) = \left(\prod_{k=1}^{m} x_k^{\gamma_k}\right) f\left[\beta^{\frac{m}{\beta}} (x_1 \ldots x_m)^{\frac{1}{\beta}}\right] \tag{3.5.5}$$

is a similarity from the hyper-Bessel integral operator (3.1.22)-(3.1.23) to the operator Λ_m of multiple integration with respect $x_1, \ldots, x_m$:

$$\Lambda_m f(x_1, \ldots, x_m) = \int_0^{x_1} \ldots \int_0^{x_m} f(\tau_1, \ldots, \tau_m)\, d\tau_1 \ldots d\tau_m, \tag{3.5.6}$$

namely:

$$XL = \Lambda_m X \tag{3.5.7}$$

and the inverse transformation has the form

$$X^{-1} f(x_1, \ldots, x_m) = \left(\frac{1}{\beta} x^{\frac{\beta}{m}}\right)^{\gamma_1 + \cdots + \gamma_m} f\left[\frac{1}{\beta} x^{\frac{\beta}{m}}, \ldots, \frac{1}{\beta} x^{\frac{\beta}{m}}\right]. \tag{3.5.8}$$

Now let $\tilde{B}$ and B be two arbitrary hyper-Bessel differential operators of order $m > 1$ whose parameters, right inverse operators, transformations (3.5.5), spaces, etc., are denoted by $\left\{\tilde{\beta}, \tilde{\gamma}_k, k = 1, \ldots, m\right\}$ and $\{\beta, \gamma_k, k = 1, \ldots, m\}$, respectively $\tilde{L}$ and $\tilde{L}$, $\tilde{\mathcal{X}}$ and $\mathcal{X}$, $C_{\tilde{\alpha}}$ and C_α, etc.

Theorem (Dimovski [70]-[71]). *The transformation*

$$T = X^{-1} Z \tilde{X}, \tag{3.5.9}$$

where Z is a correcting operator defined as a product of Riemann-Liouville fractional integrals $R_k^{\delta_k}$ with respect to the variables x_k, $k = 1, \ldots, m$:

$$Z = \prod_{k=1}^{m} R_k^{\lambda + \gamma_k - \tilde{\gamma}_k}, \tag{3.5.10}$$

λ being an arbitrary real such that

$$\lambda > \frac{\alpha}{\beta} - \frac{\tilde{\alpha}}{\tilde{\beta}}, \quad \lambda \geq \max_{1 \leq k \leq m} (\tilde{\gamma}_k - \gamma_k), \tag{3.5.11}$$

is a similarity from $\tilde{L}$ to L, i.e.

$$T\tilde{L} = LT \quad in \quad \tilde{\mathcal{X}} = C_{\tilde{\alpha}}. \tag{3.5.12}$$

The explicit integral representation of transformation (3.5.9), $T : C_{\tilde{\alpha}} \longrightarrow C_\alpha$, is found by Dimovski in [70]:

$$Tf(x) = \beta^{-m\lambda} x^{\lambda\beta} \int_0^1 \dots \int_0^1 \prod_{k=1}^m \left[\frac{(1-t_k)^{\lambda+\gamma_k-\tilde{\gamma}_k-1}}{\Gamma(\lambda+\gamma_k-\tilde{\gamma}_k)} t_k^{\tilde{\gamma}_k} \right] \\ \times f\left[\left(\frac{\tilde{\beta}}{\beta}\right)^{\frac{m}{\beta}} x^{\frac{\beta}{\tilde{\beta}}} (t_1 \dots t_m)^{\frac{1}{\tilde{\beta}}} \right] dt_1 \dots dt_m, \tag{3.5.13}$$

where $\lambda > \max\limits_{1\le k\le m} (\tilde{\gamma}_k - \gamma_k)$. If, however,

$$\max_k (\tilde{\gamma}_k - \gamma_k) > \frac{\alpha}{\beta} - \frac{\tilde{\alpha}}{\tilde{\beta}}$$

then we can choose $\lambda = \max\limits_k (\tilde{\gamma}_k - \gamma_k)$ and in this way we reduce the number of integrations in (3.5.13) by one, i.e. we represent T by means of an $(m-1)$-tuple integration. In the simplest case:

$$\tilde{\gamma}_1 - \gamma_1 = \dots = \tilde{\gamma}_m - \gamma_m = \lambda > \frac{\alpha}{\beta} - \frac{\tilde{\alpha}}{\tilde{\beta}},$$

we have simply

$$Tf(x) = x^{\lambda\beta} f\left(x^{\frac{\beta}{\tilde{\beta}}}\right).$$

Also, if some of the differences $\Delta_k = \tilde{\gamma}_k - \gamma_k$, $k = 1, \dots, s$, are equal, then we can reduce by their number s the multiplicity of integral (3.5.13).

Since the transformation T, defined by (3.5.9), (3.5.13) *generalizes the classical transformations of Poisson and Sonine, related to the Bessel operator*

$$B_\nu = \frac{d^2}{dx^2} + \frac{2\nu+1}{x}\frac{d}{dx},$$

we adopt the following definition.

Definition 3.5.4. The transformations (transmutation operators) (3.5.9),(3.5.13) which are similarities from the hyper-Bessel integral operator $\tilde{L}$ to another hyper-Bessel operator L, are said to be *Poisson-Sonine-Dimvoski (P.-S.-D.) transformations.* Their particular specifications for the cases: $\tilde{L} = l^m$ (i.e. $\tilde{B} = \left(\frac{d}{dx}\right)^m$) and arbitrary L; arbitrary $\tilde{L}$ and $L = l^m$ (i.e. $B = \left(\frac{d}{dx}\right)^m$), will be called *Poisson-Dimovski (P.-D.)* and *Sonine-Dimovski* (S.-D.) *transformations*, respectively.

Further, our aim is to find a *simpler representation* of the P.-S.-D. transformations (3.5.13) by means of single integrals with Meijer's G-functions in the kernel. Then, using these new concise representations of the transmutation operators T, we propose some new applications of the transmutation method in the theory of hyper-Bessel operators and equations.

Theorem 3.5.5. *Let Ξ be the transformation*

$$\Xi : f(x) \longrightarrow \tilde{f}(x) = f\left(cx^{\frac{\beta}{\tilde{\beta}}}\right) \quad with \quad c = \left(\frac{\tilde{\beta}}{\beta}\right)^{\frac{m}{\tilde{\beta}}}. \tag{3.5.14}$$

Then, the P.-S.-D. transformation (3.5.13) can be represented as a generalized operator of fractional integration

$$Tf(x) = \frac{x^{\lambda\beta}}{\beta^{m\lambda}} I_{\tilde{\beta},m}^{(\tilde{\gamma}_k),(\lambda+\gamma_k-\tilde{\gamma}_k)} \Xi f(x), \tag{3.5.15}$$

having the explicit integral representation

$$Tf(x) = \frac{x^{\lambda\beta}}{\beta^{m\lambda}} \int_0^1 G_{m,m}^{m,0}\left[\sigma \left| \begin{matrix} (\lambda+\gamma_k)_1^m \\ (\gamma_k)_1^m \end{matrix} \right.\right] f\left[\left(\frac{\tilde{\beta}}{\beta}\right)^{\frac{m}{\tilde{\beta}}} x^{\frac{\beta}{\tilde{\beta}}} \sigma^{\frac{1}{\tilde{\beta}}}\right] d\sigma. \tag{3.5.16}$$

Proof. The theorem follows from (3.5.13) and Theorem 1.2.10. It finds exclusively many applications. For instance, we are able to obtain immediately the following inversion formula.

Theorem 3.5.6. *(Inversion formula for the P.-S.-D. transformation). Let us denote*

$$\delta_k = \lambda - \Delta_k = \lambda + \gamma_k - \tilde{\gamma}_k > 0, \quad \eta_k = \begin{cases} [\delta_k]+1, & \text{if } \delta_k \text{ is non integer} \\ \delta_k, & \text{if } \delta_k \text{ is integer} \end{cases}, \quad k = 1, \ldots, m. \tag{3.5.17}$$

If $f \in C_{\tilde{\alpha}}$, then $g = Tf \in C_{\alpha}^{(\eta_1,\ldots,\eta_m)}$ and the following inversion formula for transformation T holds:

$$\begin{aligned} f(x) = T^{-1}g(x) &= \tilde{\beta}^{m\lambda} D_{\eta}^{(\tilde{\gamma})} I_{\tilde{\beta},m}^{(\lambda+\gamma_k),(\eta_k-\lambda-\gamma_k+\tilde{\gamma}_k)} \left\{ x^{-\lambda\tilde{\beta}} g\left(\left(\left(\frac{\beta}{\tilde{\beta}}\right)^{\frac{m}{\beta}} x^{\frac{\beta}{\tilde{\beta}}}\right)\right)\right\} \\ &= \tilde{\beta}^{m\lambda} x^{-\tilde{\beta}\lambda} D_{\eta}^{(\tilde{\gamma}-\lambda)} I_{\tilde{\beta},m}^{(\gamma_k),(\eta_k-\lambda-\gamma_k-\tilde{\gamma}_k)} \left\{ g\left(\left(\left(\frac{\beta}{\tilde{\beta}}\right)^{\frac{m}{\beta}} x^{\frac{\beta}{\tilde{\beta}}}\right)\right)\right\}, \end{aligned} \tag{3.5.18}$$

where the differential operator $D^{(\tilde{\gamma})}$ is the following polynomial of the Euler differential operator $\delta_x = x\frac{d}{dx}$:

$$D^{(\tilde{\gamma})} = \prod_{k=1}^{m} \prod_{j=1}^{\eta_k} \left(\frac{1}{\tilde{\beta}} x \frac{d}{dx} + \tilde{\gamma}_k + j\right). \tag{3.5.19}$$

Otherwise, formula (3.5.18) takes the explicit differintegral form:

$$
\begin{aligned}
f(x) =&\tilde{\beta}^{m\lambda} x^{-\tilde{\beta}\lambda} \prod_{k=1}^{m} \prod_{j=1}^{\eta_k} \left(\frac{1}{\beta} x \frac{d}{dx} + \tilde{\gamma}_k + j \right) \\
&\times \int_0^1 G_{m,m}^{m,0} \left[\sigma \left| \begin{matrix} (\eta_k + \tilde{\gamma}_k - \lambda)_1^m \\ (\gamma_k)_1^m \end{matrix} \right. \right] g\left(\left(\frac{\beta}{\tilde{\beta}} \right)^{\frac{m}{\beta}} x^{\frac{\beta}{\tilde{\beta}}} \sigma^{\frac{1}{\beta}} \right) d\sigma.
\end{aligned} \tag{3.5.20}
$$

Proof. The obvious inversion of the Ξ-transform is:

$$\Xi^{-1} g(x) = g\left[\left(\frac{x}{c} \right)^{\frac{\tilde{\beta}}{\beta}} \right].$$

Also, if $f \in C_{\tilde{\alpha}}$, then $\tilde{f} = \Xi f \in C_{\tilde{\alpha}\frac{\beta}{\tilde{\beta}}}$. The operator $I_{\beta,m}^{(\tilde{\gamma}_k),(\lambda+\gamma_k-\tilde{\gamma}_k)}$ preserves the power function and therefore, since $\lambda > \frac{\alpha}{\beta} - \frac{\tilde{\alpha}}{\tilde{\beta}}$,

$$Tf = \frac{x^{\lambda\beta}}{\beta^{m\lambda}} I_{\beta,m}^{(\tilde{\gamma}_k),(\lambda+\gamma_k-\tilde{\gamma}_k)} \tilde{f} \in C_{\tilde{\alpha}\frac{\beta}{\tilde{\beta}}+\lambda\beta}^{(\eta_1+\cdots+\eta_m)} \subset C_{\alpha}^{(\eta_1+\cdots+\eta_m)}.$$

Due to Lemma 1.3.3 with $\omega = \frac{\beta}{\tilde{\beta}} > 0$, equality (3.5.15) can be written in the form:

$$
\begin{aligned}
g(x) = Tf(x) &= \frac{x^{\lambda\beta}}{\beta^{m\lambda}} I_{\beta=\tilde{\beta}\omega,m}^{(\tilde{\gamma}_k),(\delta_k)} \Xi f(x) \\
&= \frac{x^{\lambda\beta}}{\beta^{m\lambda}} \Xi I_{\tilde{\beta}_m}^{(\tilde{\gamma}_k),(\delta_k)} f(x), \quad \text{where} \quad \delta_k := \lambda + \gamma_k - \tilde{\gamma}_k, k = 1, \ldots, m.
\end{aligned}
$$

Then,

$$
\begin{aligned}
f(x) = T^{-1} g(x) &= \left\{ I_{\tilde{\beta},m}^{(\tilde{\gamma}_k),(\delta_k)} \right\}^{-1} \left\{ \Xi^{-1} \left[\frac{\beta^{m\lambda}}{x^{\lambda\beta}} g(x) \right] \right\} \\
&= \left(\tilde{\beta} \right)^{m\lambda} \left\{ I_{\tilde{\beta},m}^{(\tilde{\gamma}_k),(\delta_k)} \right\}^{-1} \left\{ x^{-\lambda\tilde{\beta}} g\left[\left(\frac{x}{c} \right)^{\frac{\tilde{\beta}}{\beta}} \right] \right\} \\
&= \left(\tilde{\beta} \right)^{m\lambda} \left\{ I_{\tilde{\beta},m}^{(\tilde{\gamma}_k),(\delta_k)} \right\}^{-1} \{G(x)\},
\end{aligned}
$$

where

$$G(x) = x^{-\lambda\tilde{\beta}} g\left[\left(\frac{\tilde{\beta}}{\beta} \right)^{-\frac{m}{\beta}} x^{\frac{\tilde{\beta}}{\beta}} \right].$$

Then, according to the inversion formula for the generalized fractional integrals (1.5.26), we have

$$\tilde{\beta}^{m\lambda}\left\{I_{\tilde{\beta},m}^{(\tilde{\gamma}_k),(\delta_k)}\right\}^{-1}G(x) = \tilde{\beta}^{m\lambda} I_{\tilde{\beta},m}^{(\lambda+\gamma_k),(\Delta_k-\lambda)}G(x)$$

$$= \tilde{\beta}^{m\lambda}\left[\prod_{k=1}^{m}\prod_{j=1}^{\eta_k}\left(\frac{1}{\tilde{\beta}}x\frac{d}{dx}+\tilde{\gamma}_k+j\right)\right] I_{\tilde{\beta},m}^{(\lambda+\gamma_k),(\eta_k-\lambda-\gamma_k-\tilde{\gamma}_k)}G(x)$$

which gives (3.4.18). Let us note that the multiplier $x^{-\tilde{\beta}\lambda}$ in $G(x)$ can be moved in front of the differintegral operator, in accordance with Lemmas 1.3.1 and 1.6.2:

$$I_{\tilde{\beta},m}^{(\lambda+\gamma_k),(\Delta_k-\lambda)}x^{-\tilde{\beta}\lambda} = x^{-\tilde{\beta}\lambda}I_{\tilde{\beta},m}^{(\gamma_k),(\Delta_k-\lambda)}$$

$$= x^{-\tilde{\beta}\lambda}\left[\prod_{k=1}^{m}\prod_{j=1}^{\eta_k}\left(\frac{1}{\tilde{\beta}}x\frac{d}{dx}+\tilde{\gamma}_k+j\right)\right] I_{\tilde{\beta},m}^{(\gamma_k),(\eta_k-\lambda-\gamma_k+\tilde{\gamma}_k)}.$$

Hence, (3.5.20) is also obtained.

Let us consider the cases when *one of the hyper-Bessel differential operators $\tilde{B}$, B is the simplest operator of m-th order*, for instance, $\tilde{B} = D^m = \left(\frac{d}{dx}\right)^m$, (3.3.f)-(3.3.f′). Then, its linear right inverse operator is the operator of m-fold integration, represented (see 3.3, Example f), (3.3.1)-(3.3.2)) either in the form

$$\tilde{L}f(x) = l^m f(x) = \left(\frac{x}{m}\right)^m I_{m,m}^{\left(\frac{k}{m}-1\right),(1)}f(x), \tag{3.5.21}$$

or equivalently,

$$\tilde{L}f(x) = l^m f(x) = x^m I_1^{0,m}f(x). \tag{3.5.21′}$$

Then, the corresponding Poisson-Dimovski and Sonine-Dimovski transformations (Definition 3.5.4) follow from the general form (3.5.15), (3.5.16) of the P.-S.-D. transformation by a suitable choice of the parameters $\tilde{\beta}$, $\tilde{\gamma}_k$, $k = 1,\dots,m$ of $\tilde{B}$ (or β, γ_k, $k = 1,\dots,m$ of B).

3.5.i. Poisson-Dimovski transformation

Let us choose

$$\begin{cases} \tilde{B} = D^m = \left(\frac{d}{dx}\right)^m & \text{with parameters} \\ \tilde{\beta} = m,\ \tilde{\gamma}_k = \frac{k}{m}-1,\ k = 1,\dots,m, & \text{so that} \\ \tilde{\gamma}_1 < \tilde{\gamma}_2 < \dots < \tilde{\gamma}_m = 0 < \tilde{\gamma}_1 + 1 \Rightarrow \tilde{\alpha} = -1. \end{cases} \tag{3.5.22}$$

and

$$\begin{cases} B = x^{-\beta}\prod_{k=1}^{m}\left(x\frac{d}{dx}+\beta\gamma_k\right) & \text{arbitrary, with parameters} \\ \beta > 0,\ \gamma_1 \ge \gamma_2 \ge \dots \ge \gamma_m \Rightarrow \alpha = -\beta(\gamma_m+1). \end{cases} \tag{3.5.23}$$

Since

$$\max_k \Delta_k = \max_k (\tilde{\gamma}_k - \gamma_k) = -\gamma_m > \frac{\alpha}{\beta} - \frac{\tilde{\alpha}}{\tilde{\beta}} = \frac{\alpha}{\beta} + \frac{1}{m},$$

we can choose for convencience,

$$\lambda = \max_k \Delta_k, \quad \text{i.e.} \quad \lambda = -\gamma_m. \tag{3.5.24}$$

Then, the generalized Poisson type transformation, proposed by Dimovski [70], has the multiple integral form:

$$\begin{aligned} Pf(x) =& c\left(\frac{x^\beta}{\beta^m}\right)^{\lambda} \int_0^1 \cdots \int_0^1 \prod_{k=1}^m \left[\frac{(1-t_k)^{\lambda-\Delta_k-1}\, t_k^{\tilde{\gamma}_k}}{\Gamma(\lambda-\Delta_k)}\right] \\ &\times f\left[\frac{m}{\beta} x^{\frac{\beta}{m}} (t_1 \dots t_m)^{\frac{1}{m}}\right] dt_1 \dots dt_m \\ =& c\left(\frac{x^\beta}{\beta^m}\right)^{-\gamma_m} \int_0^1 \cdots \int_0^1 \prod_{k=1}^m \left[\frac{(1-t_k)^{\gamma_k-\gamma_m-\frac{k}{m}}\, t_k^{\frac{k}{m}-1}}{\Gamma\left(\gamma_k-\gamma_m-\frac{k}{m}+1\right)}\right] \\ &\times f\left[\frac{m}{\beta} x^{\frac{\beta}{m}} (t_1 \dots t_m)^{\frac{1}{m}}\right] dt_1 \dots dt_m, \end{aligned} \tag{3.5.25}$$

where c stands for the constant

$$c = \sqrt{\frac{m}{(2\pi)^{m-1}}} \prod_{k=1}^m \Gamma(\gamma_k + 1). \tag{3.5.26}$$

This transformation is a similarity between the integral operators $\tilde{L} = l^m$ (3.5.21)-(3.5.21$'$) and L (the linear right inverse of B, (3.5.23)):

$$Pl^m = LP. \tag{3.5.27}$$

Denote

$$\Delta_k = \tilde{\gamma}_k - \gamma_k = \frac{k}{m} - 1 - \gamma_k; \quad p_k = \lambda - \Delta_k = \gamma_k - \gamma_m + 1 - \frac{k}{m} > 0;$$

$$\eta_k = \begin{cases} [p_k] + 1, & \text{if } p_k \text{ is non integer} \\ p_k, & \text{if } p_k \text{ is integer} \end{cases}, \quad k = 1, \dots, m.$$

Then, from Theorems 3.5.5 and 3.5.6 we obtain immediately:

Theorem 3.5.7. *The Poisson-Dimovski transformation (3.5.25),*

$$P : C_{-1} \longrightarrow C_\alpha = C_{-\beta(\gamma_m+1)}, \tag{3.5.28}$$

is a generalized fractional integral of multiorder $(p_1, \dots, p_m) = (p_1, \dots, p_{m-1}, 0)$, *namely:*

$$\begin{aligned} Pf(x) &= c\left(\frac{x^\beta}{\beta m}\right)^{\lambda} I_{\beta,m}^{\left(\frac{k}{m}-1\right),(\lambda-\Delta_k)} f\left(\frac{m}{\beta}x^{\frac{\beta}{m}}\right) \\ &= c\left(\frac{x^\beta}{\beta m}\right)^{-\gamma_m} I_{\beta,m-1}^{\left(\frac{k}{m}-1\right),(p_k)} f\left(\frac{m}{\beta}x^{\frac{\beta}{m}}\right) \\ &= c\left(\frac{x^\beta}{\beta m}\right)^{-\gamma_m} \int_0^1 G_{m-1,m-1}^{m-1,0}\left[\sigma \middle| \begin{matrix} (\gamma_k-\gamma_m)_1^{m-1} \\ \left(\frac{k}{m}-1\right)_1^{m-1} \end{matrix}\right] f\left(\frac{m}{\beta}x^{\frac{\beta}{m}}\right) d\sigma, \end{aligned} \tag{3.5.29}$$

acting from C_{-1} *onto* $C_{-\beta\left(\gamma_m+\frac{1}{m}\right)}^{(n_1+\cdots+n_{m-1})} \subset C_\alpha^{(n_1+\cdots+n_{m-1})}$. *The corresponding inversion formula is:*

$$\begin{aligned} f(x) = P^{-1}g(x) &= \frac{1}{c}\beta^{-m\gamma_m} D_n I_{m,m-1}^{(\gamma_k-\gamma_m),\left(n_k+\gamma_m-\gamma_k+\frac{k}{m}-1\right)} \\ &\times \left\{\left(\frac{m}{\beta}\right)^{-m\gamma_m} x^{\beta\gamma_m} g\left[\left(\frac{\beta}{m}\right) x^{\frac{m}{\beta}}\right]\right\}, \end{aligned} \tag{3.5.30}$$

where

$$\begin{aligned} D_n &= \prod_{k=1}^{m}\prod_{j=1}^{\eta_k}\left(\frac{1}{m}x\frac{d}{dx}+\frac{k}{m}+j-1\right) \\ &= \prod_{k=1}^{m}\left[x^{m-k}\frac{d^{n_k}}{(dx^m)^{n_k}}x^{k+mn_k-m}\right]. \end{aligned} \tag{3.5.31}$$

Let us recall now Lemmas 1.3.3 and 1.6.3 according to which, by means of of transformation Ξ (i.e. a substitution) we can consider generalized fractional integrals (derivatives) with a parameter $\beta_1 > 0$ instead of a parameter $\beta > 0$.

Then, for simplicity of notation and expressions, *we can consider hyper-Bessel operators with parameter* $\beta = m$.

The P.-D. transformation (3.5.29):

$$P : C_{-1} \longrightarrow C_{-m\gamma_m-1}^{(n_1+\cdots+n_{m-1})} \subset C_{-m(\gamma_m+1)},$$

then takes the simpler form:

$$\begin{aligned} Pf(x) &= c\left(\frac{x}{m}\right)^{-m\gamma_m} I_{m,m-1}^{\left(\frac{k}{m}-1\right),(p_k)} f(x) \\ &= c\left(\frac{x}{m}\right)^{-m\gamma_m} \int_0^1 G_{m-1,m-1}^{m-1,0}\left[\sigma \middle| \begin{matrix} (\gamma_k-\gamma_m)_1^{m-1} \\ \left(\frac{k}{m}-1\right)_1^{m-1} \end{matrix}\right] f\left(x\sigma^{\frac{1}{m}}\right) d\sigma \end{aligned} \tag{3.5.32}$$

and its inversion is:

$$
\begin{aligned}
f(x) = P^{-1}g(x) &= \frac{1}{c} m^{-m\gamma_m} D_n I_{m,m-1}^{(\gamma_k-\gamma_m),(n_k-p_k)} \{x^{m\gamma_m} g(x)\} \\
&= \frac{1}{c} m^{-m\gamma_m} \left[\prod_{k=1}^{m} \prod_{j=1}^{n_k} \left(\frac{1}{m} x \frac{d}{dx} + \gamma_m + \frac{k}{m} + j - 1 \right) \right] I_{m,m-1}^{(\gamma_k-\gamma_m),(n_k-p_k)} g(x).
\end{aligned}
\tag{3.5.33}
$$

Due to (3.5.27), the P.-D. transformation transmutes $\tilde{B} = D^m = \left(\frac{d}{dx}\right)^m$ into the hyper-Bessel differential operator B with parameters as in (3.5.23′). Namely, we have the following relationship.

Lemma 3.5.8. *In the space $\mathcal{X}_{\tilde{F}} = \operatorname{span}] \left\{x^k\right\}_0^{m-1} \oplus C_{m-1}^{(m)}$ the transmutation relation holds:*

$$
P \left(\frac{d}{dx} \right)^m = BP - BP\tilde{F}, \tag{3.5.34}
$$

where $\tilde{F}$ is the initial operator (defining projector) (3.3.3) of $\tilde{L} = l^m$, i.e. the Taylor polynomial:

$$
\tilde{F} f(x) = \sum_{k=0}^{m-1} \frac{f^{(k)}(0)}{k!} x^k. \tag{3.5.35}
$$

Corollary 3.5.9. *On the subspace $\left[\mathcal{X}_{\tilde{B}} = C_{m-1}^{(m)}\right] \subset \mathcal{X}_{\tilde{F}}$ there is a similarity relation between both differential operators:*

$$
P \left(\frac{d}{dx} \right)^m = BP. \tag{3.5.36}
$$

In particular, (3.5.36) holds for the so-called *m-even functions* (see Klučantčev [216]-[217]):

$$
f \in C_{m-1}^{(m)} \subset C^{(m)}, \quad f'(0) = \cdots = f^{(m-1)}(0) = 0. \tag{3.5.37}
$$

In the literature (e.g. Delsarte [61, I]-[62]) by a *Poisson transformation* the following transformation is meant:

$$
\begin{aligned}
P_\nu f(x) &= \frac{2}{\sqrt{\pi}} \cdot \frac{\Gamma(\nu+1)}{\Gamma(\nu+\frac{1}{2})} x^{-2\nu} \int_0^x \frac{f(\tau)}{(x^2-\tau^2)^{\frac{1}{2}-\nu}} d\tau \\
&= \frac{2}{\sqrt{\pi}} \cdot \frac{\Gamma(\nu+1)}{\Gamma(\nu+\frac{1}{2})} \int_0^1 (1-\sigma^2)^{\nu-\frac{1}{2}} f(x\sigma)\, d\sigma, \quad \nu > -\frac{1}{2}.
\end{aligned}
\tag{3.5.38}
$$

This transformation transmutes $\tilde{B} = \left(\frac{d}{dx}\right)^2$ into the classical Bessel differential operator $B = B_\nu = \frac{d^2}{dx^2} + \frac{2\nu+1}{x}\frac{d}{dx}$ (see Section 3.3, Example a)). It can be obtained as a special case of the Poisson-Dimovski transformation (3.5.29) by taking $m = \beta = 2$, $\gamma_1 = \frac{\nu}{2}$, $\gamma_2 = -\frac{\nu}{2}$.

3.5.ii. Sonine-Dimovski transformation

Now we consider a generalization of the *Sonine transformation*

$$\begin{aligned} S_\nu f(x) &= \frac{2\left(\frac{x}{2}\right)^{\nu+1}}{\Gamma\left(\nu+\frac{1}{2}\right)} \int_0^1 \left(1-\sigma^2\right)^{\nu-\frac{1}{2}} \sigma^{1-\nu} f(x\sigma) d\sigma \\ &= \frac{\left(\frac{x}{2}\right)^{\nu+1}}{\Gamma\left(\nu+\frac{1}{2}\right)} \int_0^1 (1-t)^{\nu-\frac{1}{2}} t^{-\frac{\nu}{2}} f(x\sqrt{t}) dt, \end{aligned} \tag{3.5.39}$$

related to the Bessel operator B_ν (see Delsarte [61, I]-[62] again).

We specialize the Poisson-Sonine-Dimovski transformation (3.5.13), (3.5.15) when:

$$\begin{cases} \tilde{B} = x^{-\beta} \prod_{k=1}^{m} \left(x\frac{d}{dx} + \beta\gamma_k\right) \quad \text{is arbitrary with} \\ \tilde{\beta} > 0,\ \tilde{\gamma}_1 \le \tilde{\gamma}_2 \le \cdots \le \tilde{\gamma}_m,\ \tilde{\alpha} = -\tilde{\beta}\left(\tilde{\gamma}_1 + 1\right) \end{cases} \tag{3.5.40}$$

and

$$\begin{cases} B = D^m = \left(\frac{d}{dx}\right)^m \quad \text{is with } \textit{rearranged} \text{ (!) parameters:} \\ \beta = m,\ \gamma_k = \frac{m-k+1}{m} - 1 = -\frac{k-1}{m},\ k = 1, \dots, m, \quad \text{so that} \\ \gamma_1 \ge \gamma_2 \ge \cdots \ge \gamma_m \ \Rightarrow\ \alpha = -1. \end{cases} \tag{3.5.41}$$

For this choice of the parameters, $\max_k \Delta_k = \max_k \left(\tilde{\gamma}_k + \frac{k-1}{m}\right) = \tilde{\gamma}_m + \frac{m-1}{m}$ which is greater than $\frac{\alpha}{\beta} - \frac{\tilde{\alpha}}{\tilde{\beta}} = -\frac{1}{m} + \tilde{\gamma}_1 + 1$ and we can choose

$$\lambda = \max_k \Delta_k = \tilde{\gamma}_m + \frac{m-1}{m}. \tag{3.5.42}$$

Further, for convenience we write only B, β, γ_k, α, ... instead of $\tilde{B}$, $\tilde{\beta}$, $\tilde{\gamma}_k$, $\tilde{\alpha}$, ..., i.e. for an arbitrary hyper-Bessel differential operator B, (3.5.40) we assume

$$\beta > 0,\ \gamma_1 \le \gamma_2 \le \cdots \le \gamma_m,\ \alpha = -\beta\left(\gamma_1 + 1\right). \tag{3.5.43}$$

In this way the P.-S.-D. transformation (3.5.13) reduce to the Sonine-Dimovski transformation (Definition 3.5.4) introduced by Dimovski [70]:

$$\begin{aligned} &S : C_\alpha \longrightarrow C_{-1}, \\ &Sf(x) = \left(\frac{x}{m}\right)^{m(\gamma_m+1)-1} \int_0^1 \dots \int_0^1 \prod_{k=1}^{m-1} \left[\frac{(1-t_k)^{\lambda_k - 1} t_k^{\gamma_k}}{\Gamma(\lambda_k)}\right] \\ &\qquad \times f\left[\left(\frac{\beta}{m}\right)^{\frac{m}{\beta}} x^{\frac{m}{\beta}} \left(t_1 \dots t_{m-1}\right)^{\frac{1}{\beta}}\right] dt_1 \dots dt_{m-1}, \end{aligned} \tag{3.5.44}$$

where

$$\lambda_k = \gamma_m - \gamma_k + \frac{k}{m} > 0, \; k = 1, \dots, m-1. \tag{3.5.45}$$

This transformation is a similarity from the hyper-Bessel integral operator L (corresponding to (3.5.43)) to the m-fold integration l^m:

$$SL = l^m S \;\text{ in }\; C_\alpha. \tag{3.5.46}$$

Up to a constant multiplier in the argument of $f(x)$, the S.-D. transformation (3.5.44) coincides with the isomorphism $\varphi f(x)$, used by Dimovski [68]-[69] to transfer the results from the *Heaviside-Mikusinski operational calculus* (for l^m) to an operational calculus for the hyper-Bessel operators B, L, namely:

$$\begin{aligned}\varphi f(x) = \frac{x^{m(\gamma_k+1)-1}}{\prod\limits_{k=1}^{m-1} \Gamma(\lambda_k)} \int_0^1 \dots \int_0^1 \prod_{k=1}^{m-1} \left[(1-t_k)^{\lambda_k - 1} t_k^{\gamma_k}\right] \\ \times f\left[x^{\frac{m}{\beta}} (t_1 \dots t_{m-1})^{\frac{1}{\beta}}\right] dt_1 \dots dt_{m-1}.\end{aligned} \tag{3.5.47}$$

The relation between (3.5.44) and (3.5.47) is:

$$\varphi f(x) = m^{m(\gamma_m+1)-1} S f\left(\frac{m}{\beta} x\right). \tag{3.5.48}$$

Due to Lemmas 1.3.3 and 1.6.3 again, without loss of generality one can consider only the *simpler case of hyper-Bessel operators with $\beta = m$*.

Let us use then the representation of the S.-D. transformation S as a generalized fractional integral and write down its inversion formula. An inversion formula for the alternative transformation (3.5.47) was found by Dimovski [68]-[69] but of the form (1.2.34) (Chapter 1), viz.

$$\begin{aligned}f(x) = \varphi^{-1} g\left[(x_1 \dots x_{m-1})^{\frac{1}{\beta}}\right] = \left(\prod_{k=1}^{m-1} x^{-\gamma_k}\right) \frac{\partial^{\lambda_1 + \dots + \lambda_{m-1}}}{\partial x_1^{\lambda_1} \dots \partial x_{m-1}^{\lambda_{m-1}}} \\ \times \left\{ g\left[(x_1 \dots x_{m-1})^{\frac{1}{m}}\right] \prod_{k=1}^{m-1} x_k^{\frac{k+1}{m}} \right\}.\end{aligned} \tag{3.5.49}$$

Theorem 3.5.10. *The Sonine-Dimovski transformation (3.5.44) is an $(m-1)$-tuple generalized fractional integral of multiorder $(\lambda_1, \dots, \lambda_{m-1})$ and for $\beta = m$ it has the form:*

$$\begin{aligned}Sf(x) &= \left(\frac{x}{m}\right)^{m(\gamma_m+1)-1} I_{m,m-1}^{(\gamma_k),(\lambda_k)} f(x) \\ &= \left(\frac{x}{m}\right)^{m(\gamma_m+1)-1} \int_0^1 G_{m-1,m-1}^{m-1,0}\left[\sigma \left| \begin{matrix} \left(\gamma_m + \frac{k}{m}\right) \\ (\gamma_k) \end{matrix} \right.\right] f\left(x\sigma^{\frac{1}{m}}\right) d\sigma\end{aligned} \tag{3.5.50}$$

with the λ_k's defined by (3.5.45). If we denote

$$\eta_k = \begin{cases} [\lambda_k]+1, & \textit{if } \lambda_k \textit{ is non integer} \\ \lambda_k, & \textit{if } \lambda_k \textit{ is integer} \end{cases}, \quad k=1,\ldots,m$$

then $S: C_\alpha \longrightarrow C^{(\eta_1+\cdots+\eta_{m-1})}_{m(\gamma_m-\gamma_1)-1} \subset C_{-1}$ and the inversion formula is:

$$f(x) = S^{-1}g(x) = \left(\frac{x}{m}\right)^{-m(\gamma_m+1)+1} D_\eta I^{(\frac{k+1}{m}-1),(\eta_k-\lambda_k)}_{m,m-1} g(x), \tag{3.5.51}$$

where

$$D_\eta = \prod_{k=1}^{m}\prod_{j=1}^{\eta_k}\left(\frac{1}{m}x\frac{d}{dx}+\gamma_k-\gamma_m+\frac{1}{m}-j\right).$$

Note. The corresponding results for arbitrary $\beta>0$ are obtained (according to Lemma 1.3.3) by means of the transformation

$$\Xi: f(x) \longrightarrow \tilde{f}(x) = f\left[\left(\frac{\beta}{m}x\right)^{\frac{m}{\beta}}\right] \quad \text{with} \quad \omega=\frac{m}{\beta}>0,$$

as follows:

$$\begin{aligned} Sf(x) &= \left(\frac{x}{m}\right)^{m(\gamma_m+1)-1} I^{(\gamma_k),(\lambda_k)}_{m,m-1}\{\Xi f(x)\} \\ &= \left(\frac{x}{m}\right)^{m(\gamma_m+1)-1} \Xi\left\{I^{(\gamma_k),(\lambda_k)}_{\beta,m-1} f(x)\right\}. \end{aligned}$$

Hence, for instance:

$$\varphi f(x) = \Xi\left\{x^{\beta(\gamma_m+\frac{m-1}{m})} I^{(\gamma_k),(\lambda_k)}_{\beta,m-1} f(x)\right\}, \tag{3.5.52}$$

or

$$\varphi f\left(x^{\frac{\beta}{m}}\right) = x^{\beta(\gamma_m+\frac{m-1}{m})} I^{(\gamma_k),(\lambda_k)}_{\beta,m-1} f(x). \tag{3.5.52'}$$

The same representation is obtained in a direct way in Dimovski and Kiryakova [79] and an inversion formula for φ is found there in the symbolic form

$$f(x) = \varphi^{-1}g(x) = x^{-\beta(\gamma_m+\frac{m-1}{m})} I^{(\frac{k-m+1}{m}),(-\lambda_k)}_{\beta,m-1} g\left(x^{\frac{\beta}{m}}\right). \tag{3.5.53}$$

According to Definition 1.5.3 we have to interpret *this fractional integral of negative multiorder* in the following way:

$$\begin{aligned} f(x) = \varphi^{-1}g(x) &= x^{-\beta(\gamma_m+\frac{m-1}{m})} D^{(\gamma_k-\gamma_m-\frac{m-1}{m}),(\gamma_k)}_{\beta,m-1} g\left(x^{\frac{\beta}{m}}\right) \\ &= x^{-\beta(\gamma_m+\frac{m-1}{m})} D_\eta I^{(\frac{k-m+1}{m}),(\eta_k-\lambda_k)}_{\beta,m-1} g\left(x^{\frac{\beta}{m}}\right) \end{aligned} \tag{3.5.54}$$

with

$$D_\eta = \prod_{k=1}^{m}\prod_{j=1}^{\eta_k}\left(\frac{1}{m}x\frac{d}{dx} + \gamma_k - \gamma_m + \frac{m-1}{m} + j\right),$$

i.e. *as a generalized fractional derivative.*

By Definition 3.5.4, the S.-D. transformation S transmutes the hyper-Bessel differential operator B (with $\gamma_1 \le \gamma_2 \le \cdots \le \gamma_m$) into $\left(\frac{d}{dx}\right)^m$. Namely, from (3.5.46) we obtain the following relationship.

Lemma 3.5.11. *In the subspace $\mathcal{X}_F = \ker B \oplus C^m_{(\alpha+\beta)}$ the transmutation relation holds:*

$$SB = \left(\frac{d}{dx}\right)^m S - \left(\frac{d}{dx}\right)^m SF, \tag{3.5.55}$$

where $F = I - LB$ stands for the initial operator (defining projector) (3.2.12) of L. In particular, if

$$f \in C^{(m)}_{\alpha+\beta} \subset \mathcal{X}_F \Rightarrow SBf = \left(\frac{d}{dx}\right)^m Sf. \tag{3.5.56}$$

3.6. Applications of the transmutation method to the hyper-Bessel operational calculus.

We illustrate some applications of the Poisson-Sonine-Dimovski transformations for finding some elements of the operational calculi for the hyper-Bessel operators: convolutions, representations of the fractional powers of the basic operators, etc. In [64]-[65], [69]-[70] Dimovski developed such operational calculi based on a family of convolutions of the hyper-Bessel integral operators L in the space C_α.

Theorem (Dimovski [69]). *For $f, g \in C_\alpha$ consider the auxiliary operation*

$$\begin{aligned}(f \circ g)(x) = x^\beta \underbrace{\int_0^1 \cdots \int_0^1}_{(m)} f\left[x(t_1 \dots t_m)^{\frac{1}{\beta}}\right] g\left[x((1-t_1)\dots(1-t_m))^{\frac{1}{\beta}}\right] \\ \times \prod_{k=1}^{m} [t_k(1-t_k)]^{\gamma_k} dt_1 \dots dt_m,\end{aligned} \tag{3.6.1}$$

and the "correcting" operators $T : C_{2\alpha+\beta} \longrightarrow C_\alpha$, defined by

$$Tf(x) = x^{\beta\gamma_m} \underbrace{\int_0^1 \cdots \int_0^1}_{(s)} f\left[x(t_1 \dots t_m)^{\frac{1}{\beta}}\right] \prod_{k=1}^{s}\left[t_k^{2\gamma_k}\frac{(1-t_k)^{\gamma_m-\gamma_k-1}}{\Gamma(\gamma_m-\gamma_k)}\right] dt_1 \dots dt_s, \tag{3.6.2}$$

if $\gamma_1 \leq \gamma_2 \leq \cdots \leq \gamma_s < \gamma_{s+1} = \cdots = \gamma_m$, *or by*

$$T_\lambda f(x) = x^{\beta\lambda} \underbrace{\int_0^1 \cdots \int_0^1}_{(m)} f\left[x\left(t_1 \ldots t_m\right)^{\frac{1}{\beta}}\right] \prod_{k=1}^{m} \left[t_k^{2\gamma_k} \frac{\left(1-t_k\right)^{\lambda-\gamma_k-1}}{\Gamma\left(\lambda-\gamma_k\right)}\right] dt_1 \ldots dt_m, \quad (3.6.3)$$

with $\lambda > \max\limits_k \gamma_k = \gamma_m$. *Then, the operations*

$$(f * g)(x) = T(f \circ g)(x), \quad \left(f \overset{\lambda}{*} g\right)(x) = T_\lambda (f \circ g)(x), \quad (3.6.4)$$

are convolutions of L *in* C_α *without divisors of zero.*

Obviously, T and T_λ are generalized (s- and m-tuple) fractional integrals:

$$T = x^{\beta\gamma_m} I_{\beta,s}^{(2\gamma_k),(\gamma_m-\gamma_k)}, \quad T_\lambda = x^{\beta\lambda} I_{\beta,m}^{(2\gamma_k),(\lambda-\gamma_k)}$$

and T follows from T_λ for $\lambda = \gamma_{s+1} = \cdots = \gamma_m$, i.e. both operators (3.6.2), (3.6.3) are represented by

$$T_\lambda = x^{\beta\lambda} I_{\beta,m}^{(2\gamma_k),(\lambda-\gamma_k)} \quad \text{with} \quad \lambda \geq \max_k \gamma_k = \gamma_m. \quad (3.6.5)$$

Thus we can state the following theorem.

Theorem 3.6.1. *The operation*

$$\left(f \overset{\lambda}{*} g\right)(x) = T_\lambda (f \circ g)(x) = x^{\beta\lambda} I_{\beta,m}^{(2\gamma_k),(\lambda-\gamma_k)} \left\{(f \circ g)(x)\right\}, \quad (3.6.6)$$

where $(\circ)$ *is the auxiliary operation (3.6.1),* T_λ *is the generalized fractional integral (3.6.5),* $\lambda \geq \max\limits_k \gamma_k$, *is a convolution of the hyper-Bessel integral operator*

$$L = \frac{x^\beta}{\beta^m} I_{\beta,m}^{(\gamma_k),(1)} \quad (3.6.7)$$

in C_α, *without divisors of zero.*

The new representation (3.6.6) for these convolutions has some advantages, namely, in their representations by means of $(m+1)$-tuple integrals only instead of $2m$-tuple integrals (3.6.4), and also in using the common rules for the generalized fractional integrals from Chapter 1.

For example, once calculating that

$$x^p \circ x^q = x^{p+q+\beta} \prod_{k=1}^{m} \frac{\Gamma\left(\gamma_k + \frac{p}{\beta} + 1\right) \Gamma\left(\gamma_k + \frac{q}{\beta} + 1\right)}{\Gamma\left(2\gamma_k + \frac{p+q}{\beta} + 2\right)},$$

from Lemma 1.2.1 we find immediately the convolutional product

$$
\begin{aligned}
x^p \overset{\lambda}{*} x^q &= x^{p+q+\beta+\beta\lambda} \prod_{k=1}^{m} \frac{\Gamma\left(\gamma_k + \frac{p}{\beta} + 1\right)\Gamma\left(\gamma_k + \frac{q}{\beta} + 1\right)\Gamma\left(2\gamma_k + \frac{p+q}{\beta} + 2\right)}{\Gamma\left(2\gamma_k + \frac{p+q}{\beta} + 2\right)\Gamma\left(\gamma_k + \lambda + \frac{p+q}{\beta} + 2\right)} \\
&= x^{p+q+\beta(1+\lambda)} \prod_{k=1}^{m} \frac{\Gamma\left(\gamma_k + \frac{p}{\beta} + 1\right)\Gamma\left(\gamma_k + \frac{q}{\beta} + 1\right)}{\Gamma\left(\gamma_k + \lambda + \frac{p+q}{\beta} + 2\right)}.
\end{aligned} \tag{3.6.8}
$$

One can find another explicit expression for a convolution of L in C_α by means of the Sonine type transformation φ, written in the form (3.5.52), i.e.

$$
\varphi f(x) = x^{m\left(\gamma_m + \frac{m-1}{m}\right)} \int_0^1 G_{m,m}^{m,0}\left[\sigma \left| \begin{matrix} \left(\gamma_m + \frac{k}{m}\right)_1^m \\ (\gamma_k)_1^m \end{matrix} \right.\right] f\left(x^{\frac{m}{\beta}} \sigma^{\frac{1}{\beta}}\right) d\sigma. \tag{3.6.9}
$$

To this end, the following *theorem for the convolutions of similar operators can be used.*

Theorem (Dimovski [73]). *If $T : \mathcal{X} \longrightarrow \tilde{\mathcal{X}}$ is a similarity from L in $\mathcal{X}$ to $\tilde{L}$ in $\tilde{\mathcal{X}}$, i.e. $L = T^{-1}\tilde{L}T$ in $\mathcal{X}$ and $\tilde{*} : \tilde{\mathcal{X}} \times \tilde{\mathcal{X}} \to \mathcal{X}$ is a convolution of $\tilde{L}$ in $\tilde{\mathcal{X}}$, then the operation*

$$
f * g = T^{-1}\left(Tf \tilde{*} Tg\right) \tag{3.6.10}
$$

is a convolution of L in $\mathcal{X}$.

In particular, we take $T = \varphi$ to be the Sonine-Dimovski transformation (3.6.9), use the similarity relation

$$
\varphi L = \left(\frac{m}{\beta} l\right)^m \varphi \ \text{ in } \ \mathcal{X} = C_\alpha, \tag{3.6.11}
$$

and the *Duhamel convolution* $(\tilde{*})$ of the integration operator l (also of its powers like $\left(\frac{m}{\beta} l\right)^m$):

$$
(f \tilde{*} g)(x) = \int_0^x f(x-t) g(t) dt = x \int_0^1 f\left[x(1-\sigma)\right] g(x\sigma) d\sigma. \tag{3.6.12}
$$

Then, the following result holds.

Theorem (Dimovski [68], [73]). *The operation*

$$
f \circledast g = \varphi^{-1}\left[(\varphi f) \tilde{*} (\varphi g)\right], \tag{3.6.13}
$$

where $(\tilde{*})$ denotes the Duhamel convolution (3.6.12), is defined for all $f, g \in C_\alpha$ and is a convolution of the hyper-Bessel integral operator L in C_α.

However, let us note that instead of the multiple integral representation (3.5.47) of transfomation φ and the same kind of inversion formula (3.5.49), we use in (3.6.13) the simpler representation (3.6.9) and inversion formulas (3.5.53)-(3.5.54).

By using convolution (3.6.13), it is possible to build an operational calculus for L similar to the Mikusinski operational calculus. In the corresponding field of quotients in C_α, we can consider rational expressions of the element $S := \frac{1}{L}$. The integer powers (positive, negative or zero) of L are well defined in this field. Now we resume *the application of convolution (3.6.13) for representing the fractional powers of L.*

Definition 3.6.2. (Dimovski [65], [69]). Let λ be an arbitrary real number. The fractional power L^λ is defined by means of the following convolutional product of elements of the corresponding operational (quotient) field:

$$L^\lambda f := L^{[\lambda]-s} \left\{ \frac{x^{\beta(\{\lambda\}+s-\delta-1)}}{\prod\limits_{k=1}^{m} \Gamma(\{\lambda\}+s-\delta+\gamma_k)} \right\} \circledast f, \tag{3.6.14}$$

where $s = \left[\frac{\alpha}{\beta}+\delta+2\right]$, $\delta \geq \max\limits_k \gamma_k$ and $\circledast$ is the convolution (3.6.13).

If $\lambda > \frac{\alpha}{\beta}+\delta+1$, then the power L^λ has a simpler representation by means of a function $l_\lambda \in C_\alpha$, namely:

$$L^\lambda = \left\{ \frac{x^{\beta(\lambda-\delta-1)}}{\prod\limits_{k=1}^{m} \Gamma(\lambda-\delta+\gamma_k)} \right\} = \{l_\lambda\}, \quad \text{i.e.} \quad L^\lambda f = \{l_\lambda\} \circledast f. \tag{3.6.15}$$

Furthermore, if $\gamma_1 < \gamma_2 < \cdots < \gamma_m < \gamma_1 + 1$, then we can choose $\delta = \gamma_m$ and therefore, $\frac{\alpha}{\beta}+\delta+1 > 0$.

Definition 3.6.2 is justified by the following theorem.

Theorem (Dimovski [65]). *If λ, μ are arbitrary real numbers, then*

$$L^\lambda L^\mu = L^{\lambda+\mu},$$

and for integer $\lambda = n > 0$: $L^n = \underbrace{L.L.\ldots.L}_{n \text{ times}}$.

We prove here the following result.

Theorem 3.6.3. *If $\lambda > 0$ is arbitrary, then the fractional power L^λ of the hyper-Bessel integral operator $L = \frac{x^\beta}{\beta^m} I_{\beta,m}^{(\gamma_k),(1)}$ is also a generalized fractional integral but of multiorder $(\lambda, \ldots, \lambda)$, viz.*

$$L^\lambda = \left(\frac{x^\beta}{\beta^m}\right)^\lambda I_{\beta,m}^{(\gamma_k),(\lambda)} \tag{3.6.16}$$

and has the single integral representation

$$L^\lambda f(x) = \left(\frac{x^\beta}{\beta^m}\right)^\lambda \int_0^1 G^{m,0}_{m,m}\left[\sigma \left| \begin{matrix}(\gamma_k+\lambda)_1^m \\ (\gamma_k)_1^m\end{matrix}\right.\right] f\left(x\sigma^{\frac{1}{\beta}}\right) d\sigma. \tag{3.6.17}$$

Proof. A direct proof can be found in Dimovski and Kiryakova [79]. There, representation (3.6.9) of φ is used and the expression $\varphi^{-1}\left(\frac{m}{\beta}l\right)^m \varphi$ is calculated by means of formula (A.29) for integrals of the product of two G-functions, used several times. It is the same technique as that used in Chapter 1 for deriving the common properties of the generalized fractional integrals.

Here we establish (3.6.16) on the basis that these properties are already proved. For simplicity, we consider the case

$$\lambda > 0,\ \lambda > \frac{\alpha}{\beta} + \delta + 1 = \gamma_m - \gamma_1,$$

when L^λ is represented by (3.6.15), i.e.

$$L^\lambda f(x) = \Xi\{l_\lambda\} \circledast f(x) = \varphi^{-1}\{(\varphi l_\lambda) * \varphi f(x)\}.$$

Due to Lemma 1.2.1, we evaluate

$$\varphi l_\lambda = \left\{ x^{\beta\left(\gamma_m + \frac{m-1}{m}\right)} I^{(\gamma_k),(\lambda_k)}_{\beta,m-1}\left(\frac{x^{\beta(\lambda-\gamma_m-1)}}{\prod\limits_{k=1}^m \Gamma(\lambda-\gamma_m+\gamma_k)}\right)\right\}$$

$$= \left(\frac{m}{\beta}\right)^{m\lambda} \frac{x^{m\lambda-1}}{\Gamma(m\lambda)}.$$

Further,

$$\frac{x^{m\lambda-1}}{\Gamma(m\lambda)} * \varphi f(x) = l^{m\lambda}\{\varphi f(x)\},$$

whence

$$L^\lambda f(x) = \varphi^{-1}\left(\frac{m}{\beta}l\right)^{m\lambda} \varphi f(x).$$

It is convenient now to use an alternative representation of the Riemann-Liouville fractional integral $l^{m\lambda}$ as an m-tuple Erdélyi-Kober operator (cf. the alternative representations (3.3.1), (3.3.2) of l^m in Section 3.3, Example f),

$$l^{m\lambda} = \left(\frac{x}{m}\right)^{m\lambda} I^{\left(\frac{k}{m}-1\right),(\lambda)}_{m,m} = \left(\frac{x}{m}\right)^{m\lambda} I_m^{\frac{1}{m}-1,\lambda} I^{\left(\frac{k+1}{m}-1\right)_1^{m-1},(\lambda)}_{m,m-1}.$$

Further, we use repeatedly Lemma 1.3.3 for the transformation

$$\xi : f(x) \longrightarrow f\left(x^{\frac{m}{\beta}}\right) \quad \text{with} \quad \omega = \frac{m}{\beta} > 0,$$

and also Lemma 1.3.1, property (1.3.3). Operators φ and φ^{-1} are used in the forms (3.5.52), (3.5.54), namely:

$$\varphi = \Xi x^{\beta(\gamma_m + \frac{m-1}{m})} I_{\beta,m-1}^{(\gamma_k),(\lambda_k)},$$

$$\varphi^{-1} = x^{-\beta(\gamma_m + \frac{m-1}{m})} D_{\beta,m-1}^{(\gamma_k - \gamma_m - \frac{m-1}{m}),(\lambda_k)} \Xi^{-1}$$

$$= x^{-\beta(\gamma_m + \frac{m-1}{m})} I_{\beta,m-1}^{(\frac{k+1}{m}-1),(-\lambda_k)} \Xi^{-1}.$$

Then,

$$L^\lambda f(x) = x^{-\beta(\gamma_m + \frac{m-1}{m})} I_{\beta,m-1}^{(\frac{k+1}{m}-1),(-\lambda_k)} \Xi^{-1} \frac{1}{\beta m\lambda} x^{m\lambda} I_{m,m-1}^{(\frac{k+1}{m}-1),(\lambda)}$$

$$\times I_{m,1}^{(\frac{1}{m}-1),(\lambda)} \Xi x^{\beta(\gamma_m + \frac{m-1}{m})} I_{\beta,m-1}^{(\gamma_k),(\lambda_k)} f(x)$$

$$= \frac{1}{\beta m\lambda} x^{\beta(-\gamma_m - \frac{m-1}{m})} I_{\beta,m-1}^{(\frac{k+1}{m}-1),(-\lambda_k)} \Xi^{-1} x^{m\lambda} \Xi x^{\beta(\gamma_m + \frac{m-1}{m})}$$

$$\times I_{\beta,m-1}^{(\gamma_m + \frac{k}{m}),(\lambda)} I_{\beta,1}^{(\gamma_m),(\lambda)} I_{\beta,m-1}^{(\gamma_k),(\lambda_k)} f(x)$$

$$= \frac{1}{\beta m\lambda} x^{\beta(-\gamma_m - \frac{m-1}{m})} I_{\beta,m-1}^{(\frac{k+1}{m}-1),(-\lambda_k)} x^{\beta(\gamma_m + \lambda + 1 - \frac{1}{m})}$$

$$\times I_{\beta,m-1}^{(\gamma_m + \frac{k}{m}),(\lambda)} I_{\beta,m-1}^{(\gamma_k),(\lambda_k)} I_{\beta,1}^{\gamma_m,\lambda} f(x).$$

Now, it remains to use the index law (1.3.10) (Theorem 1.3.8) and relation $\gamma_m + \frac{k}{m} = \gamma_k + \lambda_k$, $k = 1, \ldots, m-1$:

$$L^\lambda f(x) = \left(\frac{x^\beta}{\beta m}\right)^\lambda \left[I_{\beta,m-1}^{(\gamma_m + \frac{k}{m}),(\lambda - \lambda_k)} I_{\beta,m-1}^{(\gamma_k),(\lambda_k)} \right] I_\beta^{\gamma_m,\lambda} f(x)$$

$$= \left(\frac{x^\beta}{\beta m}\right)^\lambda I_{\beta,m-1}^{(\gamma_k),(\lambda - \lambda_k + \lambda_k)} I_\beta^{\gamma_m,\lambda} f(x)$$

$$= \left(\frac{x^\beta}{\beta m}\right)^\lambda I_{\beta,m}^{\gamma_m,\lambda} f(x).$$

This ends the proof.

Note. It is possible to consider also *hyper-Bessel integral operators of Weyl type* (as in McBride [289], Kiryakova [191], Dimovski and Kiryakova [79]), namely:

$$\begin{aligned} Mf(x) &= \left(\frac{x^\beta}{\beta m}\right) W_{\beta,m}^{(-\gamma_k-1),(1)} f(x) \\ &= \frac{x^\beta}{\beta m} \int_1^\infty G_{m,m}^{m,0}\left[\frac{1}{\sigma}\left| \begin{matrix} (-\gamma_k+1)_1^m \\ (-\gamma_k)_1^m \end{matrix} \right.\right] f\left(x\sigma^{\frac{1}{\beta}}\right) d\sigma, \end{aligned} \tag{3.6.18}$$

defined in $C^*_{\alpha^*}$, $\alpha^* \leq \min_k \beta\gamma_k$ (see Section 1.4, (1.4.3)). For the fractional powers of integral operator M we find analogously the representation

$$\begin{aligned} M^\lambda f(x) &= \left(\frac{x^\beta}{\beta m}\right)^\lambda W_{\beta,m}^{(-\gamma_k-1),(\lambda)} f(x) \\ &= \left(\frac{x^\beta}{\beta m}\right)^\lambda \int_1^\infty G_{m,m}^{m,0}\left[\frac{1}{\sigma}\left| \begin{matrix} (-\gamma_k+\lambda)_1^m \\ (-\gamma_k)_1^m \end{matrix} \right.\right] f\left(x\sigma^{\frac{1}{\beta}}\right) d\sigma, \ \lambda > 0. \end{aligned} \tag{3.6.19}$$

Note. Integral representations of fractional powers of operators L, M of the same forms (3.6.17), (3.6.19) were found earlier by McBride [289] but without using convolutions and notions of the generalized fractional calculus. He used the classical Erdélyi-Kober operators and the *theory of Mellin multipliers* which became a basis of his *general approach* of finding fractional powers of linear operators (see McBride [288], [290]). These results were extended further by Lamb [243]-[245], Lamb and McBride [246] and McBride and Spratt [292]-[294].

Kiryakova [190], Dimovski and Kiryakova [75]-[76], McBride [289], Kiryakova [192]-[193], Dimovski and Kiryakova [79] seem to be the first to use Meijer's G-functions for dealing with Bessel type operators and equations. Recently, this approach has found extension in the papers of other authors (see Dimovski and Kiryakova [80], Adamchik and Marichev [5], Adamchik [1]-[2], Kiryakova and Spirova [213], Kiryakova and McBride [212], Hayek and Hernandez [127]-[129], etc.).

3.7. Application of the Poisson-Sonine-Dimovski transmutations to the hyper-Bessel differential equations and functions.

3.7.i. Obtaining generalized Poisson type integrals for the hyper-Bessel functions

It is seen that the Poisson-Dimovski transformation P (3.5.29), (3.5.32) has its simplest representation for operators (3.5.23) with $\beta = m$ and $\gamma_m = \min_k \gamma_k = 0$. We shall use such a transformation to obtain a new integral representation for the hyper-Bessel functions of Delerue $J^{(m-1)}_{\nu_1,\ldots,\nu_{m-1}}(x)$, see (D.3), Appendix. To this end, we consider the m-th order Bessel type differential operator

$$B = x^{-m}\left[\prod_{k=1}^{m-1}\left(x\frac{d}{dx} + m\nu_k\right)\right] x\frac{d}{dx} \tag{3.7.1}$$

with $\beta = m$; $\gamma_k = \nu_k$, $k = 1, \ldots, m-1$, $\gamma_m = 0$, supposing that *all the ν_k's are non negative*, for example:

$$\nu_1 \geq \nu_2 \geq \cdots \geq \nu_{m-1} \geq 0. \tag{3.7.2}$$

Then, for the corresponding Poisson-Dimovski transformation denoted by $\mathbf{P_0}$, the following corollary of Theorem 3.5.7 its true.

Corollary 3.7.1. *The Poisson-Dimovski transformation (3.5.25), (3.5.29) corresponding to operator (3.7.1) maps the space C_{-1} as well as each its subspace $C^{(l)}_{\varepsilon} \subseteq C_{-1}$, $\varepsilon \geq -1$, $l \geq 0$ into itself, and has the form*

$$\begin{aligned}\mathbf{P_0} f(x) &= c.I^{\left(\frac{k}{m}-1\right),\left(\nu_k-\frac{k}{m}+1\right)}_{m,m-1} f(x)\\ &= c.\int_0^1 G^{m-1,0}_{m-1,m-1}\left[\sigma \left| \begin{matrix}(\nu_k)_1^{m-1}\\ \left(\frac{k}{m}-1\right)_1^{m-1}\end{matrix}\right.\right] f\left(x\sigma^{\frac{1}{m}}\right) d\sigma\end{aligned} \tag{3.7.3}$$

with $c = \sqrt{\frac{m}{(2\pi)^{m-1}}}\prod_{k=1}^{m-1}\Gamma(\nu_k+1)$. This transform keeps the values of the functions at the origin:

$$(\mathbf{P_0} f)(0) = f(0) \tag{3.7.4}$$

as well as the other initial-value conditions, up to constant multipliers $c.d_j$, namely:

$$(\mathbf{P_0} f)^{(j)}(0) = c.d_j f^{(j)}(0), \quad j = 1, \ldots, l, \tag{3.7.4'}$$

where $f \in C^{(l)}_{\varepsilon}$ and $d_j = \prod_{k=1}^{m-1}\frac{\Gamma\left(\frac{k-j}{m}\right)}{\Gamma\left(\nu_k+\frac{j}{m}+1\right)}$.

Representation (3.7.3) and $\mathbf{P_0} : C_{-1} \to C_\alpha = C_{-m}$ follow from Theorem 3.5.7. Up to the constant c, $\mathbf{P_0}$ is the multiple Erdélyi-Kober fractional integral $I^{\left(\frac{k}{m}-1\right),\left(\nu_k-\frac{k}{m}+1\right)}_{m,m-1}$

and therefore,

$$\mathbf{P}_0 : C_{-1} \to C_{-1}^{(\eta)} \subset C_{-1}, \text{ also } \mathbf{P}_0 : C_\varepsilon^{(l)} \to C_\varepsilon^{(l+\eta)} \subset C_\varepsilon^{(l)},$$

(see (1.2.33)). Relationships (3.7.4), (3.7.4′) follow from (1.3.4) (Lemma 1.3.2).

Let us note that as a generalized fractional integral (3.7.3), or written alternatively as multiple integrals:

$$\mathbf{P}_0 f(x) = c. \int_0^1 \dots \int_0^1 \prod_{k=1}^{m-1} \left[\frac{(1-t_k)^{\nu_k - \frac{k}{m}} t_k^{\frac{k}{m}-1}}{\Gamma\left(\nu_k - \frac{k}{m} + 1\right)} \right] f\left(x\sigma^{\frac{1}{m}}\right) d\sigma, \tag{3.7.5}$$

transmutation $\mathbf{P}_0$ is defined also under the *weaker condition* (cf. with (3.7.2))

$$\nu_k \geq \frac{k}{m} - 1, \quad k = 1, \dots, m-1 \tag{3.7.6}$$

and according to the *principle of analytical continuation*, it remains then a similarity operator from $\tilde{L} = l^m$ to $L : \mathbf{P}_0 l^m = L\mathbf{P}_0$.

Consider the hyper-Bessel functions of Delerue [60] of multiindex $(\nu_1, \dots, \nu_{m-1})$ (see Appendix, Section D.i):

$$J_{\nu_1,\dots,\nu_{m-1}}^{(m-1)}(x) = \frac{\left(\frac{x}{m}\right)^{\nu_1+\dots+\nu_{m-1}}}{\Gamma(\nu_1+1)\dots\Gamma(\nu_{m-1}+1)} {}_0F_{m-1}\left((\nu_k+1); -\left(\frac{x}{m}\right)^m\right), \tag{3.7.7}$$

where

$$\begin{aligned} j_\nu(x) = j_{\nu_1,\dots,\nu_{m-1}}^{(m-1)}(x) &= {}_0F_{m-1}\left((\nu_k+1); -\left(\frac{x}{m}\right)^m\right) \\ &= \sum_{j=0}^{\infty} \left[\prod_{k=1}^{m-1} \frac{\Gamma(\nu_k+1)}{\Gamma(\nu_k+j+1)} \right] \frac{(-1)^j}{j!} \left(\frac{x}{m}\right)^{mj} \end{aligned} \tag{3.7.8}$$

are the so-called *"normalized" hyper-Bessel functions* (Klučantčev [218]-[219]), or *Bessel-Clifford functions* (Hayek [126], Hayek and Hernandez [127]-[129]).

The latter functions (3.7.8), denoted also by $C_{\nu_1,\dots,\nu_{m-1}}$ (D.8) satisfy the initial value conditions (D.7):

$$j_\nu(0) = 1, \quad j_\nu'(0) = \dots = j_\nu^{(m-1)}(0) = 0. \tag{3.7.9}$$

Our aim is to use transformation $\mathbf{P}_0$ in order to find a generalization of the well-known *Poisson integral representations of the Bessel functions* $J_\nu(x)$, $I_\nu(x)$, $\nu > -\frac{1}{2}$, viz.:

$$\begin{aligned} J_\nu(x) &= \frac{2\left(\frac{x}{2}\right)^\nu}{\sqrt{\pi}\Gamma\left(\nu+\frac{1}{2}\right)} \int_0^{\frac{\pi}{2}} \cos(x\sin\theta)(\cos\theta)^{2\nu} d\sigma \\ &= \frac{2\left(\frac{x}{2}\right)^\nu}{\sqrt{\pi}\Gamma\left(\nu+\frac{1}{2}\right)} \int_0^1 (1-t^2)^{\nu-\frac{1}{2}} \cos xt dt = \frac{\left(\frac{x}{2}\right)^\nu}{\Gamma(\nu+1)} P_\nu\{\cos x\} \end{aligned} \tag{3.7.10}$$

and

$$I_\nu(x) = \frac{\left(\frac{x}{2}\right)^\nu}{\Gamma(\nu+1)} P_\nu \{\operatorname{ch} x\}. \tag{3.7.11}$$

Instead of $\cos(x)$, $\operatorname{ch}(x)$ we now use the generalized trigonometric functions from Section D.i (Appendix) and especially the generalized cosine function and hyperbolic functions of order m:

$$\cos_m(z) = {}_0F_{m-1}\left(\left(\frac{k}{m}\right)_1^{m-1}; -\left(\frac{x}{m}\right)^m\right) = \sum_{j=0}^\infty \frac{(-1)^j x^{mj}}{(mj)!}, \tag{3.7.12}$$

$$h_{1,m}(z) = {}_0F_{m-1}\left(\left(\frac{k}{m}\right)_1^{m-1}; \left(\frac{x}{m}\right)^m\right) = \sum_{j=0}^\infty \frac{x^{mj}}{(mj)!}. \tag{3.7.13}$$

Theorem 3.7.2. *The Poisson-Dimovski transformation* $\mathbf{P_0}$*, defined by (3.7.3), (3.7.5) for* $\nu_k \geq \frac{k}{m} - 1$*,* $k = 1, \ldots, m-1$*, maps the generalized cosine function (3.7.12) of order* $m \geq 2$ *into a normalized hyper-Bessel function of order* $(m-1)$*:*

$$\begin{aligned} j_\nu(x) &= j^{(m-1)}_{\nu_1,\ldots,\nu_{m-1}}(x) = \mathbf{P_0}\{\cos_m x\} \\ &= \sqrt{\frac{m}{(2\pi)^{m-1}}} \prod_{k=1}^{m-1} \Gamma(\nu_k+1) \int_0^1 G^{m-1,0}_{m-1,m-1}\left[\sigma \left| \begin{matrix} (\nu_k)_1^{m-1} \\ \left(\frac{k}{m}-1\right)_1^{m-1} \end{matrix} \right.\right] \cos_m\left(x\sigma^{\frac{1}{m}}\right) d\sigma. \end{aligned} \tag{3.7.14}$$

Therefore, the following integral representation of Poisson type holds for the hyper-Bessel functions (3.7.7) under conditions (3.7.6):

$$J^{(m-1)}_{\nu_1,\ldots,\nu_{m-1}}(x) = \sqrt{\frac{m}{(2\pi)^{m-1}}} \left(\frac{x}{m}\right)^{\sum \nu_k} \int_0^1 G^{m-1,0}_{m-1,m-1}\left[\sigma \left| \begin{matrix} (\nu_k)_1^{m-1} \\ \left(\frac{k}{m}-1\right)_1^{m-1} \end{matrix} \right.\right] \cos_m\left(x\sigma^{\frac{1}{m}}\right) d\sigma. \tag{3.7.15}$$

Analogously, for the modified hyper-Bessel functions (D.5), (D.6) the generalized Poisson-type integrals have the forms:

$$\begin{aligned} i^{(m-1)}_{\nu_1,\ldots,\nu_{m-1}}(x) = \mathbf{P_0}\{h_{1,m}(x)\} &= \sqrt{\frac{m}{(2\pi)^{m-1}}} \prod_{k=1}^{m-1} \Gamma(\nu_k+1) \\ &\times \int_0^1 G^{m-1,0}_{m-1,m-1}\left[\sigma \left| \begin{matrix} (\nu_k)_1^{m-1} \\ \left(\frac{k}{m}-1\right)_1^{m-1} \end{matrix} \right.\right] h_{1,m}\left(x\sigma^{\frac{1}{m}}\right) d\sigma \end{aligned} \tag{3.7.16}$$

and

$$I^{(m-1)}_{\nu_1,\ldots,\nu_{m-1}}(x) = \sqrt{\frac{m}{(2\pi)^{m-1}}} \left(\frac{x}{m}\right)^{\sum \nu_k} \int_0^1 G^{m-1,0}_{m-1,m-1}\left[\sigma \left| \begin{matrix} (\nu_k)_1^{m-1} \\ \left(\frac{k}{m}-1\right)_1^{m-1} \end{matrix} \right.\right] h_{1,m}\left(x\sigma^{\frac{1}{m}}\right) d\sigma, \tag{3.7.17}$$

under the same conditions (3.7.6).

Proof. Consider the initial value problems (3.4.23), (3.4.26), namely:

$$Bu(x) = \lambda u(x); \quad u(0) = 1,\ u'(0) = \cdots = u^{(n-1)}(0) = 0; \tag{3.7.18}$$

and

$$\left(\frac{d}{dx}\right)^m \tilde{u}(x) = \lambda \tilde{u}(x); \quad \tilde{u}(0) = 1,\ \tilde{u}'(0) = \cdots = \tilde{u}^{(m-1)}(0) = 0, \tag{3.7.19}$$

where $\lambda = \pm 1$ and B is the hyper-Bessel differential operator (3.7.1) with condition (3.7.6). As we have already shown in Examples 3.4.6 and 3.4.7, their solutions are the functions:

$$u(x) = j^{(m-1)}_{\nu_1,\ldots,\nu_{m-1}}(x), \quad \tilde{u}(x) = \cos_m(x) \ \text{ if } \lambda = -1, \tag{3.7.20}$$

$$u(x) = i^{(m-1)}_{\nu_1,\ldots,\nu_{m-1}}(x), \quad \tilde{u}(x) = h_{1,m}(x) \ \text{ if } \lambda = 1. \tag{3.7.21}$$

Now we establish that the Poisson-Dimovski transmutation $\mathbf{P}_0$, (3.7.3), (3.7.5) transforms the solution $\tilde{u}(x)$ of (3.7.19) into the solution $u(x)$ of (3.7.18), i.e. $u(x) = \mathbf{P}_0 \tilde{u}(x)$. Indeed, from Lemma 3.5.8,

$$\mathbf{P}_0 \left(\frac{d}{dx}\right)^m \tilde{u}(x) = B\mathbf{P}_0 \tilde{u}(x) - B\mathbf{P}_0 \left\{\tilde{F}\tilde{u}(x)\right\},$$

where $\tilde{F}$ is the Taylor polynomial:

$$\tilde{F}\tilde{u}(x) = \sum_{k=0}^{m-1} \frac{\tilde{u}^{(k)}(0)}{k!} x^k.$$

Let us consider the term $B\mathbf{P}_0 \left\{\tilde{F}\tilde{u}(x)\right\}$. The initial value conditions in (3.7.19) yield

$$\tilde{u}(x) = 1 + \tilde{f} \ \text{ with } \ \tilde{f} \in C^{(m)}_{m-1}$$

and from Corollary 3.5.9, for the m-even function $\tilde{f}$ we have $\tilde{F}\tilde{f} = 0$, whence $\tilde{F}\tilde{u}(x) = \tilde{F}\{1\} = 1$. Further, $\mathbf{P}_0\{1\} = 1$ and then, since $\frac{d}{dx}\{1\} = 0$,

$$B\mathbf{P}_0\tilde{F}\tilde{u}(x) = B\{1\} = \left[x^{-m} \prod_{k=1}^{m-1} \left(x\frac{d}{dx} + m\nu_k\right)\right] \frac{d}{dx}\{1\} = 0,$$

i.e.

$$\mathbf{P}_0 \left(\frac{d}{dx}\right)^m \tilde{u} = B\mathbf{P}_0 \tilde{u}.$$

Then, $\left(\frac{d}{dx}\right)^m \tilde{u} = \lambda \tilde{u}$ yields $Bu = \lambda u$ for $u = \mathbf{P}_0 \tilde{u}$.

Now let us see how $\mathbf{P_0}$ transforms the initial conditions for $\tilde{u}(x)$ in (3.7.19). From Corollary 3.7.1, (3.7.4) and (3.7.4′) give:

$$u(0) = (\mathbf{P_0}\tilde{u})(0) = \tilde{u}(0) = 1,$$
$$u^{(j)}(0) = (\mathbf{P_0}\tilde{u})^{(j)}(0) = cd_j\tilde{u}^{(j)}(0) = 0, \quad j = 1, \ldots, m-1,$$

i.e. $u = \mathbf{P_0}\tilde{u}$ is the solution of the initial value problem (3.7.18).

Combining this result with (3.7.20), respectively (3.7.21), we find

$$j_\nu^{(m-1)} = \mathbf{P_0}\left\{\cos_m(x)\right\}, \quad i_\nu^{(m-1)} = \mathbf{P_0}\left\{h_{1,m}(x)\right\},$$

i.e. representations (3.7.14), (3.7.16) for the normalized hyper-Bessel functions. Multiplying by $\left[\prod_{k=1}^{m-1}\Gamma(\nu_k+1)\right]^{-1}\left(\frac{x}{m}\right)^{\sum\nu_k}$, we obtain the generalized Poisson integrals (3.7.15), (3.7.17).

Corollary 3.7.3. *Using integral representation (3.7.5) of* $\mathbf{P_0}$, *we obtain multiple generalized Poisson type integral representations of the form:*

$$J_{\nu_1,\ldots,\nu_{m-1}}^{(m-1)}(x) = \sqrt{\frac{m}{(2\pi)^{m-1}}}\left(\frac{x}{m}\right)^{\sum\nu_k}\int_0^1\cdots\int_0^1\left[\prod_{k=1}^{m-1}\frac{(1-t_k)^{\nu_k-\frac{k}{m}}t_k^{\frac{k}{m}-1}}{\Gamma\left(\nu_k-\frac{k}{m}+1\right)}\right] \tag{3.7.22}$$
$$\times\cos_m\left[x(t_1\ldots t_{m-1})^{\frac{1}{m}}\right]dt_1\ldots dt_{m-1},$$

$$I_{\nu_1,\ldots,\nu_{m-1}}^{(m-1)}(x) = \sqrt{\frac{m}{(2\pi)^{m-1}}}\left(\frac{x}{m}\right)^{\sum\nu_k}\int_0^1\cdots\int_0^1\left[\prod_{k=1}^{m-1}\frac{(1-t_k)^{\nu_k-\frac{k}{m}}t_k^{\frac{k}{m}-1}}{\Gamma\left(\nu_k-\frac{k}{m}+1\right)}\right] \tag{3.7.23}$$
$$\times h_{1,m}\left[x(t_1\ldots t_{m-1})^{\frac{1}{m}}\right]dt_1\ldots dt_{m-1}.$$

Note. Such a generalization of the Poisson integral has been proposed by Delerue [60, p.259] in the form:

$$J_{\nu_1,\ldots,\nu_{m-1}}^{(m-1)}(x) = \frac{m^{m-\frac{1}{2}}}{(2\pi)^{\frac{m-1}{2}}}\left(\frac{x}{m}\right)^{\nu_1+\cdots+\nu_{m-1}+k-m+1}$$
$$\times\int_0^1\underset{(m-1)}{\cdots}\int_0^1\prod_{i=1}^{m-k-1}\left[\frac{(1-\zeta_i)^{\nu_i-1-\frac{i}{m}}\zeta_i^{k+i}}{\Gamma\left(\nu_i-\frac{i}{m}\right)}\right]$$
$$\times\prod_{i=m-k}^{m-1}\left[\frac{(1-\zeta_i)^{\nu_i-\frac{i}{m}}\zeta_i^{k+i-m}}{\Gamma\left(\nu_i-\frac{i}{m}+1\right)}\right]f_{k+1}^{(m)}(x\zeta_1\ldots\zeta_{m-1})\,d\zeta_1\ldots d\zeta_{m-1}, \tag{3.7.24}$$

where $f_m^{(m)}(x) = \cos_m(x)$; $f_{k+1}^{(m)}(x) = \sin_{n,k+1}(x)$, $k = 0, \ldots, m-2$.

The multiple integral formula (3.7.22) was also obtained by Klučantčev [218, p. 55] (with a misprint in the factor with the Γ-functions). However, *single integral representations involving G-functions, did not appear in the literature until* Dimovski and Kiryakova [80]-[81].

Let us consider *some special cases of the generalized Poisson integrals* (3.7.15), (3.7.17). Naturally, for $m = 1$ formulas (3.7.15), (3.7.17) turn into Poisson integrals (3.7.10), (3.7.11).

Corollary 3.7.4. *According to (D.21), from Theorem 3.7.2 we obtain the following Poisson type integral representation for the n-Bessel functions of Agarwal (D.20), $\nu > -\frac{1}{2}$:*

$$A_{\nu,n}\left\{2^{1-n}\left(\frac{z}{n}\right)^n\right\} = \frac{\sqrt{n\pi^{1-2n}}}{2^{n(\nu+1)-1}\left[\Gamma(\nu+1)\right]^n} \times \int_0^1 G_{2n-1,2n-1}^{2n-1,0}\left[\sigma\left|\begin{matrix}\underbrace{\nu,\nu,\ldots,\nu}_{n},\underbrace{0,\ldots,0}_{n-1}\\ \left(\frac{k}{2n}-1\right)_1^{2n-1}\end{matrix}\right.\right]\cos_{2n}\left(z\sigma^{\frac{1}{2}n}\right)d\sigma$$
$$= \frac{\sqrt{n\pi^{1-2n}}}{2^{n(\nu+1)-1}\left[\Gamma(\nu+1)\right]^n}\int_0^1\cdots\int_0^1\prod_{k=1}^{n}\left[\frac{(1-\zeta_k)^{\nu-\frac{k}{2n}}}{\Gamma\left(\nu-\frac{k}{2n}+1\right)}\right]$$
$$\times\prod_{k=n+1}^{2n-1}\left[\frac{(1-\zeta_k)^{-\frac{k}{2n}}}{\Gamma\left(1-\frac{k}{2n}\right)}\right]\prod_{k=1}^{2n-1}\left[\zeta_k^{\frac{k}{2n}-1}\right]$$
$$\times\cos_{2n}\left[z\left(\zeta_1\ldots\zeta_{2n-1}\right)^{\frac{1}{2}n}\right]d\zeta_1\ldots d\zeta_{2n-1}. \tag{3.7.25}$$

Corollary 3.7.5. *For the Bessel-Clifford functions of n-th order*

$$C^{(n)}_{\lambda_1,\lambda_2,\ldots,\lambda_n}(x) = \sum_{j=0}^{\infty}\frac{(-1)^j x^j}{\Gamma(\lambda_1+j+1)\ldots\Gamma(\lambda_n+j+1)\,j!}, \tag{3.7.26}$$

closely related to functions (3.7.8), Hayek and Hernandez [129] found from Theorem 3.7.2 the Poisson type integral representation

$$C^{(n-1)}_{\lambda_1,\lambda_2,\ldots,\lambda_{n-1}}(x) = \sqrt{\frac{n}{(2\pi)^{n-1}}}\int_0^1 G_{n-1,n-1}^{n-1,0}\left[t\left|\begin{matrix}\lambda_1,\lambda_2,\ldots,\lambda_{n-1}\\ \frac{1}{n}-1,\frac{2}{n}-1,\ldots,\frac{n-1}{n}-1\end{matrix}\right.\right] \times\cos_n\left[n(zt)^{\frac{1}{n}}\right]dt \tag{3.7.27}$$

with $\lambda_k > \frac{k}{n}-1$, $k = 1,\ldots,n-1$; and in particular, for the Bessel-Clifford functions *of third order ($n = 3$), these representations in [128] have the form*

$$C_{m,n} = \frac{9\sqrt{3}}{2\pi\Gamma\left(m+\frac{1}{3}\right)}\Gamma\left(n+\frac{2}{3}\right)\int_0^1\int_0^1 \left(1-\xi^3\right)^{m-\frac{2}{3}}\left(1-\eta^3\right)^{n-\frac{1}{3}}\xi \times f_1(3\xi\eta\sqrt[3]{z})d\xi d\eta, \tag{3.7.28}$$

$$C_{m,n} = \frac{9\sqrt{3}z^{-\frac{2}{3}}}{2\pi\Gamma\left(m-\frac{1}{3}\right)}\Gamma\left(n-\frac{2}{3}\right)\int_0^1\int_0^1 \left(1-\xi^3\right)^{m-\frac{4}{3}}\left(1-\eta^3\right)^{n-\frac{5}{3}}\xi\eta^2 \times f_2(3\xi\eta\sqrt[3]{z})d\xi d\eta, \tag{3.7.29}$$

$$C_{m,n} = \frac{9\sqrt{3}}{2\pi\Gamma\left(m+\frac{1}{3}\right)}\Gamma\left(n-\frac{1}{3}\right)\int_0^1\int_0^1 \left(1-\xi^3\right)^{m-\frac{2}{3}}\left(1-\eta^3\right)^{n-\frac{4}{3}}\eta^2 \times f_3(3\xi\eta\sqrt[3]{z})d\xi d\eta, \tag{3.7.30}$$

where $f_i(x)$, $i = 1, 2, 3$ stand for the generalized trigonometric functions of third order:

$$\begin{aligned} f_1(x) &= \frac{1}{3}\left(e^{-x}+e^{-\omega x}+e^{-\omega^2 x}\right), \quad f_2(x) = \frac{1}{3}\left(e^{-x}+\omega e^{-\omega x}+\omega^2 e^{-\omega^2 x}\right) \\ f_3(x) &= \frac{1}{3}\left(e^{-x}+\omega^2 e^{-\omega x}+\omega e^{-\omega^2 x}\right), \quad \textit{with} \quad \omega^3 = 1. \end{aligned} \tag{3.7.31}$$

3.7.ii. Applications of a differential Poisson-Dimovski transformation for solving hyper-Bessel differential equations

Consider again the Poisson-Sonine-Dimovski transformation (3.5.29)

$$Pf(x) = c\left(\frac{x^\beta}{\beta m}\right)^\lambda I_{\beta,m}^{\left(\frac{k}{m}-1\right),(\lambda-\Delta_k)} f\left(\frac{m}{\beta}x^{\frac{\beta}{m}}\right), \tag{3.7.32}$$

corresponding to the hyper-Bessel operators $\tilde{B} = \left(\frac{d}{dx}\right)^m$ with $\tilde{\gamma}_k = \frac{k}{m} - 1$, $k = 1, \ldots, m$, (see (3.5.22)) and $B = x^{-\beta}\prod_{k=1}^m \left(x\frac{d}{dx} + \beta\gamma_k\right)$ with arbitrary $\beta > 0$; $\gamma_1, \ldots, \gamma_m$. The choice $\lambda = \max_k \Delta_k = \max_k (\tilde{\gamma}_k - \gamma_k)$, i.e. $\lambda - \Delta_k \geq 0$, $k = 1, \ldots, m$, and $\lambda > \frac{\alpha}{\beta} - \frac{\tilde{\alpha}}{\beta} = \frac{\alpha}{\beta} + \frac{1}{m}$, made in Section 3.5.i, ensure that (3.7.32) is a generalized fractional integral (of positive multiorder) and $P : C_{-1} \to C_\alpha$.

Now, instead of taking $\lambda = \max_{1\leq k\leq m} \Delta_k$, we can consider a Poisson-Dimovski transformation with

$$\lambda = \min_{1\leq k\leq m} \Delta_k = \min_k (\tilde{\gamma}_k - \gamma_k) > \frac{\alpha}{\beta} + \frac{1}{m}. \tag{3.7.33}$$

Then, $\lambda - \Delta_k \leq 0$, $k = 1, \ldots, m$, and the *multiorder of "integration"* in (3.7.32) *becomes negative*. This means the symbol $I_{\beta,m}^{\left(\frac{k}{m}-1\right),(\lambda-\Delta_k)}$ should be interpreted as a generalized fractional derivative, according to Definitions 1.5.3 and 1.5.4, see (1.5.15).

The special case of the so-called *spherical hyper-Bessel differential operator* B deserves attention. Namely, suppose B is a hyper-Bessel operator with parameters $\beta > 0$, γ_k, $k = 1, \dots, m$, chosen as follows:

$$\begin{aligned} \beta = m; \quad \gamma_k = \tilde{\gamma}_k - \eta_k = \frac{k}{m} - 1 - \eta_k, \quad k = 1, \dots, m-1; \\ \gamma_m = -\eta_m = 0, \quad \eta_k \geq 0, \quad k = 1, \dots, m \text{ being } \textit{integers}. \end{aligned} \tag{3.7.34}$$

Then, all the differences $\Delta_k = \tilde{\gamma}_k - \gamma_k = \eta_k$ are non negative integers and at least one of them is zero ($\Delta_m = 0$). In accordance with (3.7.33), we choose $\lambda = \min\limits_k \Delta_k = 0$, whence

$$\lambda - \Delta_k = -\Delta_k = \gamma_k - \tilde{\gamma}_k = -\eta_k \leq 0, \quad k = 1, \dots, m-1; \quad \lambda - \Delta_m = 0,$$

i.e. the Poisson-Dimovski transformation (3.7.32) turns into a *generalized "fractional" differentiation operator* of integer multiorder $(\eta_1, \dots, \eta_m)$, since

$$I_{m,m-1}^{(\tilde{\gamma}_k),(-\eta_k)} = D_{m,m-1}^{(\gamma_k),(\eta_k)} = D_\eta I_{m,m-1}^{(\tilde{\gamma}_k),(-\eta_k+\eta_k)} = D_\eta,$$

where

$$D_\eta = \prod_{k=1}^{m-1} \prod_{j=1}^{\eta_k} \left(\frac{1}{m} x \frac{d}{dx} + \tilde{\gamma}_k - \eta_k + j \right) = \prod_{k=1}^{m-1} \prod_{j=1}^{\eta_k} \left(\frac{1}{m} x \frac{d}{dx} + \gamma_k + j \right).$$

It is seen now that *the special choice of a spherical hyper-Bessel differential operator B leads to a Poisson-Dimovski type transformation, represented by pure a differential operator.* We will use it as a transmutation operator for solving corresponding hyper-Bessel differential equations.

Definition 3.7.6. The differential operator $D_\eta : C_a^{(\eta_1 + \cdots + \eta_{m-1})} \to C_a$, defined by

$$D_\eta f(x) = \left[\prod_{k=1}^{m-1} \prod_{j=1}^{\eta_k} \left(x^m \frac{d}{d(x^m)} + \gamma_k + j \right) \right] f(x), \tag{3.7.35}$$

is said to be a *differential Poisson transformation* corresponding to the *spherical hyper-Bessel differential operator* B with parameters (3.7.34).

In particular, $D_\eta : C_a^{(\infty)} \to C_a^{(\infty)}$. Let $\alpha = \max\limits_k [-\beta(\gamma_k + 1)] = \max\limits_k (m\eta_k - k)$ and $\tilde{\mathcal{X}} = C_{-1}^{(\infty)}$, $\mathcal{X} = C_\alpha^{(\infty)}$. Then, for $a = \max(\alpha, -1)$ and $\hat{\mathcal{X}} = C_a^{(\infty)} = \tilde{X} \cap \mathcal{X}$, we have that D_η maps $\hat{\mathcal{X}}$ into itself:

$$D_\eta : \hat{\mathcal{X}} \to \hat{\mathcal{X}}.$$

Lemma 3.7.7. *The differential transform (3.7.35) is a similarity in $C_{\max(\alpha,-1)}^{(\eta_1 + \cdots + \eta_{m-1})}$ from $\tilde{L} = l^m$ to the spherical hyper-Bessel integral operator L with parameters (3.7.34).*

Proof. We use the symbolic representations:

$$D_\eta = I_{m,m-1}^{(\tilde{\gamma}_k),(-\eta_k)}; \quad l^m = \left(\frac{x}{m}\right)^m I_{m,m-1}^{(\tilde{\gamma}_k),(1)}; \quad L = \left(\frac{x}{m}\right)^m I_{m,m-1}^{(\gamma_k),(1)}$$

and similarly to the proof of Theorem 3.6.3, the product rules (1.3.3), (1.3.10), (1.3.11) for the generalized fractional (differ)integrals to obtain

$$I_{m,m-1}^{(\tilde{\gamma}_k),(-\eta_k)}\left(\frac{x}{m}\right)^m I_{m,m-1}^{(\tilde{\gamma}_k),(1)} = \left(\frac{x}{m}\right)^m I_{m,m-1}^{(\tilde{\gamma}_k+1),(-\eta_k)} I_{m,m-1}^{(\tilde{\gamma}_k),(1)} = \left(\frac{x}{m}\right)^m I_{m,m-1}^{(\gamma_k),(1)} I_{m,m-1}^{(\tilde{\gamma}_k),(-\eta_k)},$$

which means:

$$D_\eta l^m = L D_\eta \text{ in } C_{\max(\alpha,-1)}^{(\eta_1+\cdots+\eta_{m-1})}. \tag{3.7.36}$$

Corollary 3.7.8. *If $\tilde{u} \in C_{m+\max(\alpha,-1)}^{(\infty)}$ is a solution of the differential equation $\tilde{u}^{(m)}(x) = \lambda\tilde{u}(x)$, then the image $u = D_\eta\tilde{u} \in C_{m+\max(\alpha,-1)}^{(\infty)}$ is a solution of the spherical hyper-Bessel differential equation (with parameters (3.7.34)): $Bu(x) = \lambda u(x)$, $\lambda = \text{const}$.*

EXAMPLE 3.7.9. We look for a solution of the m-th order ordinary differential equation

$$xy^{(m)}(x) - n.m.y^{(m-1)}(x) + axy(x) = 0, \quad 0 < x < \infty, \tag{3.7.37}$$

where $m \geq 2$, $n > 0$ are integers, $a = \text{const}$.

Divided by $x \neq 0$, the equation takes the form

$$By(x) = -ay(x),$$

where

$$B = x^{-m}\left(x^m \frac{d^m}{dx^m} - nmx^{m-1}\frac{d^{m-1}}{dx^{m-1}}\right) = x^{-m} Q_m\left(x\frac{d}{dx}\right)$$

is a hyper-Bessel differential operator. Since

$$x^m \frac{d^m}{dx^m} = \delta(\delta-1)\dots(\delta-m+1); \quad x^{m-1}\frac{d^{m-1}}{dx^{m-1}} = \delta(\delta-1)\dots(\delta-m+2),$$

where $\delta = x\frac{d}{dx}$, then

$$Q_m(\delta) = \delta(\delta-1)\dots(\delta-m+2)(\delta-mn-m+1)$$

and its zeros are:

$$\mu_k = k-1, \ k = 1,\dots,m-1; \mu_m = m(n+1)-1.$$

Thus, the rearranged parameters of B are:

$$\gamma_1 = \frac{1}{m} - (n+1); \quad \gamma_k = \frac{k}{m} - 1 = \tilde{\gamma}_k, \ k = 2,\dots,m \ \ (\text{i.e. } \gamma_m = 0);$$

$\Delta_1 = \eta_1 = n > 0$ is an integer and $\Delta_k = \eta_k = 0$, $k = 2, \ldots, m$. The corresponding differential transformation of Poisson type (3.7.35) takes the form

$$D_\eta f(x) = \left[\prod_{k=1}^{n}\left(x^m \frac{d}{d(x^m)} + \gamma_1 + j\right)\right] f(x)$$
$$= \left[x^{(n+1)m-1}\left(x^{1-m}\frac{d}{dx}\right)^n x^{1-m}\right] f(x)$$

and if $\tilde{y}(x)$ is a solution of the equation

$$\tilde{y}^{(m)}(x) + a\tilde{y}(x) = 0,$$

then its image

$$y(x) = D_\eta \tilde{y}(x) = \left[x^{(n+1)m-1}\left(x^{1-m}\frac{d}{dx}\right)^n x^{1-m}\right]\tilde{y}(x) \tag{3.7.38}$$

is a solution of equation (3.7.37). *The same relation between both solutions was found by other methods* by I. Zbornik (1957), see Kamke [155, p. 483, problem 5.6].

EXAMPLE 3.7.10. Consider the third order hyper-Bessel differential equation

$$x^2 y'''(x) - 3(p+q)xy''(x) + 3p(3q+1)y'(x) - x^2 y(x) = 0, \quad 0 < x < \infty, \tag{3.7.39}$$

where $p > 0$, $q > 0$ are integers.

We divide by $x^2 \neq 0$ and obtain the equation $By = y$, where $B = x^{-3}Q_3\left(x\frac{d}{dx}\right)$ with $m = \beta = 3$. Since $x^3\frac{d^3}{dx^3} = \delta(\delta-1)(\delta-2)$, $x^2\frac{d^2}{dx^2} = \delta(\delta-1)$, $x\frac{d}{dx} = \delta$, then $Q_3(\delta) = \delta(\delta^2 + b\delta + c)$ with coefficients $b = -3(p+q+1)$, $c = (2+3q)(1+3p)$. The zeros of $Q_3(\mu)$ are $\mu_1 = 3q+2$, $\mu_2 = 3p+1$, $\mu_3 = 0$, therefore: $\gamma_1 = -q-\frac{2}{3}$, $\gamma_2 = -p-\frac{1}{3}$, $\gamma_3 = 0$. If $p < q$, then $\gamma_1 < \gamma_2 < \gamma_3 = 0$. The differences Δ_k are: $\Delta_1 = q > 0$, $\Delta_2 = p > 0$, $\Delta_3 = 0$ and we choose $\lambda = \min\limits_{1\le k\le 3} \Delta_k = 0$. The Poisson type differential transform, corresponding to this spherical hyper-Bessel operator B, is:

$$D_\eta = 3^{(p+q)}\prod_{k=1}^{3}\prod_{j'=1}^{\eta_k}\left(x^3\frac{d^3}{dx^3} + \gamma_k + j'\right)$$
$$= 3^{(p+q)}\prod_{j=0}^{p-1}\left(\frac{1}{3}x\frac{d}{dx} - j - \frac{1}{3}\right)\prod_{i=0}^{q-1}\left(\frac{1}{3}x\frac{d}{dx} - i - \frac{2}{3}\right)$$
$$= \prod_{j=0}^{p-1}\left(x\frac{d}{dx} - 3j - 1\right)\prod_{i=0}^{q-1}\left(x\frac{d}{dx} - 3i - 2\right).$$

If $\tilde{y}(x)$ is a solution of the simplest equation

$$\tilde{y}''' = \tilde{y}, \quad \text{i.e. } \tilde{y}(x) = \sum_{k=1}^{3} c_k e^{\omega_k x} \quad \text{with } \omega_k^3 = 1,$$

then we find a solution of the spherical hyper-Bessel equation (3.7.39) of the form

$$y(x) = D_\eta \tilde{y}(x) = \left[\prod_{j=0}^{p-1} \left(x\frac{d}{dx} - 3j - 1 \right) \prod_{i=0}^{q-1} \left(x\frac{d}{dx} - 3i - 2 \right) \right] \tilde{y}(x), \tag{3.7.40}$$

as found in [155, p. 466].

3.7.iii. Sonine-Dimovski transmutation operators of Weyl type and their use for the explanation of Stokes phenomenon

For solving hyper-Bessel differential equations by the transmutation method, *one can use also the Sonine-Dimovski transformations* S (3.5.44), (3.5.50), or φ (3.5.47), (3.5.52). Then, *the following scheme* is used in general.

The hyper-Bessel equation $Bu = \lambda u$ (or $Bu = f$) is transformed by the Sonine type transformation S into a simpler equation of the form:

$$\left(\frac{d}{dx}\right)^m \tilde{u} = \lambda \tilde{u} + (SFu)^{(m)} \quad \left(\text{or } \left(\frac{d}{dx}\right)^m \tilde{u} = \tilde{f} + (SFu)^{(m)}\right).$$

Depending on the original initial value conditions determining the projector Fu, we could solve the above equation explicitly and then the required solution is found by the inverse Sonine transformation: $u = S^{-1}\tilde{u}$.

Poisson-Sonine-Dimovski transformations can be also considered in spaces other then those of functions defined in a neighbourhood of the origin $x = 0$. If instead of the Riemann-Liouville type fractional integrations (3.5.15) we take their analogues of Weyl type (see Section 1.4), then we obtain transmutation operators between the m-fold integration of Weyl type (near to infinity), see Example (1.1.a*):

$$\begin{aligned} l_*^m f(x) = W^m f(x) &= \int_x^\infty dx_1 \int_{x_1}^\infty dx_2 \ldots \int_{x_{m-1}}^\infty f(x_m)\, dx_m \\ &= \frac{1}{(m-1)!} \int_x^\infty (\tau - x)^{m-1} f(\tau) d\tau \end{aligned} \tag{3.7.41}$$

and the *hyper-Bessel integral operator of Weyl type* (3.6.18):

$$
\begin{aligned}
Mf(x) &= \frac{x^\beta}{\beta m} W_{\beta,m}^{(-\gamma_k-1),(1)} f(x) \\
&= \frac{x^\beta}{\beta m} \int_1^\infty \cdots \int_1^\infty \left(\prod_{k=1}^m \sigma_k^{\gamma_k} \right) f\left[x\left(\sigma_1 \ldots \sigma_m\right)^{\frac{1}{\beta}} \right] d\sigma_1 \ldots d\sigma_m \\
&= \frac{x^\beta}{\beta m} \int_1^\infty G_{m,m}^{m,0} \left[\frac{1}{\sigma} \middle| \begin{matrix} (-\gamma_k+1)_1^m \\ (-\gamma_k)_1^m \end{matrix} \right] d\sigma,
\end{aligned} \tag{3.7.42}
$$

defined in the space $C^*_{\alpha^*}$, $\alpha^* \le \min_k (\beta\gamma_k)$.

In [79], [196] we have shown that there exists a transmutation operator of Weyl type, analogous to the Sonine-Dimovski transformation φ, (3.5.47), (3.5.52). Namely, we have the following result.

Theorem 3.7.11. *The generalized Weyl type fractional integral*

$$
\begin{aligned}
\psi f(x) &= x^{m(\gamma_m+1)-1} \int_1^\infty \cdots \int_1^\infty \left[\prod_{k=1}^{m-1} \frac{(\sigma_k - 1)^{\lambda_k-1} \sigma_k^{\gamma_k}}{\Gamma(\lambda_k)} \right] \\
&\quad \times f\left[x^{\frac{m}{\beta}} \left(\sigma_1 \ldots \sigma_{m-1}\right)^{\frac{1}{\beta}} \right] d\sigma_1 \ldots d\sigma_{m-1} \\
&= \Xi \left[x^{\beta\left(\gamma_m + \frac{m-1}{m}\right)} W_{\beta,m-1}^{(-\gamma_k-\lambda_k),(\lambda_k)} \right] f(x),
\end{aligned} \tag{3.7.43}
$$

where $\Xi : f(x) \to f\left[\left(\frac{\beta}{m} x \right)^{\frac{m}{\beta}} \right]$, *is a transmutation operator* $\psi : C^*_{\alpha^*} \to C^*_{-1}$ *from the m-th order hyper-Bessel operator* B *to* $\tilde{B} = \left(\frac{d}{dx} \right)^m$. *Also, in* $C^{*(m)}_{\alpha^*+\beta} \subset C^*_{\alpha^*}$ *it is a similarity between them:*

$$
\psi B = \left(\frac{\beta}{m} \cdot \frac{d}{dx} \right)^m \psi. \tag{3.7.44}
$$

We state also the following *inversion formula*: if $g(x) = \psi f(x)$, then

$$
\begin{aligned}
f\left[\left(x_1 \ldots x_{m-1}\right)^{\frac{1}{\beta}} \right] &= (-1)^{[\lambda_1]+\cdots+[\lambda_{m-1}]+m-1} \left(\prod_{k=1}^{m-1} x_k^{-\gamma_k} \right) \\
&\quad \times \frac{\partial^{*\ \lambda_1+\cdots+\lambda_{m-1}}}{\partial^* x_1^{\lambda_1} \ldots \partial^* x_{m-1}^{\lambda_{m-1}}} \left\{ g\left[\left(x_1 \ldots x_m\right)^{\frac{1}{m}} \right] \prod_{k=1}^{m-1} x_k^{\frac{k+1}{m}-1} \right\},
\end{aligned} \tag{3.7.45}
$$

where

$$\frac{\partial^{*\lambda_k}}{\partial^* x^{\lambda_k}} c(x_1, \ldots, x_{m-1}) := \frac{\partial^{[\lambda_k]+1}}{\partial x_k^{[\lambda_k]+1}} \cdot \frac{1}{\Gamma(1-\{\lambda_k\})} \times \int_{x_k}^{\infty} \frac{c(x_1, \ldots, \tau_k, \ldots, x_{m-1})}{(\tau_k - x_k)^{\{\lambda_k\}}} d\tau_k.$$

The same considerations hold not only for continuous type spaces $C^*_{\alpha^*}$ but also for thc spaces:

$$\mathfrak{H}^*_{\alpha^*}(\Omega) = \left\{ f(z) = z^q \tilde{f}(z);\ q < \alpha^*,\ \tilde{f} \in \mathfrak{H}^*(\Omega) \right\} \tag{3.7.46}$$

where $\mathfrak{H}^*(\Omega)$ denotes the space of *analytic functions in a domain* Ω*, starlike with respect to infinity point* $z = \infty$. Then, the similarity is:

$$\psi B u(z) = \left(\frac{\beta}{m} \cdot \frac{d}{dz} \right)^m \psi u(z) \text{ for } u(z) \in \mathfrak{H}^*_{\alpha^*+\beta}. \tag{3.7.47}$$

and the same inversion formula (3.7.45) holds.

We are now going to demonstrate an application of the Weyl type Sonine-Dimovski transformations to solutions of differential equations and their different asymptotic behaviour in parts of the complex plane.

An explanation of the Stokes phenomenon for the Airy equation

The Stokes phenomenon concerns the linear combinations of *asymptotic solutions* that represent given continuous solutions of ordinary differential equations of the form

$$\frac{d^2u}{dz^2} + k^2 q(z) u = 0, \quad k = \text{const},\ q(z) \text{ is an entire function of } z, \tag{3.7.48}$$

in different parts of the complex plane. The zeros of $q(z)$ are said to be *turning (transition) points* and the main terms of the asymptotic expansions of the solutions as $|z| \to \infty$ are called *WKB-solutions.* Recently, interest in this phenomenon has been stimulated by its applications in quantum mechanics and the theory of propagation of radiowaves, etc. In this circle of problems, the simplest non trivial case (the so-called model equation) is the *Airy differential equation*

$$\frac{d^2u}{dz^2} - zu = 0. \tag{3.7.49}$$

Important contributions to studying (3.7.49) and some of its generalizations have been made by Heading [130]-[133], Olver [341], Millington [308], Kohno[224], Paris [356], Paris and Wood [357] and others.

Let us consider the Airy equation in the complex domain

$$\Omega = \{\delta < \arg z < 2\pi + \delta, \quad |z| > 0\}, \delta > 0,$$

which is starlike with respect to the infinity point $z = \infty$ and is obtained from $\mathbb{C}$ by putting a cut from $z = 0$ to $z = \infty$. Since $z \neq 0$, equation (3.7.49) can be rewritten as a hyper-Bessel differential equation

$$\frac{1}{z}.\frac{d^2}{dz^2}u = u, \qquad \text{i.e.} \quad Bu = u, \tag{3.7.49$'$}$$

with parameters of B as follows: $m = 2,\ \beta = 3,\ \gamma_1 = -\frac{1}{3},\ \gamma_2 = 0,\ \alpha^* = -1$. *The transmutation operator* (3.7.43) *of Weyl type* then has the form

$$\tilde{u}(z) = \psi u(z) = z\int_1^\infty \frac{(\tau-1)^{-\frac{1}{6}}}{\Gamma(\frac{5}{6})}\tau^{-\frac{1}{3}}u\left(z^{\frac{2}{3}}\tau^{\frac{1}{3}}\right)d\tau, \tag{3.7.50}$$

or

$$\tilde{u}\left(\sqrt{z}\right) = \int_z^\infty \frac{(\zeta - z)^{-\frac{1}{6}}}{\Gamma(\frac{5}{6})}\zeta^{-\frac{1}{3}}u\left(\zeta^{\frac{1}{3}}\right)d\zeta = W^{\frac{5}{6}}\left\{\zeta^{-\frac{1}{3}}u\left(\zeta^{\frac{1}{3}}\right);\ z\right\},$$

where $W^\alpha\{u(\zeta);\ z\}$ denotes the Weyl fractional integral of order $\alpha > 0$, (1.1.b*):

$$W^\alpha u(z) = W^\alpha\{u(\zeta);\ z\} = \frac{1}{\Gamma(\alpha)}\int_z^\infty (\zeta - z)^{\alpha-1}u(\zeta)d\zeta. \tag{3.7.51}$$

On the real positive half-line $\{\arg z = 2\pi, |z| > 0\}$ the operator ψ, as a Sonine-Dimovski type transmutation, transforms the Airy equation (3.7.49$'$) into the simpler equation with constant coefficients

$$\frac{d^2\tilde{u}}{dz^2} - \frac{4}{9}\tilde{u} = 0 \tag{3.7.52}$$

in such a way that its exponentially decreasing solution $\tilde{u}_1(z) = \exp\left(-\frac{2}{3}z\right)$ corresponds to the solution $u_1(z)$ of (3.7.49$'$), exponentially decreasing on $\{\arg z = 2\pi,\ |z| > 0\}$. The latter can be obtained by using the inversion formula for ψ, as a particular case of (3.7.45), namely:

$$u\left(z^{\frac{1}{3}}\right) = \psi^{-1}\tilde{u}(z) = -z^{\frac{1}{3}}\frac{d}{dz}\int_z^\infty \frac{(\zeta - z)^{-\frac{5}{6}}}{\Gamma(\frac{1}{6})}\tilde{u}\left(\sqrt{\zeta}\right)d\zeta. \tag{3.7.53}$$

Thus we find

$$u_1(z) = \frac{2}{3^{\frac{4}{3}}\sqrt{\pi}}z^{\frac{1}{2}}K_{\frac{1}{3}}\left(\frac{2}{3}z^{\frac{3}{2}}\right) = \left(2\sqrt{\pi}\right)3^{\frac{5}{6}}A_i(z),$$

where $K_\nu(z)$ and $A_i(z)$ stand respectively for the Macdonald function (Bessel function of third kind) (C.31) and for the *Airy function of first kind*, (C.32):

$$Ai(z) = \frac{1}{\pi}\sqrt{\frac{z}{3}}K_{\frac{1}{3}}\left(\frac{2}{3}z^{\frac{3}{2}}\right). \tag{3.7.54}$$

Indeed, the Airy function of the first kind $Ai(z)$ is the entire solution of (3.7.49) which is exponentially decreasing as $|z| \to \infty$ on the real positive half-line.

To extend our considerations to the whole domain Ω and also, for the exponentially increasing solutions $u_2(z)$ of the Airy equation and respectively $\tilde{u}_2(z) = \exp\left(\frac{2}{3}z\right)$ of the transformed equation (3.7.52), we are to use the *following relation between the Weyl fractional integrals* (3.7.51) (see [107, II, p. 201, (1)]):

$$W^{\alpha}\left\{u(\gamma\zeta); z\right\} = \gamma^{-\alpha} W^{\alpha}\left\{u(\zeta); \gamma z\right\}, \tag{3.7.55}$$

where γ is an *arbitrary complex constant.*

So, if for example $\gamma = e^{2\pi i}$, then the inverse operator ψ^{-1} is analytically continued like

$$\begin{aligned} u\left(z^{\frac{1}{3}}\right) &= -z^{\frac{1}{3}}\frac{d}{dz}W^{\frac{1}{6}}\left\{\tilde{u}\left(\zeta^{\frac{1}{2}}\right); e^{2\pi i}z\right\} \\ &= e^{-\frac{\pi i}{3}} z^{\frac{1}{3}}\frac{d}{dz}W^{\frac{1}{6}}\left\{\tilde{u}\left(-\zeta^{\frac{1}{2}}\right); z\right\} \end{aligned} \tag{3.7.53$'$}$$

and transforms $\tilde{u}_2(z)$ into the second entire solution of the Airy equation, exponentially increasing for $|z| \to \infty$, $\arg z = 2\pi$, namely:

$$u_2(z) = \frac{2e^{\frac{4\pi i}{3}}}{3^{\frac{4}{3}}\sqrt{\pi}} z^{\frac{1}{2}} K_{\frac{1}{3}}\left(-\frac{2}{3}z^{\frac{3}{2}}\right). \tag{3.7.56}$$

In general, if we consider the set of the entire solutions of both equations (3.7.49) and (3.7.52) in the whole complex plane, then we are to extend analogously transmutation operators ψ and ψ^{-1} by suitable choices of the complex constant γ, depending on the corresponding values of $\arg z$.

For the transformed equation (3.7.52) the pair $\tilde{u}_1(z)$, $\tilde{u}_2(z)$ of linearly independent solutions near the infinity point coincides with the pair of its WKB-solutions. However, the general solution $\tilde{u}(z) = A\tilde{u}_1(z) + B\tilde{u}_2(z)$ (where A, B are arbitrary constants) has an essentially different asymptotic behaviour in different parts of the complex plane. The following notions are accepted: the solution exponentially inceasing in some domain is said to be a *dominant solution* and the exponentially decreasing solution there a *subdominant solution*, while a solution whose modulus of the exponential multiplier is 1 is said to be a *neutral solution.*

Our transmutation formula (3.7.50) relating the corresponding solutions $u(z)$ and $\tilde{u}(z)$ shows the following phenomenon. If we consider the Airy equation in the domain Ω (with the necessary cut to keep the single-valued solutions), i.e. for $\delta < \arg z < 2\pi + \delta$, then equation (3.7.52) should be considered in the corresponding domain $\tilde{\Omega} : \frac{3}{2}\delta < \arg z < 3\pi + \frac{3}{2}\delta$. This latter domain is separated into four sections depending on the values of $\arg z$, namely:

$$(\tilde{\mathrm{I}}) = \left(\frac{3}{2}\delta, \frac{\pi}{2}\right), \ (\widetilde{\mathrm{II}}) = \left(\frac{\pi}{2}, \frac{3\pi}{2}\right), \ (\widetilde{\mathrm{III}}) = \left(\frac{3\pi}{2}, \frac{5\pi}{2}\right), \ (\widetilde{\mathrm{IV}}) = \left(\frac{5\pi}{2}, 2\pi + \frac{3}{2}\delta\right).$$

So, in the domains $\left(\widetilde{\mathrm{I}}\right)$, $\left(\widetilde{\mathrm{II}}\right)$ the solution $\tilde{u}_1(z)$ is subdominant and $\tilde{u}_2(z)$ is dominant, and conversely, in $\left(\widetilde{\mathrm{III}}\right)$, $\left(\widetilde{\mathrm{IV}}\right)$ $\tilde{u}_1(z)$ is dominant and $\tilde{u}_2(z)$ – subdominant. In *Poincaré sense*, in each subdomain we should retain from the linear combination $\tilde{u}(z) = A\tilde{u}_1(z) + B\tilde{u}_2(z)$ the dominant solution only. When crossing the lines $\{\Re z = 0\}$, i.e. $\left\{\arg z = \frac{\pi}{2}, \frac{3\pi}{2}, \frac{5\pi}{2}\right\}$, the property of dominance is transferred from one solution to the other, i.e. they change roles. Such lines are called *anti-Stokes lines* (or *conjugated Stokes lines*). Since none of the solutions is dominant on them, we retain both solutions in the case. On the other hand, the dominant or subdominant properties of the solutions are best revealed on the so-called *Stokes lines*, the lines $\{\operatorname{Im} z = 0\}$, i.e. $\{\arg z = \pi, 3\pi, 5\pi\}$. If we denote by $(*)$ the dominant solution in each subdomain, then we can express the asymptotic behaviour of the general solution of equation (3.7.52) in the following way:

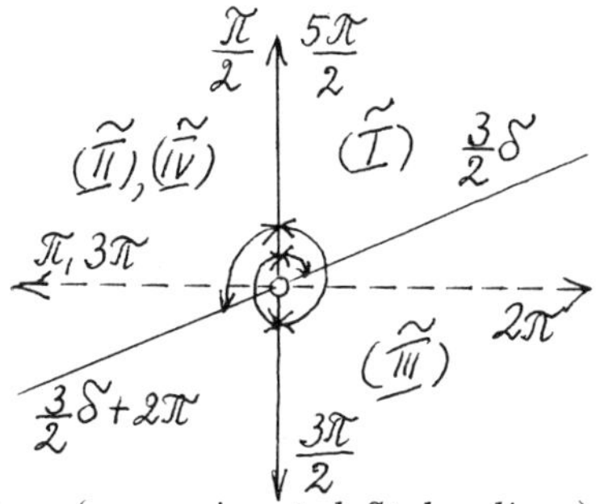

$$\begin{aligned}
\left(\widetilde{\mathrm{I}}\right) &: A\tilde{u}_1(z) + B\tilde{u}_2^*(z), \quad \text{i.e. } \tilde{u}(z) \sim B\tilde{u}_2^*(z),\\
\left(\widetilde{\mathrm{II}}\right) &: A\tilde{u}_1^*(z) + B\tilde{u}_2(z), \quad \text{i.e. } \tilde{u}(z) \sim A\tilde{u}_1^*(z),\\
\left(\widetilde{\mathrm{III}}\right) &: A\tilde{u}_1(z) + B\tilde{u}_2^*(z), \quad \text{i.e. } \tilde{u}(z) \sim B\tilde{u}_2^*(z),\\
\left(\widetilde{\mathrm{IV}}\right) &: A\tilde{u}_1^*(z) + B\tilde{u}_2(z), \quad \text{i.e. } \tilde{u}(z) \sim A\tilde{u}_1^*(z),
\end{aligned}$$

while on the three anti-Stokes lines we have:

$$\tilde{u}(z) \sim A\tilde{u}_1(z) + B\tilde{u}_2(z).$$

Further, we consider the asymptotic behaviour of the solutions $u_1(z)$, $u_2(z)$ and $u(z) = Au_1(z) + Bu_2(z)$ of the Airy equation in the corresponding four subdomains of $\Omega = \{\delta < \arg z < 2\pi + \delta\}$. For this equation, the Stokes lines correspond to $\left\{\operatorname{Im} z^{\frac{3}{2}} = 0\right\}$, i.e. they are $\left\{\arg z = \frac{2\pi}{3}, \frac{4\pi}{3}, 2\pi\right\}$ and the anti-Stokes lines are $\left\{\arg z = \frac{\pi}{3}, \pi, \frac{5\pi}{3}\right\}$ when $\left\{\Re z^{\frac{3}{2}}\right\} = 0$.

Using transmutation formula (3.7.50), we find that in the subdomains of Ω:

$$(\mathrm{I}) = \left(\delta, \frac{\pi}{3}\right), \ (\mathrm{II}) = \left(\frac{\pi}{3}, \frac{\pi}{3}\right), \ (\mathrm{III}) = \left(\pi, \frac{5\pi}{3}\right), \ (\mathrm{IV}) = \left(\frac{5\pi}{3}, 2\pi + \delta\right)$$

the general solution $u(z)$ of the Airy equation has the following asymptotic behaviour for large values of $|z|$:

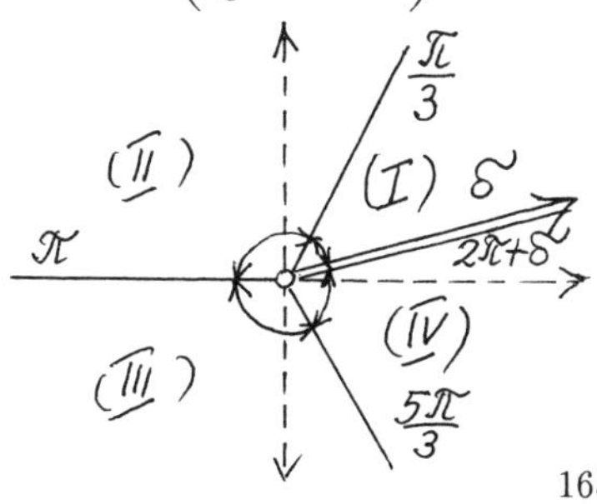

$$\begin{aligned}
(\mathrm{I}) &: \ u(z) \sim M\mathfrak{U}_+^*,\\
(\mathrm{II}) &: \ u(z) \sim N\mathfrak{U}_-^*,\\
(\mathrm{III}) &: u(z) \sim (M + iN)\mathfrak{U}_+^*,\\
(\mathrm{IV}) &: u(z) \sim iM\mathfrak{U}_-^*,
\end{aligned}$$

where M and N are new arbitrary constants and

$$\mathfrak{U}_{\pm} = z^{-\frac{1}{4}} \exp\left(\pm\frac{2}{3}z^{\frac{3}{2}}\right) \tag{3.7.57}$$

are the *WKB-solutions of the Airy equation.* Therefore, on the anti-Stokes lines, we have:

$$\begin{aligned} \left\{\arg z = \frac{\pi}{3}\right\} &: \ u(z) \sim M\mathfrak{U}_{+} + N\mathfrak{U}_{-}, \\ \{\arg z = \pi\} &: \ u(z) \sim (M+iN)\mathfrak{U}_{+} + N\mathfrak{U}_{-}, \\ \left\{\arg z = \frac{5\pi}{3}\right\} &: \ u(z) \sim (M+iN)\mathfrak{U}_{+} + iM\mathfrak{U}_{-}. \end{aligned}$$

Thus, the Stokes phenomenon can be better explained for the case of the Airy equation, if we consider seven smaller subdomains (subsectors) to which the domain Ω is divided by all the six Stokes and anti-Stokes lines and by the cut $\{\arg z = \delta,\ \arg z = 2\pi + \delta\}$. We denote them correspondingly by (1), (2), ... , (7).

Then, we obtain *the following asymptotic formulas*:

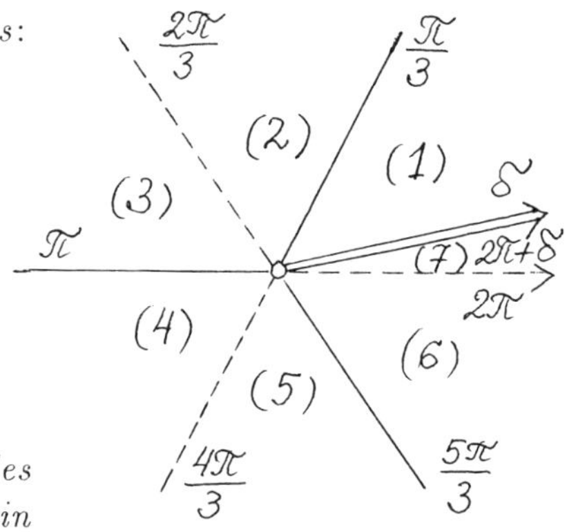

$$\begin{aligned} (1) &= \left(\delta, \tfrac{\pi}{3}\right): \ u \sim M\mathfrak{U}_{+}^{*} + N\mathfrak{U}_{-}, \\ (2) &= \left(\tfrac{\pi}{3}, \tfrac{2\pi}{3}\right): \ u \sim M\mathfrak{U}_{+} + N\mathfrak{U}_{-}^{*}, \\ (3) &= \left(\tfrac{2\pi}{3}, \pi\right): \ u \sim (M+iN)\,\mathfrak{U}_{+} + N\mathfrak{U}_{-}^{*}, \\ (4) &= \left(\pi, \tfrac{4\pi}{3}\right): \ u \sim (M+iN)\,\mathfrak{U}_{+}^{*} + N\mathfrak{U}_{-}, \\ (5) &= \left(\tfrac{4\pi}{3}, \tfrac{5\pi}{3}\right): \ u \sim (M+iN)\,\mathfrak{U}_{+}^{*} + iM\mathfrak{U}_{-}, \\ (6) &= \left(\tfrac{5\pi}{3}, 2\pi\right): \ u \sim (M+iN)\,\mathfrak{U}_{+} + iM\mathfrak{U}_{-}^{*}, \\ (7) &= (2\pi, 2\pi+\delta): \ u \sim (M+iN)\,\mathfrak{U}_{+} + N\mathfrak{U}_{-}^{*}. \end{aligned}$$

From these formulas we can derive the *following rules for determining the leaping changed of the constants in the general asymptotic solution.*

When crossing a Stokes line, let us denote by N_s and N_s' the old and the new coefficients associated with the subdominant solution and by N_d the coefficient associated with the dominant solution. Then, $N_s' = N_s + c.N_d$, where c is the so-called *Stokes multiplier.* In particular, for the Airy equation the Stokes multipliers corresponding to each of the three Stokes lines are given by $c_1 = c_2 = c_3 = i$. Therefore, on each Stokes line $\left\{\arg z = \frac{2\pi}{3}, \frac{4\pi}{3}, 2\pi\right\}$ the change of the coefficient of the subdominant solution is according to the formula $N_s' = N_s + iN_d$. On the other hand, on crossing the cut $\{\arg z = \delta,\ \arg z = 2\pi + \delta\}$, the rule is the following: the coefficients of the dominant and subdominant solutions, multiplied by $(-i)$, change their places. *This scheme, described here with the help of the transmutation method, coincides with the scheme for the change of the coefficients given earlier by Heading,* see [130]-[132]. To keep the continuity of the asymptotic expansion of the solution in the different subdomains, a change of coefficient always takes place in the subdomain where the corresponding solution is subdominant and therefore, *its modulus is less than the error* $\exp\left(\frac{2}{3}z^{\frac{3}{2}}\right)\mathcal{O}\left(|z|^{-\frac{7}{4}}\right)$, $|z| \to \infty$, of the asymptotic approximation of the dominant solution. In this way, *only a formal leap in the coefficient of the subdominant solution appears, characterizing the so-called Stokes phenomenon.*

Some generalizations of the Airy equation, being also hyper-Bessel differential equations (and therefore the same Sonine-Dimovski transmutation technique is applicable), like (cf. Example 3.4.10):

$$\frac{d^n u(z)}{dz^n} = (-1)^n z^m u(z), \quad z^n \frac{d^n u(z)}{dz^n} = z^q u(z)$$

have been considered by other techniques by Heading [133], Kohno [224], Paris [356], Paris and Wood [357] and other authors. *The following general problem can be stated.*

OPEN PROBLEM 3.7.12. By using suitable analogues of the Poisson-Sonine-Dimovski transformations of Weyl type (like (3.7.43)) and the transmutation method, give a description and explanation of the corresponding Stokes phenomenon for the general hyper-Bessel differential equations of m-th order: $Bu(z) = \lambda u(z)$. In this case, the Stokes phenomenon will concern the change of the coefficients in the linear combinations of Delerue's hyper-Bessel functions (3.4.19) (or Meijer's $G^{1,0}_{0,m}$-functions), representing the general asymptotic solution as $|z| \to \infty$.

3.8. Solution of the non homogeneous hyper-Bessel equation

Having in mind the results of Section 3.4 (Corollary 3.4.5), it is seen that *the only problem that remains open* in order to solve the general Cauchy (initial value) problem

$$\begin{cases} By(x) = \lambda y(x) + f(x), \quad \lambda \neq 0, \quad f \neq 0 \\ \lim\limits_{x \to +0} B_k y(x) = b_k, \ k = 1, \ldots, m \ \text{ of the form (3.4.4)}, \end{cases} \tag{3.8.1}$$

is to find an explicit *particular solution* of the non homogeneous hyper-Bessel differential equation $By(x) = \lambda y(x) + f(x)$, for instance, *satisfying zero initial conditions.* We show that the G-functions play an important role again.

Theorem 3.8.1. *Suppose the parameters of the hyper-Bessel operator (3.1.3) are arranged in a decreasing (increasing) order, for example*

$$\beta > 0 \; ; \; \gamma_1 \geq \gamma_2 \geq \ldots \geq \gamma_m \; , \quad \text{i.e. } \alpha := \max_k[-\beta(\gamma_k + 1)] = -\beta(\gamma_m + 1). \tag{3.8.2}$$

Then the initial value problem

$$\begin{cases} By(x) = \lambda y(x) + f(x) \; , \qquad f \in C_\alpha, \\ y(0) = y'(0) = \ldots = y^{(m-1)}(0) = 0 \end{cases} \tag{3.8.3}$$

has a solution $y \in C^{(m)}_{\alpha+\beta}$, *given by the series*

$$y(x) = \frac{x^\beta}{\beta^m} \sum_{r=0}^{\infty} \left(\frac{\lambda x^\beta}{\beta^m} \right)^r . \; G_r(x) \; , \quad \textit{convergent for } \; 0 \leq x < \infty, \tag{3.8.4}$$

with $G_r(x)$, $r = 0, 1, 2, ...,$ *standing for the integrals of G-functions:*

$$G_r(x) = \int_0^1 G_{m,m}^{m,0}\left[\sigma \left| \begin{matrix} (\gamma_k + r + 1)_1^m \\ (\gamma_k)_1^m \end{matrix} \right.\right] f\left(x\sigma^{\frac{1}{\beta}}\right) d\sigma. \tag{3.8.5}$$

Proof. To solve the problem (3.8.3) we use *the transmutation method* (see Section 3.5). In this special case we are interested in a transmutation operator, transforming the simplest m-th order hyper-Bessel differential operator $\tilde{B} = D^m = \left(\frac{d}{dx}\right)^m$ into the general operator B of the form (3.1.3). We shall use the Poisson-Dimovski transformation $P: C_{-1} \to C_\alpha$ which is a similarity between the integral operators l^m and L, right inverses of $\tilde{B}$ and B, respectively:

$$\begin{aligned} l^m \tilde{f}(x) &= \int_0^x \frac{(x-t)^{(m-1)}}{\Gamma(m)} \tilde{f}(t) dt, \\ Lf(x) &= \frac{x^\beta}{\beta^m} \int_0^1 G_{m,m}^{m,0}\left[\sigma \left| \begin{matrix} (\gamma_i + 1)_1^m \\ (\gamma_i)_1^m \end{matrix} \right.\right] f(x\sigma^{\frac{1}{\beta}}) d\sigma, \end{aligned} \tag{3.8.6}$$

namely:

$$P l^m \tilde{f}(x) = L P \tilde{f}(x) \quad \text{for each} \quad \tilde{f} \in C_{-1}; \tag{3.8.7}$$

see Sections 3.5 and 3.6. As shown there, this *Poisson-Dimovski transform* can be represented more simply as a generalized fractional integral, i.e. as an integral transform with a G-function kernel:

$$Pf(x) = \sqrt{\frac{m}{(2\pi)^{m-1}}} \left[\prod_{k=1}^m \Gamma(\gamma_k + 1)\right] I_{\beta,m-1}^{(-1+\frac{k}{m}),\left(1+\gamma_k-\gamma_m-\frac{k}{m}\right)} f\left(\frac{m}{\beta} x^{\frac{\beta}{m}}\right). \tag{3.8.8}$$

Since P transforms the simpler initial value problem for $\tilde{y} \in C_{-1}$:

$$\begin{cases} D^m \tilde{y}(x) = \tilde{y}^{(m)}(x) = \lambda \tilde{y}(x) + \tilde{f}(x), & \tilde{f} \in C_{-1}, \\ \tilde{y}(0) = \tilde{y}'(0) = ... = \tilde{y}^{(m-1)}(0) = 0 \end{cases} \tag{3.8.9}$$

into initial value problem (3.8.3) for $y \in C_\alpha$ (cf. the proof of Theorem 3.7.2, or Kiryakova [196] and Dimovski and Kiryakova [80]), *then* $P\tilde{y}(x) = y(x)$ *will be the required solution.*

By the techniques of the operational calculus and the Laplace transform one can find the solution of (3.8.9) *as the Duhamel convolution*

$$\tilde{y}(x) = h(x) * \tilde{f}(x) = \int_0^x h(x-t) \tilde{f}(t) dt, \tag{3.8.10}$$

where $h(x)$ is a hyperbolic function $h_{m,m}(x)$ of order m, see (D.13) (if $\lambda > 0$), or a generalized trigonometric function $k_{m,m}(x)$, see (D.10), (if $\lambda < 0$) but in both cases it is represented by means of the $G^{1,0}_{0,m}$-function, and therefore by the power series:

$$\begin{aligned} h(x) &= \sqrt{\frac{(2\pi)^{m-1}}{m}} G^{1,0}_{0,m}\left[-\lambda\left(\frac{x}{m}\right)^m \middle| \left(\frac{i}{m}\right)_0^{m-1}\right] \\ &= \frac{x^{m-1}}{(m-1)!}\,{}_0F_{m-1}\left[\left(1+\frac{i}{m}\right)_1^{m-1};\ \lambda\left(\frac{x}{m}\right)^m\right] \\ &= \sum_{r=0}^{\infty}\frac{\lambda^r x^{mr+m-1}}{(mr+m-1)!}, \qquad \text{convergent for } |x|<\infty. \end{aligned} \tag{3.8.11}$$

Therefore, the solution of the simpler problem (3.8.9) can be written down by the series

$$\tilde{y}(x) = \sum_{r=0}^{\infty}\lambda^r\left[\int_0^x \frac{(x-t)^{mr+m-1}}{(mr+m-1)!}\tilde{f}(t)\,dt\right] = \sum_{r=0}^{\infty}\lambda^r\ l^{m(r+1)}\ \tilde{f}(x) \tag{3.8.12}$$

and according to (3.8.10) and [307], it belongs to $C^{(m)}_{m-1} \subset C_{-1}$. Then, the solution $y(x)$ as a P-image of (3.8.12) with $\tilde{f}(x) = P^{-1}f(x)$ is:

$$y(x) = P\tilde{y}(x) = \sum_{r=0}^{\infty}\lambda^r\left[Pl^{m(r+1)}P^{-1}\right]\ f(x) = \sum_{r=0}^{\infty}\lambda^r\ L^{r+1}\ f(x)\ , \tag{3.8.13}$$

since $Pl^{m(r+1)}P^{-1} = L^{r+1}$, due to (3.8.7), and putting P under the sign of the series is justified by the absolute convergence of the series $h(x)$ and the integral operator P in C_{-1}. To find $y(x)$ in the form (3.8.4) it remains only to use the integral representation for the powers L^{r+1}, $r = 0, 1, 2, \ldots$ of the hyper-Bessel integral operator L (see (3.8.6)), found by McBride [289], [291] and later by Dimovski and Kiryakova [79], namely:

$$\begin{aligned} L^{\delta}f(x) &= \left(\frac{x^{\beta}}{\beta m}\right)^{\delta} I^{(\gamma_k),(\delta,\ldots,\delta)}_{\beta,m} f(x) \\ &= \left(\frac{x^{\beta}}{\beta m}\right)^{\delta}\int_0^1 G^{m,0}_{m,m}\left[\sigma\middle|\begin{matrix}(\gamma_k+\delta)_1^m\\(\gamma_k)_1^m\end{matrix}\right] f(x\sigma^{\frac{1}{\beta}})\,dt\ ,\quad \delta>0\ . \end{aligned} \tag{3.8.14}$$

The absolute convergence of series (3.8.4) for all $x \geq 0$ can be derived from conditions (3.8.2), $f \in C_{\alpha}$ and the asymptotic behaviour of the G-functions involved in $G_r(x)$, as the same has been made for the generalized fractional integrals $I^{(\gamma_k),(\delta_k)}_{\beta,m}$ in C_{α} (see Chapter 1). The same result can however be shown alternatively, to illustrate the advantages of the transmutation method. Let us go back to (3.8.10) and denote $g(x) = Ph(x)$, i.e. $h(x) = P^{-1}g(x)$ and $\tilde{f}(x) = P^{-1}f(x)$. The function $g(x)$ can be evaluated as a P-image of the $G^{1,0}_{0,m}$-function, according to the general formula (A.28) and this gives:

$$g(x) = \left[(-\lambda)^{\gamma_m+\frac{1}{m}-1}\prod_{k=1}^{m}\Gamma(\gamma_k+1)\right] G^{1,0}_{0,m}\left[-\lambda\frac{x^{\beta}}{\beta m}\middle| -\gamma_m;\ (-\gamma_k)_1^{m-1}\right], \tag{3.8.15}$$

therefore

$$g(x)=\mathcal{O}\left(x^{q}\right) \quad \text{as} \quad x \to 0 \quad \text{with} \quad q=\max_{k}\left(-\beta \gamma_{k}\right)=-\beta \gamma_{m}=\alpha+\beta,$$

i.e. $g(x)=Ph(x) \in C_{\alpha+\beta}^{(m)}$. Further, we obtain

$$y(x)=P\tilde{y}(x)=P\left[\left(P^{-1} f\right) *\left(P^{-1} g\right)\right](x):=f(x) \tilde{*} g(x). \tag{3.8.16}$$

Since the convolution $(*)$ is a convolution of l^m, and P is a similarity from l^m to L, then according to the theorem for convolutions of similar operators (Dimovski [72, p. 36]; see Chapter 2, (2.2.47) with $T=P^{-1}$), the new operation $(\tilde{*})$ is a convolution of the hyper-Bessel integral operator L in C_α, i.e. $(\tilde{*}) : C_\alpha \times C_\alpha \longrightarrow C_\alpha$. In particular, for $f \in C_\alpha$, $g \in C_{\alpha+\beta}^{(m)}$, by the arguments in Dimovski [72], Bozhinov [38] and Bozhinov and Dimovski [39], if follows that

$$y(x)=P\tilde{y}(x)=f(x) \tilde{*} g(x) \in C_{\alpha+\beta}^{(m)}.$$

The same conclusion holds for the equivalent series representation (3.8.4) of $y(x)$. This completes the proof.

Examples. Solutions to various special cases of hyper-Bessel ODEs can be obtained from the above general results.

EXAMPLE 3.8.2. Most of the elementary and special functions of mathematical physics are only special cases of Meijer's G-function. Thus, let us consider the case when $f(x)$ is an *arbitrary G-function* in C_α, that is:

$$f(x)=G_{\sigma,\tau}^{\mu,\nu}\left[x \left| \begin{matrix} (c_\alpha)_1^\sigma \\ (d_\beta)_1^\tau \end{matrix} \right.\right], \quad 0 \leq \sigma \leq \tau, \tag{3.8.17}$$
$$\min_{1 \leq \beta \leq \mu} d_\beta+\min_{1 \leq k \leq m} \gamma_k>-1,$$

where γ_k, $k=1,\ldots,m$, are the parameters of operator B (3.1.1) and for brevity we assume $\beta=m$. Using general formula (A.28) one can evaluate the integrals $G_r(x)$, $r=0,1,\ldots$, in (3.8.4), namely:

$$G_r(x)=A.G_{m\sigma+m,m\tau+m}^{m\mu,m\mu+m}\left[\left(\frac{x}{m^{\sigma-\tau}}\right) \left| \begin{matrix} \Delta\left(m, c_\alpha\right)_1^\nu ;\ \left(-\gamma_k\right)_1^m ;\ \Delta\left(m, c_\alpha\right)_{\nu+1}^\sigma \\ \Delta\left(m, d_\beta\right)_1^\mu ;\ \left(-\gamma_k-r-1\right)_1^m ;\ \Delta\left(m, d_\beta\right)_{\mu+1}^\tau \end{matrix} \right.\right], \tag{3.8.18}$$

where the constant A stands for

$$A=(2\pi)^{(1-m)\left[\mu+\nu-\frac{\sigma+\tau}{2}\right]} m^{\sum d_\beta-\sum c_\alpha+1+\frac{\sigma-\tau}{2}}$$

and the symbols $\Delta(m,c)$, $\Delta\left(m, c_i\right)_k^l$ denote

$$\Delta(m, c)=\left(\frac{c}{m}, \frac{c+1}{m}, \ldots, \frac{c+m-1}{m}\right),$$
$$\Delta\left(m, c_i\right)_k^l=\left(\Delta\left(m, c_k\right), \Delta\left(m, c_{k+1}\right), \ldots, \Delta\left(m, c_l\right)\right).$$

Each of the G-functions (3.8.18) belong to C_α, and the solution $y(x)$ takes the form of a series, convergent for all $x \geq 0$:

$$y(x) = A\left(\frac{x}{m}\right)^m \sum_{r=0}^{\infty} \left[\lambda\left(\frac{x}{m}\right)^m\right]^r \tag{3.8.19}$$
$$\times\, G^{m\mu,m\nu+m}_{m\sigma+m,m\tau+m}\left[\lambda\left(\frac{x}{m^{\sigma-\tau}}\right)\middle| \begin{matrix}\Delta\left(m,c_\alpha\right)_1^\nu;\ (-\gamma_k)_1^m;\ \Delta\left(m,c_\alpha\right)_{\nu+1}^\sigma \\ \Delta\left(m,d_\beta\right)_1^\mu;\ (-\gamma_k-r-1)_1^m;\ \Delta\left(m,d_\beta\right)_{\mu+1}^\tau\end{matrix}\right].$$

More details on the series in G-functions can be found in Luke [26, I] and some numerical computational methods are discused in Luke [26, II] and Mathai and Saxena [286]-[287].

EXAMPLE 3.8.3. The *simplest* but very *common* example in the case of the equation $By(x) = \lambda y(x) + f(x)$ of arbitrary order $m > 1$ is with a right-hand side function

$$f(x) = x^p, \quad p > \alpha \Rightarrow f \in C_\alpha. \tag{3.8.20}$$

Then integrals (3.8.5) turn into: $G_r(x) = b_{p,r}x^p$, $r = 0, 1, 2, \dots$, with constants $b_{p,r}$ (see formula (B.4)):

$$b_{p,r} = \int_0^1 G^{m,0}_{m,m}\left[\sigma \middle| \begin{matrix}\left(\gamma_k + r + 1 + \frac{p}{\beta}\right)_1^m \\ \left(\gamma_k + \frac{p}{\beta}\right)_1^m\end{matrix}\right] d\sigma = \left[\prod_{k=1}^m \left(\gamma_k + 1 + \frac{p}{\beta}\right)\cdot\left(\gamma_k + 2 + \frac{p}{\beta}\right)_r\right]^{-1}$$

and series (3.8.4) takes the form

$$\begin{aligned} y(x) &= \left[\prod_{k=1}^m (\beta\gamma_k + p + \beta)\right]^{-1} x^{p+\beta} \sum_{r=0}^\infty \frac{(1)_r}{\prod\limits_{k=1}^m \left(\gamma_k + 2 + \frac{p}{\beta}\right)_r} \frac{\left(\frac{\lambda x^\beta}{\beta^m}\right)^r}{r!} \\ &= \left[\prod_{k=1}^m (\beta\gamma_k + p + \beta)\right]^{-1} x^{p+\beta}\, {}_1F_m\left[1;\ \left(\gamma_k + 2 + \frac{p}{\beta}\right)_1^m;\ \lambda\frac{x^\beta}{\beta^m}\right],\quad 0 \leq x < \infty, \end{aligned} \tag{3.8.21}$$

that is, the solution of the particular problem

$$By = \lambda y + x^p, \quad y(0) = y'(0) = \cdots = y^{(m-1)}(0) = 0$$

is the *generalized hypergeometric function* $x^{p+\beta}{}_1F_m\left(\frac{\lambda x^\beta}{\beta^m}\right)$ in (3.8.21), and it is also a $G^{1,1}_{1,m+1}$-function of Meijer.

In the following we confine ourselves to some well-known examples related to ODEs involving the classical *second-order operator of Bessel* B_ν (3.3.a) with $\beta = m = 2$, $\gamma_1 = \frac{\nu}{2}$, $\gamma_2 = -\frac{\nu}{2}$, namely to equations of the form

$$x^2y'' + xy' - \left(\nu^2 \mp x^2\right)y = F(x), \tag{3.8.22}$$

i.e.

$$B_\nu y = \lambda y + f \quad \text{with } \lambda = \pm 1 \text{ and } f(x) = x^{-2}F(x). \tag{3.8.22$'$}$$

It is worth pointing out that the corresponding Poisson-Dimovski transformation (3.5.25), (3.5.29), transmuting D^2 into B_ν, is the *well-known Poisson transformation* (3.5.38):

$$P_\nu f(x) = \left[2\frac{\left(\frac{x}{2}\right)^\nu}{\sqrt{\pi}}\Gamma\left(\nu + \frac{1}{2}\right)\right]\int_0^1 (1-\sigma^2)^{\nu-\frac{1}{2}} f(x\sigma)\, d\sigma, \quad \nu > -\frac{1}{2}.$$

In this case Theorem 3.8.1 provides the following solutions:

EXAMPLE 3.8.4. Consider equation (3.8.22′) with $\lambda = +1$, $f(x) = x^{\mu-1}$, that is, $F(x) = x^{\mu+1}$, assuming $\nu > \mu + 1 > 0$. Series (3.8.4) turns into

$$\begin{aligned} y(x) &= \left(\frac{x}{2}\right)^2 \frac{x^{\mu-1}}{4} \sum_{r=0}^{\infty} \frac{(1)_r}{\left(\frac{\mu+\nu+3}{2}\right)_r \left(\frac{\mu-\nu+3}{2}\right)_r} \cdot \frac{\left(\frac{-x^2}{4}\right)_r}{r!} \\ &= [(\mu+\nu+1)(\mu-\nu+1)]^{-1}\, x^{\mu+1}\, {}_1F_2\left(1; \frac{\mu-\nu+3}{2}, \frac{\mu+\nu+3}{2}; -\frac{x^2}{4}\right) \\ &= s_{\mu,\nu}(x), \qquad \text{the Lommel function (C.8),} \end{aligned} \tag{3.8.23}$$

known to be a solution $y(x)$ with $y(0) = y'(0) = 0$ of the equation

$$x^2y'' + xy' + \left(x^2 - \nu^2\right) y = x^{\mu+1} \tag{3.8.24}$$

(compare with [272, I, p. 217-218, (1), (16)], [106, II, 7.5.5, (68)].

EXAMPLE 3.8.5. Analogously (even as a special case of (3.8.23)), the solution $y(x)$ of equation (3.8.22′) with $\lambda = +1$, $f(x) = \left[\Gamma\left(\frac{1}{2}\right)\Gamma\left(\nu+\frac{1}{2}\right)\right]^{-1}\left(\frac{x}{2}\right)^{\nu-1}$, that is, of equation

$$x^2y'' + xy' + \left(x^2 - \nu^2\right) y = \frac{4\left(\frac{x}{2}\right)^{\nu+1}}{\Gamma\left(\frac{1}{2}\right)\Gamma\left(\nu+\frac{1}{2}\right)}, \qquad \nu > 0, \tag{3.8.25}$$

is obtained as a *Struve function* $H_\nu(x)$ (see [272, I, p. 217-218, (3), (21)], [106, II, 7.5.4, (62)]:

$$\begin{aligned} y(x) &= \left[\Gamma\left(\frac{3}{2}\right)\Gamma\left(\nu+\frac{3}{2}\right)\right]^{-1}\left(\frac{x}{2}\right)^{\nu+1} {}_1F_2\left(1;\ \frac{3}{2};\ \nu+\frac{3}{2};\ -\frac{x^2}{4}\right) \\ &= \left[\pi 2^{\nu-1}\left(\frac{1}{2}\right)_\nu\right]^{-1} s_{\nu,\nu}(x) = H_\nu(x)\ . \end{aligned} \tag{3.8.26}$$

EXAMPLE 3.8.6. An example of the Bessel equation (3.8.22′) with $\lambda = -1$ and another kind of right-hand side $f(x) = \exp(-x)x^{\mu-1}$ is the following:

$$x^2y'' + xy' - \left(x^2 + \nu^2\right) y = \exp(-x).x^{\mu+1}, \quad \nu > \mu + 1 > 0. \tag{3.8.27}$$

It has the solution (as a special case of series (3.8.4) in Theorem 3.8.1):

$$\begin{aligned} y(x) = & [(\mu-\nu+1)(\mu+\nu+1)]^{-1}\exp(-x)x^{\mu+1} \\ & \times {}_2F_2\left(1,\mu+\frac{3}{2};\mu-\nu+2,\mu+\nu+2;2x\right) = h_{\mu,\nu}(x), \end{aligned} \tag{3.8.28}$$

the so-called *"associated Bessel function"* (see [272, I, p. 219, (25), (27)].

It is easily seen that the solutions (3.8.23), (3.8.26) and (3.8.28), related to the Bessel differential operator B_ν, belong to the subspace

$$C_\nu^{(\infty)} \subset C_\nu^{(2)} \text{ of the space } C_{\alpha+\beta}^{(m)} = C_{-\beta\gamma_m}^{(m)} = C_\nu^{(2)},$$

as suggested by Theorem 3.8.1.

3.9. The Obrechkoff integral transform. Relation to hyper-Bessel operators, operational properties, inversion formulas, Abelian theorems

One of the most commonly used mathematical techniques for justifying the Heaviside operational calculus for operator the $D = \frac{d}{dx}$ is based on the *Laplace transform*

$$F(z) = \mathfrak{L}\{f(x);z\} = \int_0^\infty \exp(-zx)f(x)dx. \tag{3.9.1}$$

In this sense, the key role is played by its property to algebraize the operator $D = \frac{d}{dx}$, its powers and inverse integration operator $lf(x) = \int_0^x f(\tau)d\tau$, namely:

$$\begin{aligned} \mathfrak{L}\left\{\left(\frac{d}{dx}\right)^m f(x);z\right\} = & z^m\mathfrak{L}\{f(x);z\} - z^{m-1}f(0) - \dots \\ & - zf^{(m-2)}(0) - f^{(m-1)}(0), \quad m=1,2,\dots, \end{aligned} \tag{3.9.2}$$

$$\mathfrak{L}\{lf(x);z\} = \frac{1}{z}\mathfrak{L}\{f(x);z\}. \tag{3.9.3}$$

Relations (3.9.2) allow us to reduce initial value problems for ordinary differential equations with constant coefficients to an algebraical problem for solving linear equations. The solution of the original equation is then found by the *complex inversion formula of Riemann-Mellin*:

$$f(x) = \frac{1}{2\pi i}\int_{\rho-i\infty}^{\rho+i\infty} \exp(xz)F(z)dz, \tag{3.9.4}$$

or by the *real inversion formula of Post-Widder*:

$$f(x) = \lim_{k\to\infty} \frac{(-1)^k}{k!}\left(\frac{k}{x}\right)^{k+1} F^{(k)}\left(\frac{k}{x}\right). \tag{3.9.5}$$

More about the theory of Laplace transform can be found for example, in [510], [107], [88], [90]. An integral transform of this kind, corresponding to the classical Bessel differential operator B_ν, (3.3.a) is the *Meijer integral transform*

$$F_\nu(z) = \mathfrak{K}_\nu\{f(x); z\} = 2\int_0^\infty (zx)^{\frac{\nu}{2}} K_\nu\left(2\sqrt{zx}\right) f(x)dx, \tag{3.9.6}$$

where K_ν stands for the Bessel function of third kind (Macdonald's function) (C.31).

Many authors have introduced integral transforms, suitable in developing operational calculi for Bessel type differential operators of special kinds, as those mentioned in Examples (3.3.c), (3.3.d), (3.3.g), (3.3.h): Ditkin [85], Meller [298], Prudnikov [366], Ditkin and Prudnikov [86], [89], Botashev [37], Krätzel [235]-[239], etc. These are integral transforms generalizing the Laplace transform (3.9.1) and Meijer transform (3.9.6) but all of them turn out to be quite special cases of *an integral transformation introduced and investigated by the Bulgarian mathematician N.Obrechkoff* [339] *in 1958*, earlier than the other authors mentioned above. For details see Section 3.10.

The original form of this transform is:

$$F(t) = \int_0^\infty \Phi(tx)f(x)dx \tag{3.9.7}$$

with a kernel-function ($p \geq 1$)

$$\Phi(x) = \int_0^\infty \dots \int_0^\infty u_1^{\beta_1} u_2^{\beta_2} \dots u_p^{\beta_p} \exp\left(-u_1 - \dots - u_p - \frac{x}{u_1 \dots u_p}\right) du_1 \dots du_p. \tag{3.9.8}$$

Obrechkoff himself did not consider it as an integral transform but only as a formula for representing real functions $F(t)$ on the real half-line, a continuation of his results from [337]-[338]. It was Dimovski [64] who acknowledged the priority of Obrechkoff in these matters. He observed that a transformation (3.9.7) can be used as a transform basis of an operational calculus for the Bessel type operator (3.3.i) of order $(p+1)$. In [65]-[70] Dimovski proposed a suitable modification of this transform to make it useful for the same purposes but for the most general hyper-Bessel differential operator (3.1.2)-(3.1.3):

$$\begin{gathered} B = x^{\alpha_0}\frac{d}{dx}x^{\alpha_1 \dots \frac{d}{dx}x^{\alpha_m}} = x^{-\beta}\prod_{k=1}^{m}\left(x\frac{d}{dx} + \beta\gamma_k\right), \\ \beta = m - (\alpha_0 + \dots \alpha_m) > 0, \quad 0 < x < \infty. \end{gathered} \tag{3.9.9}$$

He found also a convolution of this modified transform, analogous to the Duhamel convolution and other useful properties in the complex plane. We now call it simply the Obrechkoff transform.

Definition 3.9.1. (Dimovski [64]-[70]). Let $\gamma_1 \leq \gamma_2 \leq \cdots \leq \gamma_m$ be a non decreasing sequence of real numbers, $\beta > 0$ and

$$K(z) = \int_0^\infty \underset{(m-1)}{\cdots} \int_0^\infty \exp\left(-u_1 - \cdots - u_{m-1} - \frac{z}{u_1 \dots u_{m-1}}\right) \prod_{k=1}^{m-1} u_k^{\gamma_m - \gamma_k - 1} du_1 \dots du_{\mu-1}. \tag{3.9.10}$$

The integral transform of the form

$$F(z) = \mathfrak{O}\{f(x); z\} = \beta \int_0^\infty x^{\beta(\gamma_m+1)-1} K\left[(zx)^\beta\right] f(x) dx \tag{3.9.11}$$

is said to be an *Obrechkoff integral transform*, corresponding to the hyper-Bessel operator (3.9.9).

Transform (3.9.11) is defined in the subspace of the so-called *Obrechkoff transformable functions*

$$\Omega = C_\alpha^{\exp} = \left\{ f \in C_\alpha; f(x) = \mathcal{O}\left(\exp\left(\lambda x^{\frac{\beta}{m}}\right)\right), x \to \infty \right\} \subset C_\alpha, \tag{3.9.12}$$

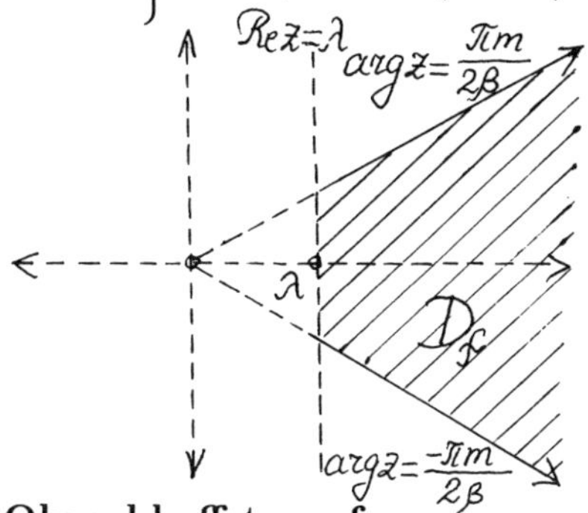

where λ is a real number, $\alpha = \max_k [-\beta(\gamma_k + 1)] = -\beta(\gamma_1 + 1)$, and C_α is the basic space (3.1.1). For $f(x) \in \Omega$, *the image* $F(z) = \mathfrak{O}\{f(x); z\}$ *is an analytic function* in the truncated angle domain

$$D_f = \left\{ z : \Re z > \lambda, \ |\arg z| < \frac{\pi m}{2} \beta \right\}.$$

3.9.i. New definition and some properties of the Obrechkoff transform

First, the Obrechkoff transform was studied by Dimovski [64]-[70] by using multiple integral representations (3.9.8), (3.9.10) of the kernel-functions $\Phi(x)$, $K(z)$.

Now, we show that the kernel-function

$$\lambda(z, x) = x^{\beta(\gamma_m+1)-1} K\left[(zx)^\beta\right] \tag{3.9.13}$$

of the Obrechkoff transform (3.9.11):

$$\mathfrak{O}\{f(x); z\} = \beta \int_0^\infty \lambda(z, x) f(x) dx \tag{3.9.11'}$$

is nothing but a Meijer's G-function.

Lemma 3.9.2. ([190], [75], [192], [196]) *The kernel-function of the Obrechkoff transform (3.9.11)-(3.9.11′) is a Meijer's G-function of order* $(m, 0; 0, m)$, *namely:*

$$\lambda(z,x) = z^{-\beta(\gamma_m+1)+1} G^{m,0}_{0,m}\left[(zx)^{\beta} \,\middle|\, \left(\gamma_k - \frac{1}{\beta} + 1\right)_1^m\right]. \tag{3.9.14}$$

Proof. To find representation (3.9.14) we shall use the techniques of *Mellin transform*

$$\mathfrak{M}\{g(x);s\} = \mathfrak{M}(s) = \int_0^\infty g(x)x^{s-1}dx$$

for functions $g(x)$ satisfying the condition

$$\int_0^\infty |g(x)|x^{\rho_0-1}dx < +\infty \quad \text{with some} \quad \rho_0 = \Re s. \tag{3.9.15}$$

Then, its complex inversion formula has the form

$$g(x) = \frac{1}{2\pi i}\int_{\rho-i\infty}^{\rho+i\infty} \mathfrak{M}(s)x^{-s}ds, \quad \rho > \rho_0. \tag{3.9.16}$$

Here, let us put $g(x) := \lambda(z,x)$ and check the validity of condition (3.9.15). We have

$$\int_0^\infty |g(x)|x^{\rho_0-1}dx = \int_0^\infty x^{\beta(\gamma_m+1)-1+\rho_0-1}K\left[(zx)^{\beta}\right]dx. \tag{3.9.17}$$

Since for $\gamma_1 \leq \cdots \leq \gamma_m$ (see [106, I, §1.1, (5)]):

$$K(0) = \prod_{k=1}^{m-1}\left[\int_0^\infty \exp(-u_k)\, u_k^{\gamma_m-\gamma_k-1}du_k\right] = \prod_{k=1}^{m-1}\Gamma(\gamma_m-\gamma_k) < \infty,$$

then in a neighbourhood of $x = 0$, $K\left[(zx)^{\beta}\right]$ is bounded and the convergence of (3.9.15) depends on the behaviour of the other multiplier $x^{\beta(\gamma_m+1)+\rho_0-2}$ as $x \to +0$. It is necessary to establish that ρ_0 can be choosen so that $\beta(\gamma_m+1)+\rho_0-2 > -1$. It is satisfied for $\rho_0 > \alpha+1$. On the other hand, from Obrechkoff's result [339] for the asymptotic behaviour of $\Phi(x)$ as $x \to \infty$, it follows that

$$K(x) \sim \frac{(\sqrt{2\pi})^{m-1}}{\sqrt{m}} x^{\sum_1^m \frac{\gamma_i}{m} - \gamma_m - \frac{m-1}{2m}} \exp\left(-m.x^{\frac{1}{m}}\right), \quad x \to \infty \tag{3.9.18}$$

and therefore,

$$|\lambda(z,x)| \leq \frac{A}{x^b}, \quad b > 1, \quad A = \text{const}, \quad \text{as } x \to \infty.$$

The established behaviour of the integrand of (3.9.17) near singular points $x = 0$, and $x = \infty$ yields that condition (3.9.15) is satisfied with $\rho_0 > \alpha + 1$. Therefore, $\lambda(z,x)$ is a Mellin transformable function and its Mellin image

$$\mathfrak{M}(s) = \mathfrak{M}\{\lambda(z,x); s\} = \int_0^\infty x^{s-1}\lambda(z,x)dx$$

is a uniformly convergent integral in $\{\Re s > \rho_0\}$ defining and analytic function there. Then, substituting $\lambda(z,x)$ with (3.9.13), (3.9.10) and changing the order of integrations, we obtain

$$\mathfrak{M}(s) = \int_0^\infty x^{\beta(\gamma_m+1)+s-2}dx$$
$$\times \left\{ \int_0^\infty \underset{(m-1)}{\cdots} \int_0^\infty \exp\left(-u_1 - \cdots - u_{m-1} - \frac{(zx)^\beta}{u_1 \ldots u_{m-1}}\right) \prod_{k=1}^{m-1} u_k^{\gamma_k - \gamma_m - 1} du_1 \ldots du_{m-1} \right\}$$
$$= \prod_{k=1}^{m-1} \left[\int_0^\infty \exp(-u_k)\, u_k^{\gamma_k - \gamma_m - 1} du_k \right] \int_0^\infty x^{\beta(\gamma_m+1)+s-2} \exp\left(-\frac{z^\beta x^\beta}{u_1 \ldots u_{m-1}}\right) dx.$$

By routine substitutions and using [106, I, §1.1, (5)] again, the inner integral in x can be easily seen to have the value

$$\left(\frac{z^\beta}{u_1 \ldots u_{m-1}}\right)^{-\left(\gamma_m+1+\frac{s-1}{\beta}\right)} \Gamma\left(\gamma_m + \frac{s-1}{\beta} + 1\right).$$

The same procedure for the integrals in u_k, $k = 1, \ldots, m-1$, shows that they are equal to

$$\Gamma\left(\gamma_k + \frac{s-1}{\beta} + 1\right), \quad k = 1, \ldots, m-1.$$

Therefore, we obtain

$$\mathfrak{M}(s) = \mathfrak{M}\{\lambda(z,x); s\} = z^{-\beta\left(\gamma_m + \frac{s-1}{\beta}+1\right)} \prod_{k=1}^{m} \Gamma\left(\gamma_k + \frac{s-1}{\beta} + 1\right)$$

and by inversion formula (3.9.16):

$$\lambda(z,x) = z^{-\beta\left(\gamma_m - \frac{1}{\beta}+1\right)} \frac{1}{2\pi i} \int\limits_{\rho-i\infty}^{\rho+i\infty} (zx)^{-s} \prod_{k=1}^{m} \Gamma\left(\gamma_k + \frac{s-1}{\beta} + 1\right) ds, \quad \rho > \rho_0 > \alpha + 1. \tag{3.9.19}$$

This is the new representation of the kernel-function by means of a contour Mellin-Barnes type integral. From Definition A.3 (see (A.7)-(A.8)) and property (A.14) (Appendix), the required representation is:

$$\lambda(z,x) = z^{-\beta(\gamma_m+1)+1} G^{m,0}_{0,m}\left[(zx)^\beta \,\middle|\, \left(\gamma_k - \frac{1}{\beta}+1\right)_{k=1}^{m}\right].$$

Corollary 3.9.3. *The original Obrechkoff kernel (3.9.8) is represented as follows:*

$$\Phi(x) = G^{p+1,0}_{0,p+1}\left[x \,\middle|\, (\beta_k+1)_{k=1}^{p}, 0\right] \tag{3.9.20}$$

and according to Corollary B.5, equation (B.11), it satisfies the hyper-Bessel ODEs of $(p+1)$-th order:

$$\begin{aligned} &x^{-1} \prod_{k=1}^{p} \left(x\frac{d}{dx} - \beta_k - 1\right)\left(x\frac{d}{dx}\right)\Phi(x) \\ &= x^{\beta_p+1}\frac{d}{dx} x^{-\beta_p+\beta_{p-1}+1}\frac{d}{dx} \dots x^{-\beta_2+\beta_1+1}\frac{d}{dx} x^{-\beta_1}\frac{d}{dx} = (-1)^{p-1}\Phi(x) \end{aligned} \tag{3.9.21}$$

(cf. the same equation in [339] obtained by rather long calculations).

Lemma 3.9.2 gives the reason for the following *new definition of the Obrechkoff transform* (3.9.11) *as a special case of the so-called G-transforms*

$$\mathfrak{G}\{f(x);z\} = A \int\limits_0^\infty G^{m,n}_{p,q}\left[c(zx)^\beta \,\middle|\, \begin{matrix} (a_k)_1^p \\ (b_j)_1^q \end{matrix}\right] f(x)dx, \tag{3.9.22}$$

considered by Rooney [403] and, recently, by other authors too.

Definition 3.9.4. The G-transformation

$$\begin{aligned} F(z) = \mathfrak{O}\{f(x);z\} &= \beta \int\limits_0^\infty \lambda(z,x) f(x)dx \\ &= \beta z^{-\beta(\gamma_m+1)+1} \int\limits_0^\infty G^{m,0}_{0,m}\left[(zx)^\beta \,\middle|\, \left(\gamma_k - \frac{1}{\beta}+1\right)_1^m\right] f(x)dx \end{aligned} \tag{3.9.23}$$

is said to be an *Obrechkoff integral transform* corresponding to the hyper-Bessel differential operator (3.9.9).

This new definition allows as to simplify most of the calculations and proofs of the already known properties of the Obrechkoff transform and also to obtain new ones easily. For example, using representation (3.9.23) one can find the Obrechkoff images of G-functions as well as of simpler functions from Ω, namely:

Lemma 3.9.5. *If* $p > -\beta\left[\min_k(\gamma_k+1)\right]$, $\lambda > 0$:

$$\mathfrak{O}\{x^p; z\} = z^{-\beta(\gamma_m+1)-p}\prod_{k=1}^{m}\Gamma\left(\gamma_k+\frac{p}{\beta}+1\right); \tag{3.9.24}$$

$$\begin{aligned}\mathfrak{O}\left\{x^p\exp\left(-\lambda x^{\frac{\beta}{m}}\right);\ z\right\} =& z^{-\beta(\gamma_m+1)-p}(2\pi)^{\frac{1-m}{2}}\sqrt{m}\\ &\times G^{m,m}_{m,m}\left[\left(\frac{m}{\lambda}\right)^m z^\beta \middle| \begin{matrix}\left(1-\frac{k}{m}\right)^{m-1}_1\\ \left(\gamma_k+\frac{p}{\beta}+1\right)^m_1\end{matrix}\right];\end{aligned} \tag{3.9.25}$$

also, if $G^{\mu,\nu}_{\sigma,\tau}\left(x^{\beta\frac{l}{k}}\right)\in\Omega$, $l>0$ *and* $k>0$ *being integers,*

$$\begin{aligned}&\mathfrak{O}\left\{G^{\mu,\nu}_{\sigma,\tau}\left[x^{\beta\frac{l}{k}}\middle|\begin{matrix}(c_i)^\sigma_1\\ (d_j)^\tau_1\end{matrix}\right];\ z\right\} = g z^{-\beta(\gamma_m+1)-1}\\ &\times G^{lm+k\nu,k\mu}_{k\tau,lm+k\sigma}\left[\frac{z^{\beta l}}{l^{lm}k^{k(\sigma-\tau)}}\middle|\begin{matrix}\Delta\left(k,1-d_j-\frac{k}{l\beta}\right)^\tau_{j=1}\\ \Delta\left(l,\gamma_k-\frac{1}{\beta}+1\right)^m_{k=1},\ \Delta\left(k,1-c_i-\frac{k}{l\beta}\right)^\sigma_{i=1}\end{matrix}\right],\end{aligned} \tag{3.9.26}$$

where $\Delta(k,c) = \left(\frac{c}{k},\frac{c+1}{k},\dots,\frac{c+k-1}{k}\right)$ *and the constant* g *is:*

$$g = (2\pi)^{\left(\mu+\nu-\frac{\sigma+\tau}{2}\right)(1-k)+\frac{m}{2}(1-l)}k^{\sum_{j=1}^{\tau}\left(d_j+\frac{k}{l\beta}-1\right)-\sum_{i=1}^{\tau}\left(c_i+\frac{k}{l\beta}-1\right)+\frac{\sigma-\tau}{2}+2}\,l^{\sum_{k=1}^{l}\left(\gamma_k-\frac{1}{\beta}+1\right)-\frac{m}{2}}.$$

Note. Formula (3.9.26) incorporates *practically all the necessary Obrechkoff images.*

The Obrechkoff transform satisfies *some integral and differential rules*, analogous to (3.9.2), (3.9.3) and given below, that allow its use as a transform basis of operational calculi for hyper-Bessel operators. First, they have been proved in Dimovski [67], Kiryakova [190] and Dimovski and Kiryakova [74] by using the original definition (3.9.11). Later, in Kiryakova [196] they found simpler proofs by means of the G-function technique and we illustrate this down.

Theorem 3.9.6. *The Obrechkoff transform (3.9.11), (3.9.23) satisfies the following operational rules related to the hyper-Bessel integral and differential operators* L, B: *if*

$f \in \Omega$ is an Obrechkoff transformable function, then

$$\mathfrak{O}\{Lf(x);z\} = \frac{1}{\beta^m z^\beta}\mathfrak{O}\{f(x);z\}, \tag{3.9.27}$$

and more generally, if $\lambda > 0$,

$$\mathfrak{O}\left\{L^\lambda f(x)\right\} = \frac{1}{\left(\beta^m z^\beta\right)^\lambda}\mathfrak{O}\{f(x);z\}. \tag{3.9.28}$$

If additionally to $f \in \Omega$,

$$f \in \mathcal{X}_F = \left(\text{span}\right]\left\{x^{-\beta\gamma_k}\right\}_1^m \oplus C_{\alpha+\beta}^{(m)}\right) \subset C_\alpha^{(m)} \subset C_\alpha$$

(cf. (3.2.11)) and also $\gamma_i - \gamma_j \neq l$, $l = 0, \pm 1, \pm 2, \dots$, for example :

$$\gamma_1 < \gamma_2 < \cdots < \gamma_m < \gamma_1 + 1, \qquad \text{(cf. (3.2.3))}$$

then the following differential property holds:

$$\begin{aligned}\mathfrak{O}\{Bf(x);z\} &= \beta^m z^\beta \mathfrak{O}\{f(x);z\} \\ &- \sum_{i=1}^m \left\{\left[\beta^i z^{\beta(\gamma_i - \gamma_m)} \prod_{j=1}^{i-1}\Gamma\left(\gamma_j - \gamma_i + 1\right) \prod_{j=i+1}^m \Gamma\left(\gamma_j - \gamma_i\right)\right] \lim_{x \to +0} B_i f(x)\right\}.\end{aligned} \tag{3.9.29}$$

Proof. To prove (3.9.28) we use representations (3.9.23) and (3.6.17) of the Obrechkoff transform and fractional powers L^λ, $\lambda > 0$. Then, changing the order of integrations, we find:

$$\begin{aligned}\mathfrak{O}\left\{L^\lambda f(x);z\right\} &= \frac{\beta z^{-\beta(\gamma_m+1)+1}}{\beta^{m\lambda}} \int_0^\infty G_{0,m}^{m,0}\left[(zx)^\beta \middle| \left(\gamma_k - \frac{1}{\beta} + 1\right)_1^m\right] x^{\beta(\lambda-1)}dx \\ &\times \int_0^x G_{m,m}^{m,0}\left[\left(\frac{\tau}{x}\right)^\beta \middle| \begin{matrix}(\gamma_k+\lambda)_1^m \\ (\gamma_k)_1^m\end{matrix}\right] f(\tau) d\left(\tau^\beta\right) \\ &= \frac{\beta z^{-\beta(\gamma_m+1)+1}}{\beta^{m\lambda}} \int_0^\infty f(\tau)d\left(\tau^\beta\right) \int_\tau^\infty x^{\beta(\lambda-1)} G_{0,m}^{m,0}\left[(zx)^\beta \middle| \left(\gamma_k - \frac{1}{\beta}+1\right)_1^m\right] \\ &\times G_{m,m}^{0,m}\left[\frac{1}{\tau^\beta}x^\beta \middle| \begin{matrix}(1-\gamma_k)_1^m \\ (1-\gamma_k-\lambda)_1^m\end{matrix}\right] dx.\end{aligned}$$

Since $G_{m,m}^{0,m}\left(\frac{x^\beta}{\tau^\beta}\right) \equiv 0$ for $x < \tau$ (see (A.15) and (A.12)), then the inner integral can be taken in limits from $x = 0$ to $x = \infty$ and evaluated according to formula (A.28). This gives

$$\mathfrak{O}\left\{L^{\lambda}f(x);z\right\} = \frac{\beta z^{-\beta(\gamma_m+1)+1}}{\beta^{m\lambda}} z^{-\beta\lambda} \int_0^\infty G^{m,0}_{0,m}\left[(z\tau)^\beta \middle| \left(\gamma_k - \frac{1}{\beta}+1\right)_1^m\right] f(\tau)d\tau$$

$$= \frac{1}{\beta^{m\lambda} z^{\beta\lambda}} \mathfrak{O}\{f(x);z\},$$

which is (3.9.28). For $\lambda = 1$, this gives (3.9.27) and means that the *Obrechkoff transformation transforms the hyper-Bessel integral operator into an algebraical operation*: multiplication by $\left(\beta^m z^\beta\right)^{-1}$.

Now let us establish the property of the Obrechkoff transform to algebraize the differential operator B of Bessel type. Since we require additionally $f \in C^m_{\alpha+\beta}$, then we can apply (3.9.27) to the function $f_1(x) := Bf(x) \in \Omega$ namely:

$$\frac{1}{\beta^m z^\beta}\mathfrak{O}\{Bf(x);z\} = \mathfrak{O}\left\{L\left(Bf(x)\right);z\right\} = \mathfrak{O}\left\{f(x);z\right\} - \left\{Ff(x);z\right\},$$

where F denotes the *defining projector (initial operator)*: $F = I - LB$, having the explicit form (3.2.12):

$$Ff(x) = \sum_{k=1}^{m} c_k(f) x^{-\beta\gamma_k}$$

with coefficients $c_k(f)$, (3.2.13) depending on the initial value conditions

$$\lim_{x\to+0} B_i f(x) = \lim_{x\to+0} x^{\alpha_i}\frac{d}{dx}x^{\alpha_i+1}\cdots\frac{d}{dx}x^{\alpha_m}f(x)$$

$$= \lim_{x\to+0}\left[x^{\beta\gamma_i}\prod_{k=i+1}^{m}\left(x\frac{d}{dx}+\beta\gamma_k\right)\right]f(x).$$

To obtain (3.9.28), it remains to use formula (3.9.24) for the images of the functions $y_k = x^{-\beta\gamma_k}$, $k = 1,\ldots,m$.

Corollary 3.9.7. *According to Corollary 3.2.4,*

$$F|_{\mathcal{X}_B} \equiv 0, \quad \textit{where} \quad \mathcal{X}_B = C^{(m)}_{\alpha+\beta} \subset \mathcal{X}_F \subset C^{(m)}_\alpha \subset C_\alpha.$$

That is why, if we suppose

$$f(x) \in C^{(m)}_{\alpha+\beta} \quad \textit{and} \quad f(x) = \mathcal{O}\left(\exp\lambda x^{\frac{\beta}{m}}\right) \quad \textit{as} \quad x \to +\infty,$$

then the initial operator F and the term involving it in (3.9.29) vanish and the differential property takes the following simpler form:

$$\mathfrak{V}\{Bf(x);z\} = \beta^m z^\beta \mathfrak{O}\{f(x);z\}, \tag{3.9.30}$$

i.e. the hyper-Bessel differential operator B goes into a multiplication by $\beta^m z^\beta$, under the Obrechkoff transform in $C^{(m)}_{\alpha+\beta} \cap \Omega$.

An interesting problem is *to find a convolution of the Obrechkoff integral transform*, i.e. an operation in Ω, playing the same role as the Duhamel convolution

$$(f * g)(x) = \int_0^x f(x-t)g(t)dt$$

for the Laplace transform:

$$\mathfrak{L}\{(f*g)(x);z\} = \mathfrak{L}\{f(x);z\}.\mathfrak{L}\{g(x);z\}.$$

This problem was solved by Dimovski [64]-[70] by showing that *the convolution* (3.6.4), (3.6.6)*: $f*g = T(f\circ g)$ of the hyper-Bessel integral operator L in C_α is also a convolution of the Obrechkoff transform in $\Omega \subset C_\alpha$*, namely:

Theorem 3.9.8. (Dimovski [64]-[70]) *The operation* $\left(\overset{\mathcal{O}}{*}\right) : \Omega \times \Omega \to \Omega$*, defined by*

$$\left(f \overset{\mathcal{O}}{\circ} g\right)(x) = T(f \circ g)(x), \tag{3.9.31}$$

with T being the correcting operator (3.6.2) and auxiliary operation $(\circ)$ as in (3.6.1), is a convolution of the Obrechkoff transform, namely:

$$\mathfrak{O}\left\{(f \overset{\mathcal{O}}{\circ} g)(x);z\right\} = \mathfrak{O}\{f(x);z\}.\mathfrak{O}\{g(x);z\}. \tag{3.9.32}$$

Note. According to Theorem 3.6.1, the operator T *can be represented also as a generalized fractional integral* (3.6.5):

$$T = x^{\beta\gamma_m} I_{\beta,s}^{(2\gamma_k),(\gamma_m-\gamma_k)} \quad (\text{if } \gamma_1 \le \gamma_2 \le \cdots \le \gamma_s < \gamma_{s+1} = \cdots = \gamma_m),$$

i.e. by means of a single integral involving a G-function and then, the expression for the operation $\left(\overset{\mathcal{O}}{*}\right)$ is written in a simpler and more concise form.

Note. A convolution of the Obrechkoff transform, satisfying (3.9.32) can be obtained also by using the transmutation method, namely this is the operation

$$(f \otimes g)(x) = \varphi^{-1}[\varphi f * \varphi g](x), \tag{3.9.33}$$

where φ is the Sonine-Dimovski transformation (3.5.47), written by means of the G-functions as (3.5.52).

It is interesting to note that the same Sonine-Dimovski transformation φ from B to $D^m = \left(\frac{d}{dx}\right)^m$ serves as a *relationship between the Obrechkoff and Laplace transforms*, as proved in Dimovski and Kiryakova [75]-[76].

Theorem 3.9.9. *If $\varphi : C_\alpha \to C_{-1}$ is the Sonine-Dimovski transform (3.5.47), (3.5.52) and $f(x) \in \Omega \subset C_\alpha$, then the following relationship holds:*

$$\mathfrak{O}\left\{f(x); \left(\frac{z}{m}\right)^{\frac{m}{\beta}}\right\} = \sqrt{(2\pi)^{m-1}m}\ \mathfrak{L}\{\varphi f(x); z\}. \tag{3.9.34}$$

Proof. A direct proof is shown in [75]-[76]. Here we propose a simpler proof, following from representation (3.5.52) of the transmutation operator φ, namely:

$$\varphi f\left(x^{\frac{\beta}{m}}\right) = x^{\beta(\gamma_m+1)-\frac{\beta}{m}} I_{\beta,m-1}^{(\gamma_k),(\lambda_k)} f(x)$$

with $\lambda_k = \gamma_m - \gamma_k + \frac{k}{m} > 0$, $k = 1, \dots, m-1$, i.e.

$$\varphi f(x) = x^{m\gamma_m - 1} \int_0^{x^{\frac{m}{\beta}}} G_{m-1,m-1}^{m-1,0}\left[\frac{\tau^\beta}{x^m}\middle| \begin{matrix}(\gamma_k+\lambda_k)_1^{m-1}\\ (\gamma_k)_1^{m-1}\end{matrix}\right] f(\tau) d\left(\tau^\beta\right). \tag{3.9.35}$$

Then,

$$\begin{aligned}
\mathfrak{L}\{\varphi f(x); z\} &= \int_0^\infty \exp(-zx)\varphi f(x) dx\\
&= \int_0^\infty \exp(-zx) x^{m\gamma_m-1} dx \int_0^{x^{\frac{\beta}{m}}} G_{m-1,m-1}^{m-1,0}\left[\frac{\tau^\beta}{x^m}\middle| \begin{matrix}(\gamma_k+\lambda_k)_1^{m-1}\\ (\gamma_k)_1^{m-1}\end{matrix}\right] f(\tau) d\left(\tau^\beta\right)\\
&= \int_0^\infty f(\tau) d\left(\tau^\beta\right) \int_{\tau^{\frac{\beta}{m}}}^\infty \exp(-zx) x^{m\gamma_m-1}\\
&\quad\times G_{m-1,m-1}^{0,m-1}\left[\frac{1}{\tau^\beta}x^m\middle| \begin{matrix}(1-\gamma_k)_1^{m-1}\\ (1-\gamma_k-\lambda_k)_1^{m-1}\end{matrix}\right] dx,
\end{aligned}$$

changing the order of integrations and using the properties of the G-functions. Now we denote the inner integral by J and substitute there

$$\exp(-zx) = G_{0,1}^{1,0}[zx|0], \quad G_{m-1,m-1}^{0,m-1}\left[\frac{x^m}{\tau^\beta}\right] \equiv 0 \text{ for } x < \tau^{\frac{\beta}{m}},$$

whence it takes the form:

$$J = \int_0^\infty x^{m\gamma_m-1} G_{0,1}^{1,0}[zx|0]\, G_{m-1,m-1}^{0,m-1}\left[\frac{1}{\tau^\beta}x^m\middle| \begin{matrix}(1-\gamma_k)_1^{m-1}\\ (1-\gamma_k-\lambda_k)_1^{m-1}\end{matrix}\right] dx$$

and can be evaluated immediately by using formula (A.28) again. Hence,

$$J=\sqrt{(2\pi)^{1-m}m}\; z^{-m\gamma_m}\; G^{0,m+(m-1)}_{m+(m-1),(m-1)}\left[\frac{m^m}{\tau^\beta z^m}\,\middle|\,\begin{matrix}(1-\gamma_k)_{k=1}^{m-1},\left(\frac{k}{m}-\gamma_m\right)_{k=1}^{m}\\(1-\gamma_j-\lambda_j)_{j=1}^{m-1}\end{matrix}\right].$$

But since $(1-\gamma_j-\lambda_j)_{j=1}^{m-1}=\left(1-\frac{j}{m}-\gamma_m\right)_{j=1}^{m-1}=\left(\frac{k}{m}-\gamma_m\right)_{k=1}^{m-1}$, according to properties (A.13′), (A.15), (A.14) of the G-function, we obtain

$$\begin{aligned}J&=\sqrt{(2\pi)^{1-m}m}\; z^{-m\gamma_m}G^{0,m}_{m,0}\left[\frac{m^m}{\tau^\beta z^m}\,\middle|\,(1-\gamma_k)_1^m\right]\\&=\sqrt{(2\pi)^{1-m}m}\; z^{-m\gamma_m}G^{m,0}_{0,m}\left[\left(\frac{z}{m}\right)^m\tau^\beta\,\middle|\,(\gamma_k)_1^m\right]\\&=\sqrt{(2\pi)^{1-m}m}\; z^{-m\gamma_m}\left[\left(\frac{z}{m}\right)^m\tau^\beta\right]^{\frac{1}{\beta}-1}G^{m,0}_{0,m}\left[\left(\frac{z}{m}\right)^m\tau^\beta\,\middle|\,\left(\gamma_k-\frac{1}{\beta}+1\right)_1^m\right].\end{aligned}$$

Then,

$$\mathfrak{L}\{\varphi f(x);z\}=\beta\int_0^\infty f(\tau)J\tau^{\beta-1}d\tau=\sqrt{\frac{(2\pi)^{1-m}}{m}}\beta\int_0^\infty[\lambda(z,\tau)]_{z\to\left(\frac{z}{m}\right)^{\frac{m}{\beta}}}f(\tau)d\tau,$$

which is equivalent to (3.9.34).

Note. The Obrechkoff transform *can be related also to the m-dimensional Laplace transform*

$$\begin{aligned}\mathfrak{L}_m\{f(x_1,\dots,x_m);z_1,\dots,z_m\}=\int_0^\infty\dots\int_0^\infty&\exp(-z_1x_1-\dots-z_mx_m)\\&\times f(x_1,\dots,x_m)\,dx_1\dots dx_m,\end{aligned}$$

namely (see Dimovski [67]): if $F(z)=\mathfrak{O}\{f(x);z\}$, then

$$F\left[(z_1\dots z_m)^{\frac{1}{\beta}}\right]=\left(\prod_{k=1}^m z_k^{\gamma_k-\gamma_m}\right)\mathfrak{L}_m\left\{f\left[(x_1\dots x_m)^{\frac{1}{\beta}}\right]\prod_{k=1}^m x_k^{\gamma_k};\;z_1,\dots,z_m\right\}.\quad(3.9.36)$$

3.9.ii. Inversion formulas for the Obrechkoff transform

Complex inversion formulas, analogous to the Riemann-Mellin inversion formula (3.9.4) but for the Obrechkoff transform, can be obtained in several different ways. According to the specific purpose and case, we can use any of the inversion formulas given below which seems to be the most suitable and useful.

Theorem 3.9.10. (Complex inversion formula *No* 1) *If* $\varphi(x) \in \Omega$ *is an* m*-times continuously differentiable function in* $[0, \infty)$ *and* $F(z) = \mathfrak{O}\{f(x); z\}$, *then*

$$f(x) = \frac{x^{-\frac{\beta}{m}(\gamma_1+\cdots+\gamma_m)}}{(2\pi i)^m} \underbrace{\int_{\sigma-i\infty}^{\sigma+i\infty} \cdots \int_{\sigma-i\infty}^{\sigma+i\infty}}_{m} \exp\left[x^{\frac{\beta}{m}}(z_1 + \cdots + z_m)\right] \times \prod_{k=1}^{m} z_k^{\gamma_k-\gamma_m} F\left[(z_1 \ldots z_m)^{\frac{1}{\beta}}\right] dz_1 \ldots dz_m, \tag{3.9.37}$$

where $\sigma > \frac{\lambda}{m}$ *is a sufficiently large constant.*

Proof. Under the hypothesis of the theorem, the function

$$f\left[(x_1 \ldots x_m)^{\frac{1}{\beta}}\right] \prod_{k=1}^{m} x_k^{\gamma_k-\gamma_m}$$

satisfies the conditions for validity of the complex inversion formula

$$f(x_1, \ldots, x_m) = \frac{1}{(2\pi i)^m} \int_{\sigma-i\infty}^{\sigma+i\infty} \cdots \int_{\sigma-i\infty}^{\sigma+i\infty} \exp(z_1 x_1 + \cdots + z_m x_m) \times \mathfrak{L}_m\{f;\ z_1, \ldots, z_m\}\, dz_1 \ldots dz_m$$

of the m-dimensional Laplace transform (see [91, p. 319]), since

$$f\left[(x_1 \ldots x_m)^{\frac{1}{\beta}}\right] \leq M \exp\left[\lambda (x_1 \ldots x_m)^{\frac{1}{m}}\right] \leq M \exp\left(\frac{\lambda}{m} x_1 + \cdots + \frac{\lambda}{m} x_m\right), \quad M = \text{const}.$$

Then, using relation (3.9.36),

$$f\left[(x_1 \ldots x_m)^{\frac{1}{\beta}}\right] = \frac{1}{(2\pi i)^m} \left[\prod_{k=1}^{m} x_k^{-\gamma_k}\right] \underbrace{\int_{\sigma-i\infty}^{\sigma+i\infty} \cdots \int_{\sigma-i\infty}^{\sigma+i\infty}}_{m} \exp(z_1 x_1 + \cdots + z_m x_m) \times \prod_{k=1}^{m} z_m^{\gamma_m-\gamma_k} F\left[(z_1 \ldots z_m)^{\frac{1}{\beta}}\right] dz_1 \ldots dz_m,$$

provided $\sigma > \frac{\lambda}{m}$. By putting here $x_1 = x_2 = \cdots = x_m = x^{\frac{\beta}{m}}$, we obtain formula (3.9.37). For more details see [190], [74]-[75].

The next formula follows from the relation (3.9.34) between the Obrechkoff transform and the one-dimensional Laplace integral transform.

Theorem 3.9.11. (Complex inversion formula *No* 2) *If* $\varphi(x) \in \Omega$ *and* $F(z) = \mathfrak{O}\{f(x); z\}$ *is its Obrechkooff image, then*

$$f(x) = \frac{1}{i\sqrt{(2\pi)^{m-1}m}} \varphi^{-1} \left\{ \int_{\sigma-i\infty}^{\sigma+i\infty} \exp(zx)\, F\left[\left(\frac{z}{m}\right)^{\frac{m}{\beta}}\right] dz, \right\} \tag{3.9.38}$$

where φ *is the Sonine-Dimovski transform (3.5.47), written as a generalized fractional integral (3.5.52) and its inversion* φ^{-1} *is given explicitly by the differintegral operator (see (3.5.54)):*

$$\varphi^{-1} g(x) = x^{-\beta\left(\gamma_m + \frac{m-1}{m}\right)} D_\eta I_{\beta,m-1}^{\left(\frac{k-m+1}{m}\right),(\eta_k - \lambda_k)} g\left(x^{\frac{\beta}{m}}\right). \tag{3.9.39}$$

Written in a compact form, (3.9.38) leads to the following
Complex inversion formula *No* 3*:*

$$\begin{aligned} f\left(x^{\frac{1}{\beta}}\right) = \frac{x^{-\left(\gamma_m + \frac{m-1}{m}\right)}}{(2\pi)^m i} \int_{\sigma-i\infty}^{\sigma+i\infty} G_{0,m}^{m,0}\left[\left(-\frac{z}{m}\right)^m x \,\middle|\, \left(\frac{m-k+1}{m} + \lambda_k\right)_1^m\right] \\ \times F\left[\left(\frac{z}{m}\right)^{\frac{m}{\beta}}\right] dz \end{aligned} \tag{3.9.40}$$

with a sufficiently large σ *and* $\lambda_k = \gamma_m - \gamma_k + \frac{k}{m}$, $k = 1, \ldots, m-1$.

Proof. Formula (3.9.38) follows from relation (3.9.34). Substituting there φ^{-1} the corresponding differential expression (3.9.39) involving the $G_{m-1,m-1}^{m-1,0}$-function and $\exp(zx) = G_{0,1}^{1,0}[-zx|0]$, after changing the order of integrations and using formula (A.28), we obtain (3.9.40). For details see [75], [79].

Further, we propose an inversion formula, found directly by using the new Definition 3.9.4 of the Obrechkoff transform as a G-transform.

Theorem 3.9.12. (Complex inversion formula *No* 4) *If* $f(x) \in \Omega$ *with Obrechkoff image* $F(z) = \mathfrak{O}\{f(x); z\}$*, then the inversion formula*

$$f(x) = \frac{1}{2\pi i} \int_{c-i\infty}^{c+i\infty} \frac{x^{-p}}{\prod_{k=1}^{m} \Gamma\left(\gamma_k - \frac{p}{\beta} + 1\right)} dp \left\{ \int_0^\infty z^{\beta(\gamma_m+1)-p-1} F(z) dz \right\} \tag{3.9.41}$$

holds provided that the integrals

$$\int_0^\infty x^{c-1} f(x) dx, \qquad \int_0^\infty z^{\beta(\gamma_m+1)-p-1} F(z) dz \tag{3.9.42}$$

are absolutely convergent for $p = c + iT$, $-\infty < T < \infty$, *and suitably chosen constant* $c < -\alpha = \beta\left[1 - \min_k \gamma_k\right]$.

Proof. We integrate the expression $z^{\beta(\gamma_m+1)-p-1}F(z)$ with respect to z on the real half-line $(0,\infty)$. Under conditions (3.9.42) we can change the order of the improper integrals (according to the Vallee-Poussin theorem):

$$\int_0^\infty z^{\beta(\gamma_m+1)-p-1}F(z)dz = \beta\int_0^\infty z^{\beta(\gamma_m+1)-p-1-\beta(\gamma_m+1)+1}dz$$
$$\times\int_0^\infty G_{0,m}^{m,0}\left[(zx)^\beta \middle| \left(\gamma_k - \frac{1}{\beta}+1\right)_1^m\right] f(x)dx$$
$$= \beta\int_0^\infty f(x)dx\int_0^\infty z^{-p}G_{0,m}^{m,0}\left[(zx)^\beta \middle| \left(\gamma_k - \frac{1}{\beta}+1\right)_1^m\right]dz.$$

The inner integral has the value

$$\frac{1}{\beta}x^{p-1}\prod_{k=1}^m \Gamma\left(\gamma_k - \frac{p}{\beta}+1\right),$$

provided $|\arg z| < \frac{m\pi}{2\beta}$ (see [107, II, p. 418]) which is satisfied for real $z \in (0,\infty)$. Then,

$$\int_0^\infty z^{\beta(\gamma_m+1)-p-1}F(z)dz = \prod_{k=1}^m \Gamma\left(\gamma_k - \frac{p}{\beta}+1\right)\int_0^\infty x^{p-1}f(x)dx,$$

i.e.

$$\frac{1}{2\pi i}\int_{c-i\infty}^{c+i\infty}\frac{x^{-p}}{\prod_{k=1}^m \Gamma\left(\gamma_k - \frac{p}{\beta}+1\right)}dp\int_0^\infty z^{\beta(\gamma_m+1)-p-1}F(z)dz$$
$$= \frac{1}{2\pi i}\int_{c-i\infty}^{c+i\infty} x^{-p}dx\int_0^\infty x^{p-1}f(x)dx = f(x),$$

according to the inversion formula for the Mellin transform. For details, see [75].

Another kind of inversion formula for the integral transform (3.9.7), namely, *a real inversion formula analogous to the Post-Widder formula* (3.9.5) *for the Laplace transform*, was found by Obrechkoff [339]. In a similar manner, we obtain

Theorem 3.9.13. (Real inversion formula, Kiryakova [190], Dimovski and Kiryakova [74]) *Let the function $f(t)$ be Lebesgue integrable on each finite interval $(0,T)$, $T>0$, and let its Obrechkoff integral transform (3.9.11):*

$$F(x) = \int_0^\infty t^{\beta(\gamma_m+1)-1} K\left[(xt)^\beta\right] f(t)dt \tag{3.9.43}$$

converge for some real $x = x_0 > 0$.

If $F(x) \in C^{(\infty)}(0,\infty)$, then we form the sequence of functions

$$F_0(x) = F(x),\ F_1(x) = \tilde{B}F(x),\ \ldots\ , F_k(x) = \tilde{B}^k F(x),\ \ldots\ , \tag{3.9.44}$$

where $\tilde{B}$ is a "conjugate" hyper-Bessel differential operator (cf. with B in (3.9.9))

$$\tilde{B} = x^{1-\beta}\frac{d}{dx}x^{\alpha_{m-1}}\frac{d}{dx}x^{\alpha_{m-2}}\ldots\frac{d}{dx}x^{\alpha_1}\frac{d}{dx}x^{\alpha_0+\alpha_m+\beta-1}. \tag{3.9.45}$$

Then, at each point of continuity $t=t_0$ of the original $f(t)$, the following real inversion formula holds:

$$f(t_0) = \lim_{k\to\infty}\left[\frac{(-1)^k}{\beta^k k!}\right]^m k^{-\sum_{i=1}^m \gamma_i}\left(\frac{k^{\frac{m}{\beta}}}{t_0}\right)^{\beta(\gamma_m+k+1)} F_k\left(\frac{k^{\frac{m}{\beta}}}{t_0}\right). \tag{3.9.46}$$

Proof. Following Obrechkoff [339] we use *the Laplace method* for asymptotic evaluation of the integrals. It is easily seen that the convergence of Obrechkoff transform for some $x = x_0 > 0$ yields its convergence for each $x > x_0$ and also, that the integrals obtained from (3.9.43) by differentiation with respect to x, are also convergent for $x > x_0$ and represent the corresponding derivatives $F'(x)$, $F''(x)$, $\ldots$. Taking into account the hyper-Bessel differential equation (3.9.21) satisfied by $\Phi(x)$, respectively by $K\left[(xt)^\beta\right]$, and substituting $x = \frac{k^{\frac{\beta}{m}}}{t_0}$, we find

$$\begin{aligned} F_k\left(\frac{k^{\frac{m}{\beta}}}{t_0}\right) &= (-1)^{mk}\beta^{mk+1}\int_0^\infty t^{\beta(\gamma_m+k+1)-1}K\left[k^m\left(\frac{t}{t_0}\right)^\beta\right] f(t)dt \\ &:= \beta(-\beta)^{mk} I_k, \quad k=0,1,2,\ldots \end{aligned} \tag{3.9.47}$$

According to (3.9.13), (3.9.14),

$$K\left[k^m\left(\frac{t}{t_0}\right)^\beta\right] = G^{m,0}_{0,m}\left[k^m\left(\frac{t}{t_0}\right)^\beta \middle| (\gamma_k-\gamma_m)_1^m\right]$$

and formulas (A.14), (B.4) (Lemma B.2) allow us to calculate the values of integrals I_k, $k = 0, 1, 2, \dots$, in $F_k\left(\frac{k^{\frac{m}{\beta}}}{t_0}\right)$, namely:

$$\begin{aligned} I_k &= \frac{1}{\beta}\left(\frac{k^{\frac{m}{\beta}}}{t_0}\right)^{-\beta(\gamma_m+k+1)} \prod_{i=1}^{m} \Gamma(\gamma_i + k + 1) \\ &\sim \frac{1}{\beta}\left(\frac{k^{\frac{m}{\beta}}}{t_0}\right)^{-\beta(\gamma_m+k+1)} k^{\sum\limits_{i=1}^{m}\gamma_i} (k!)^m := \frac{1}{g_k}, \qquad k \to \infty, \end{aligned} \tag{3.9.48}$$

following from Stirling's asymptotic formula.

Further, by the Laplace method, using asymptotic formula (3.9.18) for the kernel-function $K\left[k^m\left(\frac{t}{t_0}\right)^{\beta}\right]$ as $k \to \infty$, Stirling's asymptotic formula and boring analytical procedures, we find the limiting equalities:

$$\lim_{k\to\infty} g_k \int_0^{\alpha-\varepsilon} t^{\beta(\gamma_m+k+1)-1} K\left[k^m\left(\frac{t}{t_0}\right)^{\beta}\right] f(t)dt = 0, \tag{3.9.50}$$

and

$$\lim_{k\to\infty} g_k \int_{\alpha+\varepsilon}^{\infty} t^{\beta(\gamma_m+k+1)-1} K\left[k^m\left(\frac{t}{t_0}\right)^{\beta}\right] f(t)dt = 0 \tag{3.9.49}$$

provided $k^{\frac{m}{\beta}} > x_0 t_0$ (since $k \to \infty$) and $\varepsilon > 0$ is an arbitrary small positive number.

The above two equalities taken for $f(t) = 1$ and combined with

$$\lim_{k\to\infty} g_k \int_0^{\infty} t^{\beta(\gamma_m+k+1)-1} K\left[k^m\left(\frac{t}{t_0}\right)^{\beta}\right] dt = 1,$$

following from (3.9.48): $I_k \sim \frac{1}{g_k}$, $k \to \infty$, yield

$$\lim_{k\to\infty} g_k \int_{t_0-\varepsilon}^{t_0+\varepsilon} t^{\beta(\gamma_m+k+1)-1} K\left[k^m\left(\frac{t}{t_0}\right)^{\beta}\right] dt = 1, \tag{3.9.51}$$

for arbitrary $\varepsilon > 0$.

Since, by assumption t_0 is a continuity point of $f(t)$, for arbitrary small $\eta > 0$ we can choose $\varepsilon > 0$ so that for $|t - t_0| \le \varepsilon$ it follows that $|f(t) - f(t_0)| < \eta$. Consider then the integral

$$\begin{aligned} & g_k \int_{t_0-\varepsilon}^{t_0+\varepsilon} t^{\beta(\gamma_m+k+1)-1} K\left[k^m\left(\frac{t}{t_0}\right)^\beta\right] f(t) dt \\ & = f(t_0)\ g_k \int_{t_0-\varepsilon}^{t_0+\varepsilon} t^{\beta(\gamma_m+k+1)-1} K\left[k^m\left(\frac{t}{t_0}\right)^\beta\right] dt \\ & \quad - g_k \int_{t_0-\varepsilon}^{t_0+\varepsilon} t^{\beta(\gamma_m+k+1)-1} K\left[k^m\left(\frac{t}{t_0}\right)^\beta\right] [f(t)-f(t_0)]\, dt \\ & := f(t_0)\,\Delta_1 - \Delta_2. \end{aligned}$$

According to (3.9.51), we have $\lim\limits_{k\to\infty} \Delta_1 = 1$ and due to the choice of ε, $|\Delta_2| < \eta$, where η is arbitrary small.

Then,

$$\lim_{k\to\infty} g_k \int_{t_0-\varepsilon}^{t_0+\varepsilon} t^{\beta(\gamma_m+k+1)-1} K\left[k^m\left(\frac{t}{t_0}\right)^\beta\right] f(t) dt = f(t_0),$$

and according to (3.9.49), (3.9.50) this means,

$$\lim_{k\to\infty} g_k \int_0^\infty t^{\beta(\gamma_m+k+1)-1} K\left[k^m\left(\frac{t}{t_0}\right)^\beta\right] f(t) dt = f(t_0).$$

Due to (3.9.47), this leads to

$$f(t_0) = \lim_{k\to\infty} g_k \frac{(-1)^{mk}}{\beta^{mk+1}} F_k\left(\frac{k^{\frac{m}{\beta}}}{t_0}\right)$$

and it remains only to put here (3.9.48) for $\lim\limits_{k\to\infty} g_k$, to obtain the Post-Widder type inversion formula (3.9.46).

Note. The above theorem holds also if instead of the continuity of $f(t)$ at $t = t_0$ we suppose the condition

$$\int_{t_0}^{t_0+h} |f(t) - f(t_0)|\, dt = o(|h|), \quad h \to 0.$$

Then, for $t = t_0$ the limiting equation (3.9.46) holds again. Moreover, by the Lebesgue theorem (3.9.46) takes place for almost all $t_0 > 0$.

3.9.iii. Abelian type theorems for the Obrechkoff transform

Knowledge of Abelian type theorems for a given integral transform is of considerable importance in solving initial and boundary value problems arising in mathematical physics. For the Laplace transform (3.9.1) several different variants of the initial and final value theorems (Abelian theorems) are known, for example:

Lemma 3.9.14. ([510], [453], [90]) *Let $f(x)$ be a Lebesgue inetgrable function on $(0,\infty)$ which is $\mathcal{O}(\exp \lambda x)$ with $\lambda > 0$ as $x \to \infty$. Assume that there exists the limit*

$$\lim_{x \to +0} x^{-\rho} f(x) := f_0 \tag{3.9.52}$$

and consider the Laplace transform

$$\mathfrak{L}(s) = \mathfrak{L}\{f(x); s\} = \int_0^\infty \exp(-xs) f(x) dx, \ \ s > \lambda.$$

Then, there exists the limit

$$\lim_{s \to \infty} s^{\rho+1} \mathfrak{L}(s) = f_0 \, \Gamma(\rho + 1), \tag{3.9.53}$$

or

$$f(x) = \mathcal{O}(x^\rho), \ \ x \to +0 \Rightarrow \mathfrak{L}(s) = \mathcal{O}\left(s^{-\rho-1}\right), \ \ s \to +\infty \, . \tag{3.9.54}$$

Note. In the subspace of continuous functions

$$C_{-1}^{\exp} = \left\{ f(x) = x^\rho \tilde{f}(x), \ \rho > -1, \ \tilde{f} \in C[0,\infty); \ f(x) = \mathcal{O}(\exp \lambda x), \ x \to \infty \right\}$$

condition (3.9.52) takes the form

$$\lim_{x \to +0} x^{-\rho} f(x) = \lim_{x \to +0} \tilde{f}(x) = \tilde{f}(0) := f_0 \tag{3.9.52$'$}$$

and yields the existence of the limit

$$\lim_{s \to \infty} s^{\rho+1} \mathfrak{L}(s) = \tilde{f}(0) \, \Gamma(\rho + 1). \tag{3.9.53$'$}$$

We can state analogous Abelian theorems for the Obrechkoff transform either in the space (3.9.12) of continuous Obrechkoff transformable functions $\Omega \subset C_\alpha$, or for Lebesgue integrable functions with prescribed growth at $x = 0$ and $x = \infty$, namely (cf. Section 1.1.i, c)):

$$L_\alpha^{\exp} = \Bigg\{ f \in L(0,\infty); \ f(x) = \mathcal{O}(x^\rho), \ \rho > \alpha \ \text{ as } x \to +0; \\ f(x) = \mathcal{O}\left(\exp\left(\lambda x^{\frac{\beta}{m}}\right)\right), \ \lambda > 0 \ \text{ as } x \to +\infty \Bigg\}. \tag{3.9.55}$$

To this end, we again use the transmutation method, namely, the Sonine-Dimovski transform φ, relating the Obrechkoff transform with the Laplace transform.

Theorem 3.9.15. (Abelian initial value theorem) *Let $f \in L_\alpha^{\exp}$ be an Obrechkoff transformable function and $\rho > \alpha = \max_k [-\beta(\gamma_k + 1)] = -\beta(\gamma_1 + 1)$. Assume there exists the limit (3.9.52):*

$$\lim_{x \to +0} x^{-\rho} f(x) = f_0$$

and denote the Obrechkoff transform (3.9.11), (3.9.23) of $f(x)$ by

$$\mathfrak{O}(s) = \mathfrak{O}\{f(x); s\}, \quad s > \lambda.$$

Then, there exists the limit

$$\lim_{s \to +\infty} s^{\rho + \beta(\gamma_m + 1)} \mathfrak{O}(s) = f_0 \prod_{k=1}^{m} \Gamma\left(\gamma_k + \frac{\rho}{\beta} + 1\right). \tag{3.9.56}$$

Moreover, if $f \in \Omega = C_\alpha^{\exp} \subset L_\alpha^{\exp}$, i.e. $f(x) = x^\rho \tilde{f}(x)$, $\rho > \alpha$, $\tilde{f} \in C(0, \infty)$, then (3.9.56) takes the explicit form:

$$\lim_{s \to +\infty} s^{\rho + \beta(\gamma_m + 1)} \mathfrak{O}(s) = \tilde{f}(0) \prod_{k=1}^{m} \Gamma\left(\gamma_k + \frac{\rho}{\beta} + 1\right). \tag{3.9.56$'$}$$

Proof. For $f(x) = x^\rho \tilde{f}(x)$ with $\rho > \alpha$, $\tilde{f} \in C(0, \infty)$, condition (3.9.52$'$) is satisfied. We now use the Sonine-Dimovski transform φ, (3.5.47), (3.5.52) which according to (3.9.34) can be considered as a transmutation operator from the Laplace transform to the Obrechkoff transform. We have seen in Section 3.5 that $\varphi : C_\alpha \to C_{-1}$, and then, in C_{-1} the Laplace transform of $\varphi f(x)$ in the right-hand side of (3.9.34) can be considered. We need however to make precise the asymptotic behaviour of $\varphi f(x)$ and to this end we use some properties of the generalized fractional integrals. According to (3.5.52$'$) and (1.3.3), we have

$$\begin{aligned}
(\varphi f)\left(x^{\frac{\beta}{m}}\right) &= x^{\beta(\gamma_m + 1) - \frac{\beta}{m}} I_{\beta, m-1}^{(\gamma_k), (\gamma_m - \gamma_k + \frac{k}{m})} x^\rho \tilde{f}(x) \\
&= x^{\beta(\gamma_m + 1) - \frac{\beta}{m} + \rho} I_{\beta, m-1}^{(\gamma_k + \frac{\rho}{\beta}), (\gamma_m - \gamma_k + \frac{k}{m})} \tilde{f}(x) \\
&:= x^{\beta(\gamma_m + 1) - \frac{\beta}{m} + \rho} \hat{f}(x),
\end{aligned}$$

and hence,

$$\lim_{x \to +o} x^{-m(\gamma_m + \frac{\rho}{\beta} + 1) + 1} \varphi f(x) = \lim_{x \to +o} \hat{f}\left(x^{\frac{m}{\beta}}\right) = \hat{f}(0).$$

On the other hand,

$$\hat{f}(0) = \left(I_{\beta,m-1}^{\left(\gamma_k+\frac{\rho}{\beta}\right),\left(\gamma_m-\gamma_k+\frac{k}{m}\right)} \tilde{f}(x) \right)(0)$$

$$= \tilde{f}(0) \prod_{k=1}^{m-1} \frac{\Gamma\left(\gamma_k + \frac{\rho}{\beta} + 1\right)}{\Gamma\left(\gamma_k + \frac{k}{m} + \frac{\rho}{\beta} + 1\right)},$$

by virtue of (1.3.4) with $j = 0$. In this manner we find that condition (3.9.52′) implies the condition

$$\lim_{x \to +\infty} x^{-m\left(\gamma_m + \frac{\rho}{\beta} + 1\right) + 1} \varphi f(x) = \tilde{f}(0) \prod_{k=1}^{m-1} \frac{\Gamma\left(\gamma_k + \frac{\rho}{\beta} + 1\right)}{\Gamma\left(\gamma_k + \frac{k}{m} + \frac{\rho}{\beta} + 1\right)}, \tag{3.9.57}$$

where $\varphi f(x)$ is the Sonine type transmutation of $f(x)$. Let us now apply the initial value Lemma 3.9.14 to the Laplace transform in the right-hand side of (3.9.34). Condition (3.9.57) yields that the limit (3.9.52′) exists when ρ is substituted by $\left(m\left(\gamma_m + \frac{\rho}{\beta} + 1\right) - 1\right)$ and f_0 by $\tilde{f}_0 \prod_{k=1}^{m-1} \left(\gamma_k + \frac{\rho}{\beta} + 1\right)$, and then,

$$\lim_{s \to +\infty} s^{-m\left(\gamma_m + \frac{\rho}{\beta} + 1\right)} \mathfrak{L}\{\varphi f(x); s\}$$

$$= \tilde{f}(0) \Gamma\left(m\left(\gamma_m + \frac{\rho}{\beta} + 1\right)\right) \prod_{k=1}^{m-1} \frac{\Gamma\left(\gamma_k + \frac{\rho}{\beta} + 1\right)}{\Gamma\left(\gamma_k + \frac{k}{m} + \frac{\rho}{\beta} + 1\right)}.$$

Using the Gauss-Legendre multiplication formula for the Γ-functions ([106, I]):

$$\Gamma\left(m\left(\gamma_m + \frac{\rho}{\beta} + 1\right)\right) = (2\pi)^{-\frac{m-1}{2}} m^{-\frac{1}{2}} m^{m\left(\gamma_m + \frac{\rho}{\beta} + 1\right)} \Gamma\left(\gamma_m + \frac{\rho}{\beta} + 1\right)$$

$$\times \prod_{k=1}^{m-1} \Gamma\left(\gamma_k + \frac{k}{m} + \frac{\rho}{\beta} + 1\right),$$

we obtain

$$\lim_{s \to +\infty} s^{m\left(\gamma_m + \frac{\rho}{\beta} + 1\right)} \mathfrak{L}\{\varphi f(x); s\}$$

$$= \tilde{f}(0) (2\pi)^{-\frac{m-1}{2}} m^{-\frac{1}{2}} m^{m\left(\gamma_m + \frac{\rho}{\beta} + 1\right)} \prod_{k=1}^{m-1} \Gamma\left(\gamma_k + \frac{\rho}{\beta} + 1\right),$$

or:

$$\lim_{s\to\infty} F(s) = \tilde{f}(0) \prod_{k=1}^{m} \Gamma\left(\gamma_k + \frac{\rho}{\beta} + 1\right),$$

where we have denoted

$$F(s) = \sqrt{(2\pi)^{m-1} m} \left(\frac{s}{m}\right)^{m\left(\gamma_m + \frac{\rho}{\beta} + 1\right)} \mathfrak{L}\{\varphi f(x); s\}.$$

The later, according to (3.9.34), being equal to:

$$\left(\left(\frac{s}{m}\right)^{\frac{m}{\beta}}\right)^{\rho+\beta(\gamma_m+1)} \mathfrak{O}\left\{f(x); \left(\frac{s}{m}\right)^{\frac{m}{\beta}}\right\}.$$

Since

$$\begin{aligned}\lim_{s\to+\infty} F(s) &= \lim_{m\sigma^{\frac{\beta}{m}}\to\infty} F\left(m\sigma^{\frac{\beta}{m}}\right) \\ &= \lim_{\sigma\to+\infty} \sigma^{\rho+\beta(\gamma_m+1)} \mathfrak{O}\{f(x); \sigma\},\end{aligned}$$

we obtain the limit (3.9.56′). Let us note that the condition $f(x) = \mathcal{O}\left(\exp^{\lambda x^{\frac{\beta}{m}}}\right)$, $x \to +\infty$ for $f \in C_\alpha^{\exp} \subset C_\alpha$ is required to ensure the convergence of the Obrechkoff integral (3.9.11), (3.9.23) for $s > \lambda$, and therefore the existence of the Obrechkoff image $\mathfrak{O}(s)$. It is in the same way as the condition $f(x) = \mathcal{O}(\exp \lambda x)$, $x \to \infty$, ensures the convergence of Laplace integral (3.9.1) for $s > \lambda$. As we established at the beginning of this section, this is seen from the asymptotic behaviour (3.9.18) of the kernel-function $K(s)$ as $s \to \infty$. To prove the theorem in the most general case of Lebesgue integrable functions $f \in L_\alpha^{\exp}$, we have to make only slight modifications. It is sufficient to replace condition (3.9.52′) by the assumption that $\lim_{x\to+0} \tilde{f}(x) = f_0$ exists. Then, this imply the existence of the limit (3.9.56). This ends the proof.

Using the Sonine type transmutation operator φ again, or directly, the new definition (3.9.23) of the Obrechkoff transform and following the pattern in [510], [518], [361], we can obtain a theorem analogous to the final value theorems for the Laplace transform.

Theorem 3.9.16. (Abelian final value theorem) *Let $f \in L_\alpha^{\exp}$ be an Obrechkoff transformable function, $\mathfrak{O}(s)$ be its transform and $\rho > \alpha = -\beta(\gamma_1 + 1)$. Then the existence of the limit*

$$\lim_{x\to+\infty} x^{-\rho} f(x) = f_\infty \tag{3.9.58}$$

implies the limiting equality

$$\lim_{s\to+0} s^{\rho+\beta(\gamma_m+1)} y\mathfrak{O}(s) = f_\infty \prod_{k=1}^{m} \Gamma\left(\gamma_k + \frac{\rho}{\beta} + 1\right). \tag{3.9.59}$$

Initial and final value theorems for the Hankel and Meijer transforms were proposed by Zemanian [518]. For the general G-transformation (3.9.22), analogous and rather general results were obtained by Pathak [361]. In view of the new definition (3.9.23) of the Obrechkoff transform as a G-transformation, one can obtain the results of Theorems 3.9.15 and 3.9.16 in an alternative way, by a suitable specialization of Pathak's results [361] and specification of the list of conditions, imposed there. Examples of Abelian theorems for particular cases of the Obrechkoff transform, like the Meijer transform, will be shown in next section.

3.10. Some special cases of the Obrechkoff transform. An open problem

This section is closely related to Section 3.3. Namely, we consider some Laplace type integral transforms, related to special cases of hyper-Bessel operators, which have been introduced and investigated by different authors. Since the Obrechkoff transform is related to the most general hyper-Bessel operator (3.1.2)-(3.1.4) of arbitrary order $m \geq 2$, it is quite natural that the basic properties of these transforms, given below, follow as special cases by the results in Section 3.9. We shall use the same notation a), b), c), ... for the integral transforms, corresponding to the notation in Section 3.3 for the related hyper-Bessel operators.

a) The Laplace type integral transform corresponding to the Bessel operator B_ν, (3.3.a) is the *classical Meijer transform* ($K_\nu(z)$ is Macdonald's function (C.31)):

$$\mathfrak{K}_\nu\{f(x); z\} = \int_0^\infty \sqrt{zx} K_\nu(zx) f(x) dx, \tag{3.10.a}$$

following from the Obrechkoff transform for $m = \beta = 2$, $\gamma_{1,2} = \pm\frac{\nu}{2}$, namely:

$$\mathfrak{K}_\nu\{f(x); z\} = 2^{\nu-2} z^{\frac{1}{2}-\nu} \mathfrak{O}\left\{f(x); \frac{z}{2}\right\}.$$

Thus, we obtain from (3.9.29), (3.9.30) the differential property

$$\mathfrak{K}_\nu\left\{B_\nu f(x); z\right\} = z^2 \mathfrak{K}_\nu\left\{f(x); z\right\}, \quad f \in C_\nu^{(2)}.$$

The well-known relationship with the Laplace transform (see [510], [107, II, p.122]):

$$\mathfrak{K}_\nu\left\{f(x); z\right\} = \sqrt{\pi} 2^{-\nu} z^{\nu+\frac{1}{2}} \mathfrak{L}\left\{\int_0^t \frac{(t^2-\tau^2)^{\nu-\frac{1}{2}}}{\Gamma\left(\nu+\frac{1}{2}\right)} \tau^{-\nu+\frac{1}{2}} f(\tau) d\tau;\ z\right\} \tag{3.10.a'}$$

is a corollary of (3.9.34) and the well-known inversion formulas

$$f(x) = \frac{1}{\pi i} \int\limits_{\sigma-i\infty}^{\sigma+i\infty} \sqrt{zx} I_\mu(zx) \mathfrak{K}_\nu(z) dz,$$

$$f(x_0) = \lim_{k\to\infty} \sqrt{\frac{2}{\pi}} \frac{1}{(2k)!} \left(\frac{2k}{x_0}\right)^{2k+1} \left[S^k_{\mu,\sigma} \mathfrak{K}_\nu(\sigma)\right]_{\sigma:=\frac{2k}{x_0}},$$

where $S_{\mu,\sigma} = \sigma^{-\nu-\frac{1}{2}} \frac{d}{d\sigma} \sigma^{2\nu+1} \frac{d}{d\sigma} \sigma^{-\nu-\frac{1}{2}}$, can be derived also from (3.9.40), (3.9.46). The Abelian theorems found by Zemanian [518] for $\mathfrak{K}_\nu\{f(x); s\} = \mathfrak{K}_\nu(s)$ can be obtained as special cases of Theorems 3.9.15, 3.9.16, namely:

$$\exists \lim_{x\to+0} x^{-\rho+\frac{1}{2}} f(x) = f_0 \Rightarrow$$

$$\exists \lim_{s\to\infty} s^{\rho+\frac{1}{2}} \mathfrak{K}_\nu(s) = f_0 2^{\rho-1} \Gamma\left(\frac{\rho+1+\nu}{2}\right) \Gamma\left(\frac{\rho+1-\nu}{2}\right),$$

and

$$\exists \lim_{x\to+\infty} x^{-\rho+\frac{1}{2}} f(x) = f_\infty \Rightarrow$$

$$\exists \lim_{s\to+0} s^{\rho+\frac{1}{2}} \mathfrak{K}_\nu(s) = f_\infty 2^{\rho-1} \Gamma\left(\frac{\rho+1+\nu}{2}\right) \Gamma\left(\frac{\rho+1-\nu}{2}\right).$$

g) For the operator (3.3.g): $B_m = \frac{d}{dx} x \frac{d}{dx} \dots x \frac{d}{dx} = \frac{1}{x}\left(x \frac{d}{dx}\right)^m$, $m \geq 2$, integral transforms of Obrechkoff type were found by Ditkin and Prudnikov [86]-[87] and Botashev [36]. For the case $m = 2$: $B_2 = \frac{d}{dx} x \frac{d}{dx}$, this transform has the form

$$\mathfrak{K}\{f(x); z\} = 2z \int\limits_0^\infty K_0\left(2\sqrt{zx}\right) f(x) dx \tag{3.10.g}$$

(K_0 is the Bessel function (C.31) of third kind and zero order) and its relation with the Laplace transform has the form:

$$\mathfrak{K}\left\{f(x); \frac{z^2}{4}\right\} = \frac{\pi}{2} z^2 \mathfrak{L}\{\varphi f(x); z\}$$

with a transmutation operator, being a fractional Riemann-Liouville integral of order $\frac{1}{2}$:

$$\varphi f(x) = \frac{1}{\sqrt{\pi}} \int\limits_0^x (t-\tau)^{-\frac{1}{2}} f(x) dx.$$

In the general case $m > 2$ the corresponding transform is

$$\mathfrak{V}\{f(x); z\} = 2\int_0^\infty E_{0m}(zx)f(x)dx, \qquad (3.10.\text{g}')$$

where the kernel-function $E_{0m}(z)$ can be represented by means of $(m-1)$-tuple integrals, and the relationship

$$\mathfrak{V}\left\{f(x); \left(\frac{z}{m}\right)^m\right\} = m\mathfrak{L}\{\varphi f(x); z\}$$

holds with an operator

$$\varphi f(x) = \int_0^x \cdots \int_0^x \left[\prod_{j=1}^{m-1} (t-\tau_j)^{\frac{j}{m}-1}\right] f(\tau_1 \ldots \tau_{m-1})\, d\tau_1 \ldots d\tau_{m-1},$$

being in essence an $(m-1)$-tuple generalized fractional integral of multiorder $\left(\frac{j}{m}\right)_{j=1}^{m-1}$.

Transformations (3.10.g), (3.10.g$'$) were used by Ditkin, Prudnikov and Botashev for developing operational calculi for operator (3.3.g).

h) In a series of papers [235]-[238] Krätzel introduced and investigated an integral transformation, generalizing the Laplace and Meijer transforms and related to the hyper-Bessel differential operator (3.3.h):

$$B_{n,\nu} = \frac{d}{dx}x^{\frac{1}{n}-\nu}\left(x^{\frac{1}{n}-\nu}\frac{d}{dx}\right)^{n-1} x^{\nu+1+\frac{2}{n}},$$

namely:

$$\mathfrak{L}_\nu^{(n)}\{f(x); z\} = \int_0^\infty \lambda_\nu^{(n)}\left[n(zx)^{\frac{1}{n}}\right] f(x)dx, \qquad (3.10.\text{h})$$

where

$$\lambda_\nu^{(n)}\left[n(zx)^{\frac{1}{n}}\right] = \underbrace{\int_0^\infty \cdots \int_0^\infty}_{(n-1)} \left[\prod_{k=1}^{n-1} u_k^{\nu-1+\frac{k-1}{n}}\right]$$
$$\times \exp\left(-u_1 - \cdots - u_{n-1} - \frac{zx}{u_1 \ldots u_{n-1}}\right) du_1 \ldots du_{n-1}.$$

$\mathfrak{L}_\nu^{(n)}(z)$ can be obtained also from the corresponding Obrechkoff transform:

$$\mathfrak{O}\{f(x); z\} = z^{-\nu-1+\frac{2}{m}}\mathfrak{L}_\nu^{(m)}\{f(x); z\}.$$

Now, the differential property corresponding (3.9.29) and found in [237] is:

$$\mathfrak{L}_\nu^{(n)}\left\{B_{n,\nu}f(x); z\right\} = z\mathfrak{L}_n^{(n)}\{f(x); z\} - \gamma^{(n)}(o,\nu)f(+0), \qquad (3.10.\text{h}')$$

with $\gamma^{(n)}(0,\nu) = \prod_{r=0}^{n-2} \Gamma\left(\nu+1+\frac{r}{n}\right)$, $n \geq 2$. Krätzel [236], [238] found also a convolution of (3.10.h) and a real inversion formula, equivalent to (3.9.46), namely:

$$f(x_0) = \lim_{k\to\infty} \frac{(-1)^{nk}}{\gamma^{(n)}(k,v)} \left(\frac{s_k}{x_0}\right)^{k+1} \left[d_{n,\nu}^k \left(\mathfrak{L}_n^{(\nu)}(s)\right)^{(n)}\right]_{s:=\frac{s_k}{x_0}}, \tag{3.10.h$''$}$$

where: $\gamma^{(n)}(k,v) = \Gamma(k+1) \prod_{r=0}^{n-2} \Gamma\left(\nu+k+1+\frac{r}{n}\right)$, $n \geq 2$;

$$d_{n,\nu} = s^{-\nu-\frac{1}{n}} \left(s^{1-\frac{1}{n}} \frac{d}{ds}\right)^{n-1} s^{1-\nu} \frac{d}{ds},$$
$$s_k = n^{-n} \left\{n(k+1) + \nu(n-1) + \frac{1}{n} - 2\right\}^n.$$

From the results in Section 3.9 other integral transforms of Laplace type, corresponding to all the examples of the hyper-Bessel operators (3.3.a)-(3.3.i) and their basic properties (differential properties, convolutions, inversion formulas, Abelian theorems, etc.) can be obtained. On the other hand, some new properties of the general Obrechkoff transform can be obtained from its representation as a G-transform and the corresponding results for this kind of transform.

For other Bessel-type transformations one can see also: Koh and Deeba [223], Kljucantzev [219], Kratzel [234] and Mendez [300].

Open problem

Finally, let us state an *open problem*, related to hyper-Bessel type integral transforms. It is well known that along with the Laplace transform, there exist the Fourier transform and its modifications: the so-called cos-Fourier and sin-Fourier transforms:

$$\mathfrak{F}_c\{f(x);z\} = \sqrt{\frac{2}{\pi}} \int_0^\infty \cos zx \; f(x)dx, \quad \mathfrak{F}_s\{f(x);z\} = \sqrt{\frac{2}{\pi}} \int_0^\infty \sin zx \; f(x)dx, \tag{3.10.1}$$

suitable for operational calculi and for treating differential equations, related to operators $D = \frac{d}{dx}$, $D^m = \left(\frac{d}{dx}\right)^m$, $m > 1$. Transforms (3.10.1) have the advantage of being inverted by the formulas:

$$f(x) = \sqrt{\frac{2}{\pi}} \int_0^\infty \cos xz \; \mathfrak{F}_c(z)dz, \quad f(x) = \sqrt{\frac{2}{\pi}} \int_0^\infty \sin xz \; \mathfrak{F}_c(z)dz, \tag{3.10.2}$$

i.e. they belong to the class of so-called *symmetrical Fourier transforms*

$$\mathfrak{F}\{f;z\} = \int_0^\infty K(zy)f(y)dy, \tag{3.10.3}$$

having inversion formulas of the same kind (3.10.3). Their kernel-functions $K(x)$ are such that the identity

$$f(x) = \int_0^\infty K(zx)dz \int_0^\infty K(zy)f(y)dy \tag{3.10.4}$$

holds for each transformable function $f(x)$. It is known that (3.10.4) is equivalent to the functional equation for the Mellin image of $K(x)$:

$$\tilde{K}(s)\tilde{K}(1-s) = 1, \qquad \text{where} \quad \tilde{K}(s) = \mathfrak{M}\{K(x); s\} = \int_0^\infty x^{s-1}K(x)dx. \tag{3.10.5}$$

Another case of a symmetrical Fourier type integral transform is the *Hankel transform*

$$\mathfrak{H}_\nu(z) = \mathfrak{H}_\nu\{f(x); z\} = \int_0^\infty \sqrt{zx}\, J_\nu(zx)f(x)dx, \tag{3.10.6}$$

with $J_\nu(z)$ being the Bessel function (C.6) and which *inversion formula*

$$f(x) = \int_0^\infty \sqrt{zx}\, J_\nu(zx)\mathfrak{H}_\nu(z)dz. \tag{3.10.7}$$

This transform is closely related to the *Bessel differential operator* (3.3.6) $B_\nu = x^{-2}\left(x\frac{d}{dx} + \nu\right)\left(x\frac{d}{dx} - \nu\right)$, namely:

$$\mathfrak{H}_\nu\left\{B_\nu f(x); z\right\} = -z^2\mathfrak{H}_\nu\{f(x); z\}, \tag{3.10.8}$$

and therefore, along with the Meijer transform (3.10.a) it can also be used for the operational calculus and differential equations involving B_ν.

Open problem 3.10.1. Find a symmetrical Fourier type transform, generalizing the cos-Fourier (sin-Fourier) transform and the Hankel transform, suitable for dealing with the general hyper-Bessel differential operator (3.1.2)-(3.1.4) of order $m > 1$:

$$B = x^{-\beta}\prod_{k=1}^{m}\left(x\frac{d}{dx} + \beta\gamma_k\right).$$

This new transform, called a *generalized Hankel transform*, should generalize the Hankel transform in the same way as the Obrechkoff transform generalizes the Laplace and Meijer transforms. Find a suitable domain (functional space) for this transform and its basic properties (inversion formulas, differential properties, etc.)

CONJECTURE: Study an integral transform of the form

$$\mathfrak{H}^{(m-1)}_{\gamma_1,\dots,\gamma_m}\{f(x);z\} = \int_0^\infty (zx)^{\frac{1}{m}} J^{(m-1)}_{\gamma_1-\gamma_m,\dots,\gamma_{m-1}-\gamma_m}(zx)f(x)dx, \tag{3.10.9}$$

where the kernel-function is the Delerue's hyper-Bessel function (D.3).

Results, related to Hankel type integral transformations, their generalizations and applications can be seen, for example in Erdélyi and Kober [109], Sneddon [451]- [453], Rooney [401], Betancor [31]-[32], Mendez [300]-[301], etc.

— || —

THE MAIN RESULTS OF CHAPTER 3 HAVE BEEN PUBLISHED IN: Kiryakova [190]-[193], [196]-[197], [201], [210], Dimovski and Kiryakova [74]-[76], [79]-[80], Kiryakova and McBride [211]-[212], Kiryakova and Spirova [213].

4 Applications to the generalized hypergeometric functions

This chapter is devoted to some applications of the generalized fractional calculus to the theory of special functions. We propose new or newly written integral and differintegral representations of the generalized hypergeometric functions, the most commonly used special functions. On this basis, a suitable classification of these functions is introduced.

The study and use of the special functions is a very old branch of mathematics. The interest in them has increased incessantly, together with the discovery of their numerous applications to different problems and areas. Now, there are relatively large number of people who know a fair amount about this topic and a series of books especially devoted to it. Along with the classical handbooks of Watson [507], Erdélyi at al. [106],[108], Sneddon [452], Slater [450] and Luke [272], among the recently published books on special functions it is worth mentioning those of Olver [341], Mathai and Saxena [286], [287], Srivastava and Kashyap [468], Srivastava, Gupta and Goyal [467], Prudnikov, Brychkov and Marichev [367]-[369], Nikiforov and Uvarov [319], Rusev [414], etc.

Nevertheless, *"most of the mathematicians are totally unaware of the power of the special functions. They react to a paper which contains a Bessel function or Legendre polynomial by turning immediately to the next paper"*, confessed Askey in his lectures on orthogonal polynomials and special functions [25]. *"Hopefully these lectures will show ... how useful hypergeometric functions can be. Very few facts about them are known, but these few facts can be very useful in many different contexts. So, my advice is to learn something about hypergeometric functions: or, if this seems too hard or dull a task, get to know someone who knows something about them. And if you already know something about these functions, share your knowledge with a colleague or two, or a group of students. Every large university and research laboratory should have a person who not only find things in the Bateman Project, but can fill in a few holes in this set of books... In any case, I hope my point has been made; special functions are useful and those who need them and those who know them should start to talk to each other... The mathematical community at large needs the education on the usefulness of special functions more than most other people who could use them..."* (Askey [25]).

The most commonly encountered explicit representations for the special functions of mathematical physics are the power series and the definite or contour integrals. As an example, the Bessel function is defined by the power series

$$J_\nu(x) = \sum_{j=0}^{\infty} \frac{(-1)^j x^{2j+\nu}}{j!\Gamma(\nu+j+1)},$$

as well as by several definite integrals, one of which is the so-called *Poisson integral*

$$J_\nu(x) = \frac{2^{-\nu+1}x^\nu}{\sqrt{\pi}\Gamma\left(\nu+\frac{1}{2}\right)} \int_0^x \left(x^2-t^2\right)^{\nu-\frac{1}{2}} \cos t dt, \quad \nu > -\frac{1}{2}. \tag{4.0.1}$$

In view of the Riemann-Liouville fractional integrals, this formula is nothing but the simple relation

$$J_\nu(x) = \frac{x^{-\nu}}{2^\nu\sqrt{\pi}} R_{x^2}^{\nu+\frac{1}{2}} \left\{\frac{\cos x}{x}\right\}. \tag{4.0.1'}$$

For the classical orthogonal polynomials, the *Rodrigues formulas* are also well known and some authors (Rusev [414], Nikiforov and Uvarov [318],[319]) use these formulas as bases of their theories. As for the other special functions, formulas like those defining the orthogonal polynomials by means of repeated differentiation, are less popular and even unknown in the general case. An exception is the spherical Bessel function (see [106 ,II]):

$$J_{-n-\frac{1}{2}}(x) = \frac{2^{n+\frac{1}{2}}x^{n+\frac{1}{2}}}{\sqrt{\pi}} \frac{d^n}{(dx^2)^n} \left\{\frac{\cos x}{x}\right\}, \quad n = 0, 1, 2, \ \dots \ . \tag{4.0.2}$$

As Lavoie, Osler and Tremblay [252] noted, by using the notion of fractional derivative of arbitrary order, it is possible to generalize the above relation for Bessel functions of arbitrary order, namely:

$$\begin{aligned} J_\nu(x) = J_{-\alpha-\frac{1}{2}}(x) &= \frac{2^{\alpha+\frac{1}{2}}x^{\alpha+\frac{1}{2}}}{\sqrt{\pi}} D_{x^2}^{\alpha} \left\{\frac{\cos x}{x}\right\} \\ &= \frac{x^{-\nu}}{2^\nu\sqrt{\pi}} D_{x^2}^{-\nu-\frac{1}{2}} \left\{\frac{\cos x}{x}\right\}, \quad \nu < -\frac{1}{2}. \end{aligned} \tag{4.0.3}$$

Here, following this idea, we propose both Poisson type fractional integral represen-taions and analogues of the fractional derivative representation (4.0.3) for the generalized hypergeometric functions

$${}_pF_q\left(a_1, \dots, a_p; b_1, \dots, b_q; x\right) = \sum_{k=0}^{\infty} \frac{(a_1)_k \dots (a_p)_k}{(b_1)_k \dots (b_q)_k} \frac{x^k}{k!}, \quad 0 \le |x| \le \infty, \tag{4.0.4}$$

whit $p \le q$ or $p = q+1$ (in the latter case the condition $|x| < 1$ is also required). By $(a)_k$ the Pochhammer symbol is denoted:

$$(a)_0 = 1, \ (a)_k = \frac{\Gamma(a+k)}{\Gamma(a)} = a(a+1)\dots(a+k-1).$$

In Sections 4.2 and 4.3 we establish the fact that, depending on the cases $p < q, p = q, p = q+1$, *the functions* ${}_pF_q(x)$ *can be represented as generalized fractional integrals or derivatives* (in the sense perceived in Chapter 1) *of three different types of elementary functions*: $\cos_{q-p+1}(x), x^{a_q - 1}\exp x$ and $x^{a_q-1}(1-x)^{-a_{q+1}}$. In accordance with these results, we could assume the following *classification of the generalized hypergeometric functions (g.h.f.)* ${}_pF_q$, namely:

for $p < q$ we shall call them *"g.h.f. of Bessel type"* (the simplest examples are $J_\nu(x)$ and the hyper-Bessel functions ${}_0F_q(x)$);

for $p = q$: *"g.h.f. of confluent type"* (for example - ${}_1F_1$);

for $p = q+1$: *"g.h.f. of Gauss type"* (for example, the Gauss hypergeometric function ${}_2F_1$ and the Jacobi type orthogonal polynomials).

The starting point for these results is the fact that each ${}_pF_q$-function can be represented as a fractional differintegral of a ${}_{p-1}F_{q-1}$-function. After a finite number of steps, one reaches a hyper-Bessel function ${}_0F_{q-p}$, confluent hypergeometric function ${}_1F_1$ or Gauss function ${}_2F_1$, having corresponding differintegral representations by means of the above mentioned elementary functions. On the other hand, the compositions of the fractional integrals (derivatives) arising at subsequent steps are nothing but generalized fractional differintegrals. Interesting special cases are related to the hyper-Bessel functions, "spherical" g.h.f., n-Bessel functions, etc.

Other kinds of representations of the g.h.f. are found in Section 4.4. There, the ${}_pF_q$-functions are expressed as Laplace type G-transforms of hyper-Bessel functions, in particular, of generalized trigonometric functions. For brevity, the representations of the ${}_pF_q$-functions obtained here are for real variables but they also hold as well in suitably chosen domains of the complex plane. The corresponding specifications are a matter of technical detail.

4.1. Poisson type integral representation of the hyper-Bessel functions. Analogue of the differential representations of the spherical Bessel functions

In 1953 Delerue [60] introduced and used for the first time a generalization of the Bessel function $J_\nu(x)$ for a multiindex $\nu = (\nu_1, \dots, \nu_m)$. This is the so-called *hyper-Bessel function of Delerue of order* $m > 1$ with indices $\nu_1, \dots, \nu_m$ (see Definition D.1, Appendix):

$$J^{(m)}_{\nu_1,\dots,\nu_m}(x) = \frac{\left(\frac{x}{m+1}\right)^{\nu_1+\dots+\nu_m}}{\Gamma(\nu_1+1)\dots\Gamma(\nu_m+1)}{}_0F_m\left((\nu_k+1)_1^m; -\left(\frac{x}{m+1}\right)^{m+1}\right)$$

$$= \left[\prod_{k=1}^m \Gamma(\nu_k+1)\right]^{-1}\left(\frac{x}{m+1}\right)^{\sum\limits_{k=1}^m \nu_k}\sum_{j=0}^\infty\left(\prod_{k=1}^m\frac{\Gamma(\nu_k+1)}{\Gamma(\nu_k+j+1)}\right)\frac{(-1)^j}{j!}\left(\frac{x}{m+1}\right)^{j(m+1)}. \tag{4.1.1}$$

Analogously, the function

$$I^{(m)}_{\nu_1,\dots,\nu_m}(x) = \frac{\left(\frac{x}{m+1}\right)^{\nu_1+\dots+\nu_m}}{\Gamma(\nu_1+1)\dots\Gamma(\nu_m+1)}{}_0F_m\left((\nu_k+1)_1^m; \left(\frac{x}{m+1}\right)^{m+1}\right) \tag{4.1.2}$$

is said to be a *modified hyper-Bessel function of order m*. It generalizes the well-known function

$$I_\nu(x) = \frac{\left(\frac{x}{2}\right)^\nu}{\Gamma(\nu+1)}\, {}_0F_1\left(\nu+1; \left(\frac{x}{2}\right)^2\right).$$

Sometimes, it is more convenient to deal with the hypergeometric functions ${}_0F_m$ only, neglecting the power multipliers preceding them in (4.1.1), (4.1.2). In this manner one reaches the so-called *"normalized hypergeometric functions"* (see Klučantčev [218]) or the *"Bessel-Clifford functions of higher order"* (Hayek [126], Hayek and Hernandez [127]-[129]):

$$j(x) = j^{(m)}_{\nu_1,\ldots,\nu_m}(x) = {}_0F_m\left(\nu_1+1,\ldots,\nu_m+1; -\left(\frac{x}{m+1}\right)^{m+1}\right) \qquad (4.1.3)$$

and

$$i(x) = i^{(m)}_{\nu_1,\ldots,\nu_m}(x) = {}_0F_m\left(\nu_1+1,\ldots,\nu_m+1; \left(\frac{x}{m+1}\right)^{m+1}\right) \qquad (4.1.4)$$

being equal to 1 at $x=0$. More exactly, they satisfy the initial value conditions

$$j(0)=1, \quad j'(0) = \ \ldots \ = j^{(m)}(0) = 0 \qquad (4.1.5)$$

and

$$i(0)=1, \quad i'(0) = \ \ldots \ = i^{(m)}(0) = 0. \qquad (4.1.6)$$

As was shown in Section 3.4 (Example 2) function (4.1.3) is the unique solution of the *initial value problem*

$$By(x) = -y(x); \quad y(0)=1,\ y'(0) = \ \ldots \ = y^{(m)}(0) = 0, \qquad (4.1.7)$$

related to the hyper-Bessel differential operator of order $(m+1)$):

$$B = x^{-m+1}\left[\prod_{k=1}^{m}\left(x\frac{d}{dx} + (m+1)\nu_k\right)\right] x\frac{d}{dx} \qquad (4.1.8)$$

$$\left(\text{with parameters } \nu_k \geq \frac{k}{m+1},\quad k=1,\ldots,m,\ \nu_{m+1}=0; \beta = m+1\right).$$

In particular , taking the numbers

$$\nu_k = \frac{k}{m+1} - 1, \quad k = 1,\ldots,m,m+1 \qquad (4.1.9)$$

as components of the multiindex $\nu = (\nu_1,\ldots,\nu_m)$, we obtain the solution of the initial value problem

$$\left(\frac{d}{dx}\right)^{m+1}\tilde{y}(x) = -\tilde{y}(x); \quad \tilde{y}(0)=1, \tilde{y}'(0) = \ \ldots \ = \tilde{y}^{(m)}(0) = 0, \qquad (4.1.10)$$

related to the simplest hyper-Bessel differential operator of order $(m+1)$:

$$\tilde{B} = x^{-m+1}\left[\prod_{k=1}^{m+1}\left(x\frac{d}{dx}+k-m-1\right)\right] = \left(\frac{d}{dx}\right)^{m+1}. \tag{4.1.11}$$

This solution is the *generalized cosine function of order* $(m+1)$ (see Example 3.4.7 and Appendix, (D.9)), a special case of the hyper-Bessel function (4.1.3), namely:

$$\tilde{y}(x) = j^{(m)}_{\left(\frac{k}{m+1}-1\right)_1^m}(x) = \cos_{m+1}(x) = \sum_{j=0}^{\infty}\frac{(-1)^j x^{(m+1)j}}{((m+1)j)!}. \tag{4.1.12}$$

The *generalized Poisson transformation*, proposed by Dimovski [70], [71] corresponding to the hyper-Bessel operator (4.1.8) (see Section 3.5) has the form

$$P_0 f(x) = c\int_0^1\cdots\int_0^1\left[\prod_{k=1}^{m}\frac{(1-\sigma_k)^{\nu_k-\frac{k}{m+1}}\sigma_k^{\frac{k}{m+1}-1}}{\Gamma\left(\nu_k-\frac{k}{m+1}+1\right)}\right] f\left(x(\sigma_1\ldots\sigma_m)^{\frac{1}{m+1}}\right) d\sigma_1\ldots d\sigma_m. \tag{4.1.13}$$

According to the general Theorem 1.2.2 and the considerations in Section 3.5, it follows that this transformation can be considered also as a generalized (m-tuple) fractional integral of the form

$$\begin{aligned}P_0 f(x) &= c\, I_{m+1,m}^{\left(\frac{k}{m+1}-1\right),\left(\nu_k-\frac{k}{m+1}+1\right)} f(x)\\ &= c\int_0^1 G^{m,0}_{m,m}\left[\sigma\left|\begin{matrix}(\nu_k)_{k=1}^m\\ \left(\frac{k}{m+1}-1\right)_{k=1}^m\end{matrix}\right.\right] f\left(x\sigma^{\frac{1}{m+1}}\right) d\sigma, \text{ where } c=\sqrt{\frac{m+1}{(2\pi)^m}}\prod_{k=1}^{m}\Gamma(\nu_k+1).\end{aligned} \tag{4.1.14}$$

This transform is a transmutation operator from $\tilde{B} = \left(\frac{d}{dx}\right)^{m+1}$ to the arbitrary hyper-Bessel operator B (4.1.8) of the same order, that is:

$$P_0\left(\frac{d}{dx}\right)^{m+1}\tilde{y}(x) = BP_0\tilde{y}(x)$$

for functions $\tilde{y}(x)$ satisfying suitable initial conditions. This result is contained in Theorem 3.7.1, namely: The generalized Poisson transformation maps the solution $\tilde{y}(x)$ of the initial value problem (4.1.10) into the solution $y(x)$ of (4.1.7): $P_0\tilde{y}(x) = y(x)$, that is:

$$P_0\left\{\cos_{m+1}(x)\right\} = j^{(m)}_{\nu_1,\ldots,\nu_m}(x). \tag{4.1.15}$$

Replacing, in this transmutation formula, P_0 by (4.1.14) we obtain the following new integral representation of the normalized hyper-Bessel functions (4.1.3):

$$j_\nu(x) = j^{(m)}_{\nu_1,\dots,\nu_m}(x) = cI^{\left(\frac{k}{m+1}-1\right),\left(\nu_k-\frac{k}{m+1}+1\right)}_{m+1,m}\{\cos_{m+1}(x)\}$$
$$= \sqrt{\frac{m+1}{(2\pi)^m}}\prod_{k=1}^{m}\Gamma(\nu_k+1)\int_0^1 G^{m,0}_{m,m}\left[\sigma\left|\begin{matrix}(\nu_k)_{k=1}^m\\ \left(\frac{k}{m+1}-1\right)_{k=1}^m\end{matrix}\right.\right]\cos_{m+1}\left(x\sigma^{\frac{1}{m+1}}\right)d\sigma, \tag{4.1.16}$$

valid for $\nu_k \geq \frac{k}{m+1}-1$, $k=1,\dots,m$. In an alternative form this can be stated as the following theorem.

Theorem 4.1.1. (Poisson type integral representation of the hyper-Bessel functions) *For $\nu_k \geq \frac{k}{m+1}-1$, $k=1,\dots,m$ the following generalization of the Poisson integral (4.0.1) holds:*

$$J^{(m)}_{\nu_1,\dots,\nu_m}(x) = \sqrt{\frac{m+1}{(2\pi)^m}}\left(\frac{x}{m+1}\right)^{\sum_{k=1}^m \nu_k}\int_0^1 G^{m,0}_{m,m}\left[\sigma\left|\begin{matrix}(\nu_k)_{k=1}^m\\ \left(\frac{k}{m+1}-1\right)_{k=1}^m\end{matrix}\right.\right]\cos_{m+1}\left(x\sigma^{\frac{1}{m+1}}\right)d\sigma$$
$$= \sqrt{\frac{m+1}{(2\pi)^m}}\left(\frac{x}{m+1}\right)^{-1+\sum_{k=1}^m \nu_k}\int_0^1 G^{m,0}_{m,m}\left[\left(\frac{t}{x}\right)^{m+1}\left|\begin{matrix}\left(\nu_k+1-\frac{1}{m+1}\right)_1^m\\ \left(\frac{k-1}{m+1}\right)_1^m\end{matrix}\right.\right]\cos_{m+1}(t)dt. \tag{4.1.17}$$

In the same way, P_0 transforms the solution $\tilde{y}(x)$ of the initial value problem

$$\left(\frac{d}{dx}\right)^{m+1}\tilde{y}(x) = \tilde{y}(x); \quad \tilde{y}(0)=1, \tilde{y}'(0) = \dots = \tilde{y}^{(m)}(0) = 0, \tag{4.1.10'}$$

into the solution $y(x)$ of

$$By(x) = y(x); \quad y(0) = 1, y'(0) = \dots = y^{(m)}(0) = 0. \tag{4.1.7'}$$

Now, these solutions are:

$$\tilde{y}(x) = h_{1,m+1}(x) = \sum_{j=0}^{\infty}\frac{x^{(m+1)j}}{((m+1)j!)}, \tag{4.1.18}$$

referred to as a *generalized hyperbolic function* of order $(m+1)$ (this is a generalization of $h_{1,2}(x) = \operatorname{ch} x$; for more details about this and other generalized hyperbolic functions $h_{i,m+1}(x)$, $i=2,\dots,m+1$, see (D.13), Appendix) and

$$y(x) = i_\nu(x) = i^{(m)}_{\nu_1,\dots,\nu_m}(x) = \sum_{j=0}^{\infty}\left(\prod_{k=1}^{m}\frac{\Gamma(\nu_k+1)}{\Gamma(\nu_k+j+1)}\right)\frac{\left(\frac{x}{m+1}\right)^{(m+1)j}}{j!}, \tag{4.1.19}$$

the normalized hyper-Bessel function (4.1.4).

Therefore, the transmutation formula

$$P_0\left\{h_{1,m+1}(x)\right\} = i^{(m)}_{\nu_1,\ldots,\nu_m}(x)$$

yields an analogous result.

Theorem 4.1.1′. (Poisson type integral representation of the modified hyper-Bessel functions) *For $\nu_k \geq \frac{k}{m+1} - 1$, $k = 1,\ldots,m$, the following integral representations hold:*

$$i^{(m)}_{\nu_1,\ldots,\nu_m}(x) = \sqrt{\frac{m+1}{(2\pi)^m}}\prod_{k=1}^{m}\Gamma(\nu_k+1)\int_0^1 G^{m,0}_{m,m}\left[\sigma\left|\begin{matrix}(\nu_k)_{k=1}^m\\ \left(\frac{k}{m+1}-1\right)_{k=1}^m\end{matrix}\right.\right] h_{1,m+1}\left(x\sigma^{\frac{1}{m+1}}\right)d\sigma \tag{4.1.16′}$$

and

$$I^{(m)}_{\nu_1,\ldots,\nu_m}(x) = \sqrt{\frac{m+1}{(2\pi)^m}}\left(\frac{x}{m+1}\right)^{\sum\limits_{k=1}^m\nu_k}\int_0^1 G^{m,0}_{m,m}\left[\sigma\left|\begin{matrix}(\nu_k)_{k=1}^m\\ \left(\frac{k}{m+1}-1\right)_{k=1}^m\end{matrix}\right.\right] h_{1,m+1}\left(x\sigma^{\frac{1}{m+1}}\right)d\sigma$$

$$= \sqrt{\frac{m+1}{(2\pi)^m}}\left(\frac{x}{m+1}\right)^{-1+\sum\limits_{k=1}^m\nu_k}\int_0^1 G^{m,0}_{m,m}\left[\left(\frac{t}{x}\right)^{m+1}\left|\begin{matrix}\left(\nu_k+1-\frac{1}{m+1}\right)_1^m\\ \left(\frac{k-1}{m+1}\right)_1^m\end{matrix}\right.\right] h_{1,m+1}(t)dt. \tag{4.1.17′}$$

Generalizations of the Poisson integral for the hyper-Bessel functions (4.1.1) have been proposed *also by Delerue* [60] *and by Klučantčev* [218], but they are written *by means of repeated integrations.* For comparison, the result of Delerue ([60], p.259) has the form

$$J^{(m)}_{\nu_1,\ldots,\nu_m}(x) = \frac{(m+1)^{m+\frac{1}{2}}}{\sqrt{(2\pi)^{m+1}}}\left(\frac{x}{m+1}\right)^{\sum\limits_{k=1}^m\nu_k+j-m}\int_0^1\cdots\int_0^1\prod_{i=1}^{m-j}\left[\frac{(1-\zeta_i)^{\nu_i-1-\frac{i}{m+1}}}{\Gamma\left(\nu_i-\frac{i}{m+1}\right)}\zeta_i^{j+i}\right]$$

$$\times\prod_{i=m-j+1}^{m}\left[\frac{(1-\zeta_i)^{\nu_i-\frac{i}{m+1}}\zeta_i^{j+i-m-1}}{\Gamma\left(\nu_i-\frac{i}{m+1}+1\right)}\right] f^{(m+1)}_{j+1}(x\zeta_1\ldots\zeta_m)\,d\zeta_1\ldots d\zeta_m,$$

$$j = 0,1,\ldots,m, \tag{4.1.20}$$

where

$$f^{(m+1)}_{m+1}(x) := \cos_{m+1}(x)$$

and the other generalized trigonometric functions

$$f^{(m+1)}_{j+1}(x) := \sin_{m+1,j+1}(x),\quad j = 0,1,\ldots,m-1,$$

are defined as in Appendix, (D.10). In the case $j = m$ the same result is proved by Klučantčev [218, p. 55]:

$$J^{(m)}_{\nu_1,\ldots,\nu_m}(x) = \frac{(m+1)^{m+\frac{1}{2}}\left(\frac{x}{m+1}\right)^{\sum \nu_k}}{(2\pi)^{\frac{m}{2}} \prod\limits_{k=1}^{m} \Gamma\left(\nu_k - \frac{k}{m+1} + 1\right)}$$

$$\times \int_0^1 \ldots \int_0^1 \cos_{m+1}(x\zeta_1 \ldots \zeta_m) \prod_{i=1}^{m} \left[(1-\zeta_i)^{\nu_i - \frac{i}{m+1}} \zeta_i^{i-1}\right] d\zeta_1 \ldots d\zeta_m. \tag{4.1.21}$$

This follows also from (4.1.15), if one uses the repeated integral representation (4.1.13) of the Poisson type transformation P_0.

Let us consider *some special cases.*

Corollary 4.1.2. *For $m = 1$ Theorems 4.1.1 and 4.1.1′ give the classical Poisson integrals for the Bessel functions (see [106, II]):*

$$\begin{aligned} J_\nu(x) &= \frac{2\left(\frac{x}{2}\right)^\nu}{\sqrt{\pi}\Gamma\left(\nu+\frac{1}{2}\right)} \int_0^{\frac{\pi}{2}} \cos(x\sin\varphi)(\cos\varphi)^{2\nu} d\varphi \\ &= \frac{2\left(\frac{x}{2}\right)^\nu}{\sqrt{\pi}} \int_0^1 (1-t^2)^{\nu-\frac{1}{2}} \cos(xt)dt, \ \nu > -\frac{1}{2} \end{aligned} \tag{4.1.22}$$

and

$$I_\nu(x) = \frac{2\left(\frac{x}{2}\right)^\nu}{\sqrt{\pi}\Gamma\left(\nu+\frac{1}{2}\right)} \int_0^1 (1-t^2)^{\nu-\frac{1}{2}} \cosh(xt)dt, \ \nu > -\frac{1}{2}. \tag{4.1.23}$$

Now let us consider the so-called *n-Bessel functions* of Agarwal [6] (see also Example 3.4.11). More details about these functions are given in Appendix, Section D.ii. They are defined by the power series (D.9):

$$A_{\nu,n}(x) = \sum_{j=0}^{\infty} \frac{(-1)^j}{\{j!\}^n \{\Gamma(\nu+1+j)\}^2} \left(\frac{x}{2}\right)^{\nu+2j}.$$

It is evident (Lemma D.11) that the n-Bessel functions are special cases of hyper-Bessel functions of odd order $m = 2n-1$, namely

$$A_{\nu,n}(x) = J^{(2n-1)}_{\underbrace{\nu,\ldots,\nu}_{n},\underbrace{0,\ldots,0}_{n-1}}(z), \text{ where } z = (2n)\left(\frac{x}{2}\right)^{\frac{1}{n}}. \tag{4.1.24}$$

In particular, the "di-Bessel" functions ($n = 2$) of Exton [111] and the "tri-Bessel" functions ($n = 3$) are the following hyper-Bessel functions :

$$A_\nu(x) = A_{\nu,2}(x) = J^{(3)}_{\nu,\nu,0}\left(2^{\frac{3}{2}}\sqrt{x}\right); A_{\nu,3}(x) = J^{(5)}_{\nu,\nu,\nu,0,0}\left(3^3 2^{\frac{2}{3}}\sqrt[3]{x}\right).$$

Now, the conditions $\nu_k > \frac{k}{m+1} - 1 = \frac{k}{2n} - 1$, $k = 1, \ldots, 2n-1$, are equvalent to the unique requirement $\nu > -\frac{1}{2}$. In this case Theorem 4.1.1 gives the following result: if Ξ is the transformation

$$\Xi : f(x) \longrightarrow f\left(Nx^\omega\right), \text{ where } \omega = \frac{1}{n} > 0, N = 2^{\frac{n-1}{n}}n; \tag{4.1.25}$$

then

$$\begin{aligned} A_{\nu,n}(x) &= \Xi\left[\sqrt{\frac{2n}{(2\pi)^{2n-1}}}\left(\frac{x}{2n}\right)^{n\nu} I_{2n,2n-1}^{\left(\frac{k}{2n}-1\right),\left(\left(\nu-\frac{k}{2n}+1\right)_{k=1}^{n},\left(-\frac{k}{2n}+1\right)_{k=n+1}^{2n-1}\right)}\right]\{\cos_{2n}(x)\} \\ &= \sqrt{\frac{n}{\pi^{2n-1}}}2^{1-n}\left(\frac{x}{2}\right)^{\nu}\Xi I_{2n,2n-1}^{\left(\frac{k}{2n}-1\right),\left(\left(\nu-\frac{k}{2n}+1\right)_{k=1}^{n},\left(-\frac{k}{2n}+1\right)_{k=n+1}^{2n-1}\right)}\{\cos_{2n}(x)\} \\ &= \sqrt{\frac{n}{\pi^{2n-1}}}2^{1-n}\left(\frac{x}{2}\right)^{\nu} I_{2,2n-1}^{\left(\frac{k}{2n}-1\right),\left(\left(\nu-\frac{k}{2n}+1\right)_{k=1}^{n},\left(-\frac{k}{2n}+1\right)_{k=n+1}^{2n-1}\right)}\Xi\{\cos_{2n}(x)\}, \end{aligned}$$

that is

$$A_{\nu,n}(x) = \sqrt{\frac{n}{\pi^{2n-1}}}2^{1-n}\left(\frac{x}{2}\right)^{\nu} I_{2,2n-1}^{\left(\frac{k}{2n}-1\right),\left(\left(\nu-\frac{k}{2n}+1\right)_{k=1}^{n},\left(-\frac{k}{2n}+1\right)_{k=n+1}^{2n-1}\right)}\left\{\cos_{2n}\left(2n\left(\frac{x}{2}\right)^{\frac{1}{n}}\right)\right\}. \tag{4.1.26}$$

Written explicitly, this *generalized fractional integral representation of the n-Bessel functions*, has the following form.

Corollary 4.1.3. *For* $\nu > -\frac{1}{2}$,

$$\begin{aligned} A_{\nu,n}(x) &= \sqrt{\frac{n}{\pi^{2n-1}}}2^{1-n}\left(\frac{x}{2}\right)^{\nu}\int_0^1 G^{2n-1,0}_{2n-1,2n-1}\left[\sigma \left| \begin{matrix} \underbrace{\nu,\ldots,\nu}_{n},\underbrace{0,\ldots,0}_{n-1} \\ \left(\frac{k}{2n}-1\right)_1^{2n-1} \end{matrix}\right.\right] \\ &\times \cos_{2n}\left\{2n\left(\frac{x\sqrt{\sigma}}{2}\right)^{\frac{1}{n}}\right\} d\sigma \\ &= \sqrt{\frac{n}{\pi^{2n-1}}}2^{1-n}\left(\frac{x}{2}\right)^{\nu}\int_0^1\cdots\int_0^1 \prod_{k=1}^{n}\left[\frac{(1-\zeta_k)^{\nu-\frac{k}{2n}}}{\Gamma\left(\nu-\frac{k}{2n}+1\right)}\right]\prod_{k=n+1}^{2n-1}\left[\frac{(1-\zeta_k)^{-\frac{k}{2n}}}{\Gamma\left(1-\frac{k}{2n}\right)}\right] \\ &\times \prod_{k=1}^{2n-1}\left[\zeta_k^{\frac{k}{2n}-1}\right]\cos_{2n}\left\{2n\left(\frac{x}{2}\right)^{\frac{1}{n}}(\zeta_1\ldots\zeta_{2n-1})^{\frac{1}{2n}}\right\}d\zeta_1\ldots d\zeta_{2n-1}. \end{aligned} \tag{4.1.27}$$

In the following sections we need a modified form of the Poisson integrals (4.1.16), (4.1.17) *concerning the generalized hypergeometric function* ${}_0F_m(b_1, \ldots, b_m; -x)$ *itself.* Ones having the representation (4.1.16) of the normalized hyper-Bessel function (4.1.3), we have only to substitute

$$\nu_k + 1 \longrightarrow b_k, \ k = 1, \ldots, m, \qquad \left(\frac{x}{m+1}\right)^{m+1} \longrightarrow x.$$

Thus we get the following corollary.

Corollary 4.1.4. *For* $b_k \geq \frac{k}{m+1}$, $k = 1, \ldots, m$, *the generalized hypergeometric function of Bessel type* ${}_0F_m$, $m > 0$, *is a generalized fractional integral of the generalized cosine function:*

$${}_0F_m(b_1, \ldots, b_m; -x) = cI_{1,m}^{\left(\frac{k}{m+1}-1\right),\left(b_k - \frac{k}{m+1}\right)} \left\{\cos_{m+1}\left((m+1)(x)^{\frac{1}{m+1}}\right)\right\} \tag{4.1.28}$$

and therefore, it admits the integral representations of Poisson type:

$${}_0F_m\left((b_k)_1^m; -x\right) = c\int_0^1 G_{m,m}^{m,0}\left[\sigma \left| \begin{matrix} (b_k) \\ \left(\frac{k}{m+1}\right) \end{matrix} \right.\right] \sigma^{-1} \cos_{m+1}\left((m+1)(x\sigma)^{\frac{1}{m+1}}\right) d\sigma \tag{4.1.29}$$

and

$$\begin{aligned} {}_0F_m\left((b_k)_1^m; -x\right) = & c\int_0^1 \cdots \int_0^1 \left[\prod_{k=1}^m \frac{(1-t_k)^{b_k - \frac{k}{m+1} - 1} t_k^{\frac{k}{m+1}-1}}{\Gamma\left(b_k - \frac{k}{m+1}\right)}\right] \\ & \times \cos_{m+1}\left((m+1)(xt_1 \ldots t_m)^{\frac{1}{m+1}}\right) dt_1 \ldots dt_m \end{aligned} \tag{4.1.30}$$

with a constant

$$c = \sqrt{(m+1)(2\pi)^m} \prod_{j=1}^m \Gamma(b_j).$$

Further, we shall use this result to obtain an analogous integral representation for the more general Bessel type generalized hypergeometric function

$${}_pF_q\left(a_1, \ldots, a_p; b_1, \ldots, b_q; -x\right) \text{ with } p < q.$$

Now we procede to the consideration of *the case when the conditions* $\nu_k \geq \frac{k}{m+1} - 1$, $k = 1, \ldots, m$, that is, $b_k \geq \frac{k}{m+1}$, $k = 1, \ldots, m$, are not satisfied. *They were to ensure the positiveness of the multiorder of integration* of the generalized fractional integration operator by means of which the Poisson transformation P_0 was represented:

$$P_0 = cI_{m+1,m}^{\left(\frac{k}{m+1}-1\right),\left(\nu_k - \frac{k}{m+1}+1\right)} = cI_{m+1,m}^{\left(\frac{k}{m+1}-1\right),\left(b_k - \frac{k}{m+1}\right)}.$$

When the fully opposite conditions

$$b_k < \frac{k}{m+1} \quad \left(\nu_k < \frac{k}{m+1} - 1\right), \quad k = 1, \dots, m \tag{4.1.31}$$

are supposed, it is natural to expect that the transformation P_0 is to be considered as a generalized fractional differentiation operator (in view of Definitions 1.5.1 and 1.5.2), namely:

$$\begin{aligned} P_0 &= cI_{m+1,m}^{\left(\frac{k}{m+1}-1\right),\left(b_k-\frac{k}{m+1}\right)} = cI_{m+1,m}^{\left((b_k-1)+\left(\frac{k}{m+1}-b_k\right)\right),\left(-\left(\frac{k}{m+1}-b_k\right)\right)} \\ &= cD_{m+1,m}^{(b_k-1),\left(\frac{k}{m+1}-b_k\right)}. \end{aligned} \tag{4.1.32}$$

So, the integral representation (4.1.28) – (4.1.29) of ${}_0F_m$, $m > 0$ will be replaced by a differintegral representation:

$$\begin{aligned} {}_0F_m\left((b_k)_1^m; -x\right) &= cD_{1,m}^{(b_k-1),\left(\frac{k}{m+1}-b_k\right)} \left\{\cos_{m+1}\left((m+1)x^{\frac{1}{m+1}}\right)\right\} \\ &= cD_\eta \; I_{1,m}^{\left(\frac{k}{m+1}-1\right),\left(\eta_k+b_k-\frac{k}{m+1}\right)} \left\{\cos_{m+1}\left((m+1)x^{\frac{1}{m+1}}\right)\right\}, \end{aligned}$$

where

$$\eta_k = \begin{cases} \left[\frac{k}{m+1} - b_k\right] + 1 & \text{for non integer } \left(\frac{k}{m+1} - b_k\right), \\ \left(\frac{k}{m+1} - b_k\right) & \text{for integer } \left(\frac{k}{m+1} - b_k\right), \end{cases} \quad k = 1, \dots, m \tag{4.1.33}$$

and D_η is the differential operator (cf. (1.5.19) with $\beta = 1$, $\gamma_k \to b_k - 1$):

$$D_\eta = \prod_{k=1}^{m} \prod_{j=1}^{\eta_k} \left(x\frac{d}{dx} + b_k + j - 1\right). \tag{4.1.34}$$

Before stating and proving this result, we establish the following auxiliary proposition.

Lemma 4.1.5. *The generalized hypergeometric function of Bessel type ${}_0F_m$, $m > 0$, satisfies the differential relation ($\eta_1 > 0, \dots, \eta_m > 0$ are integers)*

$$D_\eta \frac{{}_0F_m\left((\eta_k + b_k)_1^m; -x\right)}{\prod_{j=1}^{m} \Gamma(\eta_j + b_j)} = \frac{{}_0F_m\left((b_k)_1^m; -x\right)}{\prod_{j=1}^{m} \Gamma(b_j)} \tag{4.1.35}$$

with integers $\eta_1, \dots, \eta_m > 0$ and D_η defined as in (4.1.34).

Proof. If we put $n = 1$, $p = 0$, $\sigma = b'_k$ in formula (5), p.44 of Luke [272], we obtain

$$\begin{aligned} x^{2-b'_k}\frac{d}{dx}x^{b'_k-1}{}_0F_m\left((b'_l)_{l=1}^m; -x\right) &= \left(x\frac{d}{dx} + b'_k - 1\right){}_0F_m\left((b'_k); -x\right) \\ &= (b'_k - 1)\,{}_0F_m\left(b'_1, \dots, b'_{k-1}, b'_k - 1, b'_{k+1}, \dots, b'_m; -x\right). \end{aligned}$$

Analogously, after η_k steps, we obtain for $b'_k = b_k + \eta_k$:

$$\left(x\frac{d}{dx} + b_k\right)\left(x\frac{d}{dx} + b_k + 1\right) \dots \left(x\frac{d}{dx} + b_k + \eta_k - 1\right) {}_0F_m\left(b'_1, \dots, b'_k, \dots, b'_m; -x\right)$$
$$= (b_k + \eta_k - 1)(b_k + \eta_k - 2) \dots b_k \; {}_0F_m\left(b'_1, \dots, b'_k - \eta_k = b_k, \dots, b'_m; -x\right)$$
$$= \frac{\Gamma(b_k + \eta_k)}{\Gamma(b_k)} {}_0F_m\left(b'_1, \dots, b_k, \dots, b'_m; -x\right).$$

But the function ${}_0F_m$ is symmetric in the parameters $b'_1, \dots, b'_m$ and therefore,

$$\prod_{j=1}^{m}\left[\left(x\frac{d}{dx} + b_i\right)\dots\left(x\frac{d}{dx} + b_i + \eta_i - 1\right)\right] {}_0F_m\left((\eta_k + b_k)_1^m; -x\right)$$
$$= \prod_{j=1}^{m}\left[\frac{\Gamma(b_i + \eta_i)}{\Gamma(b_i)}\right] {}_0F_m\left((b_k)_1^m; -x\right).$$

Thus the proposition is proved.

Now we are ready to state the basic result, analogous to Theorem 4.1.1.

Theorem 4.1.6. (Generalized fractional derivative representation of the hyper-Bessel functions) *Let the indices of Delerue's hyper-Bessel function $J^{(m)}_{\nu_1,\dots,\nu_m}$ satisfy conditions (4.1.31):*

$$\nu_k < \frac{k}{m+1} - 1, \quad k = 1, \dots, m.$$

Then, this function can be represented as a generalized m-tuple fractional derivative of the function $\cos_{m+1}(x)$:

$$J^{(m)}_{\nu_1,\dots,\nu_m}(x) = \sqrt{\frac{m+1}{(2\pi)^m}}\left(\frac{x}{m+1}\right)^{\sum \nu_i} D_{m+1,m}^{(\nu_k),\left(\frac{k}{m+1} - \nu_k - 1\right)}\left\{\cos_{m+1}(x)\right\} \tag{4.1.36}$$

or,

$$J^{(m)}_{\nu_1,\dots,\nu_m}(x) = \sqrt{\frac{(m+1)^3}{(2\pi)^m}}\left(\frac{x}{m+1}\right)^{1+\sum \nu_i} D_{m+1,m}^{\left(\nu_k + \frac{1}{m+1}\right),\left(\frac{k}{m+1} - \nu_k - 1\right)}\left\{\frac{\cos_{m+1}(x)}{x}\right\}, \tag{4.1.37}$$

that is, the following differintegral representation holds:

$$\begin{aligned} J^{(m)}_{\nu_1,\dots,\nu_m}(x) &= \sqrt{\frac{(m+1)}{(2\pi)^m}}\left(\frac{x}{m+1}\right)^{\sum \nu_i} \\ &\times \left[\prod_{k=1}^{m}\prod_{j=1}^{\eta_k}\left(\frac{1}{m+1}x\frac{d}{dx} + \nu_k + j\right)\right] I_{m+1,m}^{\left(\frac{k}{m+1} - 1\right),\left(\eta_k - \frac{k}{m+1} + \nu_k + 1\right)}\left\{\cos_{m+1}(x)\right\}, \end{aligned} \tag{4.1.38}$$

where

$$\eta_k = \begin{cases} \left[\frac{k}{m+1} - \nu_k - 1\right] + 1 & \textit{for non integer } \left(\frac{k}{m+1} - \nu_k - 1\right) \\ \left(\frac{k}{m+1} - \nu_k - 1\right) & \textit{for integer } \left(\frac{k}{m+1} - \nu_k - 1\right) \end{cases}, \qquad k = 1, \dots, m.$$

To prove the theorem, it is equivalent to prove the corresponding representation concerning the "normalized" hyper-Bessel function, or which is the same thing, the generalized hypergeometric function ${}_0F_m$, $m > 0$. This result can be obtained from (4.1.37) by the substitutions $\nu_k + 1 \longrightarrow b_k$, $k = 1, \dots, m$, $x \longrightarrow (m+1)x^{\frac{1}{m+1}}$.

Corollary 4.1.7. *Let the conditions (4.1.31) be satisfied:*

$$b_k < \frac{k}{m+1}, \qquad k = 1, \dots, m.$$

Then,

$$\begin{aligned} {}_0F_m(b_1, \dots, b_m; -x) &= \sqrt{\frac{m+1}{(2\pi)^m}} D_{1,m}^{(b_k - 1), \left(\frac{k}{m+1} - b_k\right)} \left\{ \cos_{m+1}\left((m+1)x^{\frac{1}{m+1}}\right)\right\} \\ &= \sqrt{\frac{(m+1)^3}{(2\pi)^m}} x^{\frac{1}{m+1}} D_{1,m}^{\left(b_k - \frac{k}{m+1}\right), \left(\frac{k}{m+1} - b_k\right)} \left\{ \frac{\cos_{m+1}\left((m+1)x^{\frac{1}{m+1}}\right)}{x^{\frac{1}{m+1}}} \right\}, \end{aligned} \tag{4.1.39}$$

that is,

$$\begin{aligned} {}_0F_m\left((b_k)_1^m; -x\right) = &\sqrt{\frac{m+1}{(2\pi)^m}} \left[\prod_{k=1}^{m} \prod_{j=1}^{\eta_k} \left(x\frac{d}{dx} + b_k + j - 1 \right) \right] \\ &\times \int_0^1 G_{m,m}^{m,0}\left[\sigma \left| \begin{matrix} (b_k + \eta_k - 1)_1^m \\ \left(\frac{k}{m+1} - 1\right)_1^m \end{matrix} \right. \right] \cos_{m+1}\left[(m+1)(x\sigma)^{\frac{1}{m+1}}\right] d\sigma. \end{aligned} \tag{4.1.40}$$

Proof. According to Definition 1.5.2,

$$D_{1,m}^{(b_k - 1), \left(\frac{k}{m+1} - b_k\right)} = D_\eta I_{1,m}^{\left(\frac{k}{m+1} - 1\right), \left(\eta_k + b_k - \frac{k}{m+1}\right)},$$

where D_η is the differential operator (4.1.34) and η_k, $k = 1, \dots, m$, are the integer numbers (4.1.33). Then $\frac{k}{m+1} - b_k > 0$, $\eta_k + b_k - \frac{k}{m+1} > 0$, $k = 1, \dots, m$, and using the generalized Poisson integral formula (4.1.28) (Corollary 4.1.4), we obtain

$$c\, D_{1,m}^{(b_k-1),\left(\frac{k}{m+1}-b_k\right)} \left\{\cos_{m+1}\left((m+1)x^{\frac{1}{m+1}}\right)\right\} = \prod_{j=1}^{m}\left[\frac{\Gamma(b_j)}{\Gamma(b_j+\eta_j)}\right]$$

$$\times D_\eta \left\{ \frac{\sqrt{m+1}}{(2\pi)^m} \left[\prod_{j=1}^{m}\Gamma(b_j+\eta_j)\right] I_{1,m}^{\left(\frac{k}{m+1}-1\right),\left(b_k+\eta_k-\frac{k}{m+1}\right)} \cos_{m+1}\left((m+1)x^{\frac{1}{m+1}}\right)\right\}$$

$$= \left[\prod_{j=1}^{m}\frac{\Gamma(b_j)}{\Gamma(b_j+\eta_j)}\right] D_\eta \left\{ {}_0F_m\left((b_k+\eta_k)_1^m; -x\right)\right\}.$$

On the other hand, by virtue of Lemma 4.1.5,

$$\left[\prod_{j=1}^{m}\frac{\Gamma(b_j)}{\Gamma(b_j+\eta_j)}\right] D_\eta \left\{ {}_0F_m\left((b_k+\eta_k)_1^m; -x\right)\right\} = {}_0F_m\left((b_k)_1^m; -x\right)$$

and thus the required representation (4.1.39) is established.

Conversely, for $b_k = \nu_k + 1$, $k = 1, \ldots, m$, and $x \longrightarrow \left(\frac{x}{m+1}\right)^{m+1}$, this expresion multiplied by $\left(\frac{x}{m+1}\right)^{\sum \nu_j}$, gives (4.1.37).

Corollary 4.1.8. *For $m = 1$ Theorem 4.1.6, especially representation (4.1.37′), reduces to the classical fractional derivative representation (4.0.3) of the Bessel function $J_\nu(x)$ [252]:*

$$J_\nu(x) = \frac{2}{\sqrt{\pi}}\left(\frac{x}{2}\right)^{1+\nu} D_{2,1}^{\nu+\frac{1}{2},-\nu-\frac{1}{2}}\left\{\frac{\cos x}{x}\right\} = \frac{x^{-\nu}}{2^\nu\sqrt{\pi}} D_{x^2}^{-\nu-\frac{1}{2}}\left\{\frac{\cos x}{x}\right\}.$$

Taking the index ν of the form $\nu = -\eta - \frac{1}{2}$, $\eta = 0, 1, 2, \ldots$, we obtain the well-known differential formula (4.0.2) for the spherical Bessel functions :

$$J_{-\eta-\frac{1}{2}}(x) = \frac{2^{\eta+\frac{1}{2}}x^{\eta+\frac{1}{2}}}{\sqrt{\pi}}\frac{d^\eta}{(dx^2)^\eta}\left\{\frac{\cos x}{x}\right\}, \eta = 0, 1, 2, \ldots$$

Let us proceed to a *multiple analogue of this result.*

Corollary 4.1.9. *Let the indices of the hyper-Bessel function $J_{\nu_1,\ldots,\nu_m}^{(m)}(x)$ be of the form*

$$\nu_k = \frac{k}{m+1} - 1 - \eta_k, \text{ with integers } \eta_k \geq 0,\ k = 1, \ldots, m. \tag{4.1.41}$$

Then, the fractional differentiation operator $D_{m+1,m}^{(\nu_k),\left(\frac{k}{m+1}-\nu_k-1\right)}$ on the right-hand side of (4.1.37) has the integers $\delta_k = \eta_k = \frac{k}{m+1} - \nu_k - 1$, $k = 1, \ldots, m$, as components of its *multiorder of differentiation. Thus in Definition (1.5.19):*

$$D_{m+1,m}^{(\nu_k),\left(\frac{k}{m+1}-\nu_k-1\right)} = D_\eta I_{m+1,m}^{\left(\nu_k+\frac{k}{m+1}-\nu_k-1\right),(\eta_k-\delta_k)}$$
$$= \left[\prod_{k=1}^{m}\prod_{j=1}^{\eta_k}\left(\frac{1}{m+1}x\frac{d}{dx}+\nu_k+j\right)\right] I_{m+1,m}^{\left(\frac{k}{m+1}-1\right),(\eta_k-\delta_k)},$$

the fractional integration operator turns into the identity operator ($\eta_k = \delta_k$, $k = 1,\ldots,m$):

$$I_{m+1,m}^{\left(\frac{k}{m+1}-1\right),(\eta_k-\delta_k)} = I_{m+1,m}^{\left(\frac{k}{m+1}-1\right),(0)} = I,$$

and hence,

$$D_{m+1,m}^{(\nu_k),\left(\frac{k}{m+1}-\nu_k-1\right)} = D_\eta = \prod_{k=1}^{m}\prod_{j=1}^{\eta_k}\left(\frac{1}{m+1}x\frac{d}{dx}+\nu_k+j\right)$$

is a purely differential operator of integer order $\eta = \eta_1 + \cdots + \eta_m$. In this manner, we obtain that if the indices of a hyper-Bessel function satisfy (4.1.41), then this function can be represented by means of a differential operator D_η of integer order of $\cos_{m+1}(x)$, namely:

$$J^{(m)}_{\left(-\eta_k+\frac{k}{m+1}-1\right)_1^m}(x) = \sqrt{\frac{m+1}{(2\pi)^m}}\left(\frac{x}{m+1}\right)^{\sum\eta_k} D_\eta\left\{\cos_{m+1}(x)\right\}$$
$$= \sqrt{\frac{m+1}{(2\pi)^m}}\left(\frac{x}{m+1}\right)^{\sum\eta_k}\left[\prod_{k=1}^{m}\prod_{j=1}^{\eta_k}\left(\frac{1}{m+1}x\frac{d}{dx}+\frac{k}{m+1}-\eta_k+j-1\right)\right]\left\{\cos_{m+1}(x)\right\}. \tag{4.1.42}$$

By analogy with the Bessel functions $J_{-\eta-\frac{1}{2}}(x)$ we call their multiple counterparts $J^{(m)}_{\left(-\eta_k+\frac{k}{m+1}-1\right)}$ *"spherical" hyper-Bessel functions* (see also Appendix, Definition D.8 and Klučantčev [218]).

If neither of the conditions

$$\nu_k \geq \frac{k}{m+1}-1, \quad \nu_k < \frac{k}{m+1}-1$$

for the indices of a hyper-Bessel function $J^{(m)}_{\nu_1,\ldots,\nu_m}$ hold simultaneously for all $k = 1,\ldots,m$, then neither Theorem 4.1.1 nor Theorem 4.1.6 can be applied. Nevertheless, a combination of such results holds. Namely, the corresponding hyper-Bessel function can be represented as a differintegral operator of $\cos_{m+1}(x)$, which is a composition of Erdélyi-Kober fractional integrals and derivatives.

Such an example concerns the *n-Bessel functions* (see also Corollary 4.1.3 and Appendix, (D.20)-(D.21)). Let us start with the fractional integral representation (4.1.26). The components $\left(-\frac{k}{2n}+1\right)$, $k = n+1,\ldots,2n-1$, of the multiorder of "integration" are always positive, while the signs of the first n components $\left(\nu - \frac{k}{2n} + 1\right)$, $k = 1, \ldots, n$, depend on the value of ν. When $\nu \geq -\frac{1}{2}$, then $\nu - \frac{k}{2n} + 1 \geq \nu - \frac{n}{2n} + 1 > 0$, $k = 1, \ldots, n$. On the contrary, *if* $\nu < \frac{1}{2n} - 1$, then $\nu - \frac{k}{2n} + 1 < 0$ for all $k = 1, \ldots, n$. In this case *we can decompose the* $(2n-1)$*-tuple* I*-operator in (4.1.26) into two operators,* n- and $(n-1)$-tuple ones, respectively:

$$I_{2,2n-1}^{\left(\frac{k}{2n}-1\right),\left(\left(\nu-\frac{k}{2n}+1\right)_1^n,\left(-\frac{k}{2n}+1\right)_{n+1}^{2n-1}\right)} = I_{2,n}^{\left(\frac{k}{2n}-1\right),\left(\nu-\frac{k}{2n}+1\right)_1^n} I_{2,n-1}^{\left(\frac{k}{2n}-1\right),\left(1-\frac{k}{2n}\right)_{n+1}^{2n-1}},$$

interpret the first one as an n-tuple fractional derivative (according to Definitions 1.5.1 and 1.5.2):

$$\begin{aligned} I_{2,n}^{\left(\frac{k}{2n}-1\right),\left(\nu-\frac{k}{2n}+1\right)_1^n} &= D_{2,n}^{(\nu)_{k=1}^n,\left(\frac{k}{2n}-\nu-1\right)_{k=1}^n} \\ &= \left[\prod_{k=1}^{m}\prod_{j=1}^{\eta_k}\left(\frac{1}{2}x\frac{d}{dx}+\nu+j\right)\right] I_{2,n}^{\left(\frac{k}{2n}-1\right),\left(\eta_k-\frac{k}{2n}+\nu+1\right)}, \end{aligned}$$

where

$$\eta_k = \begin{cases} \left[\frac{k}{2n}-\nu-1\right]+1 & \text{for non integer } \left(\frac{k}{2n}-\nu-1\right) \\ \left(\frac{k}{2n}-\nu-1\right) & \text{for integer } \left(\frac{k}{2n}-\nu-1\right) \end{cases}, \qquad \kappa = 1, \ldots, n. \tag{4.1.43}$$

Thus, we obtain

$$\begin{aligned} A_{\nu,n}(x) &= \sqrt{\frac{n}{\pi^{2n-1}}}\,2^{n-1}\left(\frac{x}{2}\right)^{\nu}\left[\prod_{k=1}^{m}\prod_{j=1}^{\eta_k}\left(\frac{1}{2}x\frac{d}{dx}+\nu+j\right)\right] \\ &\quad \times I_{2,n}^{\left(\frac{k}{2n}-1\right),\left(\eta_k-\frac{k}{2n}+\nu+1\right)} I_{2,n-1}^{\left(\frac{k}{2n}-1\right)_{n+1}^{2n-1},\left(1-\frac{k}{2n}\right)_{n+1}^{2n-1}}\left\{\cos_{2n}\left(2n\left(\frac{x}{2}\right)^{\frac{1}{n}}\right)\right\} \\ &= \sqrt{\frac{n}{\pi^{2n-1}}}\,2^{n-1}\left(\frac{x}{2}\right)^{\nu}\left[\prod_{k=1}^{m}\prod_{j=1}^{\eta_k}\left(\frac{1}{2}x\frac{d}{dx}+\nu+j\right)\right] \\ &\quad \times I_{2,2n-1}^{\left(\frac{k}{2n}-1\right),\left(\left(\eta_k-\frac{k}{2n}+\nu+1\right)_1^n,\left(1-\frac{k}{2n}\right)_{n+1}^{2n-1}\right)}\left\{\cos_{2n}\left(2n\left(\frac{x}{2}\right)^{\frac{1}{n}}\right)\right\}, \end{aligned}$$

since the composition of two generalized fractional integrals is a fractional integral too, whose multiplicity is $n + (n-1) = (2n-1)$ (see (1.3.12)). In this manner we obtain the following result.

Corollary 4.1.10. *If* $\nu < \frac{1}{2n} - 1$, *then the* n*-Bessel function (D.20) has the following* *differintegral representation:*

$$
\begin{aligned}
A_{\nu,n}(x) &= \sqrt{\frac{n}{\pi^{2n-1}}}\, 2^{n-1}\left(\frac{x}{2}\right)^{\nu}\left[\prod_{k=1}^{m}\prod_{j=1}^{\eta_k}\left(\frac{1}{2}x\frac{d}{dx}+\nu+j\right)\right] \\
&\times \int_0^1 G^{2n-1,0}_{2n-1,2n-1}\left[\sigma\left|\begin{matrix}(\eta_k+\nu)_1^n, \underbrace{(0,\dots,0)}_{n-1} \\ \left(\frac{k}{2n}-1\right)_1^{2n-1}\end{matrix}\right.\right]\cos_{2n}\left[2n\left(\frac{x\sqrt{\sigma}}{2}\right)^{\frac{1}{n}}\right]d\sigma,
\end{aligned}
\tag{4.1.44}
$$

where the integers η_k*,* $k = 1, \dots, n$*, are defined by (4.1.43).*

4.2. Poisson type integral representations of the generalized hypergeometric functions

The basic idea used here consists of a well-known fractional integral relation between the generalized hypergeometric functions ${}_{p+1}F_{q+1}$ and ${}_pF_q$. In different variants, it can be found in a series of handbooks and research papers: Erdélyi et al. [106], Askey [25, p.19], Lavoie, Osler and Tremblay [252, p.261], etc., namely:

$$
\begin{aligned}
&R^{\alpha}\left\{x^{\nu-1}\,{}_pF_q\left(a_1,\dots,a_p;b_1,\dots,b_q;\lambda x\right)\right\} \\
&\quad = \left[\frac{\Gamma(\nu)}{\Gamma(\alpha+\nu)}\right]x^{\alpha+\nu-1}\,{}_{p+1}F_{q+1}\left(\nu,a_1,\dots,a_p;\alpha+\nu,b_1,\dots,b_q;\lambda x\right),
\end{aligned}
\tag{4.2.1}
$$

where $\Re\alpha > 0$, $\Re\nu > 0$ (in particular, $\nu = 1$ can be taken), $p \le q+1$ (and $|\lambda x| < 1$, if $p = q+1$).

We shall repeatedly use this relation in the following convenient form, taking into account the symmetry of the ${}_pF_q$-functions with respect to the sets of parameters $a_1, \dots, a_p$ and $b_1, \dots, b_q$.

Lemma 4.2.1. *The Riemann-Liouville operator of integration of arbitrary order increases by 1 the indices p, q of the generalized hypergeometric function* ${}_pF_q$*. More exactly: if* $b_{q+1} > a_{p+1} > 0$ *and* $p \le q+1$*, then*

$$
\begin{aligned}
&{}_{p+1}F_{q+1}\left(a_1,\dots,a_p,a_{p+1};b_1,\dots,b_q,b_{q+1};\lambda x\right) \\
&\quad = \frac{\Gamma\left(b_{q+1}\right)}{\Gamma\left(a_{p+1}\right)}x^{1-b_{q+1}}R^{b_{q+1}-a_{p+1}}\left\{x^{a_{p+1}-1}\,{}_pF_q\left(a_1,\dots,a_p;b_1,\dots,b_q;\lambda x\right)\right\} \\
&\quad = \frac{\Gamma\left(b_{q+1}\right)}{\Gamma\left(a_{p+1}\right)}\int_0^1\frac{(1-t)^{b_{q+1}-a_{p+1}-1}}{\Gamma\left(b_{q+1}-a_{p+1}\right)}t^{a_{p+1}-1}\,{}_pF_q\left(a_1,\dots,a_p;\alpha+\nu,b_1,\dots,b_q;\lambda tx\right)dt,
\end{aligned}
\tag{4.2.2}
$$

where λ *is a constant, and in addition* $|\lambda x| < 1$ *is required, if* $p = q+1$*.*

By the way, in terms of Erdélyi-Kober operators of fractional integration (1.1.17):

$$
I_1^{a_k-1,b_l-a_k} = x^{1-b_l}R^{b_l-a_k}x^{a_k-1},
\tag{4.2.3}
$$

relation (4.2.2) has the more concise form

$$
\begin{aligned}
&{}_{p+1}F_{q+1}\left(a_1,\ldots,a_p,a_{p+1};b_1,\ldots,b_q,b_{q+1};\lambda x\right)\\
&\qquad = I_1^{a_{p+1}-1,b_{q+1}-a_{p+1}}\left\{{}_pF_q\left(a_1,\ldots,a_p;b_1,\ldots,b_q;\lambda x\right)\right\}.
\end{aligned}
\tag{4.2.2'}
$$

This result suggests that the *generalized hypergeometric function* ${}_pF_q$ *can be obtained from one of the three "basic" hypergeometric functions*

$$
\begin{cases}
{}_0F_{q-p} & \textit{for } p<q\\
{}_1F_1 & \textit{for } p=q\\
{}_2F_1 & \textit{for } p=q+1
\end{cases},
$$

by means of a finite number of Erdélyi-Kober fractional integrals (4.2.3.).

On the other hand, the functions ${}_1F_1$ and ${}_2F_1$ can be represented as fractional integrals of corresponding elementary functions, these results being well known too. A representation of the same type for the ${}_0F_{q-p}$-functions $(q-p>0)$ was proposed in Section 4.1 (see (4.1.28)); in this case a $(q-p)$-tuple generalized fractional integral was used.

Thus it remains to interpret the compositions of the Erdélyi-Kober operators (4.2.3) arising in above considerations, as generalized fractional integration operators with single integral representations involving Meijer's G-functions. In this manner, we obtain representations of the ${}_pF_q$-functions by means of definite integrals involving Meijer's G-function and some elementary functions.

4.2.i. Poisson integral representations of Bessel type generalized hypergeometric functions ${}_pF_q$, $p<q$

Denote

$$m=q-p>0.$$

For the hyper-Bessel functions ${}_0F_{q-p}={}_0F_m$ we dispose with Poisson type integral representation (4.1.28) (Corollary 4.1.4).

Further, we use Lemma 4.2.1, as well as the basic results of Chapter 1 to prove the *following intermediate result.*

Theorem 4.2.2. *Let the conditions*

$$b_{q-p+k}>a_k>0,\quad k=1,\ldots,p \tag{4.2.4}$$

be satsfied. Then, each generalized hypergeometric function ${}_pF_q$, $p<q$ *of Bessel type is an p-tuple (generalized) fractional integral of the hyper-Bessel function* ${}_0F_{q-p}$, *namely:*

$$
\begin{aligned}
&{}_pF_q\left(a_1,\ldots,a_p;b_1,\ldots,b_q;\lambda x\right)\\
&\qquad = \left(\prod_{j=1}^{p}\frac{\Gamma\left(b_{q-p+j}\right)}{\Gamma\left(a_j\right)}\right) I_{1,p}^{(\gamma_k'),(\delta_k')}\left\{{}_0F_{q-p}\left(b_1,\ldots,b_{q-p};\lambda x\right)\right\},
\end{aligned}
\tag{4.2.5}
$$

where

$$\gamma'_k = a_k - 1,\ \delta'_k = b_{q-p+k} - a_k > 0,\ k = 1, \ldots, p. \tag{4.2.6}$$

In other words, the following integral representations hold under conditions (4.2.4):

$$\begin{aligned}
&{}_pF_q\left(a_1, \ldots, a_p; b_1, \ldots, b_q; \lambda x\right) \\
&= \left(\prod_{j=1}^{p} \frac{\Gamma\left(b_{q-p+j}\right)}{\Gamma\left(a_j\right)}\right) \int_0^1 G_{p,p}^{p,0}\left[\sigma \,\middle|\, \begin{matrix} \left(b_{q-p+j}\right)_{j=1}^{p} \\ \left(a_j\right)_{j=1}^{p} \end{matrix}\right] \sigma^{-1}\, {}_0F_{q-p}(\lambda x \sigma) d\sigma \qquad (4.2.7) \\
&= \left(\prod_{j=1}^{p} \frac{\Gamma\left(b_{q-p+j}\right)}{\Gamma\left(a_j\right)}\right) \int_0^1 \cdots \int_0^1 \prod_{k=1}^{p} \left[(1 - t_k)^{b_{q-p+k} - a_k - 1}\, t_k^{a_k - 1}\right] \\
&\quad \times {}_0F_{q-p}\left(\lambda x t_1 \ldots t_p\right) dt_1 \ldots dt_p. \qquad (4.2.7')
\end{aligned}$$

under conditions (4.2.4)

Proof. Let us apply Lemma 4.2.1 to the function ${}_0F_{q-p} = {}_0F_m$:

$$\begin{aligned}
&{}_1F_{m+1}\left(a_1; b_1, \ldots, b_{m+1}; \lambda x\right) \\
&\qquad = \frac{\Gamma\left(b_{m+1}\right)}{\Gamma\left(a_1\right)} \left(x^{1-b_{m+1}} R^{b_{m+1}-a_1} x^{a_1-1}\right) {}_0F_m\left(b_1, \ldots, b_m; \lambda x\right).
\end{aligned}$$

Similarly, we obtain

$$\begin{aligned}
&{}_2F_{m+2}\left(a_1, a_2; b_1, \ldots, b_{m+1}, b_{m+2}; \lambda x\right) \\
&= \frac{\Gamma\left(b_{m+2}\right)}{\Gamma\left(a_2\right)} \left(x^{1-b_{m+2}} R^{b_{m+2}-a_2} x^{a_2-1}\right) {}_1F_{m+1}\left(a_1; b_1, \ldots, b_{m+1}; \lambda x\right). \\
&= \frac{\Gamma\left(b_{m+1}\right)\Gamma\left(b_{m+2}\right)}{\Gamma\left(a_1\right)\Gamma\left(a_2\right)} \left(x^{1-b_{m+2}} R^{b_{m+2}-a_2} x^{a_2-1}\right) \left(x^{1-b_{m+1}} R^{b_{m+1}-a_1} x^{a_1-1}\right) \\
&\quad \times {}_0F_m\left(b_1, \ldots, b_m; \lambda x\right).
\end{aligned}$$

In this manner, after p-times application of Lemma 4.2.1 we get the relation

$$\begin{aligned}
&{}_pF_q\left(a_1, \ldots, a_p; b_1, \ldots, b_q; \lambda x\right) \\
&= \prod_{j=1}^{p} \left[\frac{\Gamma\left(b_{m+j}\right)}{\Gamma\left(a_j\right)} x^{1-b_{m+j}} R^{b_{m+j}-a_j} x^{a_j-1}\right] {}_0F_m\left(b_1, \ldots, b_m; \lambda x\right) \\
&= \left(\prod_{j=1}^{p} \frac{\Gamma\left(b_{m+j}\right)}{\Gamma\left(a_j\right)}\right) \left[x^{1-b_{m+p}} R^{b_{m+p}-a_p} x^{a_p-b_{m+p-1}} R^{b_{m+p-1}-a_{p-1}} \ldots x^{a_1-1}\right] {}_0F_m(\lambda x).
\end{aligned}$$

It is more convenient to express the multipliers in the second line of this equality as Erdélyi-Kober fractional integrals (1.1.17), according to the general relation

$$I_1^{\gamma,\delta} f(x) = \left\{x^{-(\gamma+\delta)} R^{\delta} x^{\gamma}\right\} f(x), \tag{4.2.8}$$

now having the form (4.2.3). Then,

$$ {}_pF_q\left(a_1,\ldots,a_p;b_1,\ldots,b_q;\lambda x\right)=\prod_{j=1}^{p}\left[\frac{\Gamma\left(b_{m+j}\right)}{\Gamma\left(a_j\right)}I_1^{a_j-1,b_{m+j}-a_j}\right]{}_0F_m(\lambda x) $$

$$ =\prod_{j=1}^{p}\left[\frac{\Gamma\left(b_{m+j}\right)}{\Gamma\left(a_j\right)}I_{1,1}^{a_j-1,b_{m+j}-a_j}\right]{}_0F_m(\lambda x)=\left[\prod_{j=1}^{p}\frac{\Gamma\left(b_{m+j}\right)}{\Gamma\left(a_j\right)}\right]I_{1,p}^{(a_k-1),(b_{m+k}-a_k)}\,{}_0F_m(\lambda x), $$

due to the "composition" Theorem 1.2.10. Using the explicit representations (1.1.6) and (1.2.25) of the p-tuple fractional integrals

$$ I_{1,p}^{(a_k-1),(b_{m+k}-a_k)}=\prod_{k=1}^{p}I_1^{a_k-1,b_{m+k}-a_k}, \tag{4.2.9} $$

we obtain the integral representations (4.2.7), (4.2.7′).

Note. Representation (4.2.7′) can be found also in Prudnikov, Brychkov and Marichev [369, p.438,(11)]. The concept used here has been proposed in the author's papers [196], [81]. The same idea is used repeatedly in Sections 4.2.ii, 4.2.iii and 4.3 and in Kiryakova [198], [200], [209].

EXAMPLE. Let us consider the case $p=1$, $q=2$, when $m=q-p=1$. Then, for $c>a>0$, representations (4.2.7)-(4.2.7′) take the form

$$ {}_1F_2(a;b,c;\lambda x)=\frac{\Gamma(c)}{\Gamma(a)\Gamma(c-a)}\int_0^1(1-t)^{c-a-1}t^{a-1}\,{}_0F_1(b;\lambda xt)dt, $$

or (with $\lambda=-1$),

$$ {}_1F_2(a;b,c;-x)=\frac{\Gamma(b)\Gamma(c)}{\Gamma(a)\Gamma(c-a)}\int_0^1(1-t)^{c-a-1}t^{\alpha-\frac{b+1}{2}}J_{b-1}\left(2\sqrt{xt}\right)dt. $$

This is a fractional integral relation between the Bessel and ${}_1F_2$-functions. On the other hand,

$$ {}_1F_2(a;b;c;-x)=\frac{\Gamma(b)\Gamma(c)}{\Gamma(a)}G_{1,3}^{1,1}\left[x\,\middle|\,\begin{matrix}1-a\\0,1-b,1-c\end{matrix}\right], $$

and this G-function is a *"named"* special function if, for instance, $a=\frac{1}{2}$, $b=1-\nu$, $c=1+\nu$ (see [106, I, 5.6], formulas (1), (21)):

$$ G_{1,3}^{1,1}\left[x\,\middle|\,\begin{matrix}\frac{1}{2}\\0,\nu,-\nu\end{matrix}\right]=\sqrt{\pi}J_\nu\left(\sqrt{x}\right)J_{-\nu}\left(\sqrt{x}\right). $$

In this case Theorem 4.2.1, and the above example ,in particular, give the following result.

Corollary 4.2.3. *If $-\frac{1}{2} < \nu < \frac{1}{2}$, then the integral relationship*

$$J_\nu(x) J_{-\nu}(x) = \frac{x^{-\nu}}{\sqrt{\pi}\Gamma\left(\frac{1}{2}-\nu\right)} \int_0^1 (1-t)^{-\frac{1}{2}-\nu} t^{-\frac{\nu+1}{2}} J_\nu\left(2x\sqrt{t}\right) dt \tag{4.2.10}$$

holds.

Theorem 4.2.4. *Let $p < q$ and let the conditions*

$$b_k > \frac{k}{q-p+1}, \; k = 1,\ldots,q-p; \quad b_{q-p+k} > a_k > 0, \; k = 1,\ldots,p \tag{4.2.11}$$

be satisfied. Then, the ${}_pF_q$-function, $p < q$ (the generalized hypergeometric function of Bessel type) is a generalized q-tuple fractional integral of $\cos_{q-p+1}(x)$, namely:

$${}_pF_q\left(a_1,\ldots,a_p; b_1,\ldots,b_q; -x\right) = A I_{1,q}^{(\gamma_k),(\delta_k)} \left\{\cos_{q-p+1}\left((q-p+1)x^{\frac{1}{q-p+1}}\right)\right\}, \tag{4.2.12}$$

where

$$A = \sqrt{\frac{q-p+1}{(2\pi)^{q-p}}} \left[\frac{\prod_{j=1}^{q}\Gamma(b_j)}{\prod_{i=1}^{p}\Gamma(a_i)}\right]$$

and the parameters γ_k, δ_k, $k = 1,\ldots,q$ are defined as follows:

$$\gamma_k = \begin{cases} \frac{k}{q-p+1} - 1, & k = 1,\ldots,q-p \\ a_{k-q+p} - 1, & k = q-p+1,\ldots,q \end{cases}; \quad \delta_k = \begin{cases} b_k - \frac{k}{q-p+1}, & k = 1,\ldots,q-p \\ b_k - a_{k-q+p}, & k = q-p+1,\ldots,q. \end{cases} \tag{4.2.13}$$

Relation (4.2.12) means that the following Poisson type integral representations of ${}_pF_q$, $p < q$ hold:

$$\begin{aligned} {}_pF_q\left(a_1,\ldots,a_p; b_1,\ldots,b_q; -x\right) &= A \int_0^1 G_{q,q}^{q,0}\left[\sigma \left| \begin{matrix} (b_k)_{k=1}^{q} \\ \left(\frac{k}{q-p+1}\right)_{k=1}^{q-p}, (a_{k-q+p})_{k=q-p+1}^{q} \end{matrix}\right.\right] \\ &\quad \times \sigma^{-1} \cos_{q-p+1}\left[(q-p+1)(x\sigma)^{\frac{1}{q-p+1}}\right] d\sigma \end{aligned} \tag{4.2.14}$$

$$\begin{aligned} &= A \int_0^1 \underset{(q)}{\cdots} \int_0^1 \prod_{k=1}^{q-p} \left[\frac{(1-t_k)^{b_k - \frac{k}{q-p+1} - 1}}{\Gamma\left(b_k - \frac{k}{q-p+1}\right)} t_k^{\frac{k}{q-p+1}-1}\right] \\ &\quad \times \prod_{k=q-p+1}^{q} \left[\frac{(1-t_k)^{b_k - a_{k-q+p} - 1}}{\Gamma(b_k - a_{k-q+p})} t_k^{a_{k-q+p}-1}\right] \\ &\quad \times \cos_{q-p+1}\left[(q-p+1)(xt_1\ldots t_q)^{\frac{1}{q-p+1}}\right] dt_1\ldots t_q. \end{aligned} \tag{4.2.14$'$}$$

Proof. Indeed,

$$
\begin{aligned}
{}_pF_q\left(a_1,\ldots,a_p;b_1,\ldots,b_q;-x\right) &= \Gamma\ I_{1,p}^{(\gamma_k'),(\delta_k')}\left\{{}_0F_{q-p}\left(b_1,\ldots,b_{q-p};-x\right)\right\}\\
&= \Gamma I_{1,p}^{(\gamma_k'),(\delta_k')}\left\{I_{1,p}^{\left(\frac{k}{q-p+1}-1\right),\left(b_k-\frac{k}{q-p+1}\right)} cos_{q-p+1}\left((q-p+1)x^{\frac{1}{q-p+1}}\right)\right\}
\end{aligned}
$$

with $c=\sqrt{(q-p+1)(2\pi)^{q-p}}$, $\Gamma=\prod_{i=1}^{p}\left[\frac{\Gamma(b_{q-p+i})}{\Gamma(a_i)}\right]$. Further, we denote $A=c\Gamma$. Due to (1.3.12) (Theorem 1.3.8) we have that the product

$$
I_{1,p}^{(\gamma_k'),(\delta_k')}I_{1,q-p}^{\left(\frac{k}{q-p+1}-1\right),\left(b_k-\frac{k}{q-p+1}\right)}=I_{1,p}^{(\gamma_k),(\delta_k)}
$$

is a generalized fractional integral of multiplicity $p+(q-p)=q$. This operator has explicit representations of form (1.1.6) or (1.2.25), leading to (4.2.14), (4.2.14′), respectively. The proof is over.

EXAMPLE. In the case $m=q-p=1$, Theorem 4.2.4 gives a representation of the function

$$
{}_pF_{p+1}(-x)=\frac{\prod_{1}^{p+1}\Gamma\left(b_j\right)}{\prod_{1}^{p}\Gamma\left(a_i\right)}G_{p,p+2}^{1,p}\left[x\left|\begin{matrix}(1-a_k)_{k=1}^{p}\\ 0,(1-b_k)_{k=1}^{p+1}\end{matrix}\right.\right]
$$

as a generalized $(p+1)$-tuple fractional integral of the usual *cosine function.* For $p=1,2$ and for a special choice of the parameters, the $G_{p,p+2}^{1,p}$-functions reduce themselves to *Lommel and Struve functions* $S_{\mu,\nu}(x)$, $H_\nu(x)$ ((C.8)-(C.9)), *to products of Bessel functions* , etc. For example, we have (see [106, I, 5.6] and [286, p. 58]):

$$
\begin{aligned}
G_{1,3}^{1,1}\left[x\left|\begin{matrix}\frac{1}{2}\\ 0,\nu,-\nu\end{matrix}\right.\right] &= \sqrt{\pi}J_\nu\left(\sqrt{x}\right)J_{-\nu}\left(\sqrt{x}\right), G_{1,3}^{1,1}\left[x\left|\begin{matrix}\frac{1}{2}\\ 0,\nu,-\nu\end{matrix}\right.\right]\\
&= \sqrt{\pi}J_\nu^2\left(\sqrt{x}\right), G_{2,4}^{1,2}\left[x\left|\begin{matrix}\frac{\lambda}{2},\frac{\lambda+1}{2}\\ \frac{\lambda+\mu+\nu}{2},\frac{\lambda-\mu+\nu}{2},\frac{\lambda+\mu-\nu}{2},\frac{\lambda-\mu-\nu}{2}\end{matrix}\right.\right].
\end{aligned}
$$

In this way formula (4.2.14′) leads to the following special cases (corollaries of representation (4.2.14) can also be written).

Corollary 4.2.5. (Poisson type integral representation of the product $x^\lambda J_\nu(x)J_{-\nu}(x)$) *For* $-\frac{1}{2}<\nu<\frac{1}{2}$*:*

$$
x^\lambda J_\nu(x)J_{-\nu}(x)=\frac{x^\lambda}{\sqrt{\pi}}\int_0^1\int_0^1\frac{(1-t_1)^{-\nu-\frac{1}{2}}(1-t_2)^{\nu-\frac{1}{2}}}{\Gamma\left(-\nu+\frac{1}{2}\right)\Gamma\left(\nu+\frac{1}{2}\right)}(t_1t_2)^{-\frac{1}{2}}\cos\left(2x\sqrt{t_1t_2}\right)dt_1dt_2. \tag{4.2.15}
$$

Corollary 4.2.6. (Poisson type representation of $x^\lambda J_\nu^2(x)$) *For* $\nu > -\frac{1}{4}$*:*

$$x^\lambda J_\nu^2(x) = \frac{x^{\lambda+\frac{\nu}{2}}}{\sqrt{\pi}} \int_0^1 \int_0^1 \frac{(1-t_1)^{2\nu-\frac{1}{2}}(1-t_2)^{-\frac{1}{2}}}{\Gamma\left(2\nu+\frac{1}{2}\right)} t_1^{-\frac{1}{2}} t_2^{\nu-\frac{1}{2}} \cos\left(2x\sqrt{t_1 t_2}\right) dt_1 dt_2. \tag{4.2.16}$$

Corollary 4.2.7. (Poisson type integral representation of the product $x^\lambda J_\mu(x) J_\nu(x)$) *For* $\lambda > \mu > \frac{1}{2}$, $\mu + \nu > -1$*:*

$$\begin{aligned} x^\lambda J_\mu(x) J_\nu(x) = \frac{x^{\lambda+\mu+\nu}}{\sqrt{\pi}} \int_0^1 \int_0^1 \int_0^1 & \frac{(1-t_1)^{\mu-\frac{1}{2}}(1-t_2)^{\frac{\nu-\mu}{2}-1}(1-t_3)^{\frac{\mu+\nu-1}{2}}}{\Gamma\left(\mu+\frac{1}{2}\right)\Gamma\left(\frac{\nu-\mu}{2}\right)\Gamma\left(\frac{\mu+\nu-1}{2}\right)} \\ & \times t_1^{-\frac{1}{2}} t_2^{\frac{\mu+\nu}{2}} t_3^{\frac{\mu+\mu-1}{2}} \cos\left(2x\sqrt{t_1 t_2 t_3}\right) dt_1 dt_2 dt_3. \end{aligned} \tag{4.2.17}$$

Let us note that after the substitutions $t_k = sin^2\theta_k$, $k = 1, 2, \ldots$, integrals (4.2.15)-(4.2.17) acquire the *"standard" form of the Poisson type integrals*, for example:

$$\begin{aligned} x^\lambda J_\mu(x) J_\nu(x) = & \frac{8}{\sqrt{\pi}} \cdot \frac{x^{\lambda+\mu+\nu}}{\Gamma\left(\mu+\frac{1}{2}\right)\Gamma\left(\frac{\nu-\mu}{2}\right)\Gamma\left(\frac{\mu+\nu-1}{2}\right)} \\ & \times \int_0^{\frac{\pi}{2}} \int_0^{\frac{\pi}{2}} \int_0^{\frac{\pi}{2}} \cos^{2\mu}\theta_1 \cos^{\nu-\mu-1}\theta_2 \cos^{\mu+\nu}\theta_3 \sin^{\mu+\nu+1}\theta_2 \sin^{\mu+\nu}\theta_3 \\ & \times \cos\left(2x \sin\theta_1 \sin\theta_2 \sin\theta_3\right) d\theta_1 d\theta_2 d\theta_3. \end{aligned} \tag{4.2.17$'$}$$

As far as we know, representations (4.2.14)-(4.2.17) seem to be new.

4.2.ii. Poisson integral representations of confluent type generalized hypergeometric functions ${}_pF_q$

In this case, Lemma 4.2.1 takes the form

$$\begin{aligned} & {}_{k+1}F_{k+1}\left(a_1, \ldots, a_{k+1}; b_1, \ldots, b_{k+1}; \lambda x\right) \\ & = \frac{\Gamma\left(b_{k+1}\right)}{\Gamma\left(a_{k+1}\right)} x^{1-b_{k+1}} R^{b_{k+1}-a_{k+1}} x^{a_{k+1}-1} {}_kF_k\left(a_1, \ldots, a_k; b_1, \ldots, b_k; \lambda x\right) \\ & = \frac{\Gamma\left(b_{k+1}\right)}{\Gamma\left(a_{k+1}\right)} I_1^{a_{k+1}-1, b_{k+1}-a_{k+1}} {}_kF_k(\lambda x) = \frac{\Gamma\left(b_{k+1}\right)}{\Gamma\left(a_{k+1}\right)} I_{1,1}^{a_{k+1}-1, b_{k+1}-a_{k+1}} \left\{{}_kF_k(\lambda x)\right\}. \end{aligned} \tag{4.2.18}$$

On the other hand, we have the following known representation of the basic function ${}_1F_1$ (see e.g. [107, p.187, (14)] or [252, p. 261]).

Lemma 4.2.8. *For* $b > a > 0$*:*

$$ {}_1F_1(a; b; \lambda x) = \Phi(a; b; \lambda x) = \frac{\Gamma(b)}{\Gamma(a)} x^{1-b} R^{b-a} x^{a-1} \left\{e^{\lambda x}\right\}, \tag{4.2.19}$$

or in terms of Erdélyi-Kober operators:

$$
\begin{aligned}
{}_1F_1(a;b;\lambda x) &= \frac{\Gamma(b)}{\Gamma(a)} I_1^{a-1,b-a}\left\{e^{\lambda x}\right\} \\
&= \frac{\Gamma(b)}{\Gamma(a)} x^{1-a} I_1^{0,b-a}\left\{x^{a-1}e^{\lambda x}\right\},
\end{aligned} \tag{4.2.19'}
$$

if we wish to preserve the power multiplier x^{a-1} with the exponential function $e^{\lambda x}$.

Now, from (4.2.18) and (4.2.19′) we obtain the following result.

Theorem 4.2.9. (Poisson type integral representation of ${}_pF_p$) *Let $p = q$ and*

$$
b_k > a_k > 0, \quad k = 1, \ldots, p. \tag{4.2.20}
$$

Then, the generalized hypergeometric function ${}_pF_p(\lambda x)$ is an p-tuple fractional integral of the elementary function $\left\{x^{a_1-1}e^{\lambda x}\right\}$, namely:

$$
{}_pF_p\left(a_1, \ldots, a_p; b_1, \ldots, b_p; \lambda x\right) = \Gamma' x^{1-a_1} I_{1,p}^{(\gamma_k),(\delta_k)}\left\{x^{a_1-1}e^{\lambda x}\right\}, \tag{4.2.21}
$$

where $\gamma_k = a_k - a_1$, $\delta_k = b_k - a_k$, $k = 1, \ldots, p$, and $\Gamma' = \prod\limits_{j=1}^{p} \frac{\Gamma(b_j)}{\Gamma(a_j)}$. This relation yield the following integral representations:

$$
{}_pF_p(a_1, \ldots, a_p;\ b_1, \ldots, b_q;\ \lambda x) = \Gamma' \int_0^1 G_{p,p}^{p,0}\left[\sigma \left| \begin{matrix} (b_k)_{k=1}^p \\ (a_k)_{k=1}^p \end{matrix} \right.\right] \sigma^{-1} \exp(\lambda x \sigma) d\sigma \tag{4.2.22}
$$

and

$$
\begin{aligned}
&{}_pF_p\left(a_1, \ldots, a_p; b_1, \ldots, b_p; \lambda x\right) \\
&= \Gamma' \int_0^1 \underset{(p)}{\ldots} \int_0^1 \prod_{k=1}^{p}\left[\frac{(1-t_k)^{b_k-a_k-1} t_k^{a_k-1}}{\Gamma(b_k-a_k)}\right] \exp\left(\lambda x t_1 \ldots t_p\right) dt_1 \ldots dt_p.
\end{aligned} \tag{4.2.22'}
$$

Proof. Relation (4.2.18) applied subsequently $(p-1)$ times to

$$
{}_pF_p\left(a_1, \ldots, a_p; b_1, \ldots, b_p; \lambda x\right),
$$

yields:

$$
\begin{aligned}
&{}_pF_p(\lambda x) = \frac{\Gamma(b_p)}{\Gamma(a_p)} I_{1,1}^{a_p-1,b_p-a_p}\left\{{}_{p-1}F_{p-1}(\lambda x)\right\} \\
&= \frac{\Gamma(b_p)\Gamma(b_{p-1})}{\Gamma(a_p)\Gamma(a_{p-1})} I_1^{a_p-1,b_p-a_p} I_1^{a_{p-1}-1,b_{p-1}-a_{p-1}}\left\{{}_{p-2}F_{p-2}(\lambda x)\right\} \\
&= \ldots = \prod_{k=2}^{p}\left[\frac{\Gamma(b_k)}{\Gamma(a_k)} I_1^{a_k-1,b_k-a_k}\right]\left\{{}_1F_1(\lambda x)\right\} = \left[\prod_{k=2}^{p}\frac{\Gamma(b_k)}{\Gamma(a_k)}\right] I_{1,p-1}^{(a_k-1),(b_k-a_k)}\left\{{}_1F_1(\lambda x)\right\}.
\end{aligned}
$$

Further, we use (4.2.19′) and property (1.3.3) of the generalized fractional integrals, to obtain the representation

$$
\begin{aligned}
{}_pF_p(\lambda x) &= \left[\prod_{k=1}^{p} \frac{\Gamma(b_k)}{\Gamma(a_k)}\right] I_{1,p-1}^{(a_k-1),(b_k-a_k)} I_1^{0,b_1-a_1} \left\{x^{a_1-1} e^{\lambda x}\right\} \\
&= \left[\prod_{k=1}^{p} \frac{\Gamma(b_k)}{\Gamma(a_k)}\right] I_{1,p-1}^{(a_k-a_1),(b_k-a_k)} x^{1-a_1} I_1^{0,b_1-a_1} \left\{x^{a_1-1} e^{\lambda x}\right\} \\
&= \left[\prod_{k=1}^{p} \frac{\Gamma(b_k)}{\Gamma(a_k)}\right] x^{1-a_1} I_{1,p}^{(a_k-a_1),(b_k-a_k)} x^{1-a_1} \left\{x^{a_1-1} e^{\lambda x}\right\},
\end{aligned}
$$

due to the composition property (1.3.12). The same result can be written also in the *concise form*

$$
\begin{aligned}
{}_pF_p(\lambda x) &= \left[\prod_{k=1}^{p} \frac{\Gamma(b_k)}{\Gamma(a_k)}\right] I_{1,p-1}^{(a_k-1),(b_k-a_k)} I_1^{a_1-1,b_1-a_1} \left\{e^{\lambda x}\right\} \\
&= \left[\prod_{k=1}^{p} \frac{\Gamma(b_k)}{\Gamma(a_k)}\right] I_{1,p}^{(a_k-1),(b_k-a_k)} \left\{e^{\lambda x}\right\}.
\end{aligned}
\tag{4.2.23}
$$

It remains now to replace the generalized fractional integrals by their explicit integral representations.

Remark 1. Since the function ${}_pF_p$ is symmetric with respect to the parameters a_1, a_2, $\ldots$, a_p, we can replace the parameter a_1 in formula (4.2.21) by any one a_l, $l = 1, \ldots, p$.

Remark 2. Formula (4.2.21) can be seen as a formal corollary of Theorem 4.2.4 with $m = q - p = 0$, when $\cos_{m+1}\left((m+1)x^{\frac{1}{m+1}}\right) = \cos_1 x = e^{-x}$ (see Appendix, Definition D.2). *Remark 3.* Theorem 4.2.9 (in particular, Lemma 4.2.8) is not applicable to the Laguerre polynomials

$$
L_n^{\alpha}(x) = \binom{n+\alpha}{n} {}_1F_1(-n; \alpha+1; x), n = 0, 1, 2, \ldots ,
$$

since $b - a = \alpha + 1 + n > 0$ but the other condition $a = -n > 0$ is not fulfilled. However, there is another well-known relation between this confluent hypergeometric function and the elementary function $x^{-a_1+b_1-1}e^{-x} = x^{n+\alpha}e^{-x}$. This is the corresponding Rodrigues differential formula for $L_n^{(\alpha)}(x)$ (see Appendix, (C.18))

4.2.iii. Poisson integrals for the generalized hypergeometric functions of Gauss type ($p = q + 1$)

Now we use Lemma 4.2.1 in the following form:

$$\begin{aligned}
&{}_{(k+1)+1}F_{k+1}\left(a_1, \ldots, a_{k+1}, a_{k+2}; b_1, \ldots, b_{k+1}; \lambda x\right) \\
&= \frac{\Gamma\left(b_{k+1}\right)}{\Gamma\left(a_{k+2}\right)} x^{1-b_{k+1}} R^{b_{k+1}-a_{k+2}} x^{a_{k+2}-1}\ {}_{k+1}F_k\left(a_1, \ldots, a_{k+1}; b_1, \ldots, b_k; \lambda x\right) \\
&= \frac{\Gamma\left(b_{k+1}\right)}{\Gamma\left(a_{k+2}\right)} I_1^{a_{k+2}-1, b_{k+1}-a_{k+2}} \left\{ {}_{k+1}F_k(\lambda x)\right\},
\end{aligned} \tag{4.2.24}$$

provided $b_{k+1} > a_{k+2} > 0$, $k = 1, 2, \ldots$, and $|\lambda x| < 1$. For bevity we take $\lambda := \varepsilon = \pm 1$.

Suppose $b_2 > a_3 > 0, \ldots, b_q > a_{q+1} > 0$ and $|x| < 1$. Then, by $(q-1)$-times application of relation (4.2.24) we obtain, analogously to the previous considerations, that

$$\begin{aligned}
&{}_{q+1}F_q\left(a_1, \ldots, a_{q+1}; b_1, \ldots, b_q; \varepsilon x\right) \\
&\quad = \left[\prod_{k=1}^{q-1} \frac{\Gamma\left(b_{k+1}\right)}{\Gamma\left(a_{k+2}\right)}\right] I_{1,q-1}^{(a_{k+2}-1)_1^{q-1}, (b_{k+1}-a_{k+2})_1^{q-1}} \left\{ {}_2F_1(\varepsilon x)\right\},
\end{aligned} \tag{4.2.25}$$

i.e. ${}_{q+1}F_q$ is an $(q-1)$-tuple fractional integral of ${}_2F_1$.

In a modified form, this result can be found, for example, in Owa and Srivastava [352, (56)].

On the other hand, we have a known fractional integral representation of the Gauss hypergeometric function ${}_2F_1$ by means of elementary function (see [107, p.186, (9)] or [252, p. 261]). It is given by the following lemma.

Lemma 4.2.10. *Let $b_1 > a_2 > 0$ and $|x| < 1$, $\varepsilon = \pm 1$. Then,*

$${}_2F_1\left(a_1, a_2; b_1; \varepsilon x\right) = \frac{\Gamma\left(b_1\right)}{\Gamma\left(a_2\right)} x^{1-b_1} R^{b_1-a_2} x^{a_2-1} \left\{(1-\varepsilon x)^{-a_1}\right\}, \tag{4.2.26}$$

or in terms of Erdélyi-Kober operators $I_\beta^{\gamma,\delta}$,

$$\begin{aligned}
{}_2F_1\left(a_1, a_2; b_1; \varepsilon x\right) &= \frac{\Gamma\left(b_1\right)}{\Gamma\left(a_2\right)} I_1^{a_2-1, b_1-a_2} \left\{(1-\varepsilon x)^{-a_1}\right\} \\
&= \frac{\Gamma\left(b_1\right)}{\Gamma\left(a_2\right)} x^{1-a_2} I_1^{0, b_1-a_2} \left\{x^{a_2-1}(1-\varepsilon x)^{-a_1}\right\}.
\end{aligned} \tag{4.2.26'}$$

Note. Let us note that this relation can be extended analytically outside the unit disk $|x| < 1$, provided $|\arg(1-\varepsilon x)| < \pi$.

Combining (4.2.25) and (4.2.26′), we reach the final result:

$$
\begin{aligned}
{}_{q+1}F_q(\varepsilon x) &= \Gamma'' I_{1,q-1}^{(a_{k+2}-1)_1^{q-1},(b_{k+1}-a_{k+2})_1^{q-1}} I_1^{a_2-1,b_1-a_2}\left\{(1-\varepsilon x)^{-a_1}\right\} \\
&= \Gamma'' I_{1,q}^{(a_{k+2}-1)_0^{q-1},(b_{k+1}-a_{k+2})_0^{q-1}}\left\{(1-\varepsilon x)^{-a_1}\right\} = \Gamma'' I_{1,q}^{(a_{k+1}-1)_1^{q},(b_k-a_{k+1})_1^{q}}\left\{(1-\varepsilon x)^{-a_1}\right\},
\end{aligned}
\tag{4.2.27}
$$

where $\Gamma'' = \prod_{j=1}^{q}\left[\frac{\Gamma(b_j)}{\Gamma(a_{j+1})}\right]$, or in other forms:

$$
\begin{aligned}
{}_{q+1}F_q(\varepsilon x) &= \Gamma'' I_{1,q}^{(a_{k+1}-1)_2^{q},(b_k-a_{k+1})_2^{q}} x^{1-a_2} I_1^{0,b_1-a_2}\left\{x^{a_2-1}(1-\varepsilon x)^{-a_1}\right\} \\
&= \Gamma'' x^{1-a_2} I_{1,q}^{(a_{k+1}-1)_1^{q},(b_k-a_{k+1})_1^{q}}\left\{x^{a_2-1}(1-\varepsilon x)^{-a_1}\right\}.
\end{aligned}
$$

In this manner we have already proved the following theorem.

Theorem 4.2.11. *Let $p = q+1$ and let the conditions*

$$
b_k > a_k > 0, \quad k = 1, \ldots, q \tag{4.2.28}
$$

hold. Then, in the unit disk $|x| < 1$ the generalized hypergeometric function of Gauss type ${}_{q+1}F_q(\varepsilon x)$ ($\varepsilon = \pm 1$) is an q-tuple generalized fractional integral of the elementary function $x^{a_2-1}(1-\varepsilon x)^{-a_1}$, namely:

$$
\begin{aligned}
&{}_{q+1}F_q\left(a_1, \ldots, a_{q+1}; b_1, \ldots, b_q; \pm x\right) \\
&\quad = \Gamma'' x^{1-a_2} I_{1,q}^{(a_{k+1}-1)_1^{q},(b_k-a_{k+1})_1^{q}}\left\{x^{a_2-1}(1 \mp x)^{-a_1}\right\}.
\end{aligned}
\tag{4.2.29}
$$

This means that the following Poisson type integral representation holds:

$$
{}_{q+1}F_q(\pm x) = \Gamma'' \int_0^1 G_{q,q}^{q,0}\left[\sigma \,\middle|\, \begin{matrix}(b_k)\\(a_{k+1})\end{matrix}\right] \sigma^{-1}(1 \mp x\sigma)^{-a_1}\, d\sigma, \tag{4.2.30}
$$

or in terms of repeated integrals:

$$
\begin{aligned}
&{}_{q+1}F_q\left(a_1, \ldots, a_{q+1}; b_1, \ldots, b_q; \pm x\right) \\
&\quad = \left[\prod_{j=1}^{q} \frac{\Gamma(b_j)}{\Gamma(a_{j+1})\,\Gamma(b_j - a_{j+1})}\right] \int_0^1 \overset{(q)}{\cdots} \int_0^1 \prod_{j=1}^{q}\left[(1-t_k)^{b_k-a_{k+1}-1}\, t_k^{a_{k+1}-1}\right] \\
&\qquad \times \left(1 \mp x t_1 \ldots t_q\right)^{-a_1} dt_1 \ldots dt_q.
\end{aligned}
\tag{4.2.30′}
$$

Remark . The repeated integral representation (4.2.30′) can be found also in Prudnikov, Brychkov and Marichev [369, p. 438, (10)].

Corollary 4.2.12. *For $q = 1$, $\varepsilon = 1$ representations (4.2.30-30′) coincide with Lemma 4.2.10, written in the form of the known Euler formula for the Gauss function (see [106, I, p.72, (10)]):*

$$ {}_2F_1(a_1, a_2; b_1; x) = \frac{\Gamma(b_1)}{\Gamma(a_2)\Gamma(b_1 - a_2)} \int_0^1 \frac{(1-t)^{b_1-a_2-1} t^{a_2-1}}{(1-xt)^{a_1}} dt, \tag{4.2.31} $$

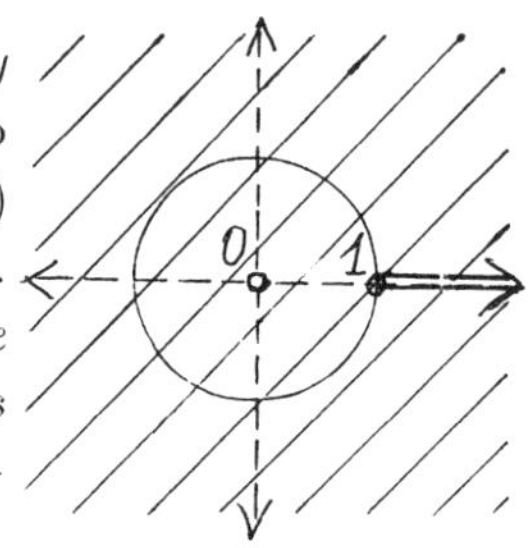

valid for $b_1 > a_2 > 0$ and $|x| < 1$. This formula *proposes a way for an analytical continuation of ${}_2F_1(x)$* outside the unit disk to the domain $|\arg(1-x)| < \pi$, where the right-hand side of (4.2.31) represents an analytical function of x (see the Remark in [106, I, p. 72]). For the same reasons, *formulas (4.2.30), (4.2.30′) can serve as analytical extension of the generalized hypergeometric functions ${}_{q+1}F_q(x)$*, $q \geq 1$ from the unit disk $|x| < 1$ to the domain $|\arg(1-x)| < \pi$.

For other Poisson type integral representations the reader is referred also to the useful works of Askey [25], Koornwinder [233], Prudnikov, Brychkov and Marichev [369]-[370], etc.

4.3. Fractional derivative representation of the generalized hypergeometric functions (Analogues of the Rodrigues formulas)

Rodrigues type formulas for defining special functions of mathematical physics and for developing their theory and applications, have been recently used by many authors like: Al-Bassam [10], [13]-[17], Al-Bassam and Kalla [22], Askey [25], Nikiforov and Uvarov [318]-[319], Rusev [414], Srivastava [460], Srivastava, Lavoie and Tremblay [469], etc. The results presented here have been inspired by the paper of Lavoie, Osler and Tremblay [252]. They are published, in parts, in the author's papers [198], [200], [202], [209].

"At times the representation by fractional derivatives is more convenient than that by power series and by definite integrals because the notation itself suggests manipulations which would otherwise not seem obvious. As an example, the definite integral (the so-called Sonine integral for the Bessel functions)

$$ \int_0^x J_{-\beta-\frac{1}{2}}(t)\left(x^2-t^2\right)^{-\alpha-1} t^{-\beta+\frac{1}{2}} dt = 2^{-a+\frac{1}{2}} x^{-\alpha-\beta-\frac{1}{2}} J_{-\alpha-\beta-\frac{1}{2}}(x), $$

in fractional derivative notation, is nothing but the obvious relation

$$ D_{x^2}^{\alpha} D_{x^2}^{\beta} \left\{\frac{\cos x}{x}\right\} = D_{x^2}^{\alpha+\beta} \left\{\frac{\cos x}{x}\right\}. $$

Mathematicians have more experience with D^n and fortunately, this experience often carries over to D^α" [252, p. 240-241].

In the above-cited paper, fractional derivative (in the classical sense) representations for the Bessel functions $J_\nu(x)$ as well as for both other basic hypergeometric functions ${}_1F_1$, ${}_2F_1$, have been mentioned, as natural extensions of the fractional integral representations in Lemmas 4.2.8, 4.2.10 for negative orders of integration. From this point of view, it might be expected that the integral representations of ${}_pF_q$ found in Section 4.2 could be extended to differintegral (differential or integro-differential) representations when the components of the corresponding multiorders of "integration" are negative. Furthermore, in Section 4.1 we have shown that such an extension holds in the case of hyper-Bessel functions ${}_0F_m$. Now, we prove this hypothesis for ${}_pF_q$ in the general case $p \leq q+1$. Some special cases are considered when the representations are purely differential and the corresponding generalized hypergeometric functions are said to be *"spherical"* hypergeometric functions.

The basic results used now as a starting point are the fractional derivative analogues of Lemmas 4.2.1, 4.2.8, 4.2.10, mentioned in [252] as well as in other papers on this subject. To illustrate the manner in which results for fractional integrals can be transferred into corresponding ones for fractional derivatives, we also give sketches of the proofs and stress on the conditions under which these analogues hold.

Lemma 4.3.1. *If* $a_{p+1} > a_{q+1} > 0$, *then the fractional derivative relation holds:*

$$\begin{aligned} {}_{p+1}F_{q+1}&\left(a_1, \ldots, a_{p+1}; b_1, \ldots, b_{q+1}; \lambda x\right) \\ &= \frac{\Gamma\left(b_{q+1}\right)}{\Gamma\left(a_{p+1}\right)} \left(x^{1-b_{q+1}} D^{a_{p+1}-b_{q+1}} x^{a_{p+1}-1}\right) \\ &\quad \times {}_pF_q\left(a_1, \ldots, a_p; b_1, \ldots, b_q; \lambda x\right), \end{aligned} \tag{4.3.1}$$

where $p \leq q+1$ *(if* $p = q+1$*, then* $|\lambda x| < 1$ *or* $|\arg(1-\lambda x)| < \pi$ *is additionally required).*

Remark . Here D^δ denotes the classical fractional derivative (1.5.7) of order $\delta > 0$. In the sense of Chapter 1, the one-tuple (single) generalized fractional derivative, i.e. the so-called Erdélyi-Kober fractional derivative $(1.6.7')$ has the form

$$D_{1,1}^{\gamma,\delta} = D_1^{\gamma,\delta} = x^{-\gamma} D^\delta x^{\gamma+\delta}, \ \delta > 0, \qquad \left(\text{cf. } I_1^{\gamma+\delta} = x^{-(\gamma+\delta)} R^\delta x^\gamma\right) \tag{4.3.2}$$

and the m-tuple fractional derivatives are compositions of m single derivatives: $D_{\beta,m}^{(\gamma_k),(\delta_k)} = D_{\beta,1}^{\gamma_m,\delta_m}\left(\ldots D_{\beta,1}^{\gamma_1,\delta_1}\right)$. Their explicit representations are given by Definition 1.5.4.

Taking into account (4.3.2), we can write down relation (4.3.1) in the concise form

$${}_{p+1}F_{q+1}(\lambda x) = \frac{\Gamma\left(b_{q+1}\right)}{\Gamma\left(a_{p+1}\right)} D_{1,1}^{b_{q+1}-1, a_{p+1}-b_{q+1}} \left\{{}_pF_q(\lambda x)\right\}. \tag{4.3.1'}$$

This result shows that ${}_{p+1}F_{q+1}$ can be obtained from ${}_pF_q$ by application of a suitable Erdélyi-Kober fractional derivative, and so, from one of the basic functions ${}_0F_{q-p}$, ${}_1F_1$ or ${}_2F_1$, by means of a composition of such derivatives.

Proof. For $a_{p+1} > b_{q+1}$ we introduce the integer

$$\eta = \begin{cases} [a_{p+1} - b_{q+1}] + 1, & \text{if } (a_{p+1} - b_{q+1}) \text{ is non integer} \\ (a_{p+1} - b_{q+1}), & \text{if } (a_{p+1} - b_{q+1}) \text{ is integer.} \end{cases}$$

Then, by the classical definition of the Riemann-Liouville fractional derivative:

$$D^{a_{p+1}-b_{q+1}} = \left(\frac{d}{dx}\right)^{\eta} R^{\eta-(a_{p+1}-b_{q+1})} \text{ with } \eta - (a_{p+1} - b_{q+1}) \geq 0.$$

Now, we can use the fractional integral representation (4.2.1):

$$D^{a_{p+1}-b_{q+1}} \left\{x^{a_{p+1}-1}\, {}_pF_q(\lambda x)\right\} = \left(\frac{d}{dx}\right)^{\eta} R^{\eta-(a_{p+1}-b_{q+1})} \left\{x^{a_{p+1}-1}\, {}_pF_q(\lambda x)\right\}$$
$$= \left(\frac{d}{dx}\right)^{\eta} \left\{\frac{\Gamma(a_{p+1})}{\Gamma(b_{q+1}+\eta)} x^{b_{q+1}+\eta-1}\, {}_{p+1}F_{q+1}\left(a_1, \ldots, a_{p+1}; b_1, \ldots, b_q, (b_{q+1}+\eta); \lambda x\right)\right\}.$$

According to Prudnikov, Brychkov and Marichev [369, p.442, (51)] (see also Luke [272, I]), the right-hand side of this equality gives

$$\frac{\Gamma(a_{p+1})}{\Gamma(b_{q+1}+\eta)} \frac{\Gamma(b_{q+1}+\eta)}{\Gamma(b_{q+1})} x^{b_{q+1}+\eta-1-\eta}\, {}_{p+1}F_{q+1}\left(a_1, \ldots, a_{p+1}; b_1, \ldots, b_{q+1}; \lambda x\right).$$

Multiplied by $\frac{\Gamma(b_{q+1})}{\Gamma(a_{p+1})} x^{1-b_{q+1}}$, this expression gives exactly ${}_{p+1}F_{q+1}(\lambda x)$.

Further, we consider separately the cases $p < q$, $p = q$ and $p = q + 1$.

4.3.i. Rodrigues type formulas for the generalized hypergeometric functions ${}_pF_q$, $p < q$

Theorem 4.3.2. *Let $p < q$ and $a_i > b_i > 0$, $i = 1, \ldots, p$. Then, the generalized hypergeometric function of Bessel type ${}_pF_q$ is an p-tuple fractional derivative of the hyper-Bessel function ${}_0F_{q-p}$, namely:*

$$\begin{aligned} &{}_pF_q\left((a_i)_{i=1}^p; (b_j)_{j=1}^p; (b_{j+p})_{j=1}^{q-p}; \lambda x\right) \\ &= \left[\prod_{i=1}^{p} \frac{\Gamma(b_i)}{\Gamma(a_i)}\right] D_{1,p}^{(b_k-1),(a_k-b_k)} \left\{{}_0F_{q-p}\left((b_{j+p})_{j=1}^{q-p}; \lambda x\right)\right\}. \end{aligned} \tag{4.3.3}$$

Proof. We apply Lemma 4.3.1 p-times to ${}_0F_m$, where $m = q - p > 0$ is denoted. For $a_i > b_i > 0$, $i = 1, \ldots, m$, we obtain in turn:

$$\begin{aligned} &{}_1F_{m+1}\left(a_p; b_p, (b_{j+p})_{j=1}^m; \lambda x\right) \\ &\quad = \frac{\Gamma(b_p)}{\Gamma(a_p)}\left(x^{1-b_p} D^{a_p-b_p} x^{a_p-1}\right) {}_0F_m\left((b_{j+p})_{j=1}^m; \lambda x\right), \\ &\quad \cdots\cdots\cdots \\ &\quad \cdots\cdots\cdots \\ &{}_pF_{m+p}\left(a_1, \ldots, a_p; b_1, \ldots, b_p, (b_{j+p})_{j=1}^m; \lambda x\right) \\ &\quad = \prod_{k=1}^{p}\left[\frac{\Gamma(b_k)}{\Gamma(a_k)} x^{1-b_k} D^{a_k-b_k} x^{a_k-1}\right] {}_0F_m\left((b_{j+p})_{j=1}^m; \lambda x\right). \end{aligned}$$

Then, we replace $D^{a_k-b_k}$ by $D_{1,1}^{0,a_k-b_k} x^{b_k-a_k}$ and use property (1.6.3) to remove all the power functions of x in front of the Erdélyi-Kober fractional derivatives. The same can be done directly using (4.3.2), that is:

$$ {}_pF_q(\lambda x) = \left[\prod_{k=1}^{p} \frac{\Gamma(b_k)}{\Gamma(a_k)} D_{1,1}^{b_k-1, a_k-b_k}\right] {}_0F_{q-p}(\lambda x). $$

This composition of single fractional derivatives can be written as an p-tuple generalized fractional derivative, which gives (4.3.3).

Now let us combine Theorem 4.3.2 with the fractional derivative representation of the hyper-Bessel functions ${}_0F_{q-p}$ found in Theorem 4.1.6, more especially in Corollary 4.1.7. In this manner we obtain the following proposition (Kiryakova, [198]).

Theorem 4.3.3. (Fractional derivative representation of the generalized hypergeometric functions of Bessel type) *Let $p < q$ and*

$$ a_k > b_k > 0, \; k = 1, \ldots, p; \quad \frac{k-p}{q-p+1} > b_k > 0, \; k = p+1, \ldots, q. \tag{4.3.4} $$

Then, the function ${}_pF_q$, $p < q$, is an q-tuple fractional derivative of $\cos_{q-p+1}(x)$, namely:

$$ \begin{aligned} {}_pF_q\left(a_1, \ldots, a_p; b_1, \ldots, b_q; -x\right) &= \sqrt{\frac{q-p+1}{(2\pi)^{q-p}}} \left[\frac{\prod_{i=1}^{q} \Gamma(b_i)}{\prod_{j=1}^{p} \Gamma(a_j)}\right] \\ &\quad \times D_{1,q}^{(\gamma_k),(\delta_k)}\left\{\cos_{q-p+1}\left((q-p+1)x^{\frac{1}{q-p+1}}\right)\right\}, \end{aligned} \tag{4.3.5} $$

where

$$ \gamma_k = b_k - 1, \; k = 1, \ldots, q; \quad \delta_k = \begin{cases} a_k - b_k & k = 1, \ldots, p \\ \frac{k-p}{q-p+1} - b_k & k = p+1, \ldots, q. \end{cases} \tag{4.3.6} $$

The explicit representation has the form (see Definition 1.5.4):

$$
\begin{aligned}
{}_pF_q(a_1,\ldots,a_p;b_1,\ldots,b_q;-x) &= \sqrt{\frac{q-p+1}{(2\pi)^{q-p}}}\left[\frac{\prod\limits_{i=1}^{q}\Gamma(b_i)}{\prod\limits_{j=1}^{p}\Gamma(a_j)}\right] \\
&\times\left[\prod_{k=1}^{q}\prod_{j=1}^{\eta_k}\left(x\frac{d}{dx}+b_k+j-1\right)\right]\int_0^1 G_{q,q}^{q,0}\left[\sigma\left|\begin{matrix}(b_k-1+\eta_k)_1^q\\ (a_k-1)_1^p,\left(\frac{k-p}{q-p+1}-1\right)_{p+1}^q\end{matrix}\right.\right] \\
&\times\cos_{q-p+1}\left[(q-p+1)(x\sigma)^{\frac{1}{q-p+1}}\right]d\sigma,
\end{aligned}
\tag{4.3.7}
$$

where the integers η_k, $k=1,\ldots,q$, are chosen in the following way:

$$
\eta_k=\begin{cases}[a_k-b_k]+1, & \text{if } (a_k-b_k) \text{ is non integer,}\\ (a_k-b_k), & \text{if } (a_k-b_k) \text{ is integer,}\end{cases}\quad \text{for } k=1,\ldots,p,
$$

and

$$
\eta_k=\begin{cases}\left[\frac{k-p}{q-p+1}-b_k\right]+1, & \text{if } \left(\frac{k-p}{q-p+1}-b_k\right) \text{ is non integer,}\\ \left(\frac{k-p}{q-p+1}-b_k\right), & \text{if } \left(\frac{k-p}{q-p+1}-b_k\right) \text{ is integer,}\end{cases}\quad \text{for } k=p+1,\ldots,q.
$$

Proof. The following property (see Section 1.6) is used:

$$
D_{1,p}^{(b_i-1),(a_i-b_i)}D_{1,q-p}^{(b_{j+p}-1),\left(\frac{j}{q-p+1}-b_{j+p}\right)}=D_{1,p+(q-p)=q}^{(\gamma_k),(\delta_k)}.
$$

Corollary 4.3.4. *Let the parameters $b_1,\ldots,b_q$ of ${}_pF_q$, $p<q$, have the form*

$$
b_k=\begin{cases}a_k-\eta_k>0 & \text{for } k=1,\ldots,p,\\ \frac{k-p}{q-p+1}-\eta_k>0 & \text{for } k=p+1,\ldots,q,\end{cases}
\tag{4.3.8}
$$

with some negative integers $\eta_1,\ldots,\eta_q$, that is: let the multiorder $(\delta_1,\ldots,\delta_q)$ of the generalized fractional derivative in (4.3.5) consist of integers $\delta_k=\eta_k$, $k=1,\ldots,q$. Then, $D_{1,q}^{(\gamma_k),(\delta_k)}$ is a purely differential operator D_η of integer order $\eta=(\eta_1+\cdots+\eta_q)$. In this manner we obtain the following differential formula for the "spherical" Bessel type generalized hypergeometric functions ${}_pF_q$ (that is, for the functions ${}_pF_q$, $p<q$ with

parameters (4.3.8)):

$$
\begin{aligned}
&{}_pF_q\left((a'_i)_{i=1}^p ; (a'_k - \eta_k)_{k=1}^q ; -x\right) \\
&= \sqrt{\frac{q-p+1}{(2\pi)^{q-p}}} \left[\frac{\prod\limits_{i=1}^{q} \Gamma(a'_i - \eta_i)}{\prod\limits_{j=1}^{p} \Gamma\left(a'_j\right)} \right] D_\eta \left\{ \cos_{q-p+1}\left((q-p+1) x^{\frac{1}{q-p+1}} \right) \right\} \\
&= \sqrt{\frac{q-p+1}{(2\pi)^{q-p}}} \left[\frac{\prod\limits_{i=1}^{q} \Gamma(a'_i - \eta_i)}{\prod\limits_{j=1}^{p} \Gamma\left(a'_j\right)} \right] \left[\prod_{k=1}^{q} \prod_{j=1}^{\eta_k} \left(x\frac{d}{dx} - a'_k - j \right) \right] \\
&\times \left\{ \cos_{q-p+1}\left((q-p+1) x^{\frac{1}{q-p+1}} \right) \right\},
\end{aligned}
\tag{4.3.9}
$$

where the auxiliary parameters $a'_k > 0$ *are defined by*

$$
a'_k = \begin{cases} a_k & \textit{for } k = 1, \dots, p \\ \frac{k-p}{q-p+1} & \textit{for } k = p+1, \dots, q. \end{cases}
$$

4.3.ii. Rodrigues type formulas for the confluent generalized hypergeometric functions $(p = q)$

Lemma 4.3.5. *If* $a_1 > b_1 > 0$*, then the confluent hypergeometric function* $\Phi(a_1; b_1; x) = {}_1F_1(x)$ *is a fractional derivative of the elementary function* $x^{a_1-1}e^x$*, viz.:*

$$
{}_1F_1(a_1; b_1; \lambda x) = \frac{\Gamma(b_1)}{\Gamma(a_1)} x^{1-b_1} D^{a_1-b_1} \left\{ x^{a_1-1} e^{\lambda x} \right\}, \tag{4.3.10}
$$

or in terms of the Erdélyi-Kober fractional derivatives:

$$
\begin{aligned}
{}_1F_1(a_1; b_1; \lambda x) &= \frac{\Gamma(b_1)}{\Gamma(a_1)} D_{1,1}^{1-b_1, a_1-b_1} \left\{ e^{\lambda x} \right\} \\
&= \frac{\Gamma(b_1)}{\Gamma(a_1)} x^{1-a_1} D_{1,1}^{b_1-a_1, a_1-b_1} \left\{ x^{a_1-1} e^{\lambda x} \right\}.
\end{aligned}
\tag{4.3.10$'$}
$$

The proof is analogous to that of Lemma 4.3.1. It uses the fractional integral representation of Lemma 4.2.8 and the differential formula:

$$
\left(\frac{d}{dx}\right)^\eta x^{b_1+\eta-1} {}_1F_1(a_1; b_1+\eta; x) = \frac{\Gamma(b_1+\eta)}{\Gamma(b_1)} x^{b_1-1} {}_1F_1(a_1; b_1; x),
$$

which is a special case of a general formula in [369, p. 442, (51)].

Theorem 4.3.6. *If $p = q$ and $a_k > b_k > 0$, $k = 1, \ldots, p$, then the "confluent" generalized hypergeometric function ${}_pF_p((a_k);(b_k);x)$ is an p-tuple generalized fractional derivative of the exponential function (with possible power weight), namely:*

$$
\begin{aligned}
{}_pF_p((a_k);(b_k);\lambda x) &= \left[\prod_{j=1}^{p} \frac{\Gamma(b_j)}{\Gamma(a_j)}\right] x^{1-a_1} D_{1,p}^{(b_k-a_1),(a_k-b_k)} \left\{x^{a_1-1} e^{\lambda x}\right\} \\
&= \left[\prod_{j=1}^{p} \frac{\Gamma(b_j)}{\Gamma(a_j)}\right] D_{1,p}^{(b_k-1),(a_k-b_k)} \left\{e^{\lambda x}\right\}.
\end{aligned}
\tag{4.3.11}
$$

Remark . Since the function ${}_pF_p$ is symmetric in the parameters $a_1, \ldots, a_p$, one can take instead of a_1 each a_l, $l = 1, \ldots, p$, in the first line of (4.3.11), i.e.

$$
{}_pF_p(\lambda x) = \left[\prod_{j=1}^{p} \frac{\Gamma(b_j)}{\Gamma(a_j)}\right] x^{1-a_l} D_{1,p}^{(b_k-a_l),(a_k-b_k)} \left\{x^{a_l-1} e^{\lambda x}\right\} \tag{4.3.11$'$}
$$

for each $l = 1, \ldots, p$.

The proof of Theorem 4.3.6 follows by p-times application of Lemma 4.3.1 to the result of Lemma 4.3.5, in a manner analogous to that in Theorem 4.3.2.

Corollary 4.3.7. *Let all the differences $a_k - b_k = \eta_k$, $k = 1, \ldots, p$, be non negative integers. Then, the differintegral operators in (4.3.11-11$'$) turn into differential operators D_η of integer order $\eta = \eta_1 + \cdots + \eta_k \geq 0$ and of form (1.5.22), namely:*

$$
\begin{aligned}
&{}_pF_p(b_1 + \eta_1, \ldots, b_p + \eta_p; b_1, \ldots, b_p; \lambda x) \\
&\quad = \left[\prod_{j=1}^{p} \frac{\Gamma(b_j)}{\Gamma(b_j + \eta_j)}\right] \left[\prod_{k=1}^{p} \prod_{j=1}^{\eta_k} \left(x\frac{d}{dx} + b_k + j - 1\right)\right] \left\{e^{\lambda x}\right\}.
\end{aligned}
\tag{4.3.12}
$$

This purely differential representation suggests that we can *call the ${}_pF_p$-functions in (4.3.12) "spherical" generalized hypergeometric functions of confluent type.*

Special case. If $b_k = a - 1$, $\eta_k = 1$, $k = 1, \ldots, p$, then from (4.3.12) and Section 3.1 ((3.1.2)-(3.1.3)), we obtain:

$$
\begin{aligned}
&{}_pF_p(a, \ldots, a; (a-1), \ldots, (a-1); x) = (a-1)^{-p} \left(x\frac{d}{dx} + a - 1\right)^p \{e^x\} \\
&= x^{1-a} \left(x\frac{d}{dx}\right)^{p-1} x^{a-1} \{e^x\}, \quad p = 1, 2, \ldots .
\end{aligned}
\tag{4.3.13}
$$

In particular, for $a = 2$:

$$\begin{aligned}{}_pF_p(2,\ldots,2;1,\ldots,1;x) &= \left(x\frac{d}{dx}+1\right)^p\{e^x\} \\ &= \frac{d}{dx}\left(x\frac{d}{dx}\right)^p\{e^x\} := D_p,\ p=1,2,\ldots\end{aligned} \tag{4.3.14}$$

Representation (4.3.14) and a representation similar to (4.3.13) can be found in Prudnikov, Brychkov and Marichev [369, p. 593, (3), (2)]. Differential formula (4.3.14) is used in [369] to find that

$$\begin{aligned}{}_pF_p(2,\ldots,2;1,\ldots,1;x) &= D_p \\ &= \sum_{k=0}^{p}\frac{(-1)^k}{(k+1)!}\sum_{l=0}^{k}(-1)^l\binom{k+1}{l+1}(l+1)^{p+1}\left\{x^ke^x\right\},\ p=0,1,2,\ldots\end{aligned}$$

that is,

$$\begin{aligned}&D_0 = e^x, \quad D_1 = (x+1)e^x,\\ &D_2 = \left(x^2+3x+1\right)e^x,\\ &D_3 = \left(x^3+6x^2+7x+1\right)e^x,\\ &D_4 = \left(x^4+10x^3+25x^2+15x+1\right)e^x,\\ &\ldots\ldots,\\ &D_7 = \left(x^7+28x^6+266x^5+1050x^4+1701x^3+966x^2+127x+1\right)e^x, \text{ etc.}\end{aligned}$$

This example shows how differential representations (4.3.12) and the other similar ones of the *"spherical" generalized hypergeometric function* ${}_pF_q(x)$ can be used for their explicit calculation (especially for $p = q$) in the form

$${}_pF_p\left(b_1+\eta_1,\ldots,b_p+\eta_p;b_1,\ \ldots,b_p;x\right) = Q_p(x)e^x, \tag{4.3.12$'$}$$

where $Q_p(x)$ is a p-degree polynomial of x.

4.3.iii. Rodrigues formulas for the Gauss type generalized hypergeometric functions ${}_{q+1}F_q$

Lemma 4.3.8. *For $a_1 > b_1 > 0$ and $|\lambda x| < 1$ the Gauss ${}_2F_1$-function is a fractional derivative of an elementary function as follows:*

$${}_2F_1\left(a_1,a_2;b_1;\lambda x\right) = \frac{\Gamma\left(b_1\right)}{\Gamma\left(a_1\right)}x^{1-b_1}D^{a_1-b_1}\left\{x^{a_1-1}(1-\lambda x)^{-a_2}\right\}. \tag{4.3.15}$$

In other words,

$$
\begin{aligned}
{}_2F_1\left(a_1, a_2; b_1; \lambda x\right) &= \frac{\Gamma\left(b_1\right)}{\Gamma\left(a_1\right)} D_{1,1}^{b_1-1, a_1-b_1}\left\{(1-\lambda x)^{-a_2}\right\} \\
&= \frac{\Gamma\left(b_1\right)}{\Gamma\left(a_1\right)} x^{a_1-1} D_{1,1}^{b_1-a_1, a_1-b_1}\left\{x^{a_1-1}(1-\lambda x)^{-a_2}\right\},
\end{aligned} \tag{4.3.15'}
$$

provided $|\lambda x|<1$ or $|\arg(1-\lambda x)|<\pi$.

Proof. For $a_1-b_1>0$ we introduce the integer:

$$
\eta=\begin{cases}\left[a_1-b_1\right]+1, & \text{when } \left(a_1-b_1\right) \text{ is non integer,} \\ \left(a_1-b_1\right), & \text{when } \left(a_1-b_1\right) \text{ is integer.}\end{cases}
$$

By the definition of the Riemann-Liouville fractional derivative,

$$
D^{a_1-b_1}=\left(\frac{d}{d x}\right)^\eta R^{\eta-\left(a_1-b_1\right)},
$$

therefore

$$
\begin{aligned}
D^{a_1-b_1}\left\{x^{a_1-1}(1-\lambda x)^{-a_2}\right\} &= \left(\frac{d}{d x}\right)^\eta R^{\eta-\left(a_1-b_1\right)}\left\{x^{a_1-1}(1-\lambda x)^{-a_2}\right\} \\
&= \frac{\Gamma\left(a_1\right)}{\Gamma\left(b_1+\eta\right)}\left(\frac{d}{d x}\right)^\eta\left\{x^{b_1+\eta-1}{}_2F_1\left(a_1, a_2; b_1+\eta; \lambda x\right)\right\},
\end{aligned}
$$

due to the fractional integral formula [107, p.186, (9)] valid for $a_1>0$, $\eta>a_1-b_1>0$. Then, the differential relation [106, I, 2.8, (22)] yields

$$
\left(\frac{d}{d x}\right)^\eta\left\{x^{b_1+\eta-1}{}_2F_1\left(a_1, a_2; b_1+\eta; \lambda x\right)\right\}=\frac{\Gamma\left(b_1+\eta\right)}{\Gamma\left(b_1\right)} x^{b_1-1}{}_2F_1\left(a_1, a_2; b_1; \lambda x\right).
$$

In this manner, we get

$$
D^{a_1-b_1}\left\{x^{a_1-1}(1-\lambda x)^{-a_2}\right\}=\frac{\Gamma\left(a_1\right) \Gamma\left(b_1+\eta\right)}{\Gamma\left(b_1+\eta\right) \Gamma\left(b_1\right)} x^{b_1-1}{}_2F_1\left(a_1, a_2; b_1; \lambda x\right),
$$

which is (4.3.15).

The q-times application of Lemma 4.3.1 to (4.3.15′) leads to:

Theorem 4.3.9. *Let $p=q+1$ and $a_k>b_k>0$, $k=1,\ldots,q$. Then, the Gauss type generalized hypergeometric function ${}_{q+1}F_q$ can be represented as an q-tuple generalized fractional derivative of each of the elementary functions $x^{a_l-1}(1-\lambda x)^{-a_{q+1}}$, $l=1,\ldots,q$,* *for example:*

$$
\begin{aligned}
{}_{q+1}F_q\left(a_1,\ldots,a_{q+1};b_1,\ldots,b_q;\lambda x\right)
&= \left[\prod_{j=1}^{q}\frac{\Gamma(b_j)}{\Gamma(a_j)}\right] x^{1-a_q} D_{1,q}^{(b_k-a_q),(a_k-b_k)}\left\{x^{a_q-1}(1-\lambda x)^{-a_{q+1}}\right\} \\
&= \left[\prod_{j=1}^{q}\frac{\Gamma(b_j)}{\Gamma(a_j)}\right] D_{1,q}^{(b_k-1),(a_k-b_k)}\left\{(1-\lambda x)^{-a_{q+1}}\right\}.
\end{aligned}
\tag{4.3.16}
$$

Corollary 4.3.10. *Let all the differences* $(a_k-b_k)=\eta_k$, $k=1,\ldots,q$, *be non negative integers. Then, the generalized fractional differentiation operator (4.3.16) is a purely differential operator* D_η *of integer order. The corresponding functions* ${}_{q+1}F_q\left((a_k)_1^{q+1};(b_l)_1^q;x\right)$ *are said to be spherical generalized hypergeometric functions of Gauss type. The following differential representation holds then:*

$$
\begin{aligned}
{}_{q+1}F_q\left(b_1+\eta_1,\ldots,b_q+\eta_q,a_{q+1};b_1,\ldots,b_q;\lambda x\right)
&= \left[\prod_{j=1}^{q}\frac{\Gamma(b_j)}{\Gamma(b_j+\varepsilon_j)}\right]\left[\prod_{k=1}^{q}\prod_{j=1}^{\eta_k}\left(x\frac{d}{dx}+b_k+j-1\right)\right]\left\{(1-\lambda x)^{-a_{q+1}}\right\} \\
&= c\,Q\left(x\frac{d}{dx}\right)\left\{(1-\lambda x)^{-a_{q+1}}\right\},
\end{aligned}
\tag{4.3.17}
$$

where $c=\text{const}$ *and* Q *is a polynomial of degree* $\eta=\eta_1+\cdots+\eta_q$.

Special case (cf. [369, p. 572, (11)]). For $b_k=1$, $\eta_k=1$, $k=1,\ldots,q$, $a_{q+1}=2$, $\lambda=1$ the previous formula (4.3.17) gives

$$
\begin{aligned}
{}_{q+1}F_q(2,2,\ldots,2;1,1,\ldots,1;x) &= \left(x\frac{d}{dx}+1\right)^q\left\{\frac{1}{1-x}\right\}^2 \\
&= x^{-1}\left(x\frac{d}{dx}\right)^q x\left\{\frac{1}{1-x}\right\}^2 = x^{-1}\left(x\frac{d}{dx}\right)^{q+1}\left\{\frac{1}{1-x}\right\} \\
&= \sum_{k=0}^{q}\frac{(k+1)(-x)^k}{(1-x)^{k+2}}\left[\sum_{l=0}^{k}(-1)^l\binom{k}{l}(l+1)^q\right].
\end{aligned}
\tag{4.3.18}
$$

This example shows, once again, the practical use of the differential formulas established here.

Corollary 4.3.11. *Let us consider the so-called hypergeometric polynomials (A.3) (see also [272, I, p. 142], [468, p. 33])*

$$ {}_{p+1}F_q\left(-n, a_1, \ldots, a_p; b_1, \ldots, b_q; x\right) = \sum_{k=0}^{n} \frac{(-n)_k (a_1)_k \ldots (a_p)_k}{(b_1)_k \ldots (b_q)_k} \frac{x^k}{k!} \tag{4.3.19} $$

in the case $p = q$. Due to the symmetry in the first group of parameters of ${}_{q+1}F_q$, we take $a_{q+1} = -n$ and suppose also that $a_k > b_k > 0$, $k = 1, \ldots, q$. Then, (4.3.16) turns into the Rodrigues type formula for these polynomials, viz.

$$ \begin{aligned} &{}_{q+1}F_q\left(-n, a_1, \ldots, a_q; b_1, \ldots, b_q; x\right) \\ &= \left[\prod_{j=1}^{q} \frac{\Gamma(b_j)}{\Gamma(a_j)}\right] x^{1-b_q} D^{a_q-b_q} x^{a_p-b_{q-1}} D^{a_{p-1}-b_{q-1}} \ldots \\ &\times x^{a_3-b_2} D^{a_2-b_2} x^{a_2-b_1} D^{a_1-b_1} \left\{x^{a_1-1}(1-x)^n\right\}, \end{aligned} \tag{4.3.20} $$

or, in terms of Erdélyi-Kober fractional derivatives,

$$ \begin{aligned} &{}_{q+1}F_q\left(-n, a_1, \ldots, a_q; b_1, \ldots, b_q; x\right) \\ &= \left[\prod_{j=1}^{q} \frac{\Gamma(b_j)}{\Gamma(a_j)}\right] x^{1-a_q} D_{1,q}^{(b_k-a_q),(a_k-b_k)} \left\{x^{a_1-1}(1-x)^n\right\} \\ &= \left[\prod_{j=1}^{q} \frac{\Gamma(b_j)}{\Gamma(a_j)}\right] D_{1,q}^{(b_k-1),(a_k-b_k)} \left\{(1-x)^n\right\}. \end{aligned} \tag{4.3.20'} $$

Note. Let us note that formula (4.3.20) for polynomials ${}_{p+1}F_q$ can be found in Misra [310] *but it is correct only for $p = q$ and under the conditions for the parameters stated here.* In this notation, for $p \neq q$, the elementary function $x^{a_1-1}(1-x)^n$ should be replaced in [310] either by $x^{a_1-1}e^x$, or by $\cos_{q-p}(x)$.

Some special cases of (4.3.20) yield the *classical Rodrigues formulas.* For example: If $p = q = 1$,

$$ {}_2F_1\left(-n, a_1; b_1; x\right) = \frac{\Gamma(b_1)}{\Gamma(a_1)} x^{1-b_1} D^{a_1-b_1} x^{a_1-1}(1-x)^n, $$

whence, for $a_1 = n+1$, $b_1 = 1$ (i.e. $a_1 > b_1 > 0$), $x \to \frac{1-x}{2}$, the *Rodrigues formula for the Legendre polynomials* follows:

$$ \begin{aligned} P_n(x) &= (-1)^n {}_2F_1\left(-n, n+1; 1; \frac{1-x}{2}\right) \\ &= \frac{(-1)^n}{n!} \frac{d^n}{dx^n} \left[\left(\frac{1-x}{2}\right)^n \left(\frac{1+x}{2}\right)^n\right] \\ &= \frac{1}{2^n n!} \frac{d^n}{dx^n} \left\{(x^2-1)^n\right\}, \end{aligned} \tag{4.3.21} $$

cf. (C. 21); analogously, formula (C.20) for the Gegenbauer polynomials C_n^ν can be derived.

If $p = q = 2$, by taking $a_1 = n+1$, $b_1 = 1$ (i.e. $a_1 > b_1 > 0$), $a_2 = \zeta$, $b_2 = p$ (provided $\zeta > p > 0$), we get the *Rodrigues formula for the Rice polynomialss*, viz.

$$\begin{aligned} R_n(x) &= {}_3F_2\left(-n, n+1, \zeta; 1, p; x\right) \\ &= \frac{\Gamma(p)}{n!\Gamma(\zeta)}\left[\frac{d^n}{dx^n}x^{1-p}\left(\frac{d}{dx}\right)^{\zeta-p}\right]\{x^n(1-x)^n\}. \end{aligned} \tag{4.3.22}$$

4.4. Integral representations of the generalized hypergeometric functions as G-transformations of hyper-Bessel and generalized trigonometric functions

Integral transforms of the formal form

$$(Gf)(x) = \int_0^\infty G_{p,q}^{m,n}\left[xt \,\middle|\, \begin{matrix} a_1, \ldots, a_p \\ b_1, \ldots, b_q \end{matrix}\right] f(t)dt \tag{4.4.1}$$

are usually referred to as *G-transforms*. Many, perhaps most, of the known integral transformations of functions on $(0, \infty)$ can be put into this form by simple changes of variables. For example, the integral transforms of Laplace, Meijer, Hankel, etc. can be represented in the form

$$(Gf)(x) = A\int_0^\infty G_{p,q}^{m,n}\left(c\,(xt)^\lambda\right) f(t)dt \tag{4.4.2}$$

($A \neq 0$, $A \in \mathbb{C}$, $c > 0$, $\lambda \neq 0$, $\lambda \in \mathbb{R}$), while the fractional integrals, the even and odd Hilbert transforms, etc. are G-transforms of the form

$$(Gf)(x) = A\int_0^\infty G_{p,q}^{m,n}\left(\left(\frac{x}{t}\right)^\lambda\right) f(t)dt \tag{4.4.3}$$

(see Rooney [403]). For more details see Section 5.6.

As shown in Section 3.9, the so-called *Obrechkoff integral transform* is also a special case of the G-transformations.

First we show that the ${}_pF_q$-functions are modified Obrechkoff transforms, that is, G-transforms of the hyper-Bessel functions ${}_0F_q$. Finally, it turns out that the ${}_pF_q$-functions are G-transforms of the form (4.4.2) of the generalized cosine function $\cos_{q+1}(x)$. This result can be compared with the similar but different result of Theorem 4.2.4; see Remark 2 at the end of this Section.

Theorem 4.4.1. *Let* $0 < x < \infty$ *and* $a_k > 0$, $k = 1, \ldots, p$. *Then, the generalized hypergeometric function* ${}_pF_q\left((a_k)_1^p; (b_l)_1^q; -x\right)$, $p \leq q+1$ *is a "modified" Obrechkoff transform* *of the hyper-Bessel function* ${}_0F_q\left((b_l)_1^q; -x\right)$, *namely:*

$$
{}_pF_q\left(a_1,\ldots,a_p;b_1,\ldots,b_q;-x\right)=\left(\prod_{k=1}^{p}\Gamma\left(a_k\right)\right)^{-1}
\times\int_0^\infty G_{0,p}^{p,0}\left[t|\left(a_k-1\right)_1^p\right]\,{}_0F_q\left((b_l)_1^q;-xt\right)dt. \tag{4.4.4}
$$

Proof. The transform

$$
\mathfrak{O}\left\{f(t);x\right\}=\left(\prod_{k=1}^{p}\Gamma\left(a_k\right)\right)^{-1}\int_0^\infty G_{0,p}^{p,0}\left[t|\left(a_k-1\right)_1^p\right]f(xt)dt
$$

is a *modification of the Obrechkoff integral transform* of order $m=p$. This is seen from Definition 3.9.4.

Representation (4.4.4) will be deduced from the more general relation

$$
G_{p,q+1}^{s,p}\left[x\left|\begin{matrix}(A_k)_1^p\\(B_l)_1^{q+1}\end{matrix}\right.\right]=\int_0^\infty G_{0,p}^{p,0}\left[t|\left(-A_k\right)_1^p\right]G_{0,q+1}^{s,0}\left[xt|\left(B_l\right)_1^{q+1}\right]dt, \tag{4.4.5}
$$

where $1\le s\le q+1$ and $p\ge 1$. This representation of the $G_{p,q+1}^{s,p}$-function in terms of the same G-function with $p=0$ is an immediate corollary of formula (A.29) for integrals of a product of two G-functions. Now, let us take $s=1$ and substitute $A_k=(1-a_k)$, $k=1,\ldots,p$; $B_1=0$, $B_l=(1-b_l)$, $l=2,\ldots,q$. Then, (4.4.5) turns into the equality

$$
G_{p,q+1}^{1,p}\left[x\left|\begin{matrix}(1-a_k)_1^p\\0,(1-b_l)_1^q\end{matrix}\right.\right]=\int_0^\infty G_{0,p}^{p,0}\left[t|\left(a_k-1\right)_1^p\right]G_{0,q+1}^{1,0}\left[xt|0,\left(1-b_l\right)_1^q\right]dt, \tag{4.4.6}
$$

that is, into (4.4.4), if one takes into account that

$$
G_{p,q+1}^{1,p}\left[x\left|\begin{matrix}(1-a_k)_1^p\\0,(1-b_l)_1^q\end{matrix}\right.\right]={}_pF_q\left((a_k);(b_l);-x\right),\quad p\ge 0.
$$

In this manner (4.4.4) is formally proved. It remains to specify the conditions under which this relation is true. Instead of using the general conditions for validity of formula (A.29) which are quite complicated, we prefer investigating the convergence of the particular integral (4.4.6), considered here.

Near the origin $t=0$ the asymptotic behaviour of the G-functions in the integrand is described simply. According to (A.21),

$$
G_{0,p}^{p,0}\left[t|\left(a_k-1\right)\right]=\mathcal{O}\left(t^{\beta_1}\right),\quad \beta_1=\min_{1\le k\le p}\left(a_k-1\right),\quad t\to+0,
$$

and

$$G^{1,0}_{0,q+1}\left[xt|0,(1-b_l)\right] = \mathcal{O}\left(t^0\right) = \mathcal{O}(1), \quad t \to +0.$$

Therefore, the integrand is $\mathcal{O}\left(t^{\beta_1}\right)$ as $t \to +0$. It is sufficient then to require $\beta_1 = \min_k (a_k - 1) > -1$, i.e. $\min_k a_k > 0$. Thus, we obtain the conditions

$$a_k > 0, \quad k = 1, \ldots, p.$$

The problem concerning the asymptotics of the integrand near infinity is not so simple but its answer is also known. In general, the G-function can increase or decrease like a power function as well as like an exponent when its variable tends to infinity. In our case, one of the multipliers, $G^{p,0}_{0,p}(t)$, decreases exponentially, while the other, $G^{1,0}_{0,q+1}(xt)$ increases exponentially as $t \to \infty$ (for $q > 1$). A description of this can be found, for example, in [106, I, 5.4.1]. However, we need the exact asymptotic formulas. To this end, the result of Obrechkoff [339, (14)] can be used. According to Lemma 3.9.1 and (3.9.18) we obtain the following asymptotic behaviour of the kernel-function $G^{p,0}_{0,p}$ of the Obrechkoff transform:

$$G^{p,0}_{0,p}\left[t|\left(a_k-1\right)_1^p\right] \sim d_p t^{\rho_1} \exp\left(-pt^{\frac{1}{p}}\right) \tag{4.4.7}$$

as $t \to \infty$, where $\rho_1 = \frac{\sum_1^p a_k}{p} - \frac{3p-1}{2p}$. The other multiplier in the integrand is the hyper-Bessel function

$$G^{1,0}_{0,q+1}\left[xt|0,(1-b_l)_1^q\right] = \left[\prod_{l=1}^{q}\Gamma\left(b_l\right)\right]^{-1} {}_0F_q\left(\left(b_l\right)_{l=1}^q; -xt\right),$$

whose asymptotics near infinity are essentially different in the cases $q = 1$ and $q > 1$.

In the particular case $q = 1$, this is the Bessel function:

$$G^{1,0}_{0,2}\left[xt|0,1-b_1\right] = \frac{1}{\Gamma\left(b_1\right)}\,{}_0F_1\left(b_1; -xt\right) = (xt)^{\frac{1-b_1}{2}} J_{b_1-1}\left(2\sqrt{xt}\right),$$

therefore (see [106, II, p. 98, (3)]):

$$G^{1,0}_{0,2}\left[xt|0,1-b_1\right] \sim \text{const}\,(xt)^{-\frac{1+b_1}{2}}\left\{\cos\left(2\sqrt{xt} - \pi\left(2b_1-1\right)\right)\right.$$
$$\left.\text{const}\,-(xt)^{-\frac{1}{2}}\sin\left(2\sqrt{xt} - \pi\left(2b_1-1\right)\right)\right\}, \quad t \to \infty,\ x > 0.$$

Due to this result, we conclude that the asymptotic behaviour of the product $G^{p,0}_{0,p}(t) \times G^{1,0}_{0,2}(xt)$ is determined by the behaviour of $G^{p,0}_{0,p}(t)$, that is, by the multiplier $\exp\left(-pt^{\frac{1}{p}}\right)$ in (4.4.7) ensuring the convergence of integral (4.4.4).

It remains to consider the case $q > 1$. Let us use Marichev's Theorem 21 from [276, p. 59-60]. Formula (4.56) in [276] for $p = 0$, $q = p+1 = 1$ and $z = t \to \infty$ with $\arg z = 0$, gives:

$$\left[\prod_{l=1}^{q} \Gamma(b_l)\right]^{-1} {}_0F_q\left((b_l)_1^q; -xt\right) \sim \frac{2\left(\sqrt{2\pi}\right)^q}{\sqrt{q+1}} \exp\left[(q+1)(xt)^{\frac{1}{q+1}} \cos\frac{\pi}{q+1}\right] \\ \times (xt)^{\rho_2} \cos\left[\pi\rho_1 + \frac{1}{q+1}(xt)^{\frac{1}{q+1}} \sin\frac{\pi}{q+1}\right], \tag{4.4.8}$$

where $\rho_2 = \frac{1}{q+1}\left[\frac{q}{2} - \sum\limits_1^q b_l\right]$. This yields that the function ${}_0F_q(-xt)$, i.e. $G^{1,0}_{0,q+1}(xt)$, increases exponentially as $t \to \infty$. So, for the integrand of (4.4.4), combining (4.4.7) and (4.4.8), we obtain:

$$\Phi(x,t) = G^{p,0}_{0,p}\left[t|\left(a_k - 1\right)\right] G^{1,0}_{0,q+1}\left[xt|0,(1-b_l)\right] \\ \sim \text{const } t^{\rho_1+\rho_2} \cos\left[\pi\rho_1 + \frac{1}{q+1}(xt)^{\frac{1}{q+1}} \sin\frac{\pi}{q+1}\right] \exp\left[\zeta(x,t)\right], \ t \to \infty, \tag{4.4.9}$$

provided $p \geq 1$, $q > 1$, $x > 0$, where

$$\zeta(x,t) = -pt^{\frac{1}{p}} + (q+1)(xt)^{\frac{1}{q+1}} \cos\frac{\pi}{q+1}. \tag{4.4.9$'$}$$

If $p \leq q$, that is, $p < q+1$, then $\frac{1}{p} > \frac{1}{q+1}$ and $t^{\frac{1}{p}} \gg t^{\frac{1}{q+1}}$ as $t \to \infty$. Therefore the dominant term in $\zeta(x,t)$ is $\left(-pt^{\frac{1}{p}}\right)$ and $\exp\left[\zeta(x,t)\right]$ is exponentially decreasing:

$$\exp\left[\zeta(x,t)\right] = \exp\left[-pt^{\frac{1}{p}} + c_{q,x} t^{\frac{1}{q+1}}\right] \sim \exp\left[-ct^{\frac{1}{p}}\right], \quad t \to \infty$$

($c_{q,x}$ and c are constants). This ensures the convergence of (4.4.4).

If $p = q+1$, then

$$\zeta(x,t) = -pt^{\frac{1}{p}} + p\left(\cos\frac{\pi}{p}\right) x^{\frac{1}{p}} t^{\frac{1}{p}} \\ = pt^{\frac{1}{p}}\left(-1 + x^{\frac{1}{p}}\cos\frac{\pi}{p}\right) < pt^{\frac{1}{p}}\left(-1 + x^{\frac{1}{p}}\right).$$

In this case, the generalized hypergeometric series ${}_pF_q(-x)$ (the left-hand side of (4.4.4)) is considered only for $|x| < 1$. This means that $0 < x^{\frac{1}{p}} < 1$, i.e. $\left(-1 + x^{\frac{1}{p}}\right) < 0$, therefore, $\zeta(x,t) < 0$ for each $t > 0$ and $0 < x < 1$. From this, we conclude that the dominant term in (4.4.9) is exponentially decreasing again and integral (4.4.4) is convergent. All these considerations confirm that both sides of (4.4.4) are well defined for $a_k > 0$, $k = 1, \ldots, p$, $p \leq q+1$ and are equal. The theorem is proved.

As *particular cases* one can obtain a series of known results, scatterred in the handbooks and research papers. Some examples are given below.

Corollary 4.4.2. *Let $p = 1$, $q = 1$. Then, Theorem 4.4.1 gives the following representation of the confluent hypergeometric function by means of the Bessel function:*

$$\Phi\left(a;\nu+1;-\frac{x^2}{4}\right) = {}_1F_1\left(a;\nu+1;-\frac{x^2}{4}\right) = \frac{2^{\nu+1}}{\Gamma(a)}x^{-\nu}\int_0^\infty e^{-t^2}t^{2a-\nu-1}J_\nu(xt)dt, \quad \textit{provided } a > 0. \tag{4.4.10}$$

This relation can be found in Luke [272, p. 115, (2)] and Prudnikov, Brychkov and Marichev [368, p. 186, (2.12.9.3)].

As a *special case* of (4.4.10), taking $a = \frac{\nu+p}{2} > 0$ and $x = \frac{\alpha}{\gamma} > 0$, we obtain the formula [106, II, p.60, (22)]:

$$\Phi\left(\frac{\nu+p}{2};\nu+1;-\frac{\alpha^2}{4\gamma^2}\right) = {}_1F_1\left(\frac{\nu+\rho}{2};\nu+1;-\frac{\alpha^2}{4\gamma^2}\right) = 2\gamma^\rho\left(\frac{\alpha}{2\gamma}\right)^{-\nu}\frac{\Gamma(\nu+1)}{\Gamma(\frac{\nu+p}{2})}\int_0^\infty J_\nu(\alpha t)e^{-\gamma^2t^2}t^{\rho-1}dt.$$

Corollary 4.4.3. *Let $p = 2$, $q = 2$. Then, Theorem 4.4.1 gives a representation of the Gauss hypergeometric function by means of the Bessel and Macdonald functions:*

$${}_2F_1\left(a,b;\nu+1;-\frac{x^2}{4}\right) = \frac{2^{\nu+2}\Gamma(\nu+1)}{\Gamma(a)\Gamma(b)}\int_0^\infty t^{a+b-\nu-1}K_{a-b}(2t)J_\nu(xt)dt, \tag{4.4.11}$$

valid for $a > 0$, $b > 0$, $|x| < 2$

This relation can be found also in Prudnikov, Brychkov and Marichev [368, p. 365, (2.16.21.1)]. Taking $a = \frac{\nu+p}{2} > 0$, $b = \frac{\nu+p+1}{2}$, $x = \frac{2\alpha}{\gamma}$, $\alpha^2 < \gamma^2$, we obtain *as its special case* the known *Hankel integral* ([106, II, p. 59, (16)])

$${}_2F_1\left(\frac{\nu+p}{2},\frac{\nu+p+1}{2};\nu+1;-\frac{\alpha^2}{\gamma^2}\right) = \left(\frac{2\gamma}{\alpha}\right)^\nu\gamma^\rho\frac{\Gamma(\nu+1)}{\Gamma(\nu+p)}\int_0^\infty e^{-\gamma t}J_\nu(\alpha t)t^{\rho-1}dt.$$

Remark 1. Theorem 4.4.1 represents the ${}_pF_q$-function as a Laplace type G-transformation of ${}_0F_q$, while Theorem 4.2.1 represents the same function as a G-transformation of the form (4.4.3) (generalized fractional integral) of ${}_0F_{q-p}$ (and for $p < q$ only).

The next result is, to a certain extent, similar to Theorem 4.2.4.

Theorem 4.4.4. *Let $x > 0$ and $a_k > 0$, $k = 1,\ldots,p$; $b_l > \frac{l}{q+1}$, $l = 1,\ldots,q$. Then, the ${}_pF_q$-function ($p \le q+1$) can be represented as a Laplace type G-transformation of the* *generalized cosine function of order* $(q+1)$*:*

$$ {}_pF_q\left((a_k);(b_l);-x\right) = \sqrt{\frac{q+1}{(2\pi)^q}} \left[\frac{\prod_{l=1}^{q} \Gamma(b_l)}{\prod_{k=1}^{p} \Gamma(a_k)} \right] \int_0^\infty G_{q,p+q}^{p+q,0}\left[y \,\middle|\, \begin{matrix} (b_l-1)_1^q \\ (a_k-1)_1^p, \left(\frac{l}{q+1}-1\right)_1^q \end{matrix} \right] \times \cos_{q+1}\left((q+1)(xy)^{\frac{1}{q+1}}\right) dy. \tag{4.4.12} $$

Proof. Let us combine Theorem 4.4.1 and Corollary 4.1.4 of Theorem 4.1.1. Due to (4.1.29), we have

$$ \begin{aligned} {}_0F_q\left((b_l);-xt\right) &= \sqrt{\frac{q+1}{(2\pi)^q}} \left[\prod_{l=1}^{q} \Gamma(b_l) \right] \int_0^1 G_{q,q}^{q,0}\left[\tau \,\middle|\, \begin{matrix} (b_l-1) \\ \left(\frac{l}{q+1}-1\right) \end{matrix} \right] \cos_{q+1}\left((q+1)(xt\tau)^{\frac{1}{q+1}}\right) d\tau \\ &= \sqrt{\frac{q+1}{(2\pi)^q}} \left[\prod_{l=1}^{q} \Gamma(b_l) \right] \int_0^\infty G_{q,q}^{q,0}\left[\tau \,\middle|\, \begin{matrix} (b_l-1) \\ \left(\frac{l}{q+1}-1\right) \end{matrix} \right] \cos_{q+1}\left((q+1)(xt\tau)^{\frac{1}{q+1}}\right) d\tau, \end{aligned} $$

since $G_{q,q}^{q,0}(\tau) \equiv 0$ for $\tau > 1$. Replace this expression for ${}_0F_q$ in (4.4.4) and interchange the order of integrations by Fubini's theorem. Then,

$$ \begin{aligned} {}_pF_q\left((a_k);(b_l);-x\right) &= \sqrt{\frac{q+1}{(2\pi)^q}} \left[\frac{\prod_{l=1}^{q} \Gamma(b_l)}{\prod_{k=1}^{p} \Gamma(a_k)} \right] \int_0^\infty G_{0,p}^{p,0}\left[t \,|\, (a_k-1)\right] dt \\ &\quad \times \int_0^\infty G_{q,q}^{q,0}\left[\tau \,\middle|\, \begin{matrix} (b_l-1) \\ \left(\frac{l}{q+1}-1\right) \end{matrix} \right] \cos_{q+1}\left((q+1)(xt\tau)^{\frac{1}{q+1}}\right) d\tau \\ &= \sqrt{\frac{q+1}{(2\pi)^q}} \left[\frac{\prod_{l=1}^{q} \Gamma(b_l)}{\prod_{k=1}^{p} \Gamma(a_k)} \right] \int_0^\infty \cos_{q+1}\left((q+1)(xy)^{\frac{1}{q+1}}\right) dy \\ &\quad \times \int_0^\infty G_{0,p}^{p,0}\left[t \,|\, (a_k-2)\right] G_{q,q}^{0,q}\left[\frac{t}{y} \,\middle|\, \begin{matrix} -\left(\frac{l}{q+1}\right) \\ (-b_l) \end{matrix} \right] dt. \end{aligned} $$

Now the inner integral contains a product of two G-functions and can be evaluated according to the known result (A.29). For $a_k > 0$, $k = 1, \ldots, p$, and $b_l < \frac{l}{q+1}$, $l = 1, \ldots, q$, this result reads:

$$\int_0^\infty G^{p,0}_{0,p}\left[t\,|\,(a_k-2)_1^p\right] G^{0,q}_{q,q}\left[\frac{t}{y}\,\middle|\,\begin{matrix}-\left(\frac{l}{q+1}\right)_1^q\\(-b_l)_1^q\end{matrix}\right]dt = G^{p+q,0}_{q,p+q}\left[y\,\middle|\,\begin{matrix}(b_l-1)_1^q\\(a_k-1)_1^p,\left(\frac{l}{q+1}-1\right)_1^q\end{matrix}\right].$$

Thus, we obtain the kernel-function of the G-transform in (4.4.12) and the theorem is proved.

Some known results can be obtained as *special cases.*

Example. For $q=1$ formula (4.4.12) takes the form

$$_pF_1\left((a_k);b;-x\right) = \frac{\Gamma(b)}{\prod\limits_k \Gamma(a_k)}\frac{1}{\sqrt{\pi}}\int_0^\infty G^{p+1,0}_{1,p+1}\left[y\,\middle|\,\begin{matrix}(b-1)\\(a_k-1),(-\frac{1}{2})\end{matrix}\right]\cos\left(2\sqrt{xy}\right)dy \qquad (4.4.13)$$

and represents the functions $_pF_1(-x)$, $0\le p\le 2$ as G-transformations of the usual cosine function. Three cases are possible, viz.:

Corollary 4.4.5. *($p=0$, $q=1$) Relation (4.4.13) turns into the classical Poisson integral (4.0.1).*

Corollary 4.4.6. *($p=1$, $q=1$) From (4.4.13) we can obtain the following relation between the confluent hypergeometric functions Φ and Ψ of Kumer and Tricomi (which seem to be new), provided $a>0$, $b>\frac{1}{2}$, $x>0$:*

$$\begin{aligned}\Phi(a;b;-x) &= \frac{\Gamma(b)}{\sqrt{\pi}\Gamma(a)}\int_0^\infty G^{2,0}_{1,2}\left[y\,\middle|\,\begin{matrix}b-1\\a-1,-\frac{1}{2}\end{matrix}\right]\cos\left(2\sqrt{xy}\right)dy\\ &= \frac{\Gamma(b)}{\sqrt{\pi}\Gamma(a)}\int_0^\infty e^{-y}y^{a-1}\Psi\left(b-\frac{1}{2};a+\frac{1}{2};y\right)\cos\left(2\sqrt{xy}\right)dy.\end{aligned} \qquad (4.4.14)$$

Corollary 4.4.7. *($p=2$, $q=1$) From (4.4.13) we get the following representation of the Gauss function, valid for $a>0$, $b>0$, $c>\frac{1}{2}$ and $0<x<1$:*

$$_2F_1(a,b;c;-x) = \frac{\Gamma(b)}{\sqrt{\pi}\Gamma(a)\Gamma(b)}\int_0^\infty G^{3,0}_{1,3}\left[y\,\middle|\,\begin{matrix}c-1\\a-1,b-1,-\frac{1}{2}\end{matrix}\right]\cos\left(2\sqrt{xy}\right)dy. \qquad (4.4.15)$$

The kernel-function $G^{3,0}_{1,3}$ of the latter integral representation is a "known" special function only for the parameters $a=\frac{1}{2}+\nu$, $b=\frac{1}{2}-\nu$, $c=1$ (see [106, I, p. 213, (30)]). Then the conditions $a>0$, $b>0$, $c>\frac{1}{2}$ are reduced to the conditions $-\frac{1}{2}<\nu<\frac{1}{2}$. In this manner we get the following *integral representation of the Gauss function*, valid for $-\frac{1}{2} < \nu < \frac{1}{2}$, $0 < x < 1$:

$$ {}_2F_1\left(\frac{1}{2}+\nu, \frac{1}{2}-\nu; 1; -x\right) = \frac{4\cos\pi\nu}{\pi^2}\int_0^\infty K_\nu^2(t)\cos\left(2\sqrt{x}t\right)dt. \tag{4.4.16} $$

Remark 2. Theorem 4.4.4 represents the ${}_pF_q$-functions ($p \leq q+1$) as Laplace type G-transformations of $\cos_{q+1}$, while Theorem 4.2.4 represents the same functions (but only for $p < q$) as other kinds of G-transformations (generalized fractional integrals) of $\cos_{q-p+1}$.

Alternative approaches to the special functions from the point of view of the fractional calculus *and related interesting results* can be seen in: Al-Bassam [10]-[21], Askey [25], Al-Saqabi [23], Al-Saqabi and Kalla [24], Campos [48]-[51], Kalla et al. [166]-[173], Koornwinder [233], Saigo and Raina [423], Srivastava [461]-[462], Srivastava and Manocha [470], R. Srivastava [483] as well as in the collected volumes of Prudnikov, Brychkov and Marichev [368]-[370].

Another good source of applications and problems, related to Special Functions, is provided by *univalent function theory*; for more details see Section 5.5. However, here we point out the recent work of de Branges [42], proposing a *solution to the famous Bieberbach conjecture.*

— || —

THE MAIN RESULTS OF CHAPTER 4 HAVE BEEN PUBLISHED IN: Kiryakova [195]-[196], [198], [200], [202], [209], Dimovski and Kiryakova [80]-[81].

5 Further generalizations and applications

5.1. Generalized fractional integrals and derivatives involving Fox's H-function

As we have seen in Chapter 1 (Section 1.2.ii), under some conditions the compositon of an arbitrary number of Erdélyi-Kober fractional integrals (1.1.17):

$$I_{\beta}^{\gamma_k,\delta_k} f(x) = \int_0^1 \frac{(1-\sigma)^{\delta_k-1}}{\Gamma(\delta_k)} \sigma^{\gamma_k} f\left(x\sigma^{\frac{1}{\beta}}\right) d\sigma, \quad k = 1,\dots,m \tag{5.1.1}$$

with the same $\beta > 0$, can be represented by means of single integrals involving Meijer's G-function, namely, by the generalized fractional integrals (1.1.6), denoted by $I_{\beta,m}^{(\gamma_k),(\delta_k)}$:

$$\begin{aligned}\left[\prod_{k=1}^{m} I_{\beta}^{\gamma_k,\delta_k} f(x)\right] &= \int_0^1 \dots \int_0^1 \prod_{k=1}^{m} \left[\frac{(1-\sigma_k)^{\delta_k-1}\sigma_k^{\gamma_k}}{\Gamma(\delta_k)}\right] f\left[x(\sigma_1\dots\sigma_m)^{\frac{1}{\beta}}\right] d\sigma_1 \dots d\sigma_m \\ &= \int_0^1 G_{m,m}^{m,0}\left[\sigma \left| \begin{matrix}(\gamma_k+\delta_k)_1^m \\ (\gamma_k)_1^m\end{matrix}\right.\right] f\left(x\sigma^{\frac{1}{\beta}}\right) d\sigma = I_{\beta,m}^{(\gamma_k),(\delta_k)} f(x).\end{aligned} \tag{5.1.2}$$

According to Lemma 1.2.9 and Theorems 1.2.10, 1.2.15, 1.2.17 and 1.2.18, the operators $I_{\beta}^{\gamma_k,\delta_k}$, $k = 1,\dots,m$, commute and relation (5.1.2) holds in the spaces:

$$\begin{aligned} &C_\alpha, && \text{if } \alpha \ge \max_k[-\beta(\gamma_k+1)], \delta_k > 0,\ k=1,\dots,m, \\ &L^p(0,\infty),\ p\ge 1, && \text{if } \beta(\gamma_k+1) > \tfrac{1}{p}, \delta_k > 0,\ k=1,\dots,m, \\ &H_\mu(\Omega),\ \mu \ge 0, && \text{if } \beta(\gamma_k+1) > -\mu, \delta_k > 0,\ k=1,\dots,m. \end{aligned} \tag{5.1.3}$$

However, one can state *the more general problem concerning compositions of arbitrary Erdélyi-Kober fractional intgrals* $I_{\beta_k}^{\gamma_k,\delta_k}$, $k=1,\dots,m$, *with different* $\beta_k > 0$, $k = 1,\dots,m$. As mentioned in Samko, Kilbas and Marichev [434, p. 159], *"we note that under corresponding conditions a commutability takes place also for products of more general operators of forms* $x^{k\beta} I_{0+;x^k}^{\alpha-\beta}$, $x^{m\beta} I_{-;x^m}^{\alpha-\beta}$ etc." (in the notation used there). Such conditions and an explicit form of the resulting generalized fractional integration operator, involving a Fox's H-function, have been derived by Kiryakova [203] (in the spaces C_α), Kalla and Kiryakova [174]-[175] (in L^p and $L_{\mu,p}$) and Kiryakova [204] (in $H_\mu(\Omega)$), cf. (5.1.3). Some of the basic properties, inversion formulas and applications have been considered in these articles too. It turned out that the generalized fractional integrals, representing the products $I_{\beta_1}^{\gamma_1,\gamma_1}\left\{I_{\beta_2}^{\gamma_2,\gamma_2}\dots\left(I_{\beta_m}^{\gamma_m,\gamma_m}\right)\right\}$ have as kernel-function Fox's $H_{m,m}^{m,0}$-function and thus, they are special cases of the *fractional integration operators considered earlier by Kalla* [161], [163]-[164]:

$$Rf(x)=\int_0^1 \sigma^{\eta}\Phi(\sigma)f(x\sigma)d\sigma, \tag{5.1.4}$$

especially with a kernel

$$\Phi(\sigma)=H_{p,q}^{m,n}\left[\sigma\left|\begin{matrix}\left(a_j+\gamma\frac{A_j}{\rho},\frac{A_j}{\rho}\right)_1^p\\ \left(b_k+\gamma\frac{B_k}{\rho},\frac{B_k}{\rho}\right)_1^q\end{matrix}\right.\right]. \tag{5.1.5}$$

Specifically, generalized fractional integration operators of the form

$$\begin{aligned}R_{\gamma,\rho,a_j,A_j,b_k,B_k}^{m,n,p,q;\,1}f(x)=&\frac{\prod\limits_{k=1}^{q}\Gamma(1-b_k)}{\prod\limits_{j=1}^{n}(a_j)\prod\limits_{j=n+1}^{p}\Gamma(1-a_j)}\\ &\times\int_0^1 H_{p,q}^{m,n}\left[\sigma\left|\begin{matrix}\left(a_j+\gamma\frac{A_j}{\rho},\frac{A_j}{\rho}\right)_1^p\\ \left(b_k+\gamma\frac{B_k}{\rho},\frac{B_k}{\rho}\right)_1^q\end{matrix}\right.\right]f(x\sigma)d\sigma,\end{aligned} \tag{5.1.6}$$

or with even more complicated arguments and parameters of the H-functions, and their Weyl type analogues, have been considered by Kalla [157]-[158], [164], Srivastava and Bushman [463], Saxena and Kumbhat [440], Saigo, Saxena and Ram [425] and recently, by other authors. In particular, if $A_j=B_k=1$, $j=1,\dots,p$, $k=1,\dots,q$, the kernel-function turns into a G-function and then, (5.1.6) become special cases of the G-transforms (3.9.22).

We now consider a special case of operators (5.1.6) with a particular kernel-function $H_{m,m}^{m,0}$ and representing products of an arbitrary number of Erdélyi-Kober operators with all with different parameters.

On the other hand, this specific choice of the kernel H-function makes possible a more detailed investigation, proper inversion by corresponding generalized fractional derivatives, finding convolutions, and a number of applications: features not available for the general case (5.1.6). Unlike in Chapter 1, where we used to emphasized on the details for the space C_α of continuous type functions, here we give the main results for *Lebesgue measurable functions* from the space

$$L_{\mu,p} = \left\{ f(x) \in L(0,\infty)\ ,\ \|f\|_{\mu,p} = \left[\int_0^\infty x^{\mu-1} |f(x)|^p dx \right]^{\frac{1}{p}} < \infty \right\}, \tag{5.1.7}$$

$$1 \le p < \infty,\ \mu \in \mathbb{R}; \qquad L_{1,p}:\ = L^p(0,\infty).$$

Definition 5.1.1. Let $m \ge 1$ be an integer; $\beta_k > 0$, $\gamma_k \in \mathbb{R}$, $\delta_k \ge 0$, $k = 1,\dots,m$. Consider $\gamma = (\gamma_1,\dots,\gamma_m)$ as a *multiweight* and respectively $\delta = (\delta_1,\dots,\delta_m)$ as a *multiorder of fractional integration.* The integral operator defined as follows:

$$I^{(\gamma_k),(\delta_k)}_{(\beta_k),m} f(x) = \begin{cases} \int_0^1 H^{m,0}_{m,m}\left[\sigma \left| \begin{matrix} \left(\gamma_k + \delta_k + 1 - \frac{1}{\beta_k}, \frac{1}{\beta_k}\right)_1^m \\ \left(\gamma_k + 1 - \frac{1}{\beta_k}, \frac{1}{\beta_k}\right)_1^m \end{matrix} \right. \right] f(x\sigma) d\sigma, & \text{if } \sum_{k=1}^m \delta_k > 0, \\ f(x), \text{ if } \delta_1 = \delta_2 = \dots = \delta_m = 0 \end{cases} \tag{5.1.8}$$

is said to be a *multiple (m-tuple) Erdélyi-Kober fractional integration operator of Riemann-Liouville type.* Then, each operator of the form

$$Rf(x) = x^{\delta_0} I^{(\gamma_k),(\delta_k)}_{(\beta_k),m} f(x),\ \delta_0 \ge 0, \tag{5.1.9}$$

is called briefly a *generalized (Riemann-Liouville) fractional integral.*

Note. We have used the same notions for the operators (1.1.6), (1.1.7), involving a $G^{m,0}_{m,m}$-function (special cases of (5.1.8), (5.1.9)), since they have essentially the same nature. The difference is only a matter of the level of generality. Operators (5.1.8) can also be put in the form:

$$I^{(\gamma_k),(\delta_k)}_{(\beta_k),m} f(x) = \frac{1}{x} \int_0^x H^{m,0}_{m,m}\left[\frac{t}{x} \left| \begin{matrix} \left(\gamma_k + \delta_k + 1 - \frac{1}{\beta_k}, \frac{1}{\beta_k}\right)_1^m \\ \left(\gamma_k + 1 - \frac{1}{\beta_k}, \frac{1}{\beta_k}\right)_1^m \end{matrix} \right. \right] f(t) dt. \tag{5.1.8$'$}$$

The special choice of the kernel H-function of the form:

$$H^{m,0}_{m,m}\left[\sigma \left| \begin{matrix} (a_k, A_k)_1^m \\ (b_k, A_k)_1^m \end{matrix} \right. \right] = \frac{1}{2\pi i} \int_{\mathfrak{L}} \left[\prod_{k=1}^m \frac{\Gamma(b_k - A_k s)}{\Gamma(a_k - A_k s)} \right] \sigma^s ds, \tag{5.1.10}$$

determines some of its specific properties (see Section E in Appendix), namely, it has three regular singular points $\sigma = 0, 1, \infty$;

$$H^{m,0}_{m,m}(\sigma) \equiv 0 \quad \text{for} \quad |\sigma| > 1, \tag{5.1.11}$$

and in $0 < |\sigma| < 1$ it is an analytic function.

If we denote the kernel-function of (5.1.8′) by $h(x,t)$, namely:

$$h(x,t)=\frac{1}{t}k\Big(\frac{x}{t}\Big),\quad \text{where}\quad k(\tau):=H_{m,m}^{m,0}\left[\frac{1}{\tau}\left|\begin{matrix}\left(\gamma_k+\delta_k+1,\frac{1}{\beta_k}\right)_1^m\\ \left(\gamma_k+1,\frac{1}{\beta_k}\right)_1^m\end{matrix}\right.\right], \tag{5.1.12}$$

then operator (5.1.8′) can be put in the form of a *convolutional type integral transform*:

$$\begin{aligned} I_{(\beta_k),m}^{(\gamma_k),(\delta_k)}f(x)&=\int_0^\infty k\Big(\frac{x}{t}\Big)f(t)\frac{dt}{t}=\int_0^\infty h(x,t)f(t)dt\\ &=\int_0^x H_{m,m}^{m,0}\left[\frac{t}{x}\left|\begin{matrix}\left(\gamma_k+\delta_k+1,\frac{1}{\beta_k}\right)_1^m\\ \left(\gamma_k+1,\frac{1}{\beta_k}\right)_1^m\end{matrix}\right.\right]f(t)\frac{dt}{t}. \end{aligned}\tag{5.1.13}$$

In terms of [434], the kernel-function $h(x,t)$, (5.1.12) is a homogeneous function of order (-1). In the we shall make use of the following general proposition, descending from Hardy, Littlewood and Polya [125]; see also Rooney [399, Lemma 3.1] and Samko, Kilbas and Marichev [434, Theorem 1.5].

We modify this approach suitably for the spaces $L_{\mu,p}$ instead of $L^p(0,\infty)$, viz.

Lemma 5.1.2. *Suppose $k(\tau)$ is measurable on $(0,\infty)$ and such that*

$$h^*=\int_0^\infty \tau^{\frac{\mu}{p}-1}\left|k(\tau)\right|d\tau<\infty. \tag{5.1.14}$$

For $f\in L_{\mu,p}$ we set

$$If(x)=\int_0^\infty k\Big(\frac{x}{t}\Big)f(t)\frac{dt}{t}. \tag{5.1.15}$$

Then, $If(x)$ exists for almost all $x\in(0,\infty)$, the integral operator I is a bounded operator from $L_{\mu,p}$ into itself and $\|I\|\le h^*$.

On the basis of (5.1.13) and Lemma 5.1.2, we are going to consider the *mapping properties of operators* (5.1.8) *in the spaces $L_{\mu,p}$, $L^p(0,\infty)$*, i.e. the conditions under which they are linear bounded operators there.

Theorem 5.1.3. *Let $1\le p<\infty$, $\mu\in\mathbb{R}$, $m>1$ be integer and let the parameters of the generalized (m-tuple) Erdélyi-Kober fractional integral (5.1.8) satisfy the conditions*

$$\beta_k(\gamma_k+1)>\frac{\mu}{p},\ \delta_k>0,\ k=1,\ldots,m. \tag{5.1.16}$$

Then, $I^{(\gamma_k),(\delta_k)}_{(\beta_k),m} f(x)$ *exists almost everywhere on* $(0,\infty)$ *and it is a bounded linear operator from the Banach space* $L_{\mu,p}$ *(5.1.7) into itself. More exactly,*

$$\left\| I^{(\gamma_k),(\delta_k)}_{(\beta_k),m} f \right\|_{\mu,p} \le h_{\mu,p} \|f\|_{\mu,p}, \quad i.e. \quad \left\| I^{(\gamma_k),(\delta_k)}_{(\beta_k),m} f \right\| \le h_{\mu,p} \tag{5.1.17}$$

with

$$h_{\mu,p} = \prod_{k=1}^{m} \frac{\Gamma\left(\gamma_k - \frac{\mu}{p\beta_k} + 1\right)}{\Gamma\left(\gamma_k + \delta_k - \frac{\mu}{p\beta_k} + 1\right)} < \infty. \tag{5.1.18}$$

Proof. The linearity of (5.1.8) is obvious. The main statement of the theorem follows from Lemma 5.1.2 applied to the kernel $k(\tau)$, defined by (5.1.12). To this end, we have to determine the corresponding conditions on the parameters γ_k, δ_k, β_k, $k = 1, \ldots, m$, under which integral (5.1.14) makes sense and to find its value $h^* := h_{\mu,p}$. Substituting $\tau = \sigma^{-1}$ in (5.1.12) and using (E.9), we have for (5.1.14):

$$\begin{aligned} h_{\mu,p} &= \int_0^\infty \sigma^{1-\frac{\mu}{p}} \left| k\left(\frac{1}{\sigma}\right) \right| d\sigma \\ &= \int_0^1 H^{m,0}_{m,m} \left[\sigma \left| \begin{matrix} \left(\gamma_k + \delta_k + 1 - \frac{\frac{\mu}{p}+1}{\beta_k}, \frac{1}{\beta_k}\right)_1^m \\ \left(\gamma_k + 1 - \frac{\frac{\mu}{p}+1}{\beta_k}, \frac{1}{\beta_k}\right)_1^m \end{matrix} \right. \right] d\sigma. \end{aligned} \tag{5.1.19}$$

The asymptotic behaviour of the kernel H-function near the singular points $\sigma = 0,\ 1$ is (see (E.16) and (E.18)):

$$\begin{aligned} H^{m,0}_{m,m} \left[\sigma \left| \begin{matrix} \left(\gamma_k + \delta_k + 1 - \frac{\frac{\mu}{p}+1}{\beta_k}, \frac{1}{\beta_k}\right)_1^m \\ \left(\gamma_k + 1 - \frac{\frac{\mu}{p}+1}{\beta_k}, \frac{1}{\beta_k}\right)_1^m \end{matrix} \right. \right] &= \mathcal{O}\left(\sigma^{\gamma}\right) \quad \text{as } \sigma \to +0, \\ H^{m,0}_{m,m} \left[\sigma \left| \begin{matrix} \left(\gamma_k + \delta_k + 1 - \frac{\frac{\mu}{p}+1}{\beta_k}, \frac{1}{\beta_k}\right)_1^m \\ \left(\gamma_k + 1 - \frac{\frac{\mu}{p}+1}{\beta_k}, \frac{1}{\beta_k}\right)_1^m \end{matrix} \right. \right] &= \mathcal{O}\left((1-\sigma)^{-\nu^*}\right) \quad \text{as } \sigma \to 1-0, \end{aligned} \tag{5.1.20}$$

where according to conditions (5.1.16):

$$\gamma = \min_{1\le k\le m} \left[\beta_k \left(\gamma_k + 1 - \frac{\frac{\mu}{p}+1}{\beta_k} \right) \right] > -1 \quad \text{for} \quad \beta_k(\gamma_k + 1) > \frac{\mu}{p},\ k = 1, \ldots, m;$$

$$-\nu^* = \sum_{k=1}^{m} \delta_k - 1 > -1 \quad \text{for} \quad \sum_{k=1}^{m} \delta_k > 0$$

(for $\sum_{k=1}^{m} \delta_k = 1, 2, \ldots$ a logarithmic term also appears in the second of formulas (5.1.20), but it does not make the convergence of (5.1.19) worse). The asymptotic formulas (5.1.20) with $\gamma > -1$, $-\nu^* > -1$ ensure the convergence of the improper integral (5.1.19), i.e. $h_{\mu,p} < \infty$. All that remains is to work out its value. To this end we use the integral formula (E.21) with

$$b_k = \gamma_k + 1 - \frac{\frac{\mu}{p} + 1}{\beta_k}, \quad a_k = b_k + \delta_k, \quad C_k = \frac{1}{\beta_k}, \quad k = 1, \ldots, m$$

and obtain (5.1.18) for $h_{\mu,p}$. This completes the proof.

It is seen that the above results are easily transferred to spaces of continuous functions like C_α (1.1.1), following the scheme of Chapter 1 (Lemma 1.2.1, Theorem 1.2.15) and the proof of Theorem 5.1.3. In particular, to evaluate the images of the functions $f(x) = x^p$, $p > \alpha$, we use property (E.9) and integral formula (E.21). Thus, we obtain (see also Kiryakova [203]):

Theorem 5.1.4. *Each multiple Erdélyi-Kober fractional integral (5.1.8) preserves the power functions in C_α, $\alpha \geq \max_k [-\beta(\gamma_k + 1)]$ up to a constant multiplier:*

$$I^{(\gamma_k),(\delta_k)}_{(\beta_k),m} \{x^p\} = c_p x^p, \; p > \alpha \quad \textit{where } c_p = \prod_{k=1}^{m} \frac{\Gamma\left(\gamma_k + \frac{p}{\beta_k} + 1\right)}{\Gamma\left(\gamma_k + \delta_k + \frac{p}{\beta_k} + 1\right)} \tag{5.1.21}$$

and maps C_α isomorphically into itself. Conversely, for a fixed index α of C_α, the conditions

$$\gamma_k \geq -\frac{\alpha}{\beta_k} - 1, \delta_k > 0, \; \eta_k := \begin{cases} [\delta_k] + 1, & \textit{for non integer } \delta_k, \\ \delta_k, & \textit{for integer } \delta_k, \end{cases} \quad k = 1, \ldots, m \tag{5.1.22}$$

ensure that (5.1.8) is an invertible mapping

$$I^{(\gamma_k),(\delta_k)}_{(\beta_k),m} : C_\alpha \longrightarrow C_\alpha^{(\eta_1 + \cdots + \eta_m)} \subset C_\alpha.$$

Note. An inversion formula of the kind of (1.2.34) can be written down, but we are proposing further an appropriate differintegral inversion formula (see (5.1.42)). The following theorem holds also in C_α, under conditions of the form (5.1.22).

For functions $f(x) \in L_{\mu,p}(0, \infty)$, the Mellin transform

$$\mathfrak{M}(s) = \mathfrak{M}\{f(x); s\} = \int_0^\infty x^{s-1} f(x) dx, \quad s \in \mathbb{C}$$

is usually considered with s restricted to the line $\Re s = \frac{\mu}{p}$. We also have in mind that for $p > 1$ the above integral is meant to be:

$$\mathfrak{M}\{f(x); s\} = \underset{X\to\infty}{\text{l.i.m.}} \int_{\frac{1}{X}}^{X} x^{s-1} f(x)dx,$$

with l.i.m. denoting the limit in mean for $L_{\mu,p}$-spaces and for $1 \le p \le 2$,

$$\mathfrak{M}(s) \in L_q\left(\frac{1}{q} - i\infty, \frac{1}{q} + i\infty\right), \quad \frac{1}{p} + \frac{1}{q} = 1.$$

Theorem 5.1.5. *In terms of the Mellin convolution*

$$(k \circ f)(x) = \int_0^x k\left(\frac{x}{t}\right) f(t)\frac{dt}{t}, \tag{5.1.23}$$

the multiple Erdélyi-Kober operator (5.1.8) has the following convolutional type representation in $L_{\mu,p}$*:*

$$I^{(\gamma_k),(\delta_k)}_{(\beta_k),m} f(x) = H^{m,0}_{m,m}\left[\frac{1}{x}\,\middle|\,\begin{matrix}\left(\gamma_k + \delta_k + 1, \frac{1}{\beta_k}\right)_1^m \\ \left(\gamma_k + 1, \frac{1}{\beta_k}\right)_1^m\end{matrix}\right] \circ f(x) \tag{5.1.24}$$

and for $1 \le p \le 2$ *its Mellin transformation is given by the equality*

$$\mathfrak{M}\left\{I^{(\gamma_k),(\delta_k)}_{(\beta_k),m} f(x); s\right\} = \left[\prod_{k=1}^{m} \frac{\Gamma\left(\gamma_k - \frac{s}{\beta_k} + 1\right)}{\Gamma\left(\gamma_k + \delta_k - \frac{s}{\beta_k} + 1\right)}\right] \mathfrak{M}\{f(x); s\}. \tag{5.1.25}$$

Proof. The convolutional type representation (5.1.24) follows immediately from (5.1.12)-(5.1.13). Then, according to the Mellin convolution theorem,

$$\begin{aligned}\mathfrak{M}\left\{I^{(\gamma_k),(\delta_k)}_{(\beta_k),m} f(x);\ s\right\} &= \mathfrak{M}\left\{(k \circ f)(x); s\right\} \\ &= \mathfrak{M}\left\{k(x); s\right\} \mathfrak{M}\left\{f(x); s\right\}.\end{aligned}$$

So, it remains to evaluate the Mellin image of the kernel-function $k(x)$ and due to formulas (E.10), (E.20) this is:

$$\begin{aligned}\mathfrak{M}\left\{k(x); s\right\} &= \mathfrak{M}\left\{H^{0,m}_{m,m}\left[x\,\middle|\,\begin{matrix}\left(-\gamma_k, \frac{1}{\beta_k}\right)_1^m \\ \left(-\gamma_k - \delta_k, \frac{1}{\beta_k}\right)_1^m\end{matrix}\right];\ s\right\} \\ &= \prod_{k=1}^{m} \frac{\Gamma\left(\gamma_k - \frac{s}{\beta_k} + 1\right)}{\Gamma\left(\gamma_k + \delta_k - \frac{s}{\beta_k} + 1\right)},\end{aligned}$$

which is the first multiplier in the right-hand side of (5.1.25).

The properties of the H-function (see Section E, Appendix) are useful in deriving *some of the basic operational properties* of the operators $I_{(\beta_k),m}^{(\gamma_k),(\delta_k)}$ that characterize them as generalizations of the classical fractional integrals. They are summarized in the following theorem.

Theorem 5.1.6. *Let $1 \leq p < \infty$, $\mu \in \mathbb{R}$ and $m > 1$, $n \geq 1$ be integers. Suppose conditions (5.1.16):*

$$\beta_k > 0,\ \beta_k(\gamma_k + 1) > \frac{\mu}{p},\ \delta_k > 0,\ k = 1, \ldots, m$$

are satisfied and analogously,

$$\varepsilon_j > 0,\ \varepsilon_j(\tau_j + 1) > \frac{\mu}{p},\ \alpha_j > 0,\ j = 1, \ldots, n.$$

Then, for functions $f \in L_{\mu,p}$ the following operational rules of the multiple Erdélyi-Kober fractional integrals (5.1.8) hold:

$$I_{(\beta_k),m}^{(\gamma_k),(\delta_k)}\{\lambda f(cx) + \eta g(cx)\} = \lambda\left\{I_{(\beta_k),m}^{(\gamma_k),(\delta_k)} f\right\}(cx) + \eta\left\{I_{(\beta_k),m}^{(\gamma_k),(\delta_k)} g\right\}(cx) \tag{5.1.26}$$

(bilinearity of (5.1.8));

$$I_{(\beta_1,\ldots,\beta_m),m}^{(\gamma_1,\ldots,\gamma_s,\gamma_{s+1},\ldots,\gamma_m),(0,\ldots,0,\delta_{s+1},\ldots,\delta_m)} f(x) = I_{(\beta_{s+1},\ldots,\beta_m),m-s}^{(\gamma_{s+1},\ldots,\gamma_m)(\delta_{s+1},\ldots,\delta_m)} f(x) \tag{5.1.27}$$

(if $\delta_1 = \delta_2 = \cdots = \delta_s = 0$, then the multiplicity reduces to $(m - s)$);

$$I_{(\beta_k),m}^{(\gamma_k),(\delta_k)} x^\lambda f(x) = x^\lambda I_{(\beta_k),m}^{\left(\gamma_k + \frac{\lambda}{\beta_k}\right),(\delta_k)} f(x), \quad \lambda \in \mathbb{R} \tag{5.1.28}$$

(generalized commutability with power functions);

$$I_{(\beta_k),m}^{(\gamma_k),(\delta_k)} I_{(\varepsilon_j),n}^{(\tau_j),(\alpha_j)} f(x) = I_{(\varepsilon_j),n}^{(\tau_j),(\alpha_j)} I_{(\beta_k),m}^{(\gamma_k),(\delta_k)} f(x) \tag{5.1.29}$$

(commutability of operators of the form (5.1.8));

$$\textit{the left-hand side of (5.1.29)} = I_{((\beta_k)_1^m,(\varepsilon_j)_1^n),m+n}^{((\gamma_k)_1^m,(\tau_j)_1^n)((\delta_k)_1^m,(\alpha_j)_1^n)} f(x) \tag{5.1.30}$$

(compositions of m-tuple and n-tuple integrals (5.1.8) are $(m+n)$-tuple integrals);

$$I_{(\beta_k),m}^{(\gamma_k+\delta_k),(\sigma_k)} I_{(\beta_k),m}^{(\gamma_k),(\delta_k)} f(x) = I_{(\beta_k),m}^{(\gamma_k),(\sigma_k+\delta_k)} f(x), \quad \textit{if } \delta_k > 0,\ \sigma_k > 0,\ k = 1, \ldots, m \tag{5.1.31}$$

(law of indices, product rule or semigroup property);

$$\left\{ I_{(\beta_k),m}^{(\gamma_k),(\delta_k)} \right\}^{-1} f(x) = I_{(\beta_k),m}^{(\gamma_k+\delta_k),(-\delta_k)} f(x) \tag{5.1.32}$$

(formal inversion formula).

Proof. For detailed proof see [174]-[175]. This results follow by using operational rules (E.8)-(E.10), (E.13) for the H-functions, in the same way as the proofs of Lemma 1.3.1 and Theorem 1.3.8 follow from the properties of the G-functions. Most of the formulas (5.1.26)-(5.1.32) are seen to be m-tuple analogues of the properties of the classical Erdélyi-Kober operators (5.1.1).

Index law (5.1.31) for $\sigma_k = -\delta_k < 0$, $k = 1, \dots, m$, and definition (5.1.8) for zero multiorder of integration yield the *formal inversion formula* (5.1.32), since:

$$I_{(\beta_k),m}^{(\gamma_k+\delta_k),(-\delta_k)} I_{(\beta_k),m}^{(\gamma_k),(\delta_k)} f(x) = I_{(\beta_k),m}^{(\gamma_k),(0,\dots,0)} f(x) = f(x).$$

But (5.1.31) is strictly applied only for $\sigma_k \geq 0$, $k = 1, \dots, m$, and on the other hand, symbols (5.1.8) have not yet been defined for negative multiorder of integration $-\delta_k < 0$, $k = 1, \dots, m$. The problem now is to propose an appropriate meaning for them and hence to avoid the divergent integral appearing in (5.1.32). The situtation is the same as in the simplest case when the Riemann-Liouville and Erdélyi-Kober operators of fractional order $\delta > 0$ are inverted by appealing to an additional differentiation of suitable integer order $\eta = [\delta] + 1$.

In Chapter 1 we have already found the corresponding *proper inversion formula* (1.5.26) for operators $I_{\beta,m}^{(\gamma_k),(\delta_k)}$ (1.1.6) (see Section 1.5). It is based on differential relation (1.5.11) for the kernel $G_{m,m}^{m,0}$-function and involves an auxiliary differential operator D_η, see (1.5.19). In the present case, the formal kernel-function of the operator $I_{(\beta_k),m}^{(\gamma_k+\delta_k),(-\delta_k)}$:

$$H_{m,m}^{m,0}\left[\sigma \,\middle|\, \begin{matrix} \left(\gamma_k + 1 - \frac{1}{\beta_k}, \frac{1}{\beta_k}\right)_1^m \\ \left(\gamma_k + \delta_k + 1 - \frac{1}{\beta_k}, \frac{1}{\beta_k}\right)_1^m \end{matrix}\right] = H_{m,m}^{m,0}\left[\frac{t}{x} \,\middle|\, \begin{matrix} \left(\gamma_k + 1, \frac{1}{\beta_k}\right)_1^m \\ \left(\gamma_k + \delta_k + 1, \frac{1}{\beta_k}\right)_1^m \end{matrix}\right], \tag{5.1.33}$$

$$\sigma := \frac{t}{x}, \quad 0 \leq t \leq x,$$

should be improved by means of a corresponding differential operator, in order to make the integral convergent.

Lemma 5.1.7. *Let $\eta_1 \geq 0, \dots, \eta_m \geq 0$ be arbitrary integers. Then, Fox's $H_{m,m}^{m,0}$-function, $m \geq 1$, satisfies the relation*

$$H_{m,m}^{m,0}\left[\frac{t}{x} \,\middle|\, \begin{matrix} \left(a_k, \frac{1}{\beta_k}\right)_1^m \\ \left(b_k, \frac{1}{\beta_k}\right)_1^m \end{matrix}\right] = D_\eta H_{m,m}^{m,0}\left[\frac{t}{x} \,\middle|\, \begin{matrix} \left(a_k + \eta_k, \frac{1}{\beta_k}\right)_1^m \\ \left(b_k, \frac{1}{\beta_k}\right)_1^m \end{matrix}\right], \tag{5.1.34}$$

where the differential operator D_η stands for the following polynomial of the Euler differential operator $\delta = x\frac{d}{dx}$ of degree $\eta = \eta_1 + \cdots + \eta_m$:

$$D_\eta = \prod_{r=1}^{m}\prod_{j=1}^{\eta_r}\left(\frac{1}{\beta_r}x\frac{d}{dx} + a_r - 1 + j\right). \tag{5.1.35}$$

This Lemma means that the parameters a_k, $k = 1, \ldots, m$, of the H-function in the upper row can be increased by arbitrary integers $\eta_k \geq 0$, $k = 1, \ldots, m$, on account of application of the operator D_η. The proof follows by a repeated use of differential formula (E.14), similarly to the proof of Lemma B.3 and Corollary B.6 for the G-functions.

Let us make use of Lemma 5.1.7 with $a_k = \gamma_k + 1 \leq \gamma_k + \delta_k + 1 = b_k$, $k = 1, \ldots, m$, and take the integers $\eta_1 \geq 0$, ..., $\eta_m \geq 0$ defined as follows:

$$\eta_k = \begin{cases} [\delta_k] + 1, & \text{if } \delta_k \text{ is non integer,} \\ \delta_k, & \text{if } \delta_k \text{ is integer,} \end{cases} \qquad k = 1, \ldots, m. \tag{5.1.36}$$

Then, we can give an appropriate meaning to the symbols $\left\{ I_{(\beta_k),m}^{(\gamma_k),(\delta_k)} \right\}^{-1} = I_{(\beta_k),m}^{(\gamma_k+\delta_k),(-\delta_k)}$ by means of the following differintegral expressions $D_{(\beta_k),m}^{(\gamma_k),(\delta_k)}$.

Definition 5.1.8. Let $m \geq 1$ be an integer, $\beta_k > 0$, γ_k, $\delta_k \geq 0$, $k = 1, \ldots, m$, be real numbers and the integers $\eta_k \geq 0$, $k = 1, \ldots, m$, be defined by (5.1.36). The differintegral operator $D_{(\beta_k),m}^{(\gamma_k),(\delta_k)}$ defined by means of the differential operator

$$D_\eta = \left[\prod_{r=1}^{m}\prod_{j=1}^{\eta_r}\left(\frac{1}{\beta_r}x\frac{d}{dx} + \gamma_r + j\right)\right] \tag{5.1.37}$$

in the following way:

$$\begin{aligned} D_{(\beta_k),m}^{(\gamma_k),(\delta_k)} f(x) &= D_\eta I_{(\beta_k),m}^{(\gamma_k+\delta_k),(\eta_k-\delta_k)} f(x) \\ &= \left[\prod_{r=1}^{m}\prod_{j=1}^{\eta_r}\left(\frac{1}{\beta_r}x\frac{d}{dx} + \gamma_r + j\right)\right] \\ &\quad \times \int_0^1 H_{m,m}^{m,0}\left[\sigma \,\middle|\, \begin{matrix} \left(\gamma_k + \eta_k + 1 - \frac{1}{\beta_k}, \frac{1}{\beta_k}\right)_1^m \\ \left(\gamma_k + 1 - \frac{1}{\beta_k}, \frac{1}{\beta_k}\right)_1^m \end{matrix}\right] f(x\sigma)d\sigma \end{aligned} \tag{5.1.38}$$

is said to be a *multiple (m-tuple) Erdélyi-Kober fractional derivative* of multiorder $\delta =$

$(\delta_1, \ldots, \delta_m)$ and in general, the differintegral operators of the form

$$Df(x) = D^{(\gamma_k),(\delta_k)}_{(\beta_k),m} x^{-\delta_0} f(x) = x^{-\delta_0} D^{\left(\gamma_k - \frac{\delta_0}{\beta_k}\right),(\delta_k)}_{(\beta_k),m} f(x) \tag{5.1.39}$$

with $\delta_0 \geq 0$ are called *generalized (multiple) fractional derivatives.*

Theorem 5.1.9. (Kalla and Kiryakova [174]-[175]) *Let $f \in L_{\mu,p}$, let conditions (5.1.16) be satisfied and*

$$g(x) = I^{(\gamma_k),(\delta_k)}_{(\beta_k),m} f(x). \tag{5.1.40}$$

Then, the following inversion formula holds with a generalized fractional derivative $D^{(\gamma_k),(\delta_k)}_{(\beta_k),m}$ defined as in (5.1.37)-(5.1.38):

$$f(x) = D^{(\gamma_k),(\delta_k)}_{(\beta_k),m} g(x), \tag{5.1.41}$$

i.e.

$$f(x) = \left\{ I^{(\gamma_k),(\delta_k)}_{(\beta_k),m} \right\}^{-1} g(x) = D^{(\gamma_k),(\delta_k)}_{(\beta_k),m} g(x) \;\; for \;\; g \in I^{(\gamma_k),(\delta_k)}_{(\beta_k),m} \left(L_{\mu,p}\right). \tag{5.1.42}$$

Proof. A detailed proof can be made following the lines of the proof of Theorem 1.5.5. However, we give here alternatively the formal manipulations, suggesting the idea of introducing the generalized fractional derivatives $D^{(\gamma_k),(\delta_k)}_{(\beta_k),m}$ of the form (5.1.37)-(5.1.38) and the proof that

$$D^{(\gamma_k),(\delta_k)}_{(\beta_k),m} I^{(\gamma_k),(\delta_k)}_{(\beta_k),m} f(x) = f(x), \quad f \in L_{\mu,p}. \tag{5.1.42'}$$

Having in mind the index law (5.1.31) and representation (5.1.8′), then putting the differential operator D_η under the integral sign and using differential relation (5.1.34), we have:

$$\begin{aligned}
D^{(\gamma_k),(\delta_k)}_{(\beta_k),m} I^{(\gamma_k),(\delta_k)}_{(\beta_k),m} f(x) &= D_\eta I^{(\gamma_k+\delta_k),(\eta_k-\delta_k)}_{(\beta_k),m} I^{(\gamma_k),(\delta_k)}_{(\beta_k),m} f(x) = D_\eta I^{(\gamma_k),(\eta_k)}_{(\beta_k),m} f(x) \\
&= D_\eta \left\{ x^{-1} \int\limits_0^x H^{m,0}_{m,m} \left[\frac{t}{x} \middle| \begin{matrix} \left(\gamma_k + \eta_k + 1 - \frac{1}{\beta_k}, \frac{1}{\beta_k}\right)_1^m \\ \left(\gamma_k + 1 - \frac{1}{\beta_k}, \frac{1}{\beta_k}\right)_1^m \end{matrix} \right] f(t) dt \right\} \\
&= D_\eta \left\{ \int\limits_0^x H^{m,0}_{m,m} \left[\frac{t}{x} \middle| \begin{matrix} \left(\gamma_k + \eta_k + 1, \frac{1}{\beta_k}\right)_1^m \\ \left(\gamma_k + 1, \frac{1}{\beta_k}\right)_1^m \end{matrix} \right] \frac{f(t)}{t} dt \right\} \\
&= \int\limits_0^x D_\eta H^{m,0}_{m,m} \left[\frac{t}{x} \middle| \begin{matrix} \left(\gamma_k + \eta_k + 1, \frac{1}{\beta_k}\right)_1^m \\ \left(\gamma_k + 1, \frac{1}{\beta_k}\right)_1^m \end{matrix} \right] \frac{f(t)}{t} dt \\
&= \int\limits_0^x H^{m,0}_{m,m} \left[\frac{t}{x} \middle| \begin{matrix} \left(\gamma_k + 1, \frac{1}{\beta_k}\right)_1^m \\ \left(\gamma_k + 1, \frac{1}{\beta_k}\right)_1^m \end{matrix} \right] \frac{f(t)}{t} dt \\
&= x^{-1} \int\limits_0^x H^{m,0}_{m,m} \left[\frac{t}{x} \middle| \begin{matrix} \left(\gamma_k + 0 + 1 - \frac{1}{\beta_k}, \frac{1}{\beta_k}\right)_1^m \\ \left(\gamma_k + 1 - \frac{1}{\beta_k}, \frac{1}{\beta_k}\right)_1^m \end{matrix} \right] f(t) dt \\
&= I^{(\gamma_k),(0,\dots,0)}_{(\beta_k),m} f(x) = f(x),
\end{aligned}$$

which is (5.1.42′).

Note. Identity (5.1.42′) *holds also in the other functional spaces* like C_α, $H_\mu(\Omega)$ etc, under the corresponding conditions for the convergence of integral (5.1.8).

Note. In the case of *integers* $\delta_k = \eta_k$, $k = 1, \dots, m$, $I^{(\gamma_k+\delta_k),(\eta_k-\delta_k)}_{(\beta_k),m}$ is the identity operator and so $D^{(\gamma_k),(\delta_k)}_{(\beta_k),m} = D_\eta$ is a *purely differential operator*, the inverse of the multiple Erdélyi-Kober integral $I^{(\gamma_k),(\delta_k)}_{(\beta_k),m}$.

Now we can consider both multiple fractional integrals (5.1.8) and multiple fractional derivatives (5.1.38) as a *united object: multiple (generalized) fractional differintegrals.* We can also use the common symbols $I^{(\gamma_k),(\delta_k)}_{(\beta_k),m}$ with *arbitrary* real $\delta_1, \dots, \delta_m$, if we adopt, in addition to (5.1.8),

$$I^{(\gamma_k),(\delta_k)}_{(\beta_k),m} = D_{\tilde\eta} I^{(\gamma_k),(\delta_k+\tilde\eta_k)}_{(\beta_k),m} \tag{5.1.43}$$

$$\text{with } \tilde\eta_k = \begin{cases} [-\delta_k] + 1, & \text{for non integer } \delta_k < 0, \\ -\delta_k, & \text{for integer } \delta_k, \\ 0, & \text{for } \delta_k \geq 0, \end{cases} \qquad k = 1, \dots, m.$$

It is seen that (5.1.43) coincides with (5.1.8) if all the $\delta_k \geq 0$, $k = 1, \ldots, m$; and with (5.1.38) if all the $\delta_k \leq 0$, $k = 1, \ldots, m$. However, if for instance

$$\delta_1 < 0, \ldots, \delta_s < 0; \quad \delta_{s+1} = \cdots = \delta_r = 0; \quad \delta_{r+1} > 0, \ldots, \delta_m > 0, \tag{5.1.44}$$

then the symbol (5.1.43) is meant as a composition of an s-tuple fractional derivative and an $(m-r)$-tuple fractional integral, namely:

$$\begin{aligned} I^{(\gamma_k),(\delta_k)}_{(\beta_k),m} &= \left[\prod_{k=1}^{s} D^{\gamma_k+\delta_k,-\delta_k}_{\beta_k}\right] \cdot \left[\underbrace{I \ldots I}_{(r-s)}\right] \cdot \left[\prod_{j=r+1}^{m} I^{\gamma_j,\delta_j}_{\beta_j}\right] \\ &= D^{(\gamma_k+\delta_k),(-\delta_k)}_{(\beta_k),s} I^{(\gamma_j),(\delta_j)}_{(\beta_j),m-r}, \end{aligned} \tag{5.1.45}$$

the latter being fully justified by the considerations in next section.

5.2. Special cases, decomposition and applications of the generalized fractional differintegrals in solving Abel type integral equations

Consider first the special case of multiple Erdélyi-Kober fractional integrals (5.1.8) with $m = 1$. The *kernel-function* $H^{1,0}_{1,1}(\sigma)$, due to formulas (E.6$'$), with $c = \frac{1}{\beta}$ and (C.10), is:

$$\begin{aligned} H^{1,0}_{1,1}\left[\sigma \left| \begin{matrix} \left(\gamma+\delta+1-\frac{1}{\beta}, \frac{1}{\beta}\right) \\ \left(\gamma+1-\frac{1}{\beta}, \frac{1}{\beta}\right) \end{matrix} \right.\right] &= \beta G^{1,0}_{1,1}\left[\sigma^\beta \left| \begin{matrix} \gamma+\delta+1-\frac{1}{\beta} \\ \gamma+1-\frac{1}{\beta} \end{matrix} \right.\right] = \beta\sigma^{\beta-1} G^{1,0}_{1,1}\left[\sigma^\beta \left| \begin{matrix} \gamma+\delta \\ \gamma \end{matrix} \right.\right] \\ &= \beta\sigma^{\beta-1} \frac{\left(1-\sigma^\beta\right)^{\delta-1}}{\Gamma(\delta)} \sigma^{\beta\gamma}, \end{aligned} \tag{5.2.1}$$

therefore

$$\begin{aligned} I^{\gamma,\delta}_{\beta,1} f(x) &= \int\limits_0^1 \frac{(1-\sigma^\beta)^{\delta-1}}{\Gamma(\delta)} \sigma^{\beta\gamma} f(x\sigma) d\sigma^\beta \\ &= \int\limits_0^1 \frac{(1-\tau)^{\delta-1}}{\Gamma(\delta)} \tau^\gamma f\left(x\tau^{\frac{1}{\beta}}\right) d\tau = I^{\gamma,\delta}_{\beta} f(x), \end{aligned} \tag{5.2.2}$$

i.e. *the 1-tuple generalized fractional integral* (5.1.8) *involving an H-function coincides with the classical Erdélyi-Kober fractional integral* (5.1.1) (and also with the 1-tuple fractional integral (1.1.6) involving Meijer's G-function).

Consider also the case $\mathbf{m} = \mathbf{2}$. In particular, if $\beta_1 = \beta_2 = \beta > 0$, due to [287, p. 11,

(1.7.11)], (E.6′) and (1.1.18):

$$\begin{aligned}&H_{2,2}^{2,0}\left[\sigma\left|\begin{matrix}\left(\gamma_1+\delta_1+1-\frac{1}{\beta},\frac{1}{\beta}\right),\left(\gamma_2+\delta_2+1-\frac{1}{\beta},\frac{1}{\beta}\right)\\ \left(\gamma_1+1-\frac{1}{\beta},\frac{1}{\beta}\right),\left(\gamma_2+1-\frac{1}{\beta},\frac{1}{\beta}\right)\end{matrix}\right.\right]\\&=\frac{\sigma^{\beta\gamma_2}(1-\sigma)^{\delta_1+\delta_2-1}}{\Gamma(\delta_1+\delta_2)}{}_2F_1\left(\gamma_2+\delta_2-\gamma_1,\ \delta_1;\ \delta_1+\delta_2;\ 1-\sigma^\beta\right)\end{aligned}\tag{5.2.3}$$

and so, we obtain again the *hypergeometric fractional integrals* (see Section 1.1.iii ($m=2$)):

$$I_{(\beta,\beta),2}^{(\gamma_1,\gamma_2),(\delta_1,\delta_2)}f(x)=Hf(x)\quad\text{of the form (1.1.19).}\tag{5.2.4}$$

For $\beta_1\neq\beta_2$, the operator $I_{(\beta_k),2}^{(\gamma_k),(\delta_k)}$ has a more complicated form, related to the Gauss hypergeometric kernel-function again.

For *arbitrary* $m>1$ let us compare the Mellin transforms of the classical Erdélyi-Kober fractional integral (5.1.1) (see (1.2.17)):

$$\mathfrak{M}\left\{I_{\beta_k}^{\gamma_k,\delta_k}f(x);s\right\}=\frac{\Gamma\left(\gamma_k-\frac{s_k}{\beta_k}+1\right)}{\Gamma\left(\gamma_k+\delta_k-\frac{s_k}{\beta_k}+1\right)}\mathfrak{M}\{f(x);s\}$$

and of the multiple Erdélyi-Kober fractional integral (5.1.8) (see (5.1.25)):

$$\mathfrak{M}\left\{I_{(\beta_k),m}^{(\gamma_k),(\delta_k)}f(x);s\right\}=\left[\prod_{k=1}^m\frac{\Gamma\left(\gamma_k-\frac{s_k}{\beta_k}+1\right)}{\Gamma\left(\gamma_k+\delta_k-\frac{s_k}{\beta_k}+1\right)}\right]\mathfrak{M}\{f(x);s\}.$$

This suggests

$$\mathfrak{M}\left\{I_{(\beta_k),m}^{(\gamma_k),(\delta_k)}f(x);s\right\}=\mathfrak{M}\left\{\left(\prod_{k=1}^m I_{\beta_k}^{\gamma_k,\delta_k}\right)f(x);s\right\},\tag{5.2.5}$$

i.e. the Mellin images of the generalized fractional integral (5.1.8) and of the product $\left(\prod\limits_{k=1}^m I_{\beta_k}^{\gamma_k,\delta_k}\right)$ of Erdélyi-Kober operators (5.1.1) coincide. The same relation follows between the corresponding kernel-functions:

$$\begin{aligned}&\mathfrak{M}\left\{H_{m,m}^{m,0}\left[\sigma\left|\begin{matrix}\left(\gamma_k+\delta_k+1-\frac{1}{\beta_k},\frac{1}{\beta_k}\right)_1^m\\ \left(\gamma_k+1-\frac{1}{\beta_k},\frac{1}{\beta_k}\right)_1^m\end{matrix}\right.\right];\ s\right\}\\&=\prod_{k=1}^m\mathfrak{M}\left\{H_{1,1}^{1,0}\left[\sigma\left|\begin{matrix}\left(\gamma_k+\delta_k+1-\frac{1}{\beta_k},\frac{1}{\beta_k}\right)_1^m\\ \left(\gamma_k+1-\frac{1}{\beta_k},\frac{1}{\beta_k}\right)_1^m\end{matrix}\right.\right];\ s\right\}.\end{aligned}\tag{5.2.6}$$

This gives the idea of the following theorem.

Theorem 5.2.1. (Composition/Decomposition theorem) *Under the conditions (5.1.16), the classical Erdélyi-Kober fractional integrals of the form (5.1.1):* $I_{\beta_k}^{\gamma_k,\delta_k}$, $k = 1, \ldots, m$, *commute in the space* $L_{\mu,p}$ *and their product*

$$\begin{aligned} I_{\beta_m}^{\gamma_m,\delta_m} \left\{ I_{\beta_{m-1}}^{\gamma_{m-1},\delta_{m-1}} \ldots \left(I_{\beta_1}^{\gamma_1,\delta_1} f(x) \right) \right\} &= \left[\prod_{k=1}^{m} I_{\beta_k}^{\gamma_k,\delta_k} \right] f(x) \\ &= \underbrace{\int_0^1 \ldots \int_0^1}_{m} \left[\prod_{k=1}^{m} \frac{(1-\sigma_k)^{\delta_k - 1} \sigma_k^{\gamma_k}}{\Gamma(\delta_k)} \right] f\left(x\sigma_1^{\frac{1}{\beta_1}} \ldots \sigma_m^{\frac{1}{\beta_m}} \right) d\sigma_1 \ldots d\sigma_m \end{aligned} \tag{5.2.7}$$

can be represented as an m*-tuple Erdélyi-Kober operator (5.1.8), i.e. by means of a single integral involving the* H*-function:*

$$\begin{aligned} \left[\prod_{k=1}^{m} I_{\beta_k}^{\gamma_k,\delta_k} \right] f(x) &= I_{(\beta_k),m}^{(\gamma_k),(\delta_k)} f(x) \\ &= \int_0^1 H_{m,m}^{m,0} \left[\sigma \left| \begin{matrix} \left(\gamma_k + \delta_k + 1 - \frac{1}{\beta_k}, \frac{1}{\beta_k} \right)_1^m \\ \left(\gamma_k + 1 - \frac{1}{\beta_k}, \frac{1}{\beta_k} \right)_1^m \end{matrix} \right. \right] f(x\sigma) d\sigma, \quad f \in L_{\mu,p}. \end{aligned} \tag{5.2.8}$$

Conversely, under the same conditions, each multiple Erdélyi-Kober operator of form (5.1.8) can be represented as a product (5.2.7).

Proof. Besides comparing the Mellin images of (5.2.7) and (5.2.8), one can use also the method of *mathematical induction*. The statement is evident for $m = 1$, due to (5.2.1). In view of (5.2.3), it can be checked for $m = 2$, after calculating the repeated integral $\prod_{k=1}^{2} I_{\beta_k}^{\gamma_k,\delta_k}$. Further, supposing that (5.2.8) holds for some $m > 1$, we can show the same for $m + 1$, following the same calculations as in the proof of Theorem 1.2.10, but with Fox's H-function. For details of the proof see Kalla and Kiryakova [175].

Similarly, let us consider a product of Weyl type Erdélyi-Kober fractional integrals (1.1.17*):

$$K_{\varepsilon_j}^{\tau_j,\alpha_j} f(x) = \int_0^1 \frac{(\sigma - 1)^{\alpha_j - 1} \sigma^{\tau_j}}{\Gamma(\alpha_j)} f\left(x\sigma^{\frac{1}{\beta}} \right) d\sigma, \quad j = 1, \ldots, n. \tag{5.2.9}$$

We introduce also the following definition.

Definition 5.2.2. Let $n \geq 1$ be integer and $\alpha_0 \geq 0$, $\alpha_j \geq 0$, τ_j, $\varepsilon_j > 0$, $j = 1, \ldots, n$, be real parameters. By *n-tuple (generalized) Erdélyi-Kober fractional integrals of Weyl type* we mean the integral operators of the form

$$\begin{aligned} Wf(x) &= x^{\alpha_0} K_{(\varepsilon_j),n}^{(\tau_j),(\alpha_j)} f(x) \\ &= x^{\alpha_0} \int_1^\infty H_{n,n}^{n,0} \left[\frac{1}{\sigma} \middle| \begin{matrix} \left(\tau_j + \alpha_j + \frac{1}{\varepsilon_j}, \frac{1}{\varepsilon_j}\right)_1^n \\ \left(\tau_j + \frac{1}{\varepsilon_j}, \frac{1}{\varepsilon_j}\right)_1^n \end{matrix} \right] f(\sigma x) dx. \end{aligned} \tag{5.2.10}$$

It is seen that for $n = 1$ and $\alpha_0 = 0$, (5.2.10) coincides with Erdélyi-Kober operator (5.2.9):

$$K_{\varepsilon,1}^{\tau,\alpha} = K_\varepsilon^{\tau,\alpha}$$

and for $\varepsilon_j = \varepsilon$, $j = 1, \ldots, n$ we obtain the multiple Weyl type fractional integrals (1.4.2) involving Meijer's $G_{n,n}^{n,0}$-function:

$$K_{(\varepsilon,\ldots,\varepsilon),n}^{(\tau_j),(\alpha_j)} = W_{\varepsilon,n}^{(\tau_j),(\alpha_j)}.$$

Theorem 5.2.3. (Kalla and Kiryakova [174]-[175]) *If $f \in L_{\mu,p}$ and*

$$\varepsilon_j > 0,\ \varepsilon_j\tau_j > -\frac{\mu}{p},\ \alpha_j > 0, \quad j = 1, \ldots, n, \tag{5.2.11}$$

then $K_{(\varepsilon_j),n}^{(\tau_j),(\alpha_j)} f \in L_{\mu,p}$ too and

$$\left\| K_{(\varepsilon_j),n}^{(\tau_j),(\alpha_j)} \right\| \leq h_{\mu,p}^* = \prod_{j=1}^n \frac{\Gamma\left(\tau_j + \frac{\mu}{p\varepsilon_j}\right)}{\Gamma\left(\tau_j + \alpha_j + \frac{\mu}{p\varepsilon_j}\right)} < \infty. \tag{5.2.12}$$

Also (cf. Section 1.4, Theorem 1.4.4) the following composition/decomposiiton property holds.

Theorem 5.2.4. ([174]-[175]) *Under conditions (5.2.11) the operators $K_{\varepsilon_j}^{\tau_j,\alpha_j}$, $j = 1, \ldots, n$, commute in $L_{\mu,p}$ and their product is representable by means of multiple Weyl type integrals (5.2.10) and conversely , i.e. for $f \in L_{\mu,p}$:*

$$\left[\prod_{j=1}^n K_{\varepsilon_j}^{\tau_j,\alpha_j} \right] f(x) = K_{(\varepsilon_j),n}^{(\tau_j),(\alpha_j)} f(x). \tag{5.2.13}$$

It is interesting then to consider products of commuting Erdélyi-Kober fractional integrals, some of them being of Riemann-Liouville type (1.1.17), and others of Weyl type (1.1.17*), namely:

$$T = \left(\prod_{k=1}^{m} I_{\beta_k}^{\gamma_k,\delta_k}\right)\left(\prod_{j=1}^{n} K_{\varepsilon_j}^{\tau_j,\alpha_j}\right). \tag{5.2.14}$$

The problem of the commutability of these multipliers and the representation of their product by means of a single integral has been considered by Samko, Kilbas and Marichev [434, Theorem 10.7] but for case of $\beta_k = \varepsilon_j = 1$, $k = 1,\dots,m$, $j = 1,\dots,n$; $L_{\mu,p}$ with $\mu = 1$ only. Under the corresponding weaker conditions, it is shown there that the operator T coincides in $L^p(0,\infty)$ with a single integral involving a $G^{m,n}_{m+n,m+n}$-function. In Kalla and Kiryakova [174]-[175] we propose the corresponding more general result.

Theorem 5.2.5. *Let $f \in L_{\mu,p}$ and let both conditions (5.1.16), (5.2.11) be satisfied. Then, the Erdélyi-Kober fractional integrals in the brackets in (5.2.14) commute and the operator T maps $L_{\mu,p}$ boundedly into itself and can be represented by the following single integral:*

$$Tf(x) = x^{-1}\int_0^\infty H^{m,n}_{m+n,m+n}\left[\frac{t}{x}\left|\begin{matrix}\left(-\tau_j+1-\frac{1}{\varepsilon_j},\frac{1}{\varepsilon_j}\right)_1^n,\left(\gamma_k+\delta_k+1-\frac{1}{\beta_k},\frac{1}{\beta_k}\right)_1^m\\ \left(\gamma_k+1-\frac{1}{\beta_k},\frac{1}{\beta_k}\right)_1^m,\left(-\tau_j-\alpha_j+1-\frac{1}{\varepsilon_j},\frac{1}{\varepsilon_j}\right)_1^n\end{matrix}\right.\right]$$
$$\times f(t)dt. \tag{5.2.15}$$

Note. By the techniques used in Chapter 1 (see also Kiryakova [203]), one can establish that the properties of the multiple Erdélyi-Kober operators (5.1.8), (5.2.10), stated in Sections 5.1 and 5.2 hold also in the spaces C_α, respectively C_α^*, under suitable conditions on the parameters.

Theorems 5.1.9 and 5.2.1 and their Weyl-type analogues concerning $K^{(\tau_j),(\alpha_j)}_{(\varepsilon_j),n}$, can be used successfully in solving certain convolutional integral equations.

Generalized Abel integral equation (explicit solution)

The famous *Abel equation*

$$\int_0^x (x-t)^{-\alpha} y(t)dt = f(x), \quad 0 < \alpha < 1, \tag{5.2.16}$$

involving the Riemann-Liouville fractional integral of order $\delta = 1-\alpha > 0$, is historically the first example of the *Volterra integral equations of the first kind* (Volterra [506]):

$$\int_0^x K(x,t)y(t)dt = f(x). \tag{5.2.17}$$

Arising from the so-called *tautochrone (isochrone) problem*, it was solved by Abel (1823) by appyling the fractional derivatives technique. The solution was so elegant that it became one of the most important motivatoins for further development of the fractional calculus. Various generalizations of the equation (5.2.16) involving generalized operators of fractional integration have been considered (for some history see e.g. Srivastava and Buschman [464]-[465]). These Volterra type integral equations find a wide range of applications, for instance, to boundary value problems for PDEs of mixed type. Naturally, we consider Abel type equations related to the multiple Erdélyi-Kober fractional integrals with $H^{m,0}_{m,m}$-functions as kernels. We can state:

Theorem 5.2.6. (Kiryakova [207]) *Consider the generalized (multiple) Abel integral equation*

$$x^{-1}\int_0^x H^{m,0}_{m,m}\left[\frac{t}{x}\left|\begin{matrix}\left(\gamma_k+\delta_k+1-\frac{1}{\beta_k},\frac{1}{\beta_k}\right)_1^m\\ \left(\gamma_k+1-\frac{1}{\beta_k},\frac{1}{\beta_k}\right)_1^m\end{matrix}\right.\right]y(t)dt=f(x), \tag{5.2.18}$$

or its equivalent multiple form:

$$\begin{aligned}&\frac{1}{\Gamma(\delta_1)\ldots\Gamma(\delta_m)}\int_0^x\left(x^{\beta_m}-t_m^{\beta_m}\right)^{\delta_m-1}t_m^{\beta_m\gamma_m}d\left(t_m^{\beta_m}\right)\ldots\\ &\qquad\times\int_0^{t_2}\left(t_2^{\beta_1}-t_1^{\beta_1}\right)^{\delta_1-1}t_1^{\beta_1\gamma_1}y(t_1)\,d\left(t_1^{\beta_1}\right)=x^{\beta_m(\gamma_m+\delta_m)}f(x)\end{aligned} \tag{5.2.19}$$

under the conditions:

$$f\in I^{(\gamma_k),(\delta_k)}_{(\beta_k),m}\left(L_{\mu,p}\right),\ \beta_k>0,\ \beta_k(\gamma_k+1)>\frac{\mu}{p},\ \delta_k>0, k=1,\ldots,m. \tag{5.2.20}$$

Then, equations (5.2.18)-(5.2.19) have unique solution of the explicit form:

$$\begin{aligned}y(x)&=D^{(\gamma_k),(\delta_k)}_{(\beta_k),m}f(x)\\ &=\left[\prod_{r=1}^m\prod_{j=1}^{\eta_r}\left(\frac{1}{\beta_r}x\frac{d}{dx}+\gamma_r+j\right)\right]x^{-1}\int_0^x H^{m,0}_{m,m}\left[\frac{t}{x}\left|\begin{matrix}\left(\gamma_k+\eta_k+1-\frac{1}{\beta_k},\frac{1}{\beta_k}\right)_1^m\\ \left(\gamma_k+1-\frac{1}{\beta_k},\frac{1}{\beta_k}\right)_1^m\end{matrix}\right.\right]f(t)dt,\end{aligned} \tag{5.2.21}$$

belonging to $L_{\mu,p}$, where the integers η_k, $k=1,\ldots,m$, are defined as in (5.1.36).

Proof. The condition $f(x)\in I^{(\gamma_k),(\delta_k)}_{(\beta_k),m}\left(L_{\mu,p}\right)$ is to ensure the existence of the solution. For a more detailed description of the space of the functions, representable by fractional integrals of functions from $L_{\mu,p}$, one can see, for example, Samko, Kilbas and Marichev [434, §6], Dinh [84], etc. Other suitable functional spaces for this purpose are the popular spaces $F_{p,\mu}$ of McBride [288]-[291]. Since integral equations (5.2.18), (5.2.19) are of convolutional type, the uniqueness of their solution follows from the theorem of Mikusinski and Ryll-Nardzewski [307], stated in the proof of Theorem

1.2.15. Both integral equations however can be written down in the operator form $I^{(\gamma_k),(\delta_k)}_{(\beta_k),m} y(x) = f(x)$ and then their solution is given by the generalized (m-tuple) fractional derivative: $y(x) = D^{(\gamma_k),(\delta_k)}_{(\beta_k),m} f(x)$, represented explicitly in the form (5.2.21).

Special case. If all the $\beta_k = \beta$, $k = 1, \dots, m$, are equal, then the generalized Abel equation (5.2.18) turns into an integral equation involving the $G^{m,0}_{m,m}$-function and its solution is expressed by $D^{(\gamma_k),(\delta_k)}_{\beta,m}$, (1.5.19) (see also Dinh [84]).

More generally, generalized Abel integral equations involving combinations of Erdélyi-Kober fractional integrals both of Riemann-Liouville and Weyl type can be considered. The simplest case concerns equations of the forms

$$\begin{gathered} u(x) \int_a^x \frac{(x-t)^{\alpha-1}}{\Gamma(\alpha)} y(t)dt + v(x) \int_x^b \frac{(t-x)^{\alpha-1}}{\Gamma(\alpha)} y(t)dt = f(x), \\ \int_a^x \frac{(x-t)^{\alpha-1}u(x)}{\Gamma(\alpha)} y(t)dt + \int_x^b \frac{(t-x)^{\alpha-1}v(x)}{\Gamma(\alpha)} y(t)dt = f(x), \end{gathered} \tag{5.2.22}$$

resolved, for example, in Samko, Kilbas and Marichev [434, §30.2-3]. One can take in particular, $a = 0$, $b = \infty$, $u(x) = x^\gamma$, $v(x) = x^\tau$ and then the Erdélyi-Kober operators (1.1.17)-(1.1.17*) appear in (5.2.22). Penzel [363] investigated systems of generalized Abel equations of the form

$$u(x) I^{\gamma,\alpha}_\beta y(x) + v(x) K^{1-\gamma-\alpha,\alpha}_\beta y(x) = f(x) \tag{5.2.23}$$

in the spaces $L_{\mu,p}$.

More about the history of solving other generalizations of the Abel integral equations can be found in [434, §34]; for related results see Sakaljuk [428]-[429], Samko [430]-[433] etc.

5.3. Solutions to dual integral equations with H- and G-functions via the generalized fractional integrals

There are many techniques available for solving mixed boundary value problems arising in mathematical physics (see Sneddon [452]). A specific approach to this class of problems is to reduce them to dual integral equations. It seems that the first and simplest example of such a pair of equations, called *Titchmarsh equations*, is the following:

$$\begin{aligned} \int_0^\infty u^{-1} J_0(xu) f(u) du &= 1, \quad 0 < x < 1, \\ \int_0^\infty J_0(xu) f(u) du &= 0, \quad x > 1. \end{aligned} \tag{5.3.1}$$

In general, the pair of equations

$$\begin{aligned} \int_0^\infty \omega(u)K(x,u)f(u)du &= \varphi(x), \quad 0<x<1, \\ \int_0^\infty K(x,u)f(u)du &= \psi(x), \quad x>1, \end{aligned} \tag{5.3.2}$$

where the kernel $K(x,u)$, the weight $\omega(u)$ and boundary conditions $\varphi(x)$, $\psi(x)$ are known functions and $f(u)$ is to be determined, is known as a *pair of dual integral equations*. Several methods for their solutions in the case $\omega(u)=u^\lambda$, $K(x,u)=J_\nu(xu)$ were developed by Weber (1873), Busbridge (1938), Titchmarsh (1948), Noble (1958), Peters (1961), Erdélyi and Sneddon [110] and others. In most of these approaches an important role is played by the operators of fractional integration. A more precise description of the history of this method can be found in [434, §39], [498], [504]. In the papers of Fox [115]-[116], Saxena [436] and Mathai and Saxena [286, p. 238-248] a general method is proposed for obtaining solutions of dual integral equations associated with an arbitrary special function having a Mellin-Barnes type integral representation. The kernel-function is taken to be a H-function of Fox's or a Meijer's G-function and compositions of the Erdélyi-Kober operators of fractional integration are applied to reduce the given equations to others with one and same kernel. Thus, in Mathai and Saxena [286] the solution is given in a closed but rather involved form, since these compositions of fractional integrals are not calculated explicitly. Moreover, no conditions on the parameters of the H- and G-functions in the kernel are stated in order for the solution to be a non formal one.

Here *we give an explicit solution to a class of dual integral equations involving H- and G-functions.* It contains as special cases the solutions of dual integral equations involving many special functions of mathematical physics. We illustrate this by the example with Bessel functions as kernels. The basic results have been published in Kiryakova [205] and Galue, Kiryakova and Kalla [119].

Consider, in general, the following pair of dual integral equations:

$$\begin{aligned} \int_0^\infty H^{m,n}_{p+n,q+m}\left[xu \left| \begin{matrix} (a_j,A_j)_1^{n+p} \\ (b_k,B_k)_1^{m+q} \end{matrix} \right.\right] f(u)du &= \varphi(x), \ 0<x<1, \\ \int_0^\infty H^{m,n}_{p+n,q+m}\left[xu \left| \begin{matrix} (c_j,C_j)_1^{n+p} \\ (d_k,D_k)_1^{m+q} \end{matrix} \right.\right] f(u)du &= \psi(x), \ x>1, \end{aligned} \tag{5.3.3}$$

where $m\geq 0$, $n\geq 0$, $p\geq 0$, $q\geq 0$ are integers and all the A_j, C_j, B_k, D_k are positive so that

$$\sum_{j=1}^{n} A_j + \sum_{k=1}^{m} B_k = \sum_{j=n+1}^{n+p} A_j + \sum_{k=m+1}^{m+q} B_k,$$
$$\sum_{j=1}^{n} C_j + \sum_{k=1}^{m} D_k = \sum_{j=n+1}^{n+p} C_j + \sum_{k=m+1}^{m+q} D_k; \tag{5.3.4}$$

$$\sum_{j=1}^{n} C_j \geq \sum_{j=1}^{n} A_j, \quad \sum_{j=n+1}^{n+p} C_j \geq \sum_{j=n+1}^{n+p} A_j,$$
$$\sum_{k=1}^{m} B_k \geq \sum_{k=1}^{m} D_k, \quad \sum_{k=m+1}^{m+q} B_k \geq \sum_{k=m+1}^{m+q} D_k; \tag{5.3.5}$$

and $\varphi(x) \in C_\alpha$, $\psi(x) \in C^*_{\alpha^*}$, $\alpha^* \leq \alpha$, C_α and $C^*_{\alpha^*}$ being spaces (1.1.1), (1.1.2) of continuous type functions.

The *main result* we achieve here by means of the generalized fractional calculus techniques gives the explicit solution as follows.

Theorem 5.3.1. (Galue, Kiryakova and Kalla [119]) *For boundary conditions (right-hand sides)*

$$\varphi(x) \in C_\alpha, \quad \psi(x) \in C^*_{\alpha^*}, \quad \alpha^* \leq \alpha, \tag{5.3.6}$$

and parameters a_j, A_j, ..., d_k, D_k, $j = 1, \dots, n+p$, $k = 1, \dots, m+q$, *satisfying conditions (5.3.4), (5.3.5) and*

$$\begin{aligned} &\alpha C_j + 1 > c_j > a_j,\ j = 1, \dots, n; \quad \alpha B_{m+k} + 1 > b_{m+k} > d_{m+k},\ k = 1, \dots, q; \\ &d_k > b_k > \alpha^* B_k,\ k = 1, \dots, m; \quad a_{n+j} > c_{n+j} > \alpha^* C_{n+j},\ j = 1, \dots, p; \end{aligned} \tag{5.3.7}$$

the explicit solution of the dual integral equations (5.3.3) has the form:

$$\begin{aligned} f(x) = &\int_0^1 H^{q,p}_{p+n,q+m}\left[xu \left| \begin{matrix} (1-a_{n+k}-A_{n+k}, A_{n+k})_1^p, (1-c_k-C_k, C_k)_1^n \\ (1-d_{m+k}-D_{m+k}, D_{m+k})_1^q, (1-b_k-B_k, B_k)_1^m \end{matrix} \right.\right] \\ &\times \left\{ \int_0^1 H^{n+q,0}_{n+q,n+q}\left[\sigma \left| \begin{matrix} (1-a_k-A_k, A_k)_1^n, (1-d_{m+k}-D_{m+k}, D_{m+k})_1^q \\ (1-c_k-C_k, C_k)_1^n, (1-b_{m+k}-B_{m+k}, B_{m+k})_1^q \end{matrix} \right.\right] \varphi(u\sigma)d\sigma \right\} du \\ &+ \int_1^\infty H^{q,p}_{p+n,q+m}\left[xu \left| \begin{matrix} (1-a_{n+k}-A_{n+k}, A_{n+k})_1^p, (1-c_k-C_k, C_k)_1^n \\ (1-d_{m+k}-D_{m+k}, D_{m+k})_1^q, (1-b_k-B_k, B_k)_1^m \end{matrix} \right.\right] \\ &\times \left\{ \int_1^\infty H^{m+p,0}_{m+p,m+p}\left[\frac{1}{\sigma} \left| \begin{matrix} (d_k+D_k, D_k)_1^m, (a_{n+k}+A_{n+k}, A_{n+k})_1^p \\ (b_k+B_k, B_k)_1^m, (c_{n+k}+C_{n+k}, C_{n+k})_1^p \end{matrix} \right.\right] \psi(u\sigma)d\sigma \right\} du. \end{aligned} \tag{5.3.8}$$

To avoid the complicated notation in the most general case (5.3.3), we illustarte below the idea and its realization in the case with G-functions as kernels, following Kiryakova [205]. The same techniques then work quite well in the case of H-functions too and the particular details can be seen in [119].

Thus, *we consider now the dual integral equations*

$$\begin{aligned} &\int_0^\infty G^{m,n}_{p+n,q+m}\left[xu \,\middle|\, \begin{matrix} a_1,\dots,a_{n+p} \\ b_1,\dots,b_{m+q} \end{matrix}\right] f(u)du = \varphi(x), \quad o < x < 1, \\ &\int_0^\infty G^{m,n}_{p+n,q+m}\left[xu \,\middle|\, \begin{matrix} c_1,\dots,c_{n+p} \\ d_1,\dots,d_{m+q} \end{matrix}\right] f(u)du = \psi(x), \quad x > 1, \end{aligned} \tag{5.3.9}$$

with condition

$$m + n = p + q, \quad \text{corresponding to (5.3.4).} \tag{5.3.10}$$

Denote the kernel-functions of (5.3.9) by

$$G_1(x) = G^{m,n}_{p+n,q+m}\left[x \,\middle|\, \begin{matrix} (a_k) \\ (b_l) \end{matrix}\right] = \frac{1}{2\pi i}\int_{\mathfrak{L}} \mathcal{X}_1(s)x^{-s}ds,$$

$$G_2(x) = G^{m,n}_{p+n,q+m}\left[x \,\middle|\, \begin{matrix} (c_k) \\ (d_l) \end{matrix}\right] = \frac{1}{2\pi i}\int_{\mathfrak{L}} \mathcal{X}_2(s)x^{-s}ds,$$

where

$$\begin{aligned} \mathcal{X}_1(s) &= \frac{\prod_{j=1}^{m}\Gamma(b_j+s)\prod_{j=1}^{n}\Gamma(1-a_j-s)}{\prod_{j=1}^{q}\Gamma(1-b_{m+j}-s)\prod_{j=1}^{p}\Gamma(a_{n+j}+s)} = \frac{\Gamma_{1,1}.\Gamma_{2,1}}{\Gamma_{3,1}.\Gamma_{4,1}}, \\ \mathcal{X}_2(s) &= \frac{\prod_{j=1}^{m}\Gamma(d_j+s)\prod_{j=1}^{n}\Gamma(1-c_j-s)}{\prod_{j=1}^{q}\Gamma(1-d_{m+j}-s)\prod_{j=1}^{p}\Gamma(c_{n+j}+s)} = \frac{\Gamma_{1,2}.\Gamma_{2,2}}{\Gamma_{3,2}.\Gamma_{4,2}}. \end{aligned} \tag{5.3.11}$$

The Mellin transform

$$\mathfrak{M}\{f(u)\} = F(s) = \int_0^\infty u^{s-1}f(u)du$$

with its inversion formula

$$\mathfrak{M}^{-1}\{F(s)\} = f(u) = \frac{1}{2\pi i}\int_{\mathfrak{L}} F(s)u^{-s}ds$$

and the known Parseval theorem, is a useful tool in solving dual integral equations. Applying this transform to equations (5.3.8), we can reduce them to a pair of equations whose kernels are the Mellin transforms of the G-functions, namely:

$$\mathfrak{M}\left\{G^{m,n}_{p+n,q+m}\left[x\left|\begin{matrix}a_1,\dots,a_{n+p}\\ b_1,\dots,b_{m+q}\end{matrix}\right.\right]\right\}=\mathcal{X}_1(s)$$
$$\mathfrak{M}\left\{G^{m,n}_{p+n,q+m}\left[x\left|\begin{matrix}c_1,\dots,c_{n+p}\\ d_1,\dots,d_{m+q}\end{matrix}\right.\right]\right\}=\mathcal{X}_2(s). \tag{5.3.12}$$

As shown in [286, §7.2], by using the Parseval theorem for the Mellin transform, the new pair of equations is

$$\frac{1}{2\pi i}\int\limits_{\mathfrak{L}}\mathcal{X}_1(s)x^{-s}F(1-s)ds=\varphi(x),\ 0<x<1,$$
$$\frac{1}{2\pi i}\int\limits_{\mathfrak{L}}\mathcal{X}_2(s)x^{-s}F(1-s)ds=\psi(x),\ x>1. \tag{5.3.13}$$

Here the new unknown function is the Mellin transform $F(s)$ of $f(u)$. For the details of the reduction of (5.3.9) to (5.3.13) we refer to [286, p. 242-243].

Next, the main idea is to transform each of the kernels $\mathcal{X}_i(s)$, $i=1,2$, of equations (5.3.13) to a common kernel $\mathcal{X}(s)$ of the same form using fractional integration operators. By an application of the generalized Riemann-Liouville type fractional integral to the first equation of (5.3.13) we try to transform the quotient $\frac{\Gamma_{2,1}}{\Gamma_{3,1}}$ of Γ-functions with $(-s)$ from the expression for $\mathcal{X}_1(s)$ in (5.3.11), into the corresponding quotient $\frac{\Gamma_2}{\Gamma_3}=\frac{\Gamma_{2,2}}{\Gamma_{3,2}}$ from the expression for $\mathcal{X}_2(s)$. Similarly, by a fractional integral of Weyl type applied to the second equation of (5.3.11) we transform the quotient $\frac{\Gamma_{1,2}}{\Gamma_{4,2}}$ of Γ-functions with $(+s)$ in $\mathcal{X}_2(s)$, (5.3.11), into the coresponding quotient $\frac{\Gamma_1}{\Gamma_4}=\frac{\Gamma_{1,1}}{\Gamma_{4,1}}$. In this way, equations (5.3.13) can be reduced to a pair of equations with a common kernel

$$\mathcal{X}(s)=\frac{\Gamma_1.\Gamma_2}{\Gamma_3.\Gamma_4}=\frac{\prod\limits_{j=1}^{m}\Gamma\left(b_j+s\right)\prod\limits_{j=1}^{n}\left(1-c_j-s\right)}{\prod\limits_{j=1}^{q}\Gamma\left(1-d_{m+j}-s\right)\prod\limits_{j=1}^{p}\Gamma\left(a_{n+j}+s\right)} \tag{5.3.14}$$

and their solution can be written by a single formula.

To this end, we consider the following Erdélyi-Kober fractional integrals (5.1.1):

$$r'_j=I_1^{-c_j,c_j-a_j},\ j=1,\dots,n;\quad r^*_k=I_1^{-b_{m+k},d_{m+k}-b_{m+k}},\ k=1,\dots,q \tag{5.3.15}$$

and form their compositions as follows:

$$\mathfrak{R}' = \prod_{j=1}^{n} r'_j, \quad \mathfrak{R}^* = \prod_{k=1}^{q} r^*_k, \quad \mathfrak{R} = \mathfrak{R}^*.\mathfrak{R}'.$$

Supposing $c_j > a_j$, $j = 1, \dots, n$, and $b_{m+k} > d_{m+k}$, $k = 1, \dots, q$. According to Theorem 1.2.10, we obtain the *multiple Erdélyi-Kober integral*

$$\mathfrak{R} = I_{1,n+q}^{(\gamma_k),(\delta_k)}; \quad \begin{array}{l} \gamma_k = -c_k, \ \delta_k = c_k - a_k, \ k = 1, \dots n, \\ \gamma_{n+k} = -b_{m+k}, \ \delta_{n+k} = b_{m+k} - d_{m+k}, \ k = 1, \dots, q, \end{array} \tag{5.3.16}$$

written by means of a $G_{n+q,n+q}^{n+q,0}$-function as a kernel.

Lemma 5.3.2. *For*

$$\varphi(x) \in C_\alpha, \ \alpha + 1 > c_j > a_j, \ j = 1, \dots, n; \ \alpha + 1 > b_{m+k} > d_{m+k}, \ k = 1, \dots, q, \tag{5.3.17}$$

the Riemann-Liouville type multiple fractional integral (5.3.16) transforms the first equation of (5.3.13) into an integral equation of the same form but with a kernel-function $\mathcal{X}(s)$, defined by (5.3.14), viz.

$$\frac{1}{2\pi i} \int\limits_{\mathfrak{L}} \mathcal{X}(s) x^{-s} F(1-s) ds = \mathfrak{R}\varphi(x), \ 0 < x < 1, \tag{5.3.18}$$

with $\mathfrak{R}\varphi(x) \in C_\alpha$.

Proof. We apply first $\mathfrak{R}' = I_{1,n}^{(-c_j),(c_j - a_j)}$ to the first equation of (5.3.13) and using (1.2.1):

$$\mathfrak{R}' \left\{ x^{-s} \right\} = x^{-s} \prod_{j=1}^{n} \frac{\Gamma(1 - c_j - s)}{\Gamma(1 - a_j - s)},$$

we obtain

$$\mathfrak{R}'\varphi(x) = \mathfrak{R}' \left\{ \frac{1}{2\pi i} \int\limits_{\mathfrak{L}} \mathcal{X}_1(s) x^{-s} F(1-s) ds \right\} = \frac{1}{2\pi i} \int\limits_{\mathfrak{L}} \mathcal{X}_1(s) \mathfrak{R}' \left\{ x^{-s} \right\} F(1-s) ds$$

$$= \frac{1}{2\pi i} \int\limits_{\mathfrak{L}} \frac{\Gamma_{1,1}.\Gamma_{2,1}}{\Gamma_{3,1}.\Gamma_{4,1}} . \frac{\Gamma_{2,2}}{\Gamma_{2,1}} x^{-s} F(1-s) ds.$$

Similarly, since

$$\mathfrak{R}^* \left\{ x^{-s} \right\} = x^{-s} \prod_{k=1}^{q} \frac{\Gamma(1 - b_{m+k} - s)}{\Gamma(1 - d_{m+k} - s)},$$

we have

$$\mathfrak{R}\varphi(x) = \mathfrak{R}^* \left\{\mathfrak{R}'\varphi(x)\right\} = \frac{1}{2\pi i}\int\limits_{\mathfrak{L}} \frac{\Gamma_{1,1}.\Gamma_{2,2}}{\Gamma_{3,1}.\Gamma_{4,1}}.\frac{\Gamma_{3,1}}{\Gamma_{3,2}} x^{-s}F(1-s)ds$$
$$= \frac{1}{2\pi i}\int\limits_{\mathfrak{L}} \mathcal{X}(s)x^{-s}F(1-s)ds,$$

as required. According to Theorem 1.2.15, for $\varphi(x) \in C_\alpha$ and the other conditions (5.3.17), it follows that $\mathfrak{R}\varphi(x) \in C_\alpha$ too.

Analogously, we consider the Weyl type multiple integral

$$\begin{aligned}\mathfrak{W} = \mathfrak{W}^*\mathfrak{W}' &= \left(\prod_{j=1}^{p} K_1^{c_{n+j},a_{n+j}-c_{n+j}}\right)\left(\prod_{k=1}^{m} K_1^{b_k,d_k-b_k}\right)\\ &= W_{1,p}^{(c_{n+j}),(a_{n+j}-c_{n+j})} W_{1,m}^{(b_k),(d_k-b_k)}\\ &= W_{1,m+p}^{(\tau_k),(\alpha_k)} \quad \text{with} \quad \begin{array}{l}\tau_k = b_k, \qquad \alpha_k = d_k - b_k,\ k=1,\dots,m\\ \tau_{m+j} = c_{n+k},\ \alpha_{m+j} = a_{n+j} - c_{n+j},\ j=1,\dots,p,\end{array}\end{aligned} \tag{5.3.19}$$

see (1.1.17*), (1.4.2), (1.4.8). Now by using formula (1.4.4), namely:

$$\mathfrak{W}'\left\{x^{-s}\right\} = x^{-s}\prod_{k=1}^{m}\frac{\Gamma(b_k+s)}{\Gamma(d_k+s)}; \quad \mathfrak{W}^*\left\{x^{-s}\right\} = x^{-s}\prod_{j=1}^{p}\frac{\Gamma(c_{n+j}+s)}{\Gamma(a_{m+j}+s)},$$

from the second equation of (5.3.13), we obtain subsequently:

$$\mathfrak{W}\psi(x) = \mathfrak{W}^*\left\{\mathfrak{W}'\psi(x)\right\} = \frac{1}{2\pi i}\int\limits_{\mathfrak{L}} \frac{\Gamma_{1,2}.\Gamma_{2,2}}{\Gamma_{3,2}.\Gamma_{4,2}}.\frac{\Gamma_{1,1}}{\Gamma_{1,2}}.\frac{\Gamma_{4,2}}{\Gamma_{4,1}} x^{-s}F(1-s)ds$$
$$= \frac{1}{2\pi i}\int\limits_{\mathfrak{L}} \frac{\Gamma_{1,1}\Gamma_{2,2}}{\Gamma_{3,2}\Gamma_{4,1}} x^{-s}F(1-s)ds = \frac{1}{2\pi i}\int\limits_{\mathfrak{L}} \mathcal{X}(s)x^{-s}F(1-s)ds.$$

Then, by Theorem 1.4.8 the next lemma follows.

Lemma 5.3.3. *For*

$$\psi(x) \in C^*_{\alpha^*},\ a_{n+j} > c_{n+j} > \alpha^*,\ j=1,\dots,p;\ d_k > b_k > \alpha^*,\ k=1,\dots,m, \tag{5.3.20}$$

the Weyl type multiple fractional integral (5.3.19) transforms the second equation of (5.3.13) into an equation with kernel $\mathcal{X}$, (5.3.14), namely:

$$\frac{1}{2\pi i}\int\limits_{\mathfrak{L}} \mathcal{X}(s)x^{-s}F(1-s)ds = \mathfrak{W}\psi(x),\ x>1, \tag{5.3.21}$$

and with $\mathfrak{W}\psi(x) \in C^*_{\alpha^*}$.

It is seen now that if we apply generalized multiple integrals (5.3.16), (5.3.19) respectively to the first and second equations (5.3.13), *we transform them into a single* integral equation with $\mathcal{X}(s)$, defined by (5.3.14) as a kernel:

$$\frac{1}{2\pi i}\int\limits_{\mathfrak{L}} \mathcal{X} x^{-s} F(1-s)ds = g(x), \quad 0 < x < \infty, \tag{5.3.22}$$

where we denote by

$$g(x) = \begin{cases} \mathfrak{R}\varphi(x), & 0 < x < 1, \\ \mathfrak{W}\psi(x), & x > 1, \end{cases} \tag{5.3.23}$$

the new right-hand side; $\mathfrak{R}\varphi \in C_\alpha$, $\mathfrak{W}\psi \in C^*_{\alpha^*}$.

According to Marichev [276, Theorem 11], an appropriate space for considering functions like $g(x)$ with Mellin transforms in the strip $\Re s = \gamma$, $-\alpha \leq \gamma \leq -\alpha^*$ is *the space*

$$C_{\alpha,\alpha^*} := \left\{ g(x) \in C[0,1] \cap C[1,\infty); \text{ and } \begin{array}{l} |g(x)| \leq Ax^{\alpha}, \ 0 < x < 1; \\ |g(x)| \leq Ax^{\alpha^*}, \ x > 1; \end{array} \right\} \text{ with } \alpha^* \leq \alpha. \tag{5.3.24}$$

Then, by the Mellin transform techniques ([286, p. 242, Lemma 7.2.2]), equation (5.3.22) can be rewritten in the form

$$f(x) = \frac{1}{2\pi i}\int\limits_{\mathfrak{L}} \frac{x^{-s}G(1-s)}{\mathcal{X}(1-s)} ds, \tag{5.3.25}$$

where $f(x)$ is the original unknown function and $G(s) = \mathfrak{M}\{g(x); s\}$. Let us denote

$$\mathcal{H}(s) = \frac{1}{\mathcal{X}(1-s)} = \frac{\prod\limits_{k=1}^{q} \Gamma(-d_{m+k}+s) \prod\limits_{j=1}^{p} \Gamma(1-(-a_{n+j})-s)}{\prod\limits_{k=1}^{m} \Gamma(1-(-b_k)-s) \prod\limits_{j=1}^{n} \Gamma(-c_j+s)} \tag{5.3.26}$$

and let $h(x) = \mathfrak{M}^{-1}\{\mathcal{H}(s)\}$ be the original of (5.3.26). This does exist for $m+n = p+q$, under the conditions on the parameters accepted here and

$$h(x) = G^{q,p}_{n+p,m+q}\left[x \left| \begin{array}{l} (-a_{n+j})_{j=1}^{p}; \ (-c_j)_{j=1}^{n} \\ (-d_{m+k})_{k=1}^{q}; \ (-b_k)_{k=1}^{m} \end{array} \right.\right]. \tag{5.3.27}$$

According to *Parseval theorem* ([286, p.242, Lemma 7.2.1]) it follows that the solution $f(x)$ of the original pair of integral equations (5.3.9) has the form

$$f(x) = \int\limits_0^\infty h(xu)g(u)du = \int\limits_0^\infty G^{q,p}_{n+p,m+q}\left[x \left| \begin{array}{l} (-a_{n+j})_{j=1}^{p}; \ (-c_j)_{j=1}^{n} \\ (-d_{m+k})_{k=1}^{q}; \ (-b_j)_{j=1}^{n} \end{array} \right.\right] g(u)du \tag{5.3.28}$$

with $g(x)$ defined by (5.3.23). Thus, we obtain that

$$f(x) = \int_0^1 h(xu)\mathfrak{R}\varphi(u)du + \int_1^\infty h(xu)\mathfrak{W}\psi(u)du. \tag{5.3.29}$$

By replacing $h(x)$ by the G-function (5.3.27) and the multiple fractional integrals $\mathfrak{R}\varphi(u)$, $\mathfrak{W}\psi(u)$ by their single integral representations via G-functions (see (1.1.6), (1.1.9)), formula (5.3.29) gives the explicit solution of the dual integral equations (5.3.9). Let us bear in mind that this was obtained by the *above formal technique*. Therefore, strictly, we have to verify it by substituting the final expression for $f(x)$ into (5.3.9) and then, working out the trivial but boring calculation of integrals involving products of G-functions (by formula (A.28)). In the most general case (5.3.3) with H-functions, this is done in Galue, Kiryakova and Kalla [119].

In this way, we finally obtain the following theorem.

Theorem 5.3.4. (Kiryakova [205]) *The function*

$$\begin{aligned} f(x) = &\int_0^1 G^{q,p}_{n+p,m+q}\left[xu \left| \begin{matrix} (-a_{n+j})_1^p;\ (-c_j)_1^n \\ (-d_{m+k})_1^q;\ (-b_k)_1^m \end{matrix} \right.\right] du \\ &\times \int_0^1 G^{n+q,0}_{n+q,n+q}\left[\nu \left| \begin{matrix} (-a_j)_1^n, (-d_{m+k})_1^q \\ (-c_j)_1^n, (-b_{m+k})_1^q \end{matrix} \right.\right] \varphi(uv)dv \\ &+ \int_1^\infty G^{q,p}_{n+p,m+q}\left[xu \left| \begin{matrix} (-a_{n+j})_1^p;\ (-c_j)_1^n \\ (-d_{m+k})_1^q;\ (-b_k)_1^m \end{matrix} \right.\right] du \\ &\times \int_1^\infty G^{m+p,0}_{m+p,m+p}\left[\frac{1}{\nu} \left| \begin{matrix} (d_k+1)_1^m, (a_{n+j}+1)_1^p \\ (b_k+1)_1^m, (c_{n+j}+1)_1^p \end{matrix} \right.\right] \psi(uv)dv \end{aligned} \tag{5.3.30}$$

satisfies the dual integral equations (5.3.9):

$$\int_0^\infty G^{m,n}_{p+n,q+m}\left[xu \left| \begin{matrix} (a_j)_1^{n+p} \\ (b_k)_1^{m+q} \end{matrix} \right.\right] f(u)du = \varphi(x), \quad 0 < x < 1$$

$$\int_0^\infty G^{m,n}_{p+n,q+m}\left[xu \left| \begin{matrix} (c_j)_1^{n+p} \\ (d_k)_1^{m+q} \end{matrix} \right.\right] f(u)du = \psi(x), \quad x > 1.$$

Note. The result obtained in [286, p. 247, (7.2.35)] is the same, but it contains compositions of fractional integrals

$$\mathfrak{R}\varphi(u) = r_1^* r_2^* \dots r_q^* \left(r_1' \dots r_n' \varphi(u)\right); \mathfrak{W}\varphi(u) = w_1^* w_2^* \dots w_p^* \left(w_1' \dots w_m' \psi(u)\right),$$

written only symbolically. Their explicit representations

$$\mathfrak{R}\varphi(u) = \int_0^1 G_{n+q,n+q}^{n+q,0}\left[\nu \left| \begin{matrix} (-a_j)_1^n, (-d_{m+k})_1^q \\ (-c_j)_1^n, (-b_{m+k})_1^q \end{matrix} \right.\right] \varphi(uv)dv$$

$$= \underbrace{\int_0^1 \cdots \int_0^1}_{(n+q)} \prod_{k=1}^{n+q} \left[\frac{(1-t_k)^{\delta_k - 1}}{\Gamma(\delta_k)} t_k^{\gamma_k}\right] \varphi\left[u\left(t_1 \ldots t_{n+q}\right)\right] dt_1 \ldots dt_{n+q}$$

and the similar ones for $\mathfrak{W}\psi(u)$ found in the above theorem allow us to write down, for the first time, the solution (5.3.30) explicitly. Moreover our conditions (combination of (5.3.17) and (5.3.20)):

$$\varphi \in C_\alpha,\ \psi \in C_{\alpha^*}^*,\ \alpha^* \leq \alpha;\ \alpha + 1 > c_j > a_j,\ j = 1, \ldots, n;\ \alpha + 1 > b_{m+k} > d_{m+k}, \\ k = 1, \ldots, q; a_{n+j} > c_{n+j} > \alpha^*,\ j = 1, \ldots, p;\ d_k > b_k > \alpha^*,\ k = 1, \ldots, m; \tag{5.3.31}$$

ensure the correct application of the fractional integration operators $\mathfrak{R}$, $\mathfrak{W}$ to the boundary conditions $\varphi(x)$ and $\psi(x)$. These conditions are missing in [286].

Note. The solutions of the dual integral equations involving various special functions, used in applied mathematics, can be derived as particular cases of (5.3.8), in particular of (5.3.30).

First of all, let us consider *the dual integral equations with H-functions, considered in Mathai and Saxena [286, §7.2].*

EXAMPLE 5.3.5.

$$\begin{aligned} &\int_0^\infty H_{p+n,q+m}^{m,n}\left[xu \left| \begin{matrix} (a_j, A_j)_1^{n+p} \\ (b_k, B_k)_1^{m+q} \end{matrix} \right.\right] f(u)du = \varphi(x), \quad 0 < x < 1, \\ &\int_0^\infty H_{p+n,q+m}^{m,n}\left[xu \left| \begin{matrix} (c_j, A_j)_1^{n+p} \\ (d_k, B_k)_1^{m+q} \end{matrix} \right.\right] f(u)du = \psi(x), \quad x > 1. \end{aligned} \tag{5.3.32}$$

This case follows from (5.3.3) if $C_j = A_j$, $j = 1, \ldots, n+p$, $D_k = B_k$, $k = 1, \ldots, m+q$, respectively. Conditions (5.3.4) reduce to condition (v), [286, p. 240]. Conditions (5.3.5) are also fulfulled with "=" signs in all of them. The additional conditions (5.3.31) have no changes. Then, Theorem 5.3.5 gives the explicit solution (5.3.8), simplified by the above substitutions.

Next, we illustrate the general results on the example of *dual integral equations involving Bessel functions of arbitrary order.*

EXAMPLE 5.3.6. (TITCHMARSH DUAL INTEGRAL EQUATIONS)

$$\begin{aligned}
&\int_0^\infty u^{-\gamma} J_\mu\left(2\sqrt{xu}\right) f(u)du = g_1(x), \quad 0 < x < 1, \\
&\int_0^\infty u^{-\delta} J_\nu\left(2\sqrt{xu}\right) f(u)du = g_2(x), \quad x > 1.
\end{aligned} \tag{5.3.33}$$

Since (see (C.6), Appendix):

$$J_\nu\left(2\sqrt{x}\right) = x^{\frac{\nu}{2}} G_{0,2}^{1,0}\left[x|0, -\nu\right],$$

the dual integral equations (5.3.33) turn into a special case of (5.3.9) with condition (5.3.10) satisfied, viz. $(m+n)-(p+q) = (1+0)-(0+1) = 1$. If we put additionally

$$\begin{aligned}
&b_1 = \tfrac{\mu}{2} - \gamma,\ b_2 = -\tfrac{\mu}{2} - \gamma,\ d_1 = \tfrac{\nu}{2} - \delta,\ d_2 = -\tfrac{\nu}{2} - \delta, \\
&\varphi(x) = x^{-\gamma} g_1(x), \qquad \psi(x) = x^{-\delta} g_2(x),
\end{aligned}$$

then equations (5.3.33) have the form:

$$\begin{aligned}
&\int_0^\infty G_{0,2}^{1,0}\left[xu|b_1, b_2\right] f(u)du = \varphi(x), \quad 0 < x < 1, \\
&\int_0^\infty G_{0,2}^{1,0}\left[xu|d_1, d_2\right] f(u)du = \psi(x), \quad x > 1.
\end{aligned} \tag{5.3.34}$$

From Theorem 5.3.5 we find the solution

$$\begin{aligned}
f(x) = &\int_0^1 G_{0,2}^{1,0}\left[xu| - d_2, -b_1\right] du \int_0^1 G_{1,1}^{1,0}\left[v \,\middle|\, \begin{matrix} -d_2 \\ -b_2 \end{matrix}\right] \varphi(uv)dv \\
&+ \int_1^\infty G_{0,2}^{1,0}\left[xu| - d_2, -b_1\right] du \int_1^\infty G_{1,1}^{1,0}\left[\frac{1}{v} \,\middle|\, \begin{matrix} d_1+1 \\ b_1+1 \end{matrix}\right] \psi(uv)dv.
\end{aligned}$$

By taking into account (C.10):

$$G_{1,1}^{1,0}\left[x \,\middle|\, \begin{matrix} \alpha+\beta \\ \alpha \end{matrix}\right] = \frac{x^\alpha(1-x)^{\beta-1}}{\Gamma(b)}, \ 0 < x < 1,$$

in the previous notation we obtain:

$$f(x) = x^{\frac{\mu+3\nu}{4}+\frac{3\delta-\gamma}{2}} \left\{ \int_0^1 u^{\frac{\mu+3\nu}{2}+3\frac{\delta-\gamma}{2}} J_\lambda\left(2\sqrt{xu}\right) du \int_0^1 \frac{v^{\frac{\mu}{2}}(1-v)^{\lambda-\mu-1}}{\Gamma(\lambda-\mu)} g_1(uv)dv \right.$$
$$\left. + \int_1^\infty u^{\frac{\mu+3\nu}{2}+\frac{\delta-\gamma}{2}} J_\lambda\left(2\sqrt{xu}\right) du \int_1^\infty \frac{v^{-\frac{\nu}{2}}(v-1)^{\nu-\lambda-1}}{\Gamma(\nu-\lambda)} g_2(uv)dv \right\} \tag{5.3.35}$$

where $\lambda = \frac{\mu+\nu}{2} + (\delta - \gamma)$.

For the right-hand sides $g_1(x) \in C_\alpha$, $g_2 \in C^*_{\alpha^*}$, the corresponding conditions (5.3.31) are:

$$\nu > \lambda > \mu, \quad \frac{\mu}{2} + \alpha + 1 > 0, \quad \frac{\mu}{2} + \delta - \gamma - \alpha^* > 0. \tag{5.3.36}$$

After a change of variables, we can obtain *the solution of the classical Titchmarsh dual integral equations* (see Sneddon [452]):

$$\begin{aligned} \int_0^\infty u^{-2\gamma} J_\mu(xu) f(u) du &= F_1(x), \quad 0 < x < 1, \\ \int_0^\infty u^{-2\delta} J_\nu(xu) f(u) du &= F_2(x), \quad x > 1, \end{aligned} \tag{5.3.37}$$

which for $\gamma = -\frac{\omega}{2}$, $\delta = 0$, $\lambda = \frac{\mu+\nu+\omega}{2}$ has the form

$$f(x) = 2^{\lambda-\nu} x^{1-\lambda+\nu} \left\{ \int_0^1 t^{1-\lambda-\nu} J_\lambda(xt) h_1(t) dt + \int_1^\infty t^{1-\lambda+\nu} J_\lambda(xt) h_2(t) dt \right\}, \tag{5.3.38}$$

found earlier by Peters ([452, p. 85-86]), where

$$h_1(t) = \frac{2^{1-\omega} t^{\nu+2\omega}}{\Gamma(\lambda-\mu)} \int_0^t \left(t^2 - \tau^2\right)^{\lambda-\mu-1} \tau^{\mu+1} F_1(\tau) d\tau$$

and

$$h_2(t) = \frac{2t^{\mu+\omega}}{\Gamma(\nu-\lambda)} \int_t^\infty \left(\tau^2 - t^2\right)^{\nu-\lambda-1} \tau^{1-\nu} F_2(\tau) d\tau.$$

Let us assume $F_1(x) \in C_{-1}$, $F_2(x) \in C^*_{-1}$, i.e. $\alpha = \alpha^* = -1$ and therefore,

$$F(x) = \left\{ \begin{array}{l} F_1(x),\ 0 < x < 1 \\ F_2(x),\ x > 1 \end{array} \right\} \in C_{-1,1} \subset L^1(0,\infty).$$

This is the most natural case usually considered, i.e. with Lebesgue integrable boundary conditions. Then, conditions (5.3.31) take the known form (see [452, p. 86] again):

$$\lambda > \mu > -1, \quad \mu + \omega > -1, \quad \nu > \lambda.$$

For other results on dual integral equations obtained by means of the fractional calculus and some alternative approaches, one can see the articles: Erdélyi and Sneddon [110], Fox [115]-[116], Love and Clements [267], Saxena [436]-[437], Virchenko [502]-[503], Virchenko and Makarenko [505] as well as the books: Sneddon [452], [455], Samko, Kilbas and Marichev [434], Ufljand [498], Virchenko [504].

5.4. Convolutions of generalized fractional integrals. Multiple Džrbashjan-Gelfond-Leontiev differintegration operators

Following Definition 2.1.1 for a *convolution of a linear operator L, mapping a linear space into itself*, we can state now the problem for finding convolutions of the multiple Erdélyi-Kober fractional integrals $L = x^{\beta_0} I^{(\gamma_k),(\delta_k)}_{(\beta_k),m}$ in spaces C_α. *Till recently, this problem has been solved in some special but important cases*, namely:

a) $m = 1$; arbitrary $\beta > 0$, γ, $\delta > 0$; $\beta_0 = \beta\delta > 0$:

$$L = x^{\beta\delta} I^{\gamma,\delta}_{\beta}, \tag{5.4.1}$$

the Erdélyi-Kober fractional integrals (2.1.4). The corresponding family of convolutions of L is found in Kiryakova [206] and Section 2.1, Theorem 2.1.2.

b) arbitrary $m \geq 2$; arbitrary γ_k, all $\delta_k = 1$, all $\beta_k = \beta > 1$, $k = 1, \ldots, m$; $\beta_0 = \beta > 0$:

$$L = x^{\beta} I^{(\gamma_k),(1,\ldots,1)}_{\beta,m} = x^{\beta} \left[\prod_{k=1}^{m} I^{\gamma_k,1}_{\beta} \right], \tag{5.4.2}$$

the hyper-Bessel integral operators (3.1.21)-(3.1.23), corresponding to the Bessel type differential operators of order $m > 1$, (3.1.2)-(3.1.3):

$$B = x^{\alpha_0} \frac{d}{dx} x^{\alpha_1} \ldots \frac{d}{dx} x^{\alpha_m} = x^{-\beta} \left[\prod_{k=1}^{m} \left(x \frac{d}{dx} + \beta\gamma_k \right) \right].$$

A one-parameter *family of convolutions of operator* (5.4.2) *was found by Dimovski* [64]-[71] *and gave rise to the development of operational calculi, integral tramsforms and the theory of hyper-Bessel operators of arbitrary order.* Later on, the same convolutions were represented briefly by means of generalized fractional integrals, see Section 3.6, Theorem 3.6.1 and related results.

c) $m = 2$; arbitrary γ_1, γ_2; integer $\delta_1 > 0$, $\delta_2 > 0$; the problem was solved for some special cases, for instance, for the operator

$$L = L_{\nu,\mu} = x^{\mu} I^{(\nu,0),(1,\mu)}_{1,2}, \quad \mu > 0 \tag{5.4.3}$$

related to the Bessel-Maitland function $J_\nu^{(\mu)}(x)$ *(E.36), see Section 1.6, Example h).* A convolution *for integer* $\mu > 0$ was found by Krätzel [239]. The problem of finding a convolution in the case of arbitrary rational $\mu > 0$ still seems to be open (*Open problem No 1*, Kiryakova [196, p. 57]).

Recently, following the general convolutional scheme of Dimovski [73], Nguen and Yakubovich [317], Yakubovich [513]-[514], Yakubovich and Luchko [515]-[516], Yakubovich and Nguen [517], Luchko and Yakubovich [270]-[271] have found some very general convolutions, related to the G- and H-functions. *Their results allow solving the problem for convolutions of almost arbitrary multiple fractional integrals* $L = x^{\beta_0} I^{(\gamma_k),(\delta_k)}_{(\beta_k),m}$. In essence, the convolutions found by Yakubovich and Luchko are related to above the operators with arbitrary γ_k, $\delta_k > 0$, $k = 1, \ldots, m$, $\beta_0 = \mu > 0$ but $\beta_k = \frac{\mu}{\delta_k}$, $k = 1, \ldots, m$, i.e. of the form $L = x^\mu I^{(\gamma_k),(\delta_k)}_{\frac{\mu}{\delta_k},m}$.

Theorem (Luchko and Yakubovich [270]-[271]). *Let* $\mu > 0$, $a_i > 0$, $\alpha_i \in \mathbb{R}$, $1 \leq i \leq n$;

$$\alpha = \max_{1 \leq i \leq n} \left(\frac{\alpha_i - 1}{a_i} \right), \quad \lambda \geq \max_{1 \leq i \leq n} \frac{1 - \alpha_i}{a_i}. \tag{5.4.4}$$

Then, the operation

$$\left(f \overset{\lambda}{*} g \right)(x) = x^\lambda \left\{ \prod_{i=1}^n I_{a_i}^{1-2\alpha_i, \alpha_i + \lambda a_i - 1} (f \circ g)(x) \right\}, \tag{5.4.5}$$

where

$$\begin{aligned} (f \circ g)(x) = \int_0^1 \ldots \int_0^1 \prod_{j=1}^n [u_i(1 - u_i)]^{-\alpha_i} f \left[x \prod_{i=1}^n u_i^{a_i} \right] \\ \times g \left[x \prod_{i=1}^n (1 - u_i)^{a_i} \right] du_1 \ldots du_n, \end{aligned} \tag{5.4.6}$$

is a convolution without divisors of zero of the linear operator L_μ *in* C_α, *where:*

$$\begin{aligned} L_\mu f(x) &= x^\mu \left[\prod_{i=1}^n I_{a_i}^{-\alpha_i, \mu a_i} \right] \\ &= x^\mu \int_0^1 \ldots \int_0^1 \prod_{i=1}^n \frac{(1 - u_i)^{\mu a_i - 1} u_i^{-\alpha_i}}{\Gamma(\mu a_i)} f \left[x \prod_{i=1}^n u_i^{a_i} \right] du_1 \ldots du_n. \end{aligned} \tag{5.4.7}$$

5.4.i. Fractional hyper-Bessel differential operators and convolutions related to them

Definition 5.4.1. By a *fractional hyper-Bessel differential operator* we mean a differintegral operator of the symbolic form

$$\tilde{B} = x^{\alpha_0}\left(\frac{d}{dx}\right)^{\delta_1} x^{\alpha_1}\left(\frac{d}{dx}\right)^{\delta_2} \dots x^{\alpha_{m-1}}\left(\frac{d}{dx}\right)^{\delta_m} x^{\alpha_m} \tag{5.4.8}$$

with integer $m \geq 1$, real $\alpha_0, \alpha_1, \dots, \alpha_m$ and positive $\delta_1, \dots, \delta_m$.

As shown in Section 1.6, operators (5.4.8) can be considered as generalized (m-tuple) fractional derivatives (1.6.17):

$$\tilde{B} = D_{1,m}^{(\gamma_k),(\delta_k)} \text{ with } \begin{matrix} \alpha_0 = -\gamma_1; & \alpha_k = \gamma_k + \delta_k - \gamma_{k+1} \\ k = 1, \dots, m-1; & \alpha_m = \gamma_m + \delta_m, \end{matrix} \tag{5.4.9}$$

i.e. as a product of Erdélyi-Kober fractional derivatives $D_\beta^{\gamma,\delta}$ of the form (1.6.7):

$$D_\beta^{\gamma,\delta} f(x) = \left[x^{-\gamma}\left(\frac{d}{dx}\right)^{\delta} x^{\gamma+\delta} f\left(x^{\frac{1}{\beta}}\right)\right]_{x\to x^\beta}, \tag{5.4.10}$$

namely (Lemma 1.6.5):

$$\tilde{B} = D_\beta^{\gamma_1,\delta_1} D_\beta^{\gamma_2,\delta_2} \dots D_\beta^{\gamma_m,\delta_m}.$$

In order to use the convolution (5.4.6) for the fractional hyper-Bessel operators $\tilde{B}$ (strictly, for the corresponding integral operators $\tilde{L}$), we now represent them in an alternative way, using compositions of Erdélyi-Kober derivatives (5.4.10) with different parameters $\beta > 0$, and hence related to Fox's H-function as a kernel.

First, we should note that a composition/decomposition theorem, analogous to Lemma 1.6.5 but corresponding Theorem 5.2.1, holds: the composition of commuting Erdélyi-Kober fractional derivatives $I_{\beta_k}^{\gamma_k,\delta_k}$, $k = 1, \dots, m$, with different β_k's can be represented as a generalized (multiple) fractional derivative $D_{(\beta_k),m}^{(\gamma_k),(\delta_k)}$ of form the (5.1.38), viz.:

$$D_{\beta_1}^{\gamma_1,\delta_1} D_{\beta_2}^{\gamma_2,\delta_2} \dots D_{\beta_m}^{\gamma_m,\delta_m} = D_{(\beta_k),m}^{(\gamma_k),(\delta_k)}. \tag{5.4.11}$$

Then, for the fractional hyper-Bessel differential operator (5.4.8)

$$\tilde{B} = x^{\alpha_0}\left(\frac{d}{dx}\right)^{\delta_1} x^{\alpha_1} \dots \left(\frac{d}{dx}\right)^{\delta_m} x^{\alpha_m}$$

with

$$M = \left[\prod_{k=1}^{m}\left(\frac{\delta_k}{\mu}\right)^{\delta_k}\right], \quad \mu = \sum_{k=1}^{m}\delta_k - \sum_{i=0}^{m}\alpha_i := \Delta - A > 0, \tag{5.4.13}$$

the same δ_k's and new parameters:

$$\begin{aligned}\gamma_k &= \frac{\delta_k}{\mu}\left[\alpha_k + \cdots + \alpha_m - \delta_m - \delta_{m+1} - \cdots - \delta_{k+1}\right], \ k = 1, \ldots, m-1,\\ \gamma_m &= \frac{\delta_m}{\mu}\alpha_\mu; \quad \beta_k = \frac{\mu}{\delta_k}, \ k = 1, \ldots, m.\end{aligned} \tag{5.4.14}$$

we find the *alternative representation*:

$$\tilde{B} = \frac{1}{M} D^{(\gamma_k),(\delta_k)}_{\left(\frac{\mu}{\delta_k}\right),m} x^{-\mu} \tag{5.4.12}$$

Indeed,

$$\begin{aligned}\tilde{B} &= x^{\alpha_0}\left(\frac{d}{dx}\right)^{\delta_1} x^{\alpha_1} \ldots \left(\frac{d}{dx}\right)^{\delta_m} x^{\alpha_m}\\ &= x^{-\delta_2 - \mu\left(1 - \frac{\gamma_1}{\delta_1}\right)}\left(\frac{d}{dx}\right)^{\delta_1} x^{\delta_2 + \mu\left(\frac{\gamma_1}{\delta_1} - \frac{\gamma_2}{\delta_2}\right)}\left(\frac{d}{dx}\right)^{\delta_2} \ldots x^{\mu\frac{\gamma_m}{\delta_m}}\\ &= \frac{1}{M}\left[\left(\frac{\delta_1}{\mu}\right)^{\delta_1} x^{-\frac{\mu}{\delta_1}(\gamma_1+\delta_1)+\delta_1}\left(\frac{d}{dx}\right)^{\delta_1} x^{\frac{\mu}{\delta_1}(\gamma_1+\delta_1)}\right]\ldots\\ &\qquad\left[\left(\frac{\delta_m}{\mu}\right)^{\delta_m} x^{-\frac{\mu}{\delta_m}(\gamma_m+\delta_m)+\delta_m}\left(\frac{d}{dx}\right)^{\delta_m} x^{\frac{\mu}{\delta_m}(\gamma_m+\delta_m)}\right] x^{-\mu}\\ &= \frac{1}{M} D^{\gamma_1,\delta_1}_{\frac{\mu}{\delta_1}} D^{\gamma_1,\delta_2}_{\frac{\mu}{\delta_2}} \ldots D^{\gamma_m,\delta_m}_{\frac{\mu}{\delta_m}} x^{-\mu} = \frac{1}{M} D^{(\gamma_k),(\delta_k)}_{\left(\frac{\mu}{\delta_k}\right),m} x^{-\mu},\end{aligned}$$

in a subspace of C_α, where $D^{\gamma_k,\delta_k}_{\frac{\mu}{\delta_k}}$ commute.

By Theorem 5.1.9 (see (5.1.42′)), the linear right inverse operator of $\tilde{B}$ (5.4.14), is the generalized (m-tuple) fractional integral of the form (5.1.8):

$$\tilde{L} = M x^\mu I^{(\gamma_k),(\delta_k)}_{\left(\frac{\mu}{\delta_k}\right),m}, \tag{5.4.15}$$

considred in C_α, $\alpha = \max\limits_{1\le k\le m}\left[-\frac{\mu}{\beta_k}(\gamma_k+1)\right]$, and $\tilde{L} : C_\alpha \to C_{\alpha+\mu} \subset C_\alpha$.

This can be called a *fractional hyper-Bessel integral operator* and has the representation:

$$\tilde{L}f(x) = Mx^{\mu}\int_0^1\cdots\int_0^1\prod_{k=1}^{m}\frac{\left[(1-t_k)^{\delta_k-1}t_k^{\gamma_k}\right]}{\Gamma(\delta_k)}f\left[xt_1^{\frac{\delta_1}{\mu}}\dots t_m^{\frac{\delta_m}{\mu}}\right]dt_1\dots dt_m$$
$$= Mx^{\mu}\int_0^1 H_{m,m}^{m,0}\left[\sigma\left|\begin{matrix}\left(\gamma_k+\delta_k+1-\frac{\delta_k}{\mu},\frac{\delta_k}{\mu}\right)_1^m\\ \left(\gamma_k+1-\frac{\delta_k}{\mu},\frac{\delta_k}{\mu}\right)_1^m\end{matrix}\right.\right]f(x\sigma)d\sigma. \tag{5.4.16}$$

Then, by a suitable change of notation in (5.4.5)-(5.4.7), from the result of Luchko and Yakubovich [270] we obtain the following convolutions of $\tilde{L}$.

Theorem 5.4.2. *For* $\lambda \geq \max\limits_{1\leq k\leq m}\left[\frac{1+\gamma_k-\delta_k}{\delta_k}\right]$ *the operations*

$$\left(f\overset{\lambda}{*}g\right)(x) = x^{\mu(\lambda+1)}\left[\prod_{k=1}^{m} I_{\frac{\mu}{\delta_k}}^{2\gamma_k+1,(\lambda+1)\delta_k-\gamma_k-1}\right](f\circ g)(x)$$
$$= x^{\mu(\lambda+1)} I_{\left(\frac{\mu}{\delta_k}\right),m}^{(2\gamma_k+1),((\lambda+1)\delta_k-\gamma_k-1)} = T_\lambda(f\circ g)(x), \tag{5.4.17}$$

where $(\circ)$ *is the auxiliary operation*

$$(f\circ g)(x) = \int_0^1\cdots\int_0^1\prod_{k=1}^{m}\left[t_k(1-t_k)\right]^{\gamma_k} f\left[x\prod_{k=1}^{m}t_k^{\frac{\delta_k}{\mu}}\right] g\left[x\prod_{k=1}^{m}(1-t_k)^{\frac{\delta_k}{\mu}}\right]dt_1\dots dt_m, \tag{5.4.18}$$

and the parameters $\mu > 0$, β_k, γ_k, $\delta_k > 0$, $k = 1,\dots,m$, *are as in (5.4.12)-(5.4.13), are convolutions without divisors of zero of the fractional hyper-Bessel integral operator* $\tilde{L}$, *(5.4.15)-(5.4.16) in the space* C_α, $\alpha = \max\limits_{1\leq k\leq m}\left[\frac{-\mu(\gamma_k+1)}{\delta_k}\right]$.

Note. Along with representations by multiple integrals like (5.2.7), the generalized fractional integrals

$$T_\lambda = x^{\mu(\lambda+1)} I_{\left(\frac{\mu}{\delta_k}\right),m}^{(2\gamma_k+1),((\lambda+1)\delta_k-\gamma_k-1)} \tag{5.4.19}$$

in (5.4.17), also have single integral representations of the form (5.1.8) involving H-functions. This fact allows representations of the convolutions $\left(\overset{\lambda}{*}\right)$ by means of $(m+1)$-tuple integrals only, namely:

$$\left(f\overset{\lambda}{*}g\right)(x) = x^{\mu(\lambda+1)}\int_0^1 H_{m,m}^{m,0}\left[\sigma\left|\begin{matrix}\left(\gamma_k+1+\left(\lambda-\frac{1}{\mu}+1\right)\delta_k,\frac{\delta_k}{\mu}\right)_1^m\\ \left(2\gamma_k+2-\frac{1}{\mu}\delta_k,\frac{\delta_k}{\mu}\right)_1^m\end{matrix}\right.\right](f\circ g)(x\sigma)d\sigma, \tag{5.4.20}$$

$(f\circ g)(x)$ being an m-tuple integral (5.4.18).

By means of these convolutions *one can develop operational calculi, corresponding to the fractional hyper-Bessel differential* operators $\tilde{B}$, (5.4.8) or (5.4.12).

Special cases:

i) for $m > 1$; $\delta_1 = \cdots = \delta_m = 1$, $\mu = m - (\alpha_0 + \cdots + \alpha_m) := \beta > 0$, $\lambda \geq \max_k \gamma_k$, C_α with $\alpha = \max_k [-\beta(\gamma_k + 1)]$ we obtain the convolutions of Dimovski [64]-[71], Dimovski and Kiryakova [79] *of the "integer" hyper-Bessel operators* (see Theorem 3.6.1, (3.6.6)).

ii) for $m = 1$, $\mu = \beta\delta > 0$, the convolutions (2.1.7) of the Erdélyi-Kober integral $L = x^{\beta\delta} I_\beta^{\gamma,\delta}$ are found. In particular, the convolutions (2.2.25) of the Džrbashjan-Gelfond-Leontiev operator $l_{\rho,\mu}$, (2.2.16)-(2.2.17) follow.

Next, we show one more interesting special case of the fractional hyper-Bessel operators which is a multiple analogue of Džrbashjan-Gelfond-Leontiev operator (2.2.16).

5.4.ii. Multiple Džrbashjan-Gelfond-Leontiev operators and their convolutions

Similarly to the Džrbashjan-Gelfond-Leontiev operators of integration (see Section 2.2):

$$
\begin{aligned}
l_{\rho,\mu} f(x) = x I_\rho^{\mu-1,\frac{1}{\rho}} f(x) &= x \int_0^1 \frac{(1-\sigma)^{\frac{1}{\rho}-1}}{\Gamma\left(\frac{1}{\rho}\right)} \sigma^{\mu-1} f\left(x\sigma^{\frac{1}{\rho}}\right) d\sigma \\
&= \sum_{k=0}^{\infty} a_k \frac{\Gamma\left(\mu + \frac{k}{\rho}\right)}{\Gamma\left(\mu + \frac{k+1}{\rho}\right)} x^{k+1}, \text{ if } f(x) = \sum_{k=0}^{\infty} a_k x^k,
\end{aligned}
\tag{5.4.21}
$$

we can consider their multiple counterparts.

Definition 5.4.3. For power series $f(x) = \sum_{k=0}^{\infty} a_k x^k$, arbitrary real $\mu_1, \ldots, \mu_m$ and $\rho_1 > 0, \ldots, \rho_m > 0$ we define a *multiple Džrbashjan-Gelfond-Leontiev integration operator* by means of the series

$$
Lf(x) = L_{(\rho_k),(\mu_k)} f(x) = \sum_{k=0}^{\infty} a_k \frac{\Gamma\left(\mu_1 + \frac{k}{\rho_1}\right) \ldots \Gamma\left(\mu_m + \frac{k}{\rho_m}\right)}{\Gamma\left(\mu_1 + \frac{k+1}{\rho_1}\right) \ldots \Gamma\left(\mu_m + \frac{k+1}{\rho_m}\right)} x^{k+1}. \tag{5.4.22}
$$

It is easily seen that this operator extends to analytic funcitons in spaces $\mathfrak{H}_\alpha(\Omega)$, continuous functions of C_α, $\alpha = \max_{1\leq k\leq m}(-\rho_k)$ and even to integrable functions by means of the integral representation

$$
\begin{aligned}
L_{(\rho_k),(\mu_k)} f(x) &= x I_{(\rho_k),m}^{(\mu_k - 1),\left(\frac{1}{\rho_k}\right)} f(x) \\
&= x \int_0^1 H_{m,m}^{m,0} \left[\sigma \,\middle|\, \begin{matrix} \left(\mu_k, \frac{1}{\rho_k}\right)_1^m \\ \left(\mu_k - \frac{1}{\rho_k}, \frac{1}{\rho_k}\right)_1^m \end{matrix} \right] f(x\sigma) d\sigma.
\end{aligned}
\tag{5.4.23}
$$

Thus, this is a generalized fractional integral of the form (5.4.15) for which we dispose with a family of convolutions (5.4.17). By putting $\mu = 1$, $\gamma_k = \mu_k - 1$, $\delta_k = \frac{1}{\rho_k}$ in (5.4.17) we obtain the following result.

Theorem 5.4.4. *For* $\lambda \geq \max_k (\mu_k \rho_k - 1)$ *the operations*

$$\left(f \underset{(\rho),(\mu)}{\overset{\lambda}{*}} g\right)(x) = x^{\lambda+1} I_{(\rho_k),m}^{(2\mu_k - 1),\left(\frac{\lambda+1}{\rho_k} - \mu_k\right)} (f \circ g)(x)$$
$$= x^{\lambda+1} \int_0^1 H_{m,m}^{m,0} \left[\sigma \left| \begin{matrix} \left(\frac{\lambda}{\rho_k} + \mu_k, \frac{1}{\rho_k}\right)_1^m \\ \left(2\mu_k - \frac{1}{\rho_k}, \frac{1}{\rho_k}\right)_1^m \end{matrix} \right. \right] (f \circ g)(x\sigma) d\sigma, \tag{5.4.24}$$

where $(\circ)$ *is the auxiliary operator*

$$(f \circ g)(x) = \int_0^1 \ldots \int_0^1 \prod_{k=1}^m [t_k (1 - t_k)]^{\mu_k - 1} f \left[x \prod_{k=1}^m t_k^{\frac{1}{\rho_k}} \right]$$
$$\times g \left[x \prod_{k=1}^m (1 - t_k)^{\frac{1}{\rho_k}} \right] dt_1 \ldots dt_m, \tag{5.4.25}$$

are convolutions of the multiple Džrbashjan-Gelfond-Leontiev integration operator (5.4.23) in C_α, $\alpha = \max_k (-\mu_k \rho_k)$.

The corresponding fractional differential operator $B_{(\rho_k),(\mu_k)}$, i.e. the multiple Džrbashjan-Gelfond-Leontiev operator of differentiation can be represented in the forms

$$B_{(\rho_k),(\mu_k)} = x^{-1} D_{(\rho_k),m}^{\left(\mu_k - 1 - \frac{1}{\rho_k}\right),\left(\frac{1}{\rho_k}\right)}$$
$$= x^{\alpha_0} \left(\frac{d}{dx}\right)^{\frac{1}{\rho_1}} x^{\alpha_1} \ldots \left(\frac{d}{dx}\right)^{\frac{1}{\rho_1}} x^{\alpha_m},$$

with $\mu = \sum_{k=1}^m \frac{1}{\rho_k} - \sum_{i=0}^m \alpha_i = 1 > 0$.

In particular, in the *case* $m = 2$, our 2-tuple Džrbashjan-Gelfond-Leontiev integration operator (5.4.22) has the representation

$$L_{(\rho_1,\rho_2),(\mu_1,\mu_2)} f(x) = \sum_{k=0}^\infty a_k \frac{\Gamma\left(\mu_1 + \frac{k}{\rho_1}\right) \Gamma\left(\mu_2 + \frac{k}{\rho_2}\right)}{\Gamma\left(\mu_1 + \frac{k+1}{\rho_1}\right) \Gamma\left(\mu_2 + \frac{k+1}{\rho_2}\right)} x^{k+1} \tag{5.4.26}$$

as well as a representation of the form (5.4.23), i.e. involving the Gauss hypergeometric function in the kernel. It is seen now that (5.4.26) is nothing but *a Gelfond-Leontiev integration operator* (2.2.8) *with respect to Džrbashjan's function* (E.33):

$$\Phi_{\rho_1,\rho_2}(x;\mu_1,\mu_2)=\sum_{k=0}^{\infty}\frac{x^k}{\Gamma\left(\mu_1+\frac{k}{\rho_1}\right)\Gamma\left(\mu_2+\frac{k}{\rho_2}\right)}, \tag{5.4.27}$$

considered as a 2-index analogue of the Mittag-Leffler function $E_\rho(x;\mu)$. Operators of the form (5.4.26), related to function (5.4.27) do not seem to have been considered till now. Formula (5.4.24) with $m=2$ provides a family of convolutions for them.

Note. Convolutions (5.4.17), (5.4.24) of the fractional hyper-Bessel integral operators $\tilde{L}=x^\mu I^{(\gamma_k),(\delta_k)}_{\left(\frac{\mu}{\delta_k}\right),m}$ and multiple Džrbashjan-Gelfond-Leontiev operators $L_{(\rho_k),(\mu_k)}$ *can be used for representations of the commutants of these operators and their integer powers,* like the results shown in Section 2.3.

OPEN PROBLEM 5.4.5. Although Theorem 5.4.2 provides convolutions for the class of generalized fractional integrals of the form $L=x^\mu I^{(\gamma_k),(\delta_k)}_{\left(\frac{\mu}{\delta_k}\right),m}$, $\mu>0$ (i.e. with $\beta_k=\frac{\mu}{\delta_k}$, $k=1,\dots,m$), the problem of finding a convolution of the most general multiple fractional integral $L=x^{\delta_0}I^{(\gamma_k),(\delta_k)}_{(\beta_k),m}$, $\delta_0>0$ of the form (5.1.9) still seems to be open. In particular, it states the problem for a convolution of Krätzel's operator (5.4.3), when $\mu>0$ is an arbitrary (rational) parameter.

5.5. Generalized fractional calculus in classes of analytic functions and applications to univalent function theory

There are several approaches in defining fractional differintegrals for functions of a complex variable, each of them with its pros and cons. There has also been some controversy leading to a certain distrust in the general concept of fractional calculus, especially in the mid-nineteenth century. *"The mathematicians at that time were aiming for a plausible definition of the generalized differentiation. But, in fairness to them, one should note they lacked the tools to examine the consequences of their definitions in the complex plane"* (Davis, see [404, p. 4]). As pointed out in [434], the first successful attempts in developing fractional calculus for functions of a complex variable belong to the Russian mathematicians Sonine [456] and Letnikov [254]-[255]. They have established that the extension of the *Cauchy integral formula*

$$f^{(\rho)}(z)=\frac{\rho!}{2\pi i}\int\limits_{\mathfrak{L}}\frac{f(\zeta)}{(\zeta-z)^{\rho+1}}d\zeta,\quad \rho=0,1,2,\dots \tag{5.5.1}$$

to non integer negative values of ρ (or in general, to complex ρ with $\Re\rho<0$) coincides with Riemann's definition for integration of order $\delta = -\rho > 0$, viz.

$$R^{\delta} f(z) = D^{\rho} f(z) = \frac{1}{\Gamma(-\rho)} \int_{c}^{z} \frac{f(\zeta)\, d\zeta}{(z-\zeta)^{\rho+1}} = \frac{1}{\Gamma(\delta)} \int_{c}^{z} (z-\zeta)^{\delta-1} f(\zeta)\, d\zeta, \tag{5.5.2}$$

where c is the cross-point of the contour $\mathfrak{L}$, encircling z, and the cut in $\mathbb{C}$ put for determinig a unique branch of $(\zeta - z)^{-\rho-1}$ in (5.5.1).

Another frequently used method descends from Hadamard's paper [124]. There, the idea of a fractional differentiation of analytic functions via *term-by-term differentiation* of their Taylor series:

$$D^{\delta} \left\{ \sum_{k=0}^{\infty} a_k (z-c)^k \right\} = \sum_{k=0}^{\infty} a_k \frac{\Gamma(k+1)}{\Gamma(k-\delta+1)} (z-c)^{k-\delta}, \quad \delta \neq -1, -2, \ldots \tag{5.5.3}$$

has been effectively used and gave rise to the so-called *Hadamard approach.* In [124], also the representation

$$R^{\delta} f(z) = z^{\delta} \int_{0}^{1} \frac{(1-\sigma)^{\delta-1}}{\Gamma(\delta)} f(z\sigma) d\sigma, \ \delta > 0 \tag{5.5.4}$$

of the Riemann-Liouville fractional integral (5.5.2) has appeared. This form, following by a simple substitution is however *much more convenient than (5.5.2) when dealing with complex variables.* Then, the choice of the unique branch there depends only on the fixed point $z = 0$ while in (5.5.2) the branch-point $z = \zeta$ is variable.

There exist also several other approaches in developing fractional calculus in complex domains, for example, *Nishimoto's approach* (see [321]-[331] and especially the books [324], [326], [329], [331]). He uses contour integrals similar to (5.5.1) but allowing fractional values of ρ. Recently, the Riemann-Liouville fractional integrals and derivatives in that sense have been used repeatedly by: Nishimoto et al. [332]-[335], Nunukawa and Owa [336], Owa [344], Owa et al. [345]-[354], Srivastava and Owa [471]-[474], Srivastava et al. [477]-[482] and other authors of the same school, see the volumes [475]-[476]. There, the Riemann-Liouville fractional derivatives of order $0 < \delta < 1$ are used in the form

$$D^{\delta} f(z) = \frac{d}{dz} \int_{0}^{z} \frac{(z-\zeta)^{-\delta}}{\Gamma(1-\delta)} f(\zeta) d\zeta, \tag{5.5.5}$$

in simply connected domains Ω and requiring $\ln(z-\zeta)$ to be real for $(z-\zeta) > 0$.

Some of the other recent articles on the topic, related mainly to the univalent starlike and convex functions are: Rusheweyh [411]-[412], Mokanu [311], Owa, Altintas and Pashkuleva [346], Sekine [443], Yakubowski [150]-[152], Dimkov [63], Kapoor and Patel [183], Saitoh [427], etc. Worthy of mentioning is the recent resolution of the famous *Bieberbach conjecture* by de Branges [42].

Most of the difficulties, arising with respect to multi-valued in integrals (5.5.2), (5.5.5) or factors z^δ in (5.5.4), can be easily overcome by considering *the Erdélyi-Kober operators of fractional integration* and especially, in their form:

$$I_\beta^{\gamma,\delta} f(z) = \int_0^1 \frac{(1-\sigma)^{\delta-1}}{\Gamma(\delta)} \sigma^\gamma f\left(z\sigma^{\frac{1}{\beta}}\right) d\sigma, \ \gamma \in \mathbb{R}, \ \delta > 0, \beta > 0. \tag{5.5.6}$$

Originally, they were introduced by Kober [220] and Erdélyi [104] and used till recently by many authors in the more complicated form

$$I_\beta^{\gamma,\delta} f(z) = z^{-\beta(\gamma+\delta)} \int_0^z \frac{\left(z^\beta - \zeta^\beta\right)^{\delta-1}}{\Gamma(\delta)} \zeta^{\beta\gamma} f(\zeta) d(\zeta^\beta), \tag{5.5.6$'$}$$

its disadvantage being obvious especially when dealing in complex domains.

This is the main idea in the following development: to use the Erdélyi-Kober fractional integrals (5.5.6) *and their multiple analogues of the same form:* $I_{\beta,m}^{(\gamma_k),(\delta_k)}$ (1.1.6), $D_{\beta,m}^{(\gamma_k),(\delta_k)}$ (1.5.19), $I_{(\beta_k),m}^{(\gamma_k),(\delta_k)}$ (5.1.8), $D_{(\beta_k),m}^{(\gamma_k),(\delta_k)}$ (1.5.38), and hence the techniques of the G- and H-functions, for the purposes of complex analysis.

5.5.i. Generalized fractional differintegrals in classes of analytic functions

Definition 5.5.1. Let Ω be a complex domain, starlike with respect to the origin $z = 0$, and hence a simply connected domain. Denote by $\mathfrak{H}(\Omega)$ *the space of analytic and single-valued (i.e. holomormphic) functions in* Ω. Further, for real $\mu \geq 0$ we consider the following *classes of functions:*

$$\mathfrak{H}_\mu(\Omega) = \left\{f(z) = z^\mu \tilde{f}(z); \tilde{f}(z) \in \mathfrak{H}(\Omega)\right\}, \quad \mathfrak{H}_0(\Omega) := \mathfrak{H}(\Omega). \tag{5.5.7}$$

(in practice, most often $\mu \geq 0$ is taken to be integer). To avoid the multiplicity of the factors z^μ we determine a unique branch of $\arg z : 0 \leq \arg z \leq 2\pi$ by putting a cut in Ω along the half-line $\{\Re z \geq 0, \ \mathrm{Im} z = 0\}$ and assuming

$$z^\mu = |z|^\mu \exp(i\mu \arg z), \quad \mu \geq 0.$$

In particular, Ω can be taken the unit disk

$$U = \Delta_1 : |z| < 1, \quad \text{or the disk } \Delta_R : |z| < R, \ R > 0.$$

In the latter case, we can consider the functions $f(z)$ given by their power series

$$f(z) = z^\mu \sum_{n=0}^\infty a_n z^n = z^\mu (a_0 + a_1 z + \ldots) \in \mathfrak{H}_\mu(\Delta_R), \tag{5.5.8}$$

absolutely convergent in Δ_R with $R > 0$ defined by the *Cauchy-Hadamard formula*

$$R = \left\{ \overline{\lim_{n\to\infty}} \sqrt[n]{|a_n|} \right\}^{-1}. \tag{5.5.9}$$

In Chapter 1 and in Sections 5.1-5.4 we have studied the multiple Erdélyi-Kober fractional integrals and derivatives in the spaces C_α (continuous type functions) and L^p, $L_{\mu,p}$ (Lebesgue integrable functions). In Chapter 2 some special cases of generalized fractional differintegrals have been considered also for functions of complex variables. Now, we go deeper into the problems, related to generalized fractional calculus in the classes (5.5.7) of analytic functions.

For the most general case of multiple fractional integrals involving $H^{m,0}_{m,m}$-functions we state the following result.

Theorem 5.5.2. (Kiryakova [204]) *Let the conditions*

$$\gamma_k > -\frac{\mu}{\beta_k} - 1, \quad \delta_k > 0, \quad k = 1, \ldots, m \tag{5.5.10}$$

be satisfied. Then, the multiple Erdélyi-Kober operator $I^{(\gamma_k),(\delta_k)}_{(\beta_k),m}$ defined by (5.1.8) or (5.2.7), maps the class $\mathfrak{H}_\mu(\Omega)$ into itself, preserving the power functions up to a constant multiplier:

$$I^{(\gamma_k),(\delta_k)}_{(\beta_k),m}\{z^p\} = c_p z^p, \; p \geq \mu \;\; \textit{with} \;\; c_p = \prod_{k=1}^{m} \frac{\Gamma\left(\gamma_k + \frac{p}{\beta_k} + 1\right)}{\Gamma\left(\gamma_k + \delta_k + \frac{p}{\beta_k} + 1\right)}. \tag{5.5.11}$$

Hence, the image of power series (5.5.8) is given by the series

$$I^{(\gamma_k),(\delta_k)}_{(\beta_k),m} f(z) = z^\mu \sum_{n=0}^{\infty} \left\{ a_n \prod_{k=1}^{m} \frac{\Gamma\left(\gamma_k + \frac{n+\mu}{\beta_k} + 1\right)}{\Gamma\left(\gamma_k + \delta_k + \frac{n+\mu}{\beta_k} + 1\right)} \right\} z^n, \tag{5.5.12}$$

having the same radius of convergence $R > 0$, defined by (5.5.9) and the same signs of the coefficients. More generaly, each generalized fractional integral $\mathfrak{R} f(z) = z^{\delta_0} I^{(\gamma_k),(\delta_k)}_{(\beta_k),m}$, $\delta_0 \geq 0$ maps $\mathfrak{H}_\mu(\Omega)$ into $\mathfrak{H}_{\mu+\delta_0}(\Omega) \subseteq \mathfrak{H}_\mu(\Omega)$.

Proof. We emphasize only the specific details for $\mathfrak{H}_\mu(\Omega)$ which have not been used in the proofs of the similar theorems for spaces C_α, $L_{\mu,p}$. For instance, (5.5.11) has been calculated in Theorem 5.1.4. Then, term-by-term integration of $f(z) \in \mathfrak{H}_\mu(\Delta_R)$, (5.5.8) leads to the power series

$$I^{(\gamma_k),(\delta_k)}_{(\beta_k),m} f(z) = \sum_{n=0}^{\infty} a_n I^{(\gamma_k),(\delta_k)}_{(\beta_k),m} \{z^{\mu+n}\} = z^{\mu} \sum_{n=0}^{\infty} a_n c_{\mu+n} z^n$$
$$= z^{\mu} \sum_{n=0}^{\infty} b_n z^n,$$

where *the new coefficients are:*

$$b_n = a_n c_{\mu+n} = a_n \prod_{k=1}^{m} \frac{\Gamma\left(\gamma_k + \frac{n+\mu}{\beta_k} + 1\right)}{\Gamma\left(\gamma_k + \delta_k + \frac{n+\mu}{\beta_k} + 1\right)}$$

and so have the same signs as a_n, $n = 0, 1, 2, \dots$. The radius of convergence of the image-series, due to (5.5.9), is:

$$\tilde{R} = \left\{ \overline{\lim_{n\to\infty}} |b_n|^{\frac{1}{n}} \right\}^{-1} = \left\{ \overline{\lim_{n\to\infty}} |a_n|^{\frac{1}{n}} |c_{\mu+n}|^{\frac{1}{n}} \right\}^{-1}.$$

By repeated use of the asymptotic formula $\frac{\Gamma(x+b)}{\Gamma(x+a)} \sim x^{b-a}$, $x \to \infty$, we obtain:

$$\lim_{n\to\infty} |c_{\mu+n}|^{\frac{1}{n}} = \lim_{n\to\infty} \left[\prod_{k=1}^{m} \left(n^{\frac{1}{n}}\right)^{-\delta_k} \left(\beta_k^{\frac{1}{n}}\right)^{\delta_k} \right] = 1,$$

i.e. $\overline{\lim_{n\to\infty}} |b_n|^{\frac{1}{n}} = \overline{\lim_{n\to\infty}} |a_n|^{\frac{1}{n}} \Rightarrow \tilde{R} = R$.

The correctness of definition (5.1.8) (with $x = z \in \Omega$) *for arbitrary functions* $f(z) = z^{\mu}\tilde{f}(z) \in \mathfrak{H}_{\mu}(\Omega)$ can be justified by means of *Range's approximation theorem* applied to each compact subset of the simply connected domain Ω. Namely, we can approximate there the analytic function $\tilde{f}(z)$ by means of polynomials $P_n(z)$, and therefore $f(z)$, by the polynomials $z^{\mu}P_n(z)$, $n = 1, 2, \dots$. Another way is to establish that the improper integral $I^{(\gamma_k),(\delta_k)}_{(\beta_k),m} \left\{ z^{\mu}\tilde{f}(z) \right\}$, depending on the parameter $z \in \Omega$, is uniformly convergent in each compact subset $\tilde{\Omega} \subset \Omega$, and thus it is a function from $\mathfrak{H}_{\mu}(\Omega)$ again. To this end, we use the asymptotics of the kernel $H^{m,0}_{m,m}$-function (see the proof of Theorem 5.1.3) and conditions (5.5.10).

For the leter use, we shall consider *the simpler case* when

$$\text{the } \beta_1 = \beta_2 = \cdots = \beta_m = \beta > 0, \quad \text{i.e. the kernel is } H^{m,0}_{m,m}(\sigma) = G^{m,0}_{m,m}(\sigma). \tag{5.5.13}$$

Then, the simpler Erdélyi-Kober fractional integrals $I^{(\gamma_k),(\delta_k)}_{\beta,m}$, (1.1.6) with a complex variable $x = z \in \Omega$, considered in the classes $\mathfrak{H}_{\mu}(\Omega)$ have the same properties as stated in Theorem 5.5.2 but with the substitution $\beta_k = \beta$, $k = 1, \dots, m$; see Theorem 1.2.18 and its proof in Kiryakova [199]. The simplest (and the most useful) case is with parameters $\beta_k = \beta = 1$, $k = 1, \dots, m$. Then, we have the following statement.

Corollary 5.5.3. *Let us consider the multiple Erdélyi-Kober fractional integral (1.1.6) with $\beta = 1$:*

$$I_{1,m}^{(\gamma_k),(\delta_k)} f(z) = \int_0^1 G_{m,m}^{m,0} \left[\sigma \left| \begin{matrix} (\gamma_k + \delta_k + 1)_1^m \\ (\gamma_k + 1)_1^m \end{matrix} \right. \right] f(z\sigma) d\sigma \qquad (5.5.14)$$

in the classes of function $\mathfrak{H}_\mu(\Omega)$. If

$$\gamma_k > -\mu - 1, \quad \delta_k > 0, \quad k = 1, \ldots, m, \qquad (5.5.15)$$

then $I_{1,m}^{(\gamma_k),(\delta_k)} : \mathfrak{H}_\mu(\Omega) \to \mathfrak{H}_\mu(\Omega)$ and in particular, for $f \in \mathfrak{H}_\mu(\Delta_R) \subset \mathfrak{H}_\mu(\Omega)$:

$$I_{1,m}^{(\gamma_k),(\delta_k)} \left\{ z^\mu \sum_{n=0}^{\infty} a_n z^n \right\} = z^\mu \sum_{n=0}^{\infty} a_n c_{\mu+n} z^n \qquad (5.5.16)$$

with

$$c_{\mu+n} = \prod_{k=1}^{n} \frac{\Gamma(\gamma_k + 1 + n + \mu)}{\Gamma(\gamma_k + \delta_k + 1 + n + \mu)} > 0, \quad n = 0, 1, 2, \ldots, \quad \mu \geq 0, \qquad (5.5.17)$$

i.e. these operators transform power series into power series with the same radius of convergence and the same signs of the coefficients.

Note. For the classes of analytic functions considered here, the generalized fractional integrals $I_{(\beta_k),m}^{(\gamma_k),(\delta_k)}$ and derivatives $D_{(\beta_k),m}^{(\gamma_k),(\delta_k)}$, (5.1.38) act in almost the same manner, so the mapping and other properties of the latter are similar to those of the integrals $I_{(\beta_k),m}^{(\gamma_k),(\delta_k)}$.

All the other operational rules and properties, stated in Chapter 1 (for the case of $G_{m,m}^{m,0}$-kernels) and in Sections 5.1-5.4 (for the $H_{m,m}^{m,0}$-kernels) *are also valid for the generalized fractional differintegrals $I_{(\beta_k),m}^{(\gamma_k),(\delta_k)}$, $D_{(\beta_k),m}^{(\gamma_k),(\delta_k)}$ in the classes of functions $\mathfrak{H}_\mu(\Omega)$* and we shall use them (e.g. Theorems 5.1.6, 5.1.9, 5.2.1, etc.) *under conditions* (5.5.10), (5.5.15).

In particular, for $m = 1$, the above general results have *as consequences the properties of the clasical Erdélyi-Kober operators in spaces of analytic functions.* Some of them are well-known but present interest as a tool, commonly used in dealing with univalent (starlike, convex) functions. Let us list them in the form of a separate proposition.

Theorem 5.5.4. *If $\gamma > -\mu - 1$, $\delta > 0$, then the Erdélyi-Kober fractional integrals (5.5.6)* *(with $\beta = 1$) preserve the class $\mathfrak{H}_\mu(\Omega)$;*

$$I_1^{\gamma,\delta}\left\{z^\mu \sum_{n=0}^{\infty} a_n z^n\right\} = z^\mu \sum_{n=0}^{\infty} a_n \frac{\Gamma(\gamma+1+n+\mu)}{\Gamma(\gamma+\delta+1+n+\mu)} z^n;$$

$$I_1^{\gamma,\delta} z^\lambda = z^\lambda I_1^{\lambda+\gamma,\delta}; \quad I_1^{\gamma+\delta,\sigma} I_1^{\gamma,\delta} = I_1^{\gamma,\delta+\sigma}; \quad I_1^{\gamma,\delta} I_1^{\tau,\alpha} = I_1^{\tau,\alpha} I_1^{\gamma,\delta} = I_{1,2}^{(\gamma,\tau),(\delta,\alpha)}; \quad (5.5.18)$$

$$I_1^{\gamma,0} = I; \quad \left\{I_1^{\gamma,\delta}\right\}^{-1} = I_1^{\gamma+\delta,-\delta} := D_1^{\gamma,\delta}; \quad D_1^{\gamma,\delta} = z^{-\gamma}\left(\frac{d}{dz}\right)^\delta z^{\gamma+\delta}$$

and

$$\left(\prod_{k=1}^{m} I_1^{\gamma_k,\delta_k}\right) = I_{1,m}^{(\gamma_k),(\delta_k)}; \quad \left(\prod_{k=1}^{m} D_1^{\gamma_k,\delta_k}\right) = D_{1,m}^{(\gamma_k),(\delta_k)}. \quad (5.5.19)$$

5.5.ii. Application to some special classes of analytic functions

Definition 5.5.5. For integer $\mu = p \geq 1$ and $\Omega = \Delta_1 = \{|z| < 1\}$ we define the subclasses $\mathfrak{A}_p \subset \mathfrak{H}_p(\Delta_1)$ of functions, analytic in the unit disk Δ_1 and satisfying the conditions

$$f(0) = f'(0) = \cdots = f^{(p-1)}(0) = 0, \quad f^{(p)}(0) = p!,$$

i.e.

$$\mathfrak{A}_p = \left\{f(z) = z^p + a_1 z^{p+1} + a_2 z^{p+2} + \cdots = z^p \tilde{f}(z);\ \tilde{f} \in \mathfrak{H}(\Delta_1);\ a_0 = 1\right\}. \quad (5.5.20)$$

We are interested in *the problem of preserving these subclasses* $\mathfrak{A}_p$, which play an important role in the theory of functions of a complex variable, when acting on them by generalized fractional differintegrals $I_{(\beta_k),m}^{(\gamma_k),(\delta_k)}$, $D_{(\beta_k),m}^{(\gamma_k),(\delta_k)}$, considered in the previous theorems. To this end, we should only put there *integer* $\mu = p \geq 1$, $\Omega = \Delta_1$ and add the requirement:

$$I_{1,m}^{(\gamma_k),(\delta_k)}\left\{z^p \sum_{n=0}^{\infty} a_n z^n\right\} = z^p \sum_{n=0}^{\infty} b_n z^n \ \text{ with } a_0 = 1 \Rightarrow b_0 = 1. \quad (5.5.21)$$

Since

$$b_n = a_n c_{n+p} \Rightarrow b_0 = a_0 c_p = c_p = \prod_{k=1}^{m} \frac{\Gamma(\gamma_k+1+p)}{\Gamma(\gamma_k+\delta_k+1+p)},$$

we "normalize" the fractional integrals and derivatives by the factors c_p, respectively $\frac{1}{c_p}$. Then, from Corollary 5.5.3, we obtain the following.

Theorem 5.5.6. *Let $p \geq 1$ be an integer and*

$$\gamma_k > -p-1,\ \delta_k > 0,\ k = 1, \ldots, m. \quad (5.5.22)$$

Then, the (normalized) multiple Erdélyi-Kober fractional integrals and derivatives:

$$\begin{aligned}\tilde{I}_{1,m}^{(\gamma_k),(\delta_k)} &= \frac{1}{c_p} I_{1,m}^{(\gamma_k),(\delta_k)} \quad \textit{with} \quad \frac{1}{c_p} = \prod_{k=1}^{m} \frac{\Gamma(\gamma_k + \delta_k + 1 + p)}{\Gamma(\gamma_k + 1 + p)}, \\ \tilde{D}_{1,m}^{(\gamma_k),(\delta_k)} &= c_p D_{1,m}^{(\gamma_k),(\delta_k)} = c_p I_{1,m}^{(\gamma_k+\delta_k),(-\delta_k)}\end{aligned} \tag{5.5.23}$$

preserve the classes $\mathfrak{A}_p$, defined by (5.5.20). In particular,

$$\tilde{I}_{1,m}^{(\gamma_k),(\delta_k)} \{z^p\} = \tilde{D}_{1,m}^{(\gamma_k),(\delta_k)} \{z^p\} = z^p$$

and for functions $f(z) = z^p(1 + a_1 z + a_2 z^2 + \dots) \in \mathfrak{A}_p$ operators (5.5.23) act as follows:

$$\begin{aligned}\tilde{I}_{1,m}^{(\gamma_k,\delta_k)} f(z) &= z^p\left(1 + \tilde{b}_1 z + \tilde{b}_2 z^2 + \dots\right), \quad \tilde{b}_n = a_n g_{n,p}, \\ \tilde{D}_{1,m}^{(\gamma_k,\delta_k)} f(z) &= z^p\left(1 + \tilde{d}_1 z + \tilde{d}_2 z^2 + \dots\right), \quad \tilde{d}_n = a_n g_{n,p}^{-1}, \quad n = 1, 2, \dots\end{aligned} \tag{5.5.24}$$

where the "multiplier coefficients" are:

$$g_{n,p} = \frac{c_{n+p}}{c_p} = \prod_{k=1}^{m} \frac{\Gamma(\gamma_k + 1 + n + p)\Gamma(\gamma_k + \delta_k + 1 + p)}{\Gamma(\gamma_k + \delta_k + 1 + n + p)\Gamma(\gamma_k + 1 + p)} > 0, \tag{5.5.25}$$

therefore,

$$\operatorname{sign} \tilde{b}_n = \operatorname{sign} \tilde{d}_n = \operatorname{sign} \tilde{a}_n, \quad n = 1, 2, \dots . \tag{5.5.26}$$

Most frequently, special cases of the generalized fractional differintegrals (5.5.23) (see the examples below) are used in the classes $\mathfrak{A}_p$ with $p = 0$ and $p = 1$, namely:

$$\mathfrak{A}_0 = \left\{ f(z) = 1 + a_1 z + a_2 z^2 + \dots \ ; \text{ analytic in } \Delta_1 \right\}, \tag{5.5.27}$$

$$\mathfrak{A} = \mathfrak{A}_1 = \left\{ f(z) = z + a_1 z^2 + \dots + a_n z^{n+1} + \dots \in \mathfrak{H}(\Delta_1) \right\}. \tag{5.5.28}$$

The latter is the familiar class $\mathfrak{A}$ of functions, analytic in the unit disk and normalized by the conditions $f(0) = f'(0) - 1 = 0$.

It is now easy to specify the conditions (5.5.22) under which the generalized Erdélyi-Kober operators (5.5.23) preserve these two classes, namely:

$$\gamma_k > -1, \ \delta_k > 0, \ k = 1, \dots, m, \ \text{ for } \mathfrak{A}_0; \tag{5.5.27'}$$

$$\gamma_k > -2, \ \delta_k > 0, \ k = 1, \dots, m, \ \text{ for } \mathfrak{A} = \mathfrak{A}_1. \tag{5.5.28'}$$

Note. In terms of the so-called *Hadamard convolution (product)* ($\circ$) of power series (see (2.2.9)) *one can say that the generalized integration* $\tilde{I}_{1,m}^{(\gamma_k),(\delta_k)}$, (5.5.23), *maps a function*

$f \in \mathfrak{A}_p$ *into its Hadamard product by an analytic function* $g_p(z)$ whose Taylor coefficients are $g_{n,p}$, namely:

$$\begin{aligned}\tilde{I}_{1,m}^{(\gamma_k),(\delta_k)} f(z) = \tilde{I}_{1,m}^{(\gamma_k),(\delta_k)} \left\{ z^p \tilde{f}(z) \right\} &= \left(f \overset{p}{\circ} g_p \right)(z) \\ := z^p \left(\tilde{f} \circ g_p \right)(z) &= z^p \sum_{n=0}^{\infty} a_n g_{n,p} z^n,\end{aligned} \tag{5.5.29}$$

where

$$g_p(z) = \sum_{n=0}^{\infty} g_{n,p} z^n \quad \text{with} \quad g_{n,p} \to 0, \ n \to \infty. \tag{5.5.30}$$

Analogously, *the generalized differentiation* $\tilde{D}_{1,m}^{(\gamma_k),(\delta_k)}$ *is* nothing but *a Hadamard product with the "associated" function*

$$g_p^*(z) = \sum_{n=0}^{\infty} g_{n,p}^{-1} z^n \quad \text{with} \quad g_{n,p}^{-1} \to \infty, \ n \to \infty, \tag{5.5.31}$$

i.e.

$$\begin{aligned}\tilde{D}_{1,m}^{(\gamma_k),(\delta_k)} f(z) = \tilde{D}_{1,m}^{(\gamma_k),(\delta_k)} \left\{ z^p \tilde{f}(z) \right\} &= \left(f \overset{p}{\circ} g_p^* \right)(z) \\ := z^p \left(\tilde{f} \circ g_p^* \right)(z) &= z^p \sum_{n=0}^{\infty} \frac{a_n}{g_{n,p}} z^n.\end{aligned} \tag{5.5.32}$$

Examples of generalized Erdélyi-Kober fractional integrals used in the topic of univalent, starlike and convex functions

These operators have been introduced by different authors who use them to study analytic functions in various subclasses of the class $\mathfrak{A}$ defined by (5.5.28). Such classes include the class $\mathcal{S}$ of functions univalent in Δ_1; the clases $\mathcal{S}^*(\alpha)$ of functions starlike of order α $(0 \geq \alpha \geq 1)$ in Δ_1; the class $\mathcal{K}$ and $\mathcal{K}(\alpha)$ of convex (convex of order α) functions in Δ_1; the $\mathcal{C}(\alpha, \beta)$ (see e.g. [351]), etc. Most of the operators listed below are defined in the class $\mathfrak{A}$ and preserve it (as a corollary of Theorem 5.5.6 when $p = 1$). Their properties in the above subclasses of $\mathfrak{A}$ have been discussed in many papers and books. We shall not describe in detail the various interesting results on this topic, but let us give some examples only. We begin with the *simplest case* $m = 1$ of the *classical (single) Erdélyi-Kober operators of fractional differintegration.*

Example 1 (Biernacki [33]). The linear integral operator B is defined in $\mathfrak{A}$ by

$$Bf(z) = \int_0^z \frac{f(t)}{t} dt = \int_0^1 \sigma^{-1} f(z\sigma) d\sigma = I_1^{-1,1} f(z).$$

EXAMPLE 2 (Komatu [230]-[232]). The integral operator L is defined in $\mathfrak{A}$ by

$$Lf(z) = \int_0^1 \frac{f(z\sigma)}{\sigma} d\omega(\sigma) = \int_0^1 \sigma^{-1} f(z\sigma) d\sigma = I_1^{-1,1} f(z),$$

in the case when the probability measure on [0, 1] is taken to be $\omega(\sigma) = \sigma$. The sequence $\{L^n\}$ $(n = 0, 1, 2, \ldots)$ is then interpolated into a family of operators $\left\{L^\lambda\right\}$, depending on a continuous parameter $\lambda \geq 0$. According to our theory, the integer powers L^n can be considered as n-tuple compositions of the operator L itself:

$$L^n = \left(I_1^{-1,1}\right)^n = I_1^{-1,1}\left\{I_1^{-1,1}\ldots\left(I_1^{-1,1}\right)\right\} = I_{1,n}^{(-1)_1^n,(1)_1^n},$$

that is, as n-tuple Erdélyi-Kober fractional integrals of the form (1.1.6):

$$\begin{aligned} L^n f(z) = I_{1,n}^{(-1),(1)} f(z) &= \int_0^1 G_{n,n}^{n,0}\left[\sigma \middle| \begin{matrix} 0,\ldots,0 \\ -1,\ldots,-1 \end{matrix}\right] f(z\sigma)d\sigma \\ &= \int_0^1 G_{n,n}^{n,0}\left[\sigma \middle| \begin{matrix} 1,\ldots,1 \\ 0,\ldots,0 \end{matrix}\right] \frac{f(z\sigma)}{\sigma} d\sigma \\ &= \frac{1}{(n-1)!}\int_0^1 \left(\ln\frac{1}{\sigma}\right)^{n-1} f(z\sigma)d\sigma \end{aligned} \tag{5.5.33}$$

(see [370, p. 637, (3)]). This result coincides with the representation for integer $\lambda = n > 1$ found by Komatu [230]-[232].

EXAMPLE 3 (Libera [256]). The linear integral operator

$$If(z) = \frac{2}{z}\int_0^z f(t)dt = 2\int_0^1 f(z\sigma)d\sigma = 2I_1^{0,1} f(z).$$

EXAMPLE 4 (Owa and Srivastava [351], Owa and Obradović [349]). The *generalized Libera operator* $\mathfrak{B}_c$ is defined for $c > -1$ by

$$\begin{aligned} \mathfrak{B}_c f(z) &= \frac{c+1}{z^c}\int_0^z t^{c-1} f(t)dt = (c+1)\int_0^1 \sigma^{c-1} f(z\sigma)d\sigma \\ &= (c+1)I_1^{c-1,1} f(z) = (c+1)I_{1,1}^{c-1,1} f(z). \end{aligned} \tag{5.5.34}$$

For integer $c = n = 0, 1, 2, \ldots$, this operator is considered by Bernardi [30].

EXAMPLE 5 (Ruscheweyh [411]-[412]; Owa and Obradović [348]). The so-called *Ruscheweyh derivative* is defined by

$$\mathfrak{D}^{\alpha} f(z) = \left\{ \frac{z}{(1-z)^{1+\alpha}} \right\} \circ f(z) \quad (\alpha \geq 0), \tag{5.5.35}$$

where (as before) $f \circ g$ denotes the Hadamard product of two power series, that is,

$$(f \circ g)(z) = \sum_{n=0}^{\infty} a_n b_n z^n \tag{5.5.36}$$

$$\text{for} \quad f(z) = \sum_{n=0}^{\infty} a_n z^n \quad \text{and} \quad g(z) = \sum_{n=0}^{\infty} b_n z^n.$$

It can easily be seen that (5.5.35) has the alternative representation:

$$\begin{aligned} \mathfrak{D}^{\alpha} f(z) &= \frac{1}{\Gamma(\alpha+1)} \left[z \left(\frac{d}{dz} \right)^{\alpha} z^{\alpha-1} \right] f(z) \\ &= \frac{1}{\Gamma(\alpha+1)} D_1^{-1,\alpha} f(z) \end{aligned} \tag{5.5.37}$$

in terms of the Erdélyi-Kober fractional derivatives of order α, defined by (5.5.18). According to Theorem 5.5.4, this differintegral operator is an inverse operator of the Erdélyi-Kober integral

$$L_{\alpha} = \Gamma(\alpha+1) I_1^{-1,\alpha} \quad (\alpha \geq 0).$$

For integer $\alpha = n = 0, 1, 2, \ldots$, the Ruscheweyh derivative (5.5.35), (5.5.37) has the known form:

$$\mathfrak{D}^{n} f(z) = (L_n)^{-1} f(z) = \frac{1}{n!} z \left(z^{n-1} f(z) \right)^{(n)} = \frac{1}{n!} D_1^{-1,n} f(z).$$

In particular, for $n = 1$, we obtain the operator

$$(L_1)^{-1} f(z) = \left(I_1^{-1,1} \right)^{-1} f(z) = D_1^{-1,1} f(z) = z f'(z),$$

inverse to the Biernacki operator considered in Example 1.

Since $\mathfrak{D}^{\alpha}$ is an Erdélyi-Kober derivative of the form $D_1^{\gamma,\delta}$ with $\gamma = -1 > -2$ and $\delta = \alpha \geq 0$ (see conditions (5.5.27′) and (5.5.28′)) and $\mathfrak{D}^{\alpha}\{z\} = z$, Theorem 5.5.6 implies that $\mathcal{D}^{\alpha}$ maps the class $\mathfrak{A}$ into itself. Further, we can consider a composition of a finite number of Ruscheweyh derivatives $\mathfrak{D}^{\alpha_k}$ ($\alpha_k \geq 0$, $k = 1, \ldots, m$), that is, the m-tuple Erdélyi-Kober derivative:

$$\mathfrak{D}^{\alpha_1,\ldots,\alpha_m} f(z) := \left(\prod_{k=1}^{m} \mathfrak{D}^{\alpha_k} \right) f(z) = D_{1,m}^{(-1)_1^m,(\alpha_k)_1^m} f(z), \tag{5.5.38}$$

defined by (1.5.19). On the other hand, formula (5.5.32) provides a representation of (5.5.38) as a Hadamard product with the analytic function

$$g^*(z) = \sum_{n=0}^{\infty} \left(\frac{1}{n!} \prod_{k=1}^{m} \frac{\Gamma(\alpha_k + n + 1)}{\Gamma(\alpha_k + 1)} \right) z^n,$$

that is,

$$D^{\alpha_1,\dots,\alpha_m} f(z) = (g^* \circ f)(z). \tag{5.5.39}$$

Then, the inverse operator $\mathfrak{I}^{\alpha_1,\dots,\alpha_m} := I_{1,m}^{(-1),(\alpha_k)}$ is a product with the generalized hypergeometric function (5.5.30):

$$\begin{aligned} g(z) &= \sum_{n=0}^{\infty} n! \left(\prod_{k=1}^{m} \frac{\Gamma(\alpha_k + 1)}{\Gamma(\alpha_k + n + 1)} \right) z^n = \sum_{n=0}^{\infty} \frac{(1)_n (1)_n}{(\alpha_1 + 1)_n \dots (\alpha_m + 1)_n} \frac{z^n}{n!} \\ &= {}_2F_m\left(1, 1; (\alpha_k + 1)_1^m; z\right), \end{aligned}$$

i.e. in terms of the Hadamard product:

$$\mathfrak{I}i^{\alpha_1,\dots,\alpha_m} f(z) = {}_2F_m\left(1, 1; (\alpha_k + 1)_1^m; z\right) \circ f(z). \tag{5.5.39'}$$

We may call the a generalized fractional derivative (5.5.38) a multiple Ruscheweyh derivative.

Example 6 (Owa and Srivastava [352]; Sekine et al. [444]). The differintegral operator Ω^λ defined by

$$\Omega^\lambda = \Gamma(2-\lambda) z^\lambda \left(\frac{d}{dz}\right)^\lambda f(z) = \Gamma(2-\lambda) D_1^{-\lambda,\lambda} f(z) \quad (\lambda \neq 2, 3, 4, \dots)$$

provides another example of an Erdélyi-Kober fractional derivative (5.5.18).

Example 7 (Carlson and Shaffer [52]; Owa and Srivastava [352]; Owa et al. [354]). Consider the integral operator $\mathcal{L}(a,c)$ defined in $\mathfrak{A}$ by the Hadamard product (5.5.36) as follows:

$$\mathcal{L}(a,c) f(z) = \Phi(a, c;\ z) \circ f(z) = \{z\, {}_2F_1(1, a;\ c;\ z)\} \circ f(z). \tag{5.5.40}$$

It is easily seen that

$$\begin{aligned} \mathcal{L}(a,c) f(z) &= \frac{\Gamma(c)}{\Gamma(a)\,\Gamma(c-a)} \int_0^1 (1-\sigma)^{c-a-1} \sigma^{a-2} f(z\sigma) d\sigma \\ &= \frac{\Gamma(c)}{\Gamma(a)} I_1^{a-2,c-a} f(z) \quad (0 < a < c). \end{aligned} \tag{5.5.41}$$

Naturally, $\mathcal{L}(a,a)$ is the identity operator. The inverse operator of $\mathcal{L}(a,c)$ is formally given (cf. [52] and our inversion formula (5.5.18)) by:

$$\mathcal{L}^{-1}(a,c) = \frac{\Gamma(a)}{\Gamma(c)} I_1^{(a-2)+(c-a),\ a-c} = \frac{\Gamma(a)}{\Gamma(c)} I_1^{c-2,\ a-c} = \mathcal{L}(c,a),$$

which has to be understood as an Erdélyi-Kober fractional derivative:

$$\mathcal{L}^{-1}(a,c) = \mathcal{L}(c,a) = \frac{\Gamma(a)}{\Gamma(c)} D_1^{a-2,c-a} \quad (c-a \geq 0).$$

The property from [52]:

$$\mathcal{L}(b,c)\mathcal{L}(a,b) = \mathcal{L}(a,c)$$

follows readily from the product rule in (5.5.18); in fact we have

$$\mathcal{L}(b,c)\mathcal{L}(a,b) = \frac{\Gamma(c)}{\Gamma(b)} I_1^{b-2,c-b} \frac{\Gamma(b)}{\Gamma(a)} I_1^{a-2,b-a} = \frac{\Gamma(c)}{\Gamma(a)} I_1^{a-2,c-a} = \mathcal{L}(a,c).$$

A composition of operators (5.5.40) was considered by Owa and Srivastava [352]. Note that, according to Theorems 1.2.10, 1.2.18, we can represent such a composition as a multiple fractional integral involving G-function in the kernel; for example:

$$\begin{aligned} \mathcal{L} &= \mathcal{L}(a_1,b_1)\ldots\mathcal{L}(a_q,b_q)\,\mathcal{L}(a_{q+1},1) \\ &= \prod_{k=1}^{q+1} \frac{\Gamma(b_k)}{\Gamma(a_k)} I_{1,q+1}^{(a_k-2),(b_k-a_k)}. \end{aligned} \tag{5.5.42}$$

Then, the following result related to operator (5.5.42) (see Owa and Srivastava [352, p. 1067, (5.6)]) can be found also by the technique of the $(q+1)$-tuple fractional integrals of Erdélyi-Kober (cf. Kiryakova [200], [209]):

$$z\ {}_{q+1}F_q(a_1,\ldots,a_{q+1};\ b_1,\ldots,b_q;\ z) = \mathcal{L}\{z\ {}_2F_1(1,1;\ 1;z)\}. \tag{5.5.43}$$

For other applications of the operator $\mathcal{L}(a,c)$, see Owa and Srivastava [352].

Let us mention now two interesting special cases of the multiple Erdélyi-Kober operators $I_{1,m}^{(\gamma_k),(\delta_k)}$ with $m=2$ which have been recently used in univalent function theory.

Example 8 (Saigo [415], [420]). The hypergeometric fractional integral operator $I^{\alpha,\beta,\eta}$ defined by

$$I^{\alpha,\beta,\eta} f(z) = z^{-(\alpha+\beta)} \int_0^z \frac{(z-\zeta)^{\alpha-1}}{\Gamma(\alpha)}\ {}_2F_1\left(\alpha+\beta,-\eta;\alpha;1-\frac{\zeta}{z}\right) f(\zeta)d\zeta \tag{5.5.44}$$

is, according to the proof of Theorem 1.2.10 with $m = 2$, a double Erdélyi-Kober fractional integral:

$$I^{\alpha,\beta,\eta} = z^{-\beta} I_{1,2}^{(\eta-\beta,0),\ (-\eta,\alpha+\eta)} = z^{-\beta} I_1^{\eta-\beta,\ -\eta} I_1^{0,\alpha+\eta}. \tag{5.5.45}$$

Therefore, for $\alpha \geq -\eta \geq 0$ and $\beta < \eta + 2$ (see condition (5.5.28′)), the normalized operator

$$\tilde{I}^{\alpha,\beta,\eta} = cz^{\beta} I^{\alpha,\beta,\eta} \quad \left(c := \frac{\Gamma(2-\beta)\Gamma(2+\eta+\alpha)}{\Gamma(2+\eta-\beta)}\right)$$

preserves the class $\mathfrak{A}$.

Example 9 (Hohlov [135], [136]). Another hypergeometric fractional integral is defined in $\mathfrak{A}$ by means of the Hadamard product, namely:

$$\mathcal{F}(a,b,c)f(z) = \{z\ {}_2F_1(a,b;\ c;\ z)\} \circ f(z). \tag{5.5.46}$$

Hohlov [135], [136] uses them in solving some interesting problems concerning the univalence of functions of the class $\mathfrak{A}$.

Operators (5.5.46) also have a representation as double Erdélyi-Kober fractional integrals:

$$\begin{aligned} \mathcal{F}(a,b,c)f(z) &= \frac{\Gamma(c)}{\Gamma(a)\Gamma(b)} \int_0^1 \frac{(1-\sigma)^{c-a-b}}{\Gamma(c-a-b+1)} \sigma^{b-c} \\ &\quad \times {}_2F_1(c-a, 1-a;\ c-a-b+1;\ 1-\sigma) f(z\sigma)d\sigma \\ &= \frac{\Gamma(c)}{\Gamma(a)\Gamma(b)} I_1^{a-2,\ 1-a} I_1^{b-2,\ c-b} f(z) = \frac{\Gamma(c)}{\Gamma(a)\Gamma(b)} I_{1,2}^{(a-2,b-2),\ (1-a,c-b)}. \end{aligned} \tag{5.5.47}$$

Naturally, when $a = 1$ or $b = c$, Hohlov's operator (5.5.47) reduces to a single Erdélyi-Kober integral. For $a < 1$ and $b < c$, it is a purely integral operator; otherwise, $\mathcal{F}(a,b,c)$ has a differintegral representation. All the properties of this operator follow from Theorems 5.1.6, 5.5.2, 5.5.6 with $m = 2$. According to (5.1.32), the inverse operator has the form:

$$\begin{aligned} \mathcal{F}^{-1}(a,b,c) &= \frac{\Gamma(a)\Gamma(b)}{\Gamma(c)} I_{1,2}^{(-1,c-2),\ (a-1,b-c)} \\ &= \frac{\Gamma(a)\Gamma(b)}{\Gamma(c)} D_{1,2}^{(a-2,b-2),\ (1-a,c-b)}. \end{aligned} \tag{5.5.48}$$

Most of the operators listed until now (those of Biernacki, Libera, Ruscheweyh and others) are easily seen to be special cases of (5.5.46) and its inverse (5.5.48).

One of the problems considered by Hohlov [135], [136] was *to find the conditions for the parameters a, b and c under which the operator $\mathcal{F}(a,b,c)$ maps the class $\mathcal{S}$ of normalized univalent functions into itself and analogously, for the class of normalized convex functions $\mathcal{K}$.*

The variety of the results and the concluding remarks in [135]-[136], as well as the collection of other papers cited here, suggest the following problem for further discussion.

OPEN PROBLEM 5.5.7. Study the same problems (as those considered by Hohlov) for the generalized (multiple) Erdélyi-Kober operators of fractional differintegration and use them in the theory of univalent functions.

In fact, compositions of Erdélyi-Kober operators and integer powers of them arise frequently in analysis. Examples of these occurences are the so-called Gelfond-Leontiev generalized integrations and differentiations of power series, the hyper-Bessel differential operators of arbitrary order $m > 1$, some transmutation operators used in solving integral and differential equations and a series of other generalized integrations and differentiations, all of them having the general form proposed by Kalla [164]. Multiple Erdélyi-Kober operators of differintegration have been recently used sometimes, along the lines discussed in this section (e.g. equations (5.5.33) and (5.5.42)) but (as a rule) the repeated integral representations ($m > 2$) or the Hadamard products ($m = 2$; e.g. equation (5.5.46)) are considered. No other authors seem to have used the excellent properties of Meijer's G-function as a kernel of the corresponding single integral representation, especially in the above field (see e.g. Srivastava and Owa [475]-[476]). By this description of the G-function approach, we would like to stimulate researchers on analytic function theory to use it.

5.6. Generalized Laplace type integral transforms generated by multiple Erdélyi-Kober fractional integrals

There are many results relating Integral Transforms and Fractional Calculus (see e.g. Erdélyi [104], Widder [510], Fox [114], Zemanian [519], McBride [288], etc.)

Here we consider briefly *the two-fold connection between the generalized (multiple) fractional differintegrals and a class of integral transforms, generalizing the Laplace transform.* First, each fractional integral $x^{\lambda_0} I_{(\beta_k),m}^{(\gamma_k),(\lambda_k)}$, (5.1.9) generates such a transformation. On the other hand, these integral transformations have the property of algebraizing some corresponding fractional (or integer) integrals and derivatives.

The Laplace transform

$$\mathfrak{L}\{f(x); z\} = \int_0^\infty \exp(-zx) f(x) dx, \quad \Re z > \mu \tag{5.6.1}$$

can be considered for functions from $C_{-1} = \{f(x) = x^p \tilde{f}(x),\ p > -1;\ \tilde{f} \in C[0,\infty)\}$ which have exponential growth as $x \to \infty$, i.e. in the subspace

$$C_{-1}^{\exp} = \{f(x) \in C_{-1};\ f(x) = \mathcal{O}(\exp \mu x), x \to \infty;\ \mu \in \mathbb{R}\}. \tag{5.6.2}$$

More general integral transforms involving Meijer's G-functions, Fox's H-functions and other special functions have been considered recently by various authors, e.g. Fox

[113], Rooney [403], Srivastava [458]-[459], Srivastava and Buschman [465], Pathak [358]-[361], Tuan [495]-[496], Tuan, Marichev and Yakubovich [497], Yakubovich and Nguen [517], Nguen and Yakubovich [317], Luchko and Yakubovich [270]-[271], McBride and Spratt [292]-[294], Berkel [29], Kiryakova [196], [201]; and for generalized functions, respectively by Lighthill [257], Zemanian [519], McBride [288], Lamb [244]-[245], Pathak and Pandey [362], etc.

According to recent developments (see Marichev [276], Brychkov, Glaeske and Marichev [44], Tuan [496], Marichev and Tuan [284]), there are *mainly three kinds of generalized integral transforms*: of the type of the Fourier transform (sin-, cos-Fourier, Hankel and symmetrical Fourier transforms); of the type of the fractional integrals (1.1.6), (5.1.8) and of the type of the Laplace transform (like the Meijer, Borel-Džrbashjanand Obrechkoff transforms).

Definition 5.6.1. The G- and H-transforms of the form

$$\mathfrak{G}f(x) = \int_0^\infty G_{p,q}^{m,n}\left[zx \,\middle|\, \begin{matrix}(a_j)_1^p \\ (b_k)_1^q\end{matrix}\right] f(x)dx, \tag{5.6.3}$$

respectively,

$$\mathfrak{H}f(x) = \int_0^\infty H_{p,q}^{m,n}\left[zx \,\middle|\, \begin{matrix}(a_j, A_j)_1^p \\ (b_k, B_k)_1^q\end{matrix}\right] f(x)dx, \tag{5.6.4}$$

are said to be *generalized integral transforms of Laplace type* when

$$\delta = m + n - \frac{p+q}{2} > 0 \tag{5.6.5}$$

for (5.6.3), respectively

$$\Omega = \sum_{j=1}^{n} A_j - \sum_{j=1}^{p} A_j + \sum_{k=1}^{m} B_k - \sum_{k=m+1}^{q} B_k > 0 \tag{5.6.6}$$

for (5.6.4) (cf. (A.10), (E.4)).

Definition 5.6.2. A generalized (multiple) fractional integral (5.1.9):

$$\begin{aligned} Tf(x) &= x^{\lambda_0} I_{(\beta_k),m}^{(\gamma_k),(\lambda_k)} f(x) \\ &= x^{\lambda_0} \int_0^1 H_{m,m}^{m,0}\left[\sigma \,\middle|\, \begin{matrix}\left(\gamma_k + \lambda_k + 1 - \frac{1}{\beta_k}, \frac{1}{\beta_k}\right)_1^m \\ \left(\gamma_k + 1 - \frac{1}{\beta_k}, \frac{1}{\beta_k}\right)_1^m\end{matrix}\right] f(x\sigma)d\sigma, \quad \lambda_0 \geq 0, \end{aligned} \tag{5.6.7}$$

considered as a *transmutation operator from the Laplace transform to a generalized Laplace type transform*, is called a *generator of the latter integral transform*.

Definition 5.6.3. A composition of the Laplace transform and a generalized fractional integral (5.6.7) as a transmutation operator T:

$$\mathfrak{H}\{f(x); z\} := \mathfrak{L}\{Tf(x); z\} \tag{5.6.8}$$

is said to be a *generalized integral transform of Laplace type, generated by the transmutation operator (generator) T.*

The notions of Definitions 5.6.2-5.6.3 have been used, for example, in Kiryakova [201] and deeper investigations on this topic have been made recently by Luchko and Yakubovich [270]-[271], Yakubovich and Luchko [515]-[516], Yakubovich and Nguen [517], etc.

Here we propose only some basic ideas along the lines of our previous results on the integral transforms of Borel-Džrbashjan, Meijer and Obrechkoff (see Chapters 2, 3) and further possible generalizations of the Laplace transform.

Theorem 5.6.4. *For functions of the space*

$$C_\alpha^{\mathfrak{H}} = \Bigg\{ f(x) \in C_\alpha, \alpha = \max_k \left[-\beta_k\left(\gamma_k+1\right)\right]; \\ f(x) = \mathcal{O}\left\{ H^{1,m}_{m,m+1}\left[-\mu x \left| \begin{matrix} \left(-\gamma_k-\lambda_k, \frac{1}{\beta_k}\right)_1^m \\ (\alpha+1,1),\ \left(-\gamma_k, \frac{1}{\beta_k}\right)_1^m \end{matrix} \right. \right] \right\} \quad as\ x\to\infty,\ \mu\in\mathbb{R} \Bigg\}, \tag{5.6.9}$$

the generalized integral transform of Laplace type (5.6.8) generated by the transmutation operator (5.6.7) with $\lambda_0 = -\alpha - 1$ has the following explicit representation as an H-transform:

$$\mathfrak{H}\{f(x);\ z\} = \mathfrak{L}\{Tf(x); z\} \\ = z^{\alpha+1} \int_0^\infty H^{m+1,0}_{m,m+1}\left[zx \left| \begin{matrix} \left(\gamma_k+\lambda_k+1-\frac{1}{\beta_k}, \frac{1}{\beta_k}\right)_1^m \\ (-\alpha-1,1),\ \left(\gamma_k+1-\frac{1}{\beta_k}, \frac{1}{\beta_k}\right)_1^m \end{matrix} \right. \right] f(x)dx,\ \Re z > \mu. \tag{5.6.10}$$

The transform (5.6.8), (5.6.10) has a convolution in $C_\alpha^{\mathfrak{H}}$ of the form

$$\left(f \overset{\mathfrak{H}}{*} g\right)(x) = T^{-1}\left\{(Tf * Tg)(x)\right\}, \tag{5.6.11}$$

where $(*)$ is the Duhamel (Laplace) convolution

$$(f*g)(x) = \int_0^x f(x-t)g(t)dt, \tag{5.6.12}$$

namely, for $f,g \in C_\alpha^{\mathfrak{H}}$:

$$\mathfrak{H}\left\{\left(f \overset{\mathfrak{H}}{*} g\right)(x);\ z\right\} = \mathfrak{H}\{f(x);z\}.\mathfrak{H}\{g(x);z\}. \tag{5.6.13}$$

Proof. Formal calculations based on representation (5.6.7) of T and integral formula (E.21′) with

$$x^{\lambda_0}\exp(-zx) = G_{0,1}^{1,0}\left[zx|\lambda_0\right] = H_{0,1}^{1,0}\left[zx\,|(\lambda_0,1)\right],$$

lead to representation (5.6.10), namely:

$$\mathfrak{H}\{f(x);z\} = \mathfrak{L}\{Tf(x);z\} = \int_0^\infty \exp(-zx)Tf(x)dx$$

$$= \int_0^\infty x^{\lambda_0}\exp(-zx)dx \int_0^1 H_{m,m}^{m,0}\left[\sigma\left|\begin{matrix}\left(\gamma_k+\lambda_k+1-\frac{1}{\beta_k},\frac{1}{\beta_k}\right)\\ \left(\gamma_k+1-\frac{1}{\beta_k},\frac{1}{\beta_k}\right)\end{matrix}\right.\right] f(x\sigma)d\sigma$$

$$= \cdots = \int_0^\infty f(t)dt \int_t^\infty x^{\lambda_0-1}\exp(-zx)H_{m,m}^{m,0}\left[\frac{t}{x}\right]dx$$

$$= z^{-\lambda_0}\int_0^\infty H_{m,m+1}^{m+1,0}\left[zt\left|\begin{matrix}\left(\gamma_k+\lambda_k+1-\frac{1}{\beta_k},\frac{1}{\beta_k}\right)_1^m\\ (\lambda_0,1),\ \left(\gamma_k+1-\frac{1}{\beta_k},\frac{1}{\beta_k}\right)_1^m\end{matrix}\right.\right] f(t)dt,$$

where $\lambda_0 = -\alpha - 1$.

On the other hand operator T, (5.6.7), maps C_α into $C_{\alpha+\lambda_0}$ for the choice $\alpha = \max\limits_k\left[-\beta_k(\gamma_k+1)\right]$ (Theorem 5.1.4). For $\lambda_0 = -\alpha - 1$ this means $T: C_\alpha \to C_{-1}$. It remains to establish additionally that the subspace $C_\alpha^{\mathfrak{H}} \subset C_\alpha$ is suitably chosen so that T transforms the behaviour of $f \in C_\alpha^{\mathfrak{H}}$ as $x \to \infty$ into the condition $Tf \in C_{-1}^{\exp}$. Actually, by using formula (E.21′) again (and (E.12)-(E.13), (E.8)), we see that:

$$T\left\{H_{m,m+1}^{1,m}\left[-\mu x\left|\begin{matrix}\left(-\gamma_k-\lambda_k,\frac{1}{\beta_k}\right)_1^m\\ (-\lambda_0,1),\ (-\gamma_k)_1^m\end{matrix}\right.\right]\right\} = x^{\lambda_0}\int_0^1 H_{m,m}^{m,0}(\sigma)H_{m,m+1}^{1,m}(-\mu x\sigma)d\sigma$$

$$= x^{\lambda_0}H_{2m,2m+1}^{1,2m}\left[-\mu x\left|\begin{matrix}\left(-\gamma_k-\lambda_k,\frac{1}{\beta_k}\right)_1^m,\left(-\gamma_k,\frac{1}{\beta_k}\right)_1^m\\ (-\lambda_0,1),\ \left(-\gamma_k-\lambda_k,\frac{1}{\beta_k}\right)_1^m,\left(-\gamma_k,\frac{1}{\beta_k}\right)_1^m\end{matrix}\right.\right] \tag{5.6.14}$$

$$= x^{\lambda_0}H_{0,1}^{1,0}\left[-\mu x\,|(-\lambda_0,1)\right] = x^{\lambda_0}\exp(\mu x)x^{-\lambda_0} = \exp(\mu x).$$

Since generalized fractional integrals "preserve", in general, the asymptotics (see, for example, the proof of Theorem 3.9.15 for $x \to +0$; but Theorem 3.9.16 deals with $x \to +\infty$; details can be seen also in the article [28]), then (5.6.14) means that the new transformation (5.6.10) is well defined in $C_\alpha^{\mathfrak{H}}$ and for $\Re z > \mu$.

Using the property of the Duhamel convolution (5.6.12): $\mathfrak{L}\{f*g;z\} = \mathfrak{L}\{f\}.\mathfrak{L}\{g\}$ and representation $\mathfrak{H}\{f;z\} = \mathfrak{L}\{Tf;z\}$, we find:

$$\mathfrak{H}\left\{h \overset{\mathfrak{H}}{*} g; z\right\} = \mathfrak{H}\left\{T^{-1}(Tf*Tg);\ z\right\} = \mathfrak{L}\left\{TT^{-1}(Tf*Tg);\ z\right\}$$
$$= \mathfrak{L}\{Tf;\ z\}.\mathfrak{L}\{Tg;\ z\} = \mathfrak{H}\{f;\ z\}.\mathfrak{H}\{g;\ z\}$$

i.e. (5.6.13).

Note. By replacing operators T and T^{-1} in (5.6.11) by their (differ)integral representations via H-functions and simplifying suitably in special cases, one can find also explicit representations for the convolutions $\left(\overset{\mathfrak{H}}{*}\right)$. For instance, it is suitable to choose $Tf(x) = x^{\mu} I^{(\gamma_k),(\delta_k)}_{\left(\frac{\mu}{\delta_k}\right),m}$, $\mu = -\alpha - 1$, as in Section 5.4.

In particular, let us consider a simpler case of a transmutation operator T involving a $G^{m,0}_{m,m}$-function and the corresponding Laplace type G-transform, generated by it. As established in Kiryakova [201], we can take the Sonine-Dimovski type transformation

$$Tf := \varphi f\left(x^{\frac{\beta}{m}}\right) = x^{\beta(\gamma_m+1)-\frac{\beta}{m}} I^{(\gamma_k),(\lambda_k)}_{\beta,m-1} f(x) \tag{5.6.15}$$

in C_α, $\alpha = \max\limits_{1\le k\le m} [-\beta(\gamma_k+1)]$, and consider the corresponding new transform

$$\mathfrak{G}\{f(x);\ z\} = \mathfrak{L}\{\varphi f(x);\ z\}$$
$$= \sqrt{m(2\pi)^{1-m}} z^{-\beta\gamma_m} \int\limits_0^\infty G^{2m-1,0}_{m-1,2m-1}\left[\left(\frac{z}{m}\right)^m x^\beta \,\middle|\, \begin{matrix} (\gamma_k+\lambda_k)_1^{m-1} \\ (\gamma_k)_1^{m-1},\ \left(\gamma_k+\frac{k}{m}\right)_0^{m-1} \end{matrix}\right] f(x)dx, \tag{5.6.16}$$

for $f \in C^{\exp}_{\alpha,\frac{\beta}{m}}$ and $\Re z > \mu$, where

$$C^{\exp}_{\alpha,\frac{\beta}{m}} = \left\{f \in C_\alpha;\ f(x) = \mathcal{O}\left(\exp \mu x^{\frac{\beta}{m}}\right),\ x\to\infty;\ \mu\in\mathbb{R}\right\}. \tag{5.6.17}$$

Being a special case of the H-transform (5.6.10), the G-transform (5.6.16) is easily seen to generalize some other integral transforms, considered in previous Chapters 2, 3.

Example 1. Let $m = 1$, then transmutation φ in (5.6.15) reduces to the simple transformation $\varphi f(x) = x^\gamma f\left(x^{\frac{1}{\beta}}\right)$ and the generalized Laplace type transformation (5.6.16) takes the form:

$$\mathfrak{L}\{\varphi f(x);\ z\} = \beta\int\limits_0^\infty \exp\left(-zx^\beta\right) x^{\beta(\gamma+1)-1} f(x)dx. \tag{5.6.18}$$

This is a *modification of the Borel-Džrbashjan transform* (2.4.4):

$$\mathfrak{B}_{\rho,\mu}\{f(x);\ z\} = \rho z^{\mu\rho-1} \int_0^\infty \exp\left(-z^\rho t^\rho\right) t^{\mu\rho-1} f(t)dt. \tag{5.6.19}$$

As shown in Section 2.4, this transform turns the Džrbashjan-Gelfond-Leontiev operators of differentiation and integration into algebraic operations (Theorems 2.4.2, 2.4.3). Its convolution is (2.4.11).

EXAMPLE 2. Let $m > 1$ and let us choose $\lambda_k = \gamma_m - \gamma_k + \frac{k}{m}$, $k = 1, \dots, m-1$ in *the transmutation* $T = \varphi$, (5.6.15). Then, it *reduces to the Sonine-Dimovski transformation* (3.5.52′) related to the hyper-Bessel operators

$$L = \frac{x^\beta}{\beta^m} I_{\beta,m}^{(\gamma_k),(1)}; \quad B = x^{-\beta} \prod_{k=1}^m \left(x\frac{d}{dx} + \beta\gamma_k\right),$$

so that

$$\varphi L = \left(\frac{m}{\beta}\right)^m l^m \varphi \text{ in } C_\alpha,\ \alpha = \max_k \left[-\beta\left(\gamma_k + 1\right)\right].$$

The Laplace type G-transform (5.6.16) is then :

$$\begin{aligned}
\mathfrak{L}\{\varphi f(x);\ z\} &= \sqrt{m(2\pi)^{1-m}}\, z^{-m\gamma_m} \int_0^\infty G_{m-1,2m-1}^{2m-1,0}\left[\left(\frac{z}{m}\right)^m x^\beta \,\middle|\, \begin{matrix} \left(\gamma_k + \frac{k}{m}\right)_1^{m-1} \\ (\gamma_k)_1^m,\ \left(\gamma_k + \frac{k}{m}\right)_1^{m-1} \end{matrix}\right] \\
&\quad \times f(x)dx \\
&= \sqrt{m(2\pi)^{1-m}}\, z^{-m\gamma_m} \int_0^\infty G_{0,m}^{m,0}\left[\left(\frac{z}{m}\right)^m x^\beta \,\middle|\, (\gamma_k)_1^m\right] f(x)dx \\
&= \sqrt{m(2\pi)^{1-m}}\, \beta Z^{-\beta(\gamma_m+1)-1} \\
&\quad \times \int_0^\infty G_{0,m}^{m,0}\left[\left(Zx^\beta\right) \,\middle|\, \left(\gamma_k - \frac{1}{\beta} + 1\right)_1^m\right] f(x)dx,
\end{aligned} \tag{5.6.20}$$

i.e. up to a constant multiplier, this is the *Obrechkoff integral transform* (3.9.23), (3.9.34). Formula (5.6.11) now gives its convolution (3.9.33): $\varphi^{-1}\{(\varphi f) * (\varphi g)\}$ and Theorem 3.9.6 illustrates its properties of algebraizing the hyper-Bessel differential and integral operators.

EXAMPLE 3. In particular, for $m = \beta = 2$, $\gamma_{1,2} = \pm\frac{\nu}{2}$, then (5.6.16), respectively (5.6.20), turns into *the Meijer transformation* (3.10.a), due to relation (3.10.a′).

Needless to say, *if all the* λ_k*'s* in transmutations T (5.6.7), (5.6.15) *are taken to be zero*, then T are the identity operators and the H-, G-transformations of Laplace type (5.6.10), (5.6.16) are nothing but the Laplace transform.

Note also, that Weyl type generalized fractional integrals of the form $T^* = x^{\lambda_0} W_{(\beta_k),m}^{(\gamma_k),(\lambda_k)}$ can also be taken as transmutations, generating new integral transforms with G- and H-functions in the kernels. The most commonly used transmutations are *the two-dimensional Riemann-Liouville and Erdélyi-Kober fractional integrals*, see: Raina [384]-[388], Raina and Kalla [389], Raina and Kiryakova [390], Raina and Koul [391]-[393]; Saxena and Kiryakova [438], Saxena, Kiryakova and Davie [439], Saxena and Ram [441]-[442], etc.

5.7. Miscellaneous

We can hardly mention all the possible applications of fractional calculus and thus, the possible ones of the present generalized fractional calculus. We refer to the handbook of Samko, Kilbas and Marichev [434]. Nevertheless, let us describe a few topics.

Summation of series

There is a series of papers showing how some interesting results for series summation and the ψ-function can be established by means of fractional calculus. Their idea was initiated by Ross [408]. His paper discussed the general problem for *"serendipity in mathematics"* (i.e. the course of a mathematical discovery), by means of an excellent example of how fractional operators can simplify the solutions of complicated functional equations. In particular, Ross' result [408] showed *"how one is led to discover that*

$$\sum_{n=1}^{\infty} \frac{1.3.5.\ldots.(2n-1)}{n2^n n!} = \frac{1}{2} + \frac{3}{16} + \frac{15}{144} + \cdots = \ln r\text{''}. \tag{5.7.1}$$

This idea was extended by Ross and Kalla [409], Kalla and Al-Saqabi [166], [168], etc. In the famous Cambridge Tract entitled *"The General Theory of Dirichlet's Series"*, Hardy and Riesz showed implicity that the concept of the Riemann-Liouville fractional integral was a very useful expedient for the summation of the series of so-called typical $(R_{\mu k})$-means. Later, Mikolas (1960-1975) succeeded in clarifying the role of fractional integrals for general summation theory. After numerous more special results, in [304] he discussed the so-called (M)-method for arbitrary function series and some of its applications.

Statistics and related topics

First of all, the special functions (in general, the G- and H-functions) used as kernel-functions for generalized fractional integrals (0.10), (0.11), (5.1.6) and generalized fractional transformations (5.6.3), (5.6.4) find also wide applications in statistical distribution theory, multivariate statistical analysis, generalized probability laws, queuing theory and related stochastic processes, spares provisioning, generalized birth and death processes, etc. One can see: Mathai and Saxena [286]-[287], Srivastava and Kashyap [468], etc. It is interesting to note that *Kabe* [153] *seems to be the first to use Meijer's* $G_{m,m}^{m,0}$*-function* (the kernel-function of our generalized fractional integrals (1.1.6), (1.1.9)) *as density functions of a random variable*:

$$f(U) = \prod_{j=1}^{m} \frac{\Gamma\left(\frac{1}{2}(M-j)\right)}{\Gamma\left(\frac{1}{2}(p-j)\right)} G_{m,m}^{m,0}\left[U \left| \begin{matrix} a_1, \ldots, a_m \\ b_1, \ldots, b_m \end{matrix} \right.\right], \tag{5.7.2}$$

the special cases $m = 1$, $m = 2$ giving the so-called *"beta- and Gauss-distributions"*.

On the other hand, *fractional calculus finds its own use* in the above topics. Examples of applications of fractional differintegral operators and prospects for future use of their generalizations can be found, for example, in Laurent (in [404, p. 256-266]), Saigo and Raina [424], Ignatov and Kaishev [144]-[145], Kaishev [154], etc.

Radiation integrals

Applications of the Gauss hypergeometric function (the kernel-function of our 2-tuple Erdélyi-Kober fractional integrals (1.1.18), (1.1.19)) and some fractional integrals of special functions are shown to be useful in the environmental sciences and protection from accidental radiations. For example, the radiation field arising from a plane isotropic rectangular source can be presented by the integral (called the *Hubbell integral*):

$$f(a,b) = \int_0^b \operatorname{arctg}\left(\frac{a}{\sqrt{1+x^2}}\right)\frac{dx}{\sqrt{1+x^2}}, \quad 0 < a \le b < \infty, \tag{5.7.3}$$

investigated in detail in a series of fundamental papers by Hubbell ([142]-[143], etc.) for its various applications. Several generalizations of (5.7.3) have been given by authors like Bach and Lamkin, Glasser, Andrews, Saigo and R. Srivastava, etc.

Recently, the following generalized radiation integral has been investigated:

$$I\left[\begin{matrix} a, b, p, \lambda, \mu \\ \alpha, \beta, \gamma \end{matrix}\right] = \frac{\sigma a}{4\pi}\int_0^b x^{\lambda}\left(x^2+p\right)^{-\alpha}\left(1-\frac{x^2}{b^2}\right)^{\mu} {}_2F_1\left(\alpha, \beta;\ \gamma; -\frac{a^2}{x^2+p}\right) dx \tag{5.7.4}$$

by Kalla, Al-Saqabi and Conde [169], Kalla, Conde and Hubbell [170], Kalla, Galue and Kiryakova [172], Galue and Kiryakova [118] etc, which is suitable for dealing also with radiation fields with specific configuration of source, barrier and detector. Series expansions and representations of (5.7.4) by means of Appell's double hypergeometric functions have been found in [169], [170], especially when $\mu = 0$, and these results have been generalized in [172], [118] by introducing the above integral with arbitrary $\mu > -1$. In the latter papers, radiation integrals (5.7.4) are considered also as Erdélyi-Kober operators (1.1.17) of the Gauss function.

Fractional powers of operators and Mellin multipliers

As shown by Dimovski [65], [69], fractional powers of some linear operators L mapping a linear space $\mathcal{X}$ into itself, can be defined by means of the convolutions of these operators (in the sense of Definition 2.1.1). In building an *operational calculus* for the operator L, its *powers* L^λ, $\lambda > 0$, *can be represented as convolutional products* with corresponding elements of the operational (quotient) field: $L^\lambda f = \{l_\lambda\} * f$, $l_\lambda, f \in \mathcal{X}$ as is done for the hyper-Bessel integral operators (see Definition 3.6.2 and Theorem 3.6.3).

To represent the fractional powers of the same hyper-Bessel operators, McBride [289], [291, p. 99-139] has used another approach, based on Mellin transform theory and extendable to fractional powers of the so-called *Mellin multiplier transforms*, see also Rooney [402]. This approach provides a rigorous framework for the development of a theory of the fractional (i.e. not necessarily integer) powers L^α, $\alpha > 0$, of some classes of operators L, based on the relation

$$\mathfrak{M}\{L^\alpha f;\ s - \alpha\gamma\} = \frac{h(s-\alpha\gamma)}{h(s)}\mathfrak{M}\{f; s\}, \quad \alpha > 0, \tag{5.7.5}$$

$g(s) := \frac{h(s-\alpha)}{h(s)}$ being the Mellin multiplier, $\gamma \in \mathbb{C}$. The above papers, as well as McBride [288], Lamb [243]-[245], Lamb and McBride [246], McBride and Spratt [292]-[294], etc. propose a unified approach to fractional calculus based on a systematic use of Mellin multipliers.

For other approach see also Komatsu [225]-[229].

Operational calculus and integral transforms

The relation of this topic with fractional calculus is so close that it is not discussed here. The modern operational calculus is related to the name of Mikusinski ([306]) but as shown later (e.g. Dimovski [68], [70]-[71], [73]) many operational calculi can be developed even for a single linear operator. More references can be seen, for example in the surveys by Ditkin and Prudnikov [88] (especially relations to fractional calculus in §24) and Brychkov, Prudnikov and Shishov [45]. Recent contributions are related also to some trends like convolutional calculi (Dimovski [73]), convolutional representations of commutants and multipliers (Bozhinov [38]), discrete operational calculi (e.g. Dimovski and Kiryakova [82]), etc.

Fractional finite differences and averaged moduli of smoothness

It is well known that for a n-times differentiable function, the following formula

$$f^{(n)}(x) = \lim_{h\to 0}\frac{\left(\Delta_h^n f\right)(x)}{h^n}, \quad n = 1, 2, \ldots \tag{5.7.6}$$

holds, where $\left(\Delta_h^n f\right)(x)$ denotes the finite difference of integer order n. This equality can be used for an alternative definition of the fractional derivatives of order $\alpha > 0$, via the *generalized differences of fractional order*

$$\left(\Delta_h^\alpha f\right)(x) = \sum_{k=0}^{\infty}(-1)^k\binom{\alpha}{k}f(x-kh), \quad \alpha > 0, \tag{5.7.7}$$

namely:

$$D^\alpha f(x) = \lim_{h \to +0} \frac{(\Delta_h^\alpha f)(x)}{h^\alpha}, \quad \alpha > 0, \tag{5.7.8}$$

the so-called *Grünwald-Letnikov fractional derivatives*, originated by Liouville (1835), Gründwald (1867), Letnikov (1868), Mikolas (1963), etc. Recently, this approach to the fractional calculus has again attracted the attention of mathematicians, due to Butzer and Westphal [47] and Butzer, Dychkoff, Gorlich and Stens [46]. Their approach, from the point of view of the contemporary theory of functions, relates several classical problems of fractional calculus with the problems of functional analysis and gives also a way of dealing with generalized functions. Differences of fractional order (5.7.7) have been used to introduce *moduli of continuity of fractional order*:

$$\omega_\alpha(f,h) = \omega_{\alpha,\mathcal{X}}(f,h) = \sup_{|t|<h} \|\Delta_t^\alpha f\|_{\mathcal{X}},$$

$\mathcal{X}$ being a Banach space; fractional averaged moduli of smoothness; spaces of functions of generalized smoothness; Hausdorff approximation; etc. Related results can be seen in Timan [487], Sendov [445]-[447], Sendov and Popov [448], Popov [364], Ivanov [149], Dryanov [93]-[95], Kaljabin [179]-[181], Kaljabin and Lizorkin [182], etc.

OPEN PROBLEM 5.7.1. Find an *appropriate discrete definition of the generalized (multiple) Erdélyi-Kober fractional differintegrals* (1.1.6), (1.5.19), analogous to (5.7.8) and based on some *"fractional multiorder differences"*, generalizing (5.7.7) and related to the G-functions.

Fractal Hausdorff dimensions and fractal geometry

Fractals are geometrical objects: points, curves, planes, cubes, etc. but unlike more familiar objects, they cannot be "measured" by conventional methods. The notion was introduced by Mandelbrot [273] who discovered that many irregular shapes in the Nature that are not just random but can result from simple formulas, based on certain shapes nesting inside others. His observation of the existence of a "Geometry of Nature" has developed into a new kind of geometry, *fractal geometry.* Whereas classical geometry is used to describe the form and structure of man-made objects the fractals are models of both physical and non physical structures such as clouds, coastlines, plants, snowflakes, etc. Recently, fractal theory has been developed intensively towards applications in geometry, analysis, graph theory, etc. See Barnsley [27], Kuyosying [242] and the series of papers of Lapidus [247]-[251].

In addition, the *question of a possible relation between fractals and fractional calculus* (not only semantically) remains foggy but needs to be clarified. A fractal set consisting of one-dimensional parts in a two-dimensional setting, is adopted to have a dimension between 1 and 2, i.e. a *fractal dimension.* It is based on the idea of the Hausdorff "measure" of a compact region in an *n-dimensional space, where n can be a fraction.* Then, we need a constructive interpretation of a fractional measure. By analogy, an integral of order $1\frac{1}{2}$ can be considered as a measure of a body's volume in a $1\frac{1}{2}$-dimensional space. Since the integer-dimensional spaces are Cartesian products of one-dimensional spaces (lines) with a continuum cardinality, to construct a $1\frac{1}{2}$-dimensional space as a product of one-dimensional and $\frac{1}{2}$-dimensional spaces, we need a solution to the following problem.

Open problem 5.7.2. Give a constructive description of a set whose cardinality is a part of the continuum, for example a half of it?

Another positive answer is suggested by some results in Takev [486]. Namely, one can consider the graph $G(f)$ of a continuous function as a fractal set with Hausdorff dimension $D(G(f))$. Suppose also that the fractional derivative $D^{\alpha}f$ exists but $D^{\alpha+\varepsilon}f$ does not exist for any $\varepsilon > 0$. Then,

$$\begin{aligned} &\omega_p(f;\ \delta) \leq C\delta^{\alpha}\left\|D^{\alpha}f\right\|_p \quad \left(\frac{1}{\alpha} > p > 1\right); \\ &D(G(f)) \leq 2-\alpha, \quad 0<\alpha<1, \end{aligned} \tag{5.7.9}$$

where $\omega_p(f;\ \delta)$ denotes the average modulus of continuity and a problem arises: If for Lipschitz functions of order α, can the "$\leq$" sign be replaced by "$=$" in the latter inequality?

In view of the above-mentioned, we would like to hope that *fractal geometry might give at last a positive answer to a long standing open question*, a challenge to fractional analysts. It was posed subsequently in [404], [291], [329], still without answer, as follows ([329, p. 283]): *"Is there a geometrical representation of a fractional derivative? If not, can one prove that a graphical representation of a fractional derivative does not exist?* (by B. Ross) .... *The consensus of experts ... is that there is, in general, NO geometrical interpretation of a derivative of fractional order ... It can be asked, however at least for a geometrical meaning or a physical phenomena that can be represented by means of equations involving a derivative of a particular order such as* $\frac{1}{2}$ ..."

Let us hope that the above conjecture will fail, so we pose the question again (and even generalize it).

Open problem 5.7.3. Find geometrical or physical meanings for the generalized Erdélyi-Kober fractional differintegrals, or at least for the classical fractional operators. The answer seems to be hidden, probably, in our *fractal world*.

— || —

The main results of Chapter 5 have been published in: Kiryakova [196], [199], [201], [203]-[205], [207], Kalla and Kiryakova [174]-[175], Galue, Kiryakova and Kalla [119], Kiryakova and Srivastava [214], Raina and Kiryakova [390], Saxena and Kiryakova [438], Saxena, Kiryakova and Davie [439].

Appendix: Definitions, examples and properties of the special functions used in this book

The notation, definitions and properties of the special functions used or mentioned in this book are scattered over a number of manuals, monographs and papers, published at different times, and in different languages and places. To make our presentation self-contained and convenient for the reader, we have tried to gather the main facts about the generalized hypergeometric functions having an essential use here. Most of these facts are well-known and can be found in the books by Erdélyi et al. [106], [108], Luke [272], Mathai and Saxena [286]-[287], Marichev [276], Srivastava and Kashyap [468], Srivastava, Gupta and Goyal [467], Prudnikov, Brychkov and Marichev [368]-[370] as well as in many other lecture notes, papers, reprints, etc. To avoid overloading our presentation, usually we do not indicate the corresponding sources. On the other hand, some of the results listed below are not so well known or have been found recently. In these cases we mention their origin. In any case, the reader interested in more detailed information on the generalized hypergeometric functions and various special cases is referred to the books and papers cited.

A. Definitions and basic properties of the generalized hypergeometric functions and Meijer's G-functions

In recent decades Meijer's G-function has found various applications in different areas close to applied mathematics like: mathematical physics (hydrodynamics, theory of elasticity, potential theory, etc), theoretical physics, mathematical statistics, queuing theory, optimization theory, sinusoidal signals, generalized birth and death processes, etc. Due to the interesting and important properties of the G-functions, it became possible to represent the solutions of many problems in these fields in terms of these functions. Stated in this way, the problems gain a much more general character, due to the great freedom of choice of the orders m, n, p, q and parameters of the G-functions, in comparison with the other special functions. Simultaneously, the calculations become simpler and moore unified. Evidence for the importance of the G-function is given by the fact that the basic elementary functions and most of the special functions of mathematical physics, including the generalized hypergeometric functions, follow as its particular cases. Therefore, each result concerning the G-function becomes a key leading to numerous particular results for the Bessel functions, confluent hypergeometric functions, classical orthogonal polynomials, etc. Here we give a brief sketch of the results used repeatedly in this book.

A.i. Definitions

Since Meijer's G-functions are natural extensions of the generalized hypergeometric functions ${}_pF_q$, we first draw our attention to them. Concerning the more general notion of a function of hypergeometric type, Marichev [276,p. 46] wrote: *"The author failed to find an accurate definition of this notion in the literature. Therefore, for convenience, further we shall base ourselves on the following definition including all the linear combinations of functions mentioned in §12, except for the functions $J^{\mu}_{\nu,\lambda}(x), E_\rho(x,\mu)$ for irrational μ and ρ."*

Definition A.1. (Marichev [276]) By a *function of hypergeometric type* of the variable z we mean each function $u(z)$ representable in a neighbourhood of the point $z = 0$ in the form of a linear combination of functions

$$E((a_k),(b_l),\alpha) z^{\alpha}_{p} F_q((a_k);(b_l);hz^v),$$

(where $E((a),(b),\alpha)$ is a function of the parameters (a), (b), α, $v > 0$, $h =$const) as well as each function that can be obtained continuously by a limiting process with respect to these parameters.

Examples

The functions $z^\alpha \exp hz$, $(1-z)^{-a}$, $\ln(1+z)$, $\ln z$, $\arcsin z$, $\arctan z$, $(1+\sqrt{1-z})^{-a}$ and linear combinations of them are functions of hypergeometric type. The fact that the function $u(z) = \ln(z)$ belongs to this class follows easily from the equality $\ln z = \lim\limits_{\alpha\to 0}(z^\alpha - 1)\alpha^{-1}$ and the second part of the definition. Naturally, all the ${}_pF_q$-functions and their particular cases are functions of hypergeometric type.

In general, functions of hypergeometric type are functions whose Mellin transforms have the form

$$\text{const}\,.\prod_{i,j,k,l}\frac{\Gamma(a_i+s)\Gamma(b_j-s)}{\Gamma(c_k+s)\Gamma(d_l-s)}$$

and therefore, these functions are *Mellin-Barnes type integrals* of the form

$$\frac{1}{2\pi i}\int_{\mathcal{L}}\prod_{i,j,k,l}\frac{\Gamma(a_i+s)\Gamma(b_j-s)}{\Gamma(c_k+s)\Gamma(d_l-s)}z^{-s}ds,$$

or linear combinations of such integrals (here $\mathcal{L}$ denotes a certain contour in the complex plane). In fact, the transition from the Mellin images to the originals $u(z)$ is performed by means of Slater's theorem based upon the theory of residues (see Slater [450]).

The generalized hypergeometric function ${}_pF_q$ was known in the time of Euler and its long history was described in a number of books and articles, e.g. in Slater [450].

Let p and q be arbitrary natural numbers such that $p \le q+1$ and $a_1,\ldots,a_p$; $b_1,\ldots,b_q$ be some complex parameters. In general, the variable z is supposed to be complex too.

Definition A.2. The *generalized hypergeometric series* ${}_pF_q(z)$ is defined by the sum of the power series

$$
\begin{aligned}
{}_pF_q((a_k);(b_l);z) &= {}_pF_q(a_1,\dots,a_p;b_1,\dots,b_q;z) \\
&= {}_pF_q\left[\begin{matrix} a_1,\dots,a_p \\ b_1,\dots,b_q \end{matrix}\middle| z\right] = \sum_{i=0}^{\infty} \frac{(a_1)_i\dots(a_p)_i}{(b_1)_i\dots(b_q)_i}\frac{z^i}{i!},
\end{aligned} \tag{A.1}
$$

in the domain of its convergence, namely: $|z|<\infty$ for $p\le q$ and $|z|<1$ for $p=q+1$, where $(a)_i$ stands for Pohhamer's symbol

$$
(a)_0 = 1, \quad (a)_i = \frac{\Gamma(a+i)}{\Gamma(a)}, \quad i = 1,2,\dots.
$$

By a *generalized hypergeometric function* ${}_pF_q(z)$ we mean the sum of the series (A.1) in these domains; or, in the case $p=q+1$, also its analytical continuation in $|z|\ge 1$.

The a_k's and b_k's are called *numerator* and *denominator parameters*, respectively. The ${}_pF_q$-function is symmetric in its numerator parameters, and likewise in its denominator parameters. If at least one of the parameters $b_1,\dots,b_q$ is zero or a negative integer, then (A.1) has no meaning at all, since the denominator of the general term vanishes for a sufficiently large index. That is why we suppose that

$$
b_l \ne -j, \quad l = 1,\dots,q, \quad j = 1,2,\dots \tag{A.2}
$$

Under this condition, if some of the numerator parameters are negative integers or zero, then the series terminates and turns into a *hypergeometric polynomial* of the following form (see for example Luke [272, I, p. 142] or Srivastava and Kashyap [468, p. 33]) with $n = 0,1,2,\dots$:

$$
\begin{aligned}
{}_{p+1}F_q\left[\begin{matrix} -n;\ \alpha_1,\dots,\alpha_p \\ \beta_1,\dots,\beta_q \end{matrix}\middle| z\right] &= \sum_{k=0}^{n} \frac{(-n)_k(\alpha_1)_k\dots(\alpha_p)_k}{(\beta_1)_k\dots(\beta_q)_k}\cdot\frac{z^k}{k!} \\
= \frac{(\alpha_1)_n\dots(\alpha_p)_n}{(\beta_1)_n\dots(\beta_q)_n}(-z)^n\ {}_{q+1}F_p & \left[\begin{matrix} -n, 1-\beta_1-n,\dots,1-\beta_q-n \\ 1-\alpha_1-n,\dots,1-\alpha_p-n \end{matrix}\middle| \frac{(-1)^{p+q}}{2}\right].
\end{aligned} \tag{A.3}
$$

If we suppose that none of the numerator parameters is zero or a negative integer (otherwise the question of convergence will not arise), and with the usual restriction (A.2), the ${}_pF_q$-series in (A.1):

$$
\begin{array}{ll}
\text{converges for } |z|<\infty, & \text{if } p\le q, \\
\text{converges for } |z|<1, & \text{if } p=q+1, \\
\text{diverges for all } z, z\ne 0, & \text{if } p>q+1.
\end{array}
$$

Furthermore, if we set

$$
\omega = \sum_{l=1}^{q} b_l - \sum_{k=1}^{p} a_k, \tag{A.4}
$$

it is known that the ${}_pF_q$-series with $p = q+1$ is :

absolutely convergent for $\vert z\vert = 1$,	if $\Re(\omega) > 0$,
conditionally convergent for $\vert z\vert = 1,\ z \neq 1$,	if $-1 < \Re(\omega) \leq 0$,
divergent for $\vert z\vert = 1$,	if $\Re(\omega) \leq -1$.

Therefore the generalized hypergeometric function ${}_pF_q$ is defined by the (absolutely) convergent series (A.1) whenever:

$$p \leq q \quad \text{and} \quad |z| < \infty, \qquad \text{or} \quad p = q+1 \quad \text{and} \quad |z| < 1.$$

As in the case of the Gauss ${}_2F_1$-function, the ${}_pF_q$-function with $p = q+1$ can be continued analytically to the domain $|\arg(1-z)| < \pi$, that is, to the plane cut along the real axis from $z = 1$ to $z = \infty$, by using (A.1) in conjunction with the Mellin-Barnes contour integral representation:

$$\begin{gathered}{}_pF_q\left((a_k),(b_l),z\right) = \frac{\Gamma(b_1)\ldots\Gamma(b_q)}{\Gamma(a_1)\ldots\Gamma(a_p)}\cdot\frac{1}{2\pi i}\int_{-i\infty}^{+i\infty}\frac{\Gamma(a_1+s)\ldots\Gamma(a_p+s)\Gamma(-s)}{\Gamma(b_1+s)\ldots\Gamma(b_q+s)}.(-z)^s\,ds, \\ a_k \neq 0,-1,-2,\ldots; \quad k = 1,\ldots,p; \quad |\arg(1-zi)| < \pi. \end{gathered} \tag{A.5}$$

Here the path of integration is the imaginary axis (in the complex s-plane) which can be distorted, if necessary, in order to separate the poles of $\Gamma(a_k+s)$, $k = 1,\ldots,p$, from those of $\Gamma(-s)$.

Another integral representation of the ${}_pF_q$-function for $p = q+1$ providing an analytic continuation to the domain $|\arg(1-z)| < \pi$, is obtained in Chapter 4, see (4.2.30)-(4.2.30′) and Corollary 4.2.12.

Since Meijer's G-function considered below is a generalization of the ${}_pF_q$-function , the basic properties and operational rules concerning the latter can be derived as easily seen special cases of those for the G-function. For the same reason we shall list some of the most familiar special cases of the ${}_pF_q$-functions in Section C.

Meijer's G-function arises in mathematical analysis in attempts to give meaning to the symbols ${}_pF_q$ in the case $p > q+1$. The first of Meijer's definitions ([295]) coincides, in essence, with the definition of the so-called E-function, introduced independently by McRobert in 1937-1938 (see [106, I, §5.2]). The basic idea consists in considering, instead of the senseless ${}_pF_q$ with $p \geq q+1$, the finite series of well defined generalized hypergeometric functions ${}_{q+1}F_{p-1}$ with $q+1 \leq (p-1)+1$, namely:

$$\begin{aligned}E(p;a_k:q;b_l:z) &= \sum_{k=1}^{p}\left[\prod_{l=1}^{p}{}^*\Gamma(a_l-a_k)\prod_{i=1}^{q}\Gamma^{-1}(b_i-a_k)\right] \\ &\times \Gamma(a_k)z^{a_k}\,{}_{q+1}F_{p-1}\left[\begin{matrix}a_k, a_k-b_1+1,\ldots,a_k-b_q+1 \\ a_k-a_1,\ldots,*,\ldots,a_k-a_p+1\end{matrix}\,\middle|\,(-1)^{p+q}z\right],\end{aligned} \tag{A.6}$$

where $|z| < 1$ if $p = q+1$. The asteriks in the product mean omission of the factor $\Gamma(a_k - a_k)$, while in F they mean omitting of the parameter $(a_k - a_k + 1)$.

Later on, Meijer [296]-[297] replaced this definition by one in terms of a Mellin-Barnes type integral. The latter allows a greater freedom of choice of the values of p and q and uses the powerful tools of complex analysis, especially the theory of contour integration and the residue theorem. More about the history of Meijer's G-functions can be seen in Braaksma [41].

Definition A.3. By a *Meijer G-function* we mean a Mellin-Barnes type integral of the form

$$G_{p,q}^{m,n}(z) = G_{p,q}^{m,n}\left[z \left| \begin{matrix} a_1, \ldots, a_p \\ b_1, \ldots, b_q \end{matrix} \right.\right] = G_{p,q}^{m,n}\left[z \left| \begin{matrix} (a_j)_1^p \\ (b_k)_1^q \end{matrix} \right.\right] = \frac{1}{2\pi i} \int_{\mathcal{L}} \mathcal{G}_{p,q}^{m,n}(s) z^s ds, \qquad (A.7)$$

where $\mathcal{L}$ is a suitably chosen contour, $z \neq 0$ is a complex variable, $z^s := \exp[s \ \ln|z| + i \ \arg z]$ with a single valued branch of $\arg z$ and the integrand is defined as follows

$$\mathcal{G}_{p,q}^{m,n}(s) = \frac{\prod_{k=1}^m \Gamma(b_k - s) \ \prod_{j=1}^n \Gamma(1 - a_j + s)}{\prod_{k=m+1}^q \Gamma(1 - b_k + s) \ \prod_{j=n+1}^p \Gamma(a_j - s)}. \qquad (A.8)$$

Here an empty product is interpreted as unity, the integers m, n, p and q (known *as orders of the G-function*, or as components of *the order* $(m, n; p, q)$) are such that $0 \leq m \leq q$, $0 \leq n \leq p$; the *parameters* a_j, $j = 1, \ldots, p$, and b_k, $k = 1, \ldots, q$, are complex numbers for which

$$a_j - b_k \neq 0, 1, 2, 3, \ldots; \ j = 1, \ldots, n, \ k = 1, \ldots, m \qquad (A.9)$$

is satisfied. The latter condition ensures that none of the poles of the functions $\Gamma(b_k - s)$, $k = 1, \ldots, m$, namely, $s_{k,l} = b_k + l$, $l = 0, 1, 2, \ldots$, coincide with a pole $s'_{j,l'} = a_j - l' - 1$, $l' = 0, 1, 2, \ldots$ of the other Γ-functions in the numerator: $\Gamma(1 - a_j + s)$, $j = 1, \ldots, n$ and that these groups of poles could be separated by a contour $\mathcal{L}$.

Denote

$$\delta = m + n - \frac{1}{2}(p + q), \quad \nu = \sum_{k=1}^q b_k - \sum_{j=1}^p a_j \ . \qquad (A.10)$$

Then, the paths of integration $\mathcal{L}$ in the complex plane $\mathbb{C}$ can be of the following three types:

i) $\mathcal{L} = \mathcal{L}_{-i\infty,+i\infty}$ chosen in a manner to go from $-i\infty$ to $+i\infty$ leaving to the right all the poles $s_{k,l} = b_k + l, l = 0, 1, 2, \ldots$ of the functions $\Gamma(b_k - s)$, $k = 1, \ldots, m$ and to the left the poles $s'_{j,l'} = a_j - l' - 1$, $l' = 0, 1, 2, \ldots$ of the functions $\Gamma(1 - a_j + s), j = 1, \ldots, n$. Integral (A.7) is convergent for $\delta > 0, |\arg z| < \pi\delta$. The possibility $|\arg z| = \pi\delta, \delta \geq 0$, is also considered. Then, the integral converges absolutely when $p = q$, if $\Re(\nu) < -1$; and when $p \neq q$, if $(q - p)\Re(s) > \Re(\nu) + 1 - \frac{1}{2}(q - p)$ as $Im(s) \to \pm\infty$.

ii) $\mathcal{L} = \mathcal{L}_{+\infty}$ is a loop beginning and ending at $+\infty$ and encirling once in the negative direction all the poles of $\Gamma(b_k - s), k = 1, \ldots, m$, but none of the poles of $\Gamma(1 - a_j + s), j = 1, \ldots, n$. Integral (A.7) converges if $q \geq 1$ and either $p < q$ or $p = q$ and $|z| < 1$.

iii) $\mathcal{L} = \mathcal{L}_{-\infty}$ is a loop beginning and ending at $-\infty$, encircling in the positive direction all the poles of $\Gamma(1' - a_j + s)$, $j = 1, \dots, n$, but none of the poles of $\Gamma(b_k - s)$, $k = 1, \dots, m$ The integral is convergent if $p \geq 1$ and either $p > q$ or $p = q$ and $|z| > 1$.

Here we give an illustration of what these contours can be like (cf. Marichev [276, p. 52]). It is supposed that the parameters a_j, b_k and the variable z are such that at least one of the definitions with contours i)-iii) has meaning. When more than one of these definitions make sense, they lead to the same results, so no confusion can arise.

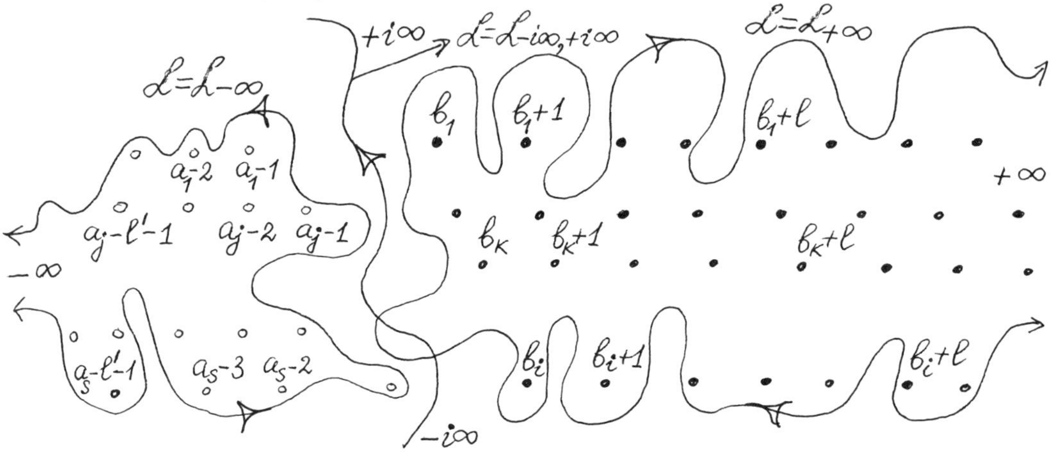

Meijer's G-function is an analytic (many-valued) function of z with a branch point at the origin. More exactly, it is representable by a function analytic in the sector $|argz| < \pi\delta$. For $\delta > 0, p = q$, the functions defined by ii) and iii) are analytical continuations of each other through the circle $|z| = 1$ in this sector. In the case $\delta \leq 0$ there is no sector of analycitity but if $p = q$ both above mentioned functions which are analytic in $|z| < 1$ and respectively in $|z| > 1$, have the same limit as $z \to 1$ along the ray $\{argz = 0\}$, if $\Re(\nu) < 0, \delta = 0$. Further details on the situation, depending on the values of the orders m, n, p, q and of the numbers δ, ν in (A.10) can be seen in the books by Marichev [276, p. 46-59], Prudnikov, Brychkov and Marichev [370, p. 618, 626], see also Marichev [280].

If we use, for example, the path of integration ii), then integral (A.7) can be evaluated as a sum of residues. So, if no two of the parameters $b_k, k = 1, ..., m$, differ by an integer or zero, all the poles of the integrand are simple and a representation in the form (A.6) is obtained, namely:

$$G_{p,q}^{m,n}\left[z \left| \begin{matrix} (a_j) \\ (b_k) \end{matrix} \right.\right] = \sum_{h=1}^{m} \frac{\prod\limits_{k=1}^{m} {}^{*}\Gamma(b_k - b_h) \prod\limits_{j=1}^{n} \Gamma(1 + b_h - a_j)}{\prod\limits_{k=m+1}^{q} \Gamma(1 + b_h - b_k) \prod\limits_{j=n+1}^{p} \Gamma(a_j - a_h)} . z^{b_h}$$
$$\times \; {}_pF_{q-1}\left[\begin{matrix} (1 + b_h - a_j) \\ (1 + b_h - b_k)^* \end{matrix} \left| (-1)^{p-m-n} z \right.\right] \text{ for } p < q \text{ or } p = q, |z| < 1. \tag{A.11}$$

If $m = 0$ and we use ii), the integrand is analytic on and within the contour and so,

$$G^{0,n}_{p,q}\left[z\left|\begin{matrix}(a_j)\\(b_k)\end{matrix}\right.\right] \equiv 0 \text{ for } p < q \text{ or } p = q, |z| < 1. \tag{A.12}$$

Analogously, if no two of the a_j, $j = 1, \dots, n$, differ by an integer or zero, and we use the contour iii), we obtain

$$G^{m,n}_{p,q}\left[z\left|\begin{matrix}(a_j)\\(b_k)\end{matrix}\right.\right] = sum^{m}_{h=1} \frac{\prod\limits_{j=1}^{n} {}^{*}\Gamma(a_h - a_j) \prod\limits_{k=1}^{m} \Gamma(b_k - a_h + 1)}{\prod\limits_{j=n+1}^{p} \Gamma(a_j - a_h + 1) \prod\limits_{k=m+1}^{q} \Gamma(a_h - b_k)} \tag{A.11'}$$

$$\times\, z^{a_h - 1}\, {}_qF_{p-1}\left[z\left|\begin{matrix}(1 + b_k - a_h)\\(1 + a_j - a_h)^*\end{matrix}\right.\right] \text{ for } p > q \text{ or } p = q, |z| > 1,$$

and for $n = 0$:

$$G^{m,0}_{p,q}\left[z\left|\begin{matrix}(a_j)\\(b_k)\end{matrix}\right.\right] \equiv 0 \text{ for } p > q \text{ or } p = q, |z| > 1. \tag{A.12'}$$

A.ii. Some basic properties of the G-functions

The G-function is *symmetric* in the groups of parameters $(a_1, \dots, a_n)$, $(a_{n+1}, \dots, a_p)$, $(b_1, \dots, b_m)$, $(b_{m+1}, \dots, b_q)$. If one of the a_j's, $j = 1, \dots, n$, is equal to some of the b_k's, $k = m + 1, \dots, q$, then the G-function reduces to one of Lower order. For example, if $n, p, q \geq 1$,

$$G^{m,n}_{p,q}\left[z\left|\begin{matrix}a_1, \dots, a_p\\b_1, \dots, b_{q-1}, a_1\end{matrix}\right.\right] = G^{m,n-1}_{p-1,q-1}\left[z\left|\begin{matrix}a_2, \dots, a_p\\b_1, \dots, b_{q-1}\end{matrix}\right.\right]. \tag{A.13}$$

Similary, if one of the a_j's, $j = n + 1, \dots, p$, is equal to some of the b_k's, $k = 1, \dots, m$, the function has its orders m, p, q reduced by 1, for example, if $m, p, q \geq 1$,

$$G^{m,n}_{p,q}\left[z\left|\begin{matrix}a_1, \dots, a_{p-1}, b_1\\b_1, b_2, \dots, b_q\end{matrix}\right.\right] = G^{m-1,n}_{p-1,q-1}\left[z\left|\begin{matrix}a_1, \dots, a_{p-1}\\b_2, \dots, b_q\end{matrix}\right.\right]. \tag{A.13'}$$

A change of variable in the integral (A.7) leads to the properties:

$$z^{\sigma} G^{m,n}_{p,q}\left[z\left|\begin{matrix}(a_j)\\(b_k)\end{matrix}\right.\right] = G^{m,n}_{p,q}\left[z\left|\begin{matrix}(a_j + \sigma)\\(b_k + \sigma)\end{matrix}\right.\right], \tag{A.14}$$

$$G^{m,n}_{p,q}\left[z\left|\begin{matrix}(a_j)_1^p\\(b_k)_1^q\end{matrix}\right.\right] = G^{n,m}_{q,p}\left[\frac{1}{z}\left|\begin{matrix}(1 - b_k)_1^q\\(1 - a_j)_1^p\end{matrix}\right.\right]. \tag{A.15}$$

The latter property allows us to transform a function with $p \geq q$ into a function with $p \leq q$. *So, without any loss of generality, we can consider only the case of G-functions with* $p \leq q$. We use this property also when passing from the domain $|z| < 1$ to domain $|z| > 1$ and vice versa.

If in the integrand of (A.7) we replace s by ks, k being a positive integer, and use the Gauss-Legendre multiplication formula for Γ-functions (see [106, I, §1.2], [272, I, p. 11], etc.), then the following *useful multiplication formula* is obtained:

$$G^{m,n}_{p,q}\left[z\left|\begin{matrix}(a_j)_1^p\\(b_k)_1^q\end{matrix}\right.\right]=(2\pi)^u k^v G^{mk,nk}_{pk,qk}\left[\frac{z^k}{k^{k(q-p)}}\left|\begin{matrix}c_{1k},\ldots,c_{pk}\\d_{1k},\ldots,d_{qk}\end{matrix}\right.\right], \tag{A.16}$$

where

$$u=(k-1)\left[\frac{1}{2}(p+q)-(m+n)\right],\quad v=\sum_{k=1}^{q}b_k-\sum_{j=1}^{p}a_j+\frac{1}{2}(p-q)+1$$

and $c_{h,k}, d_{i,k}$ stand respectively for the set of parameters

$$c_{h,k}=\Delta(k,a_h)=\left\{\frac{a_h}{k},\frac{a_h+1}{k},\ldots,\frac{a_h+k-1}{k}\right\},\quad h=1,\ldots,p,$$

$$d_{i,k}=\Delta(k,d_i)=\left\{\frac{b_i}{k},\frac{b_i+1}{k},\ldots,\frac{b_i+k-1}{k}\right\},\quad i=1,\ldots,q.$$

A.iii. Differential equations and asymptotics

Many differential relations are known for the Meijer's G-function. We list only a few of them that are in use in the book, for instance:

$$\frac{d}{dz}\left\{z^{-b_i}G^{m,n}_{p,q}\left[z\left|\begin{matrix}(a_j)\\(b_k)\end{matrix}\right.\right]\right\}=\varepsilon_i z^{-1-b_i}G^{m,n}_{p,q}\left[z\left|\begin{matrix}(a_j)_1^p\\b_1,\ldots,b_{i-1},b_i+1,b_{i+1},\ldots,b_q\end{matrix}\right.\right],$$

$$\text{with}\quad \varepsilon_i=\begin{cases}-1 & i=1,2,\ldots,m\\+1 & i=m+1,\ldots,q,\end{cases} \tag{A.17}$$

$$\frac{d}{dz}\left\{z^{1-a_h}G^{m,n}_{p,q}\left[z\left|\begin{matrix}(a_j)\\(b_k)\end{matrix}\right.\right]\right\}=\varepsilon'_h z^{-a_h}G^{m,n}_{p,q}\left[z\left|\begin{matrix}a_1,\ldots,a_{h-1},a_h+1,a_{h+1},\ldots,a_p\\(b_k)_1^q\end{matrix}\right.\right],$$

$$\text{with}\quad \varepsilon'_i=\begin{cases}+1 & h=1,2,\ldots,n\\-1 & h=n+1,\ldots,p,\end{cases} \tag{A.17$'$}$$

and

$$z^l\frac{d^l}{dz^l}\left\{G^{m,n}_{p,q}\left[z\left|\begin{matrix}(a_j)\\(b_k)\end{matrix}\right.\right]\right\}=G^{m,n+1}_{p+1,q+1}\left[z\left|\begin{matrix}0,(a_j)\\(b_k),0\end{matrix}\right.\right],\quad l=1,2,\ldots \tag{A.18}$$

See also [106, I, §5.3], [272, I, p.151-152], [286, p. 8-9], [370, p. 620-621].

Using these relations or directly definition (A.7), one can establish that *Meijer's G-function* $y(z)=G^{m,n}_{p,q}\left[z\left|\begin{matrix}(a_j)\\(b_k)\end{matrix}\right.\right]$ *satisfies the linear ordinary differential equation* of generalized hypergeometric type:

$$\left[(-1)^{p-m-n}z\prod_{j=1}^{p}\left(z\frac{d}{dz}-a_j+1\right)-\prod_{k=1}^{q}\left(z\frac{d}{dz}-b_k\right)\right]y(z)=0, \tag{A.19}$$

whose order is equal to $\max(p, q)$. Due to property (A.15), only the case $p \leq q$ need be considered only. For fixed p, q and parameters $a_1, \ldots, a_p; b_1, \ldots, b_q$ the same differential equation is satisfied also by all the $(p+1)(q+1)$ functions $G_{p,q}^{m,n}[(-1)^{m+n}z]$ with $0 \leq m \leq q, \quad 0 \leq n \leq p$.

Equation (A.19) has two (if $p \neq q$) or three (if $p = q$) singular points. If $p < q$, then $z = 0$ is a regular singular point and $z = \infty$ is an irregular one. For $p > q$ these two points change their roles. In the first case the fundamental system of q linearly independent solutions (f.s.s.) in the vicinity of $z = 0$ consists of the G-functions ($h = 1, 2, \ldots, q$):

$$G_{p,q}^{1,p}\left[z.e^{(p-m-n-1)\pi i}\,\middle|\,\begin{matrix}(a_j)\\ b_h, b_1, \ldots, b_{h_1}, b_{h+1}, \ldots, b_q\end{matrix}\right], \tag{A.20}$$

being respectively $\mathcal{O}\left(|z|^{b_h}\right)$ as $z \to 0$ (according to (A.21) below). Near the irregular singular point $z = \infty$ the variable z should be considered in a sector only. The corresponding f.s.s. is described by Meijer [297], cf. [106, I, §5.4], [279], [370, p. 621-622].

The case $p = q$ is more peculiar. Then equation (A.19) and *the* $G_{p,q}^{m,n}$*-function have three regular singular points*, namely, $z = 0, \quad z = (-1)^{p-m-n}$ and $z = \infty$. The behaviour of the G-function in the vicinity of the new singular point $(-1)^{p-m-n}$ is rather complicated and as one can read in [106, I, §5.4], [272, p. 181]: "No fundamental system for the neighbourhood of this point has been given in the literature". Fortunately, Marichev has recently solved this problem both for the ${}_pF_{p-1}$- and $G_{p,q}^{m,n}$-functions (see [279]-[280], [370, p. 621-622]).

The asymptotic behaviour of the G-function near the points $z = 0$, $z = \infty$ has been studied yet by Meijer [297]. It is well known that

$$G_{p,q}^{m,n}\left[z\,\middle|\,\begin{matrix}(a_j)\\(b_k)\end{matrix}\right] = \mathcal{O}\left(|z|^{\beta}\right), \tag{A.21}$$

where $p \leq q$ and $\beta = \min\limits_k \Re b_k$, $k = 1, \ldots, m$. Since $z = \infty$ is an irregular singular point for equation (A.19) (if $p \neq q$), the behaviour of $G(z)$ as $|z| \to \infty$ has a more complicated character: it can increase or decrease algebraically, or tend exponentially to zero or to infinity, depending of the values of m, n, p, q and $\arg z$ (see [106, I, §5.4], [272, I, p. 181-182], [370, §8.2], [286, §1.4], etc.). The behaviour of the G-function near the third singular point $(-1)^{p-m-n}$, when $p = q$, is found by Marichev [279]-[280]. For example, Meijer's $G_{m,m}^{m,0}$-function used as a kernel of our generalized fractional integrals $I_{\beta,m}^{(\gamma_k),(\delta_k)}$, has the following asymptotics:

$$\begin{aligned} &G_{m,m}^{m,0}\left[z\,\middle|\,\begin{matrix}(a_k)_a^m\\(b_k)_1^m\end{matrix}\right] \sim \frac{(1-z)^{\nu_m^*}}{\Gamma(\nu_m^*+1)} \quad \text{as} \quad z \to 1, \qquad |z| < 1, \\ &\text{where} \quad \nu_m^* = \sum_{k=1}^{m}(a_k - b_k) - 1 = -\nu - 1 \neq 0, \pm 1, \pm 2, \ldots \end{aligned} \tag{A.22}$$

In this case $m+n-p=0$, and according to Marichev [280], the $G_{m,m}^{m,0}$-function is not a function analytic simultaneously inside and outside the unit disk, since it is representable by means of different analytic formulas. Indeed, we have

$$G_{m,m}^{m,0}(z) \equiv 0 \qquad \text{for} \quad |z|>1,$$

according to (A.12′), while in $|z|<1$:$G_{m,m}^{m,0}(z)$ is continuous as $z \to 1$, provided $\Re \nu_m^* > 0$; it is bounded near $z=1$ if $\Re \nu_m^* = 0$ but $\nu_m^* \neq 0$; in the other cases ($\nu_m^* = 0, \Re \nu_m^* < 0$) it has, in general, a logarithmic singularity or a singularity of the power order $(-\nu_m^*)$.

A.iv. Integrals involving Meijer's G-functions

Here we list only the formulas useful in our considerations. For more extensive lists of various integrals involving Meijer's G-functions one can see, for example, [106, I, §5.5)], [272, I, §5.6], [286, Ch. III], [370, §2.24], etc.

Riemann-Liouville integrals of fractional order $\alpha, \Re\alpha > 0$

$$\begin{aligned} R^\alpha \left\{G_{p,q}^{m,n}(\eta z)\right\} &= \frac{1}{\Gamma(\alpha)} \int_0^z (z-\zeta)^{\alpha-1} G_{p,q}^{m,n}\left[\eta\zeta \,\middle|\, \begin{matrix}(a_j)\\(b_k)\end{matrix}\right] d\zeta \\ &= z^\alpha \int_0^1 \frac{(1-\sigma)^{\alpha-1}}{\Gamma(\alpha)} G_{p,q}^{m,n}\left[\eta z\sigma \,\middle|\, \begin{matrix}(a_j)\\(b_k)\end{matrix}\right] d\sigma = z^\alpha G_{p+1,q+1}^{m,n+1}\left[\eta z \,\middle|\, \begin{matrix}0,(a_j)\\(b_k),-\alpha\end{matrix}\right], \end{aligned} \tag{A.23}$$

provided $\Re b_k > -1, \quad k=1,\ldots,m$ and $p \le q$. If $p=q$, then the formula is valid for $|\eta z|<1$. For $p+q<2(m+n)$ we require that $|\arg(\eta z)| < \left(m+n-\frac{p+q}{2}\right)\tau$ (see [107, II, p. 200, (96)-(97)]).

Weyl integral of fractional order $\alpha, \Re\alpha > 0$

$$\begin{aligned} \mathcal{W}^\alpha \left\{G_{p,q}^{m,n}(\eta z)\right\} &= \frac{1}{\Gamma(\alpha)} \int_z^\infty (\zeta-z)^{\alpha-1} G_{p,q}^{m,n}\left[\eta\zeta \,\middle|\, \begin{matrix}(a_j)\\(b_k)\end{matrix}\right] d\zeta \\ &= z^\alpha \int_1^\infty \frac{(\sigma-1)^{\alpha-1}}{\Gamma(\alpha)} G_{p,q}^{m,n}\left[\eta z\sigma \,\middle|\, \begin{matrix}(a_j)\\(b_k)\end{matrix}\right] d\sigma = z^\alpha G_{p+1,q+1}^{m+1,n}\left[\eta z \,\middle|\, \begin{matrix}(a_j),0\\-\alpha,(b_k)\end{matrix}\right], \end{aligned} \tag{A.24}$$

provided $0 < \Re\alpha < 1 - \Re a_j, \quad j=1,\ldots,n$ and $p \ge q$. For $p=q$ the condition $|\eta z|>1$ is required additionally, and for $p+q<2(m+n)$ the corresponding condition is $|\arg(\eta z)| < \left(m+n-\frac{p+q}{2}\right)\pi$.

Mellin transform

One of the most useful integral transforms involving Meijer's G-function is the Mellin transform. This is due to the fact that it transforms the G-function into the coefficient

$\mathcal{G}^{m,n}_{p,q}(s)$ associated with the factor z^s in the integrand of (A.7). Thus,

$$\mathfrak{M}\left\{G^{m,n}_{p,q}(\eta t)\right\} = \int\limits_0^\infty t^{s-1} G^{m,n}_{p,q}\left[\eta t \left| \begin{matrix}(a_j)\\(b_k)\end{matrix}\right.\right] dt$$

$$= \eta^{-s}\mathcal{G}^{m,n}_{p,q}(-s) = \eta^{-s}\frac{\prod\limits_{k=1}^{m}\Gamma(b_k+s)\prod\limits_{j=1}^{n}\Gamma(1-a_j-s)}{\prod\limits_{k=m+1}^{q}\Gamma(1-b_k-s)\prod\limits_{j=n+1}^{p}\Gamma(a_j+s)} \tag{A.25}$$

if, for example, $1 \le n \le p < q$, $1 \le m \le q$, $\eta \ne 0$, $\quad \delta > 0$, $|\arg\eta| < \delta\pi$, $-\min\limits_{1\le k\le m}\Re b_k < \Re s < 1 - \max\limits_{1\le j\le n}\Re a_j$. For other conditions under which formula (A.25) holds true, see [272, I, p. 157-159], [286, p. 79, (3.2.1)], etc.

Let us consider a special case of the situation $p = q$, for instance, $m = p = q$, $\quad n = 0$, that is: $\delta = 0$. Then, the corresponding result has the form

$$\int\limits_0^\infty t^{s-1} G^{m,0}_{m,m}\left[\eta t \left| \begin{matrix}(a_k)\\(b_k)\end{matrix}\right.\right] dt = \prod_{k=1}^{m}\frac{\Gamma(b_k+s)}{\Gamma(a_k+s)}, \tag{A.26}$$

provided $\eta \ne 0, 2(\lambda-1)\pi < \arg\eta < 2\lambda\pi$ with some arbitrary λ (for $\lambda_1 = 1$ we obtain the condition $0 < \arg\eta < 2\pi$ and for $\lambda_2 = \frac{1}{2}$: $-\pi < \arg\eta < \pi$) and $-\min\limits_{1\le k\le m}\Re b_k < \Re s < 1 - \max\limits_{1\le j\le m}\Re a_k$ (see [272, I, p. 159, Case 7]).

Laplace transform

The image of the G-function under the Laplace transform can be obtained from the more general formula (cf. [106, I, §5.5, (8)], [286, p. 87, (3.3.12)], [272, p. 166, (1)]):

$$\int\limits_0^\infty t^{\sigma-1}e^{-st} G^{m,n}_{p,q}\left[\eta t \left| \begin{matrix}(a_j)\\(b_k)\end{matrix}\right.\right] dt = s^{-\sigma} G^{m,n+1}_{p+1,q}\left[\frac{\eta}{s} \left| \begin{matrix}1-\sigma,(a_j)\\(b_k)\end{matrix}\right.\right], \tag{A.27}$$

valid, for example, if $\Re\left(\sigma + \min\limits_{1\le k\le m} b_k\right) > 0$, $\Re s > 0$, $\delta > 0$, $|\arg\eta| < \delta\pi$. This integral is convergent also for $\delta = 0$, under some additional conditions (see e.g. [272, I, p. 166-169]).

Most of the other classical integral transforms and like the Hankel, Meijer, Laguerre and Stieltjes transforms, are defined by means of integrals whose kernels (e.g. Bessel functions of first and third kind) are quite special cases of Meijer's G-functions . So the images of the G-function under these transforms can be evaluated as integrals of products of two G-functions , according to a formula given in almost all of the mentioned books. Since we use this result repeatedly, we cite it below even with a sketch of its proof.

Integral of the product of two G-functions

The general formula has the form

$$\int_0^\infty z^{\gamma-1} G_{p,q}^{m,n}\left[\eta z \left| \begin{matrix}(a_i)\\(b_j)\end{matrix}\right.\right] G_{\sigma,\tau}^{\mu,\nu}\left[\omega z^{\frac{k}{l}} \left| \begin{matrix}(c_\alpha)\\(d_\beta)\end{matrix}\right.\right] dz = \eta^{-\gamma}(2\pi)^{\delta(1-k)+\rho(1-l)} k^{U+\gamma(q-p-1)} l^V$$
$$\times G_{l\sigma+kq,l\tau+kp}^{l\mu+kn,l\nu+km}\left[W \left| \begin{matrix}\Delta(l,c_1),\ldots,\Delta(l,c_\nu);\Delta(k,1-b_j-\gamma)|_1^q;\Delta(l,c_{\nu+1}),\ldots,\Delta(l,c_\sigma)\\ \Delta(l,d_1),\ldots,\Delta(l,d_\mu);\Delta(k,1-a_i-\gamma)|_1^p;\Delta(l,d_{\mu+1}),\ldots,\Delta(l,d_\tau)\end{matrix}\right.\right], \tag{A.28}$$

where $k, l > 0$ are integers, $\eta \neq 0$, $\omega \neq 0$, $\delta = m + n - \frac{p+q}{2}$, $\rho = \mu + \nu - \frac{\sigma+\tau}{2}$, $U = \sum_1^q b_j - \sum_1^p a_j + \frac{p-q}{2} + 1$, $V = \sum_1^\tau d_\beta - \sum_1^\sigma c_\alpha + \frac{\sigma-\tau}{2} + 1$, $W = \frac{\omega^l l^{l(\sigma-\tau)}}{\eta^k k^{k(p-q)}}$ and the symbol $\Delta(k,c)$ stands for the sequence of the k numbers: $\frac{c}{k}, \frac{c+1}{k}, \ldots, \frac{c+k-1}{k}$.

The main conditions under which (A.28) is valid can be found in [272, p. 159-164], [286, p. 80-82]. The first source contains also the manner of establishing (A.28) in some of the more specific cases. For an extensive list of conditions we recommend also the book [370, p. 346-347, 1)-35)]. Here we shall demonstrate the evaluation of the integral (A.28) in a special case to which the general one is easily reduced (by using properties (A.14), (A.16)):

$$\begin{aligned}\int_0^\infty & G_{p,q}^{m,n}\left[\eta z \left| \begin{matrix}(a_i)\\(b_j)\end{matrix}\right.\right] G_{\sigma,\tau}^{\mu,\nu}\left[\omega z \left| \begin{matrix}(c_\alpha)\\(d_\beta)\end{matrix}\right.\right] dz \\ &= \frac{1}{\eta} G_{q+\sigma,p+\tau}^{n+\mu,m+\nu}\left[\frac{\omega}{\eta} \left| \begin{matrix}-b_1,\ldots,-b_m;(c_\alpha)_1^\sigma;-b_{m+1},\ldots,-b_q\\ -a_1,\ldots,-a_n;(d_\beta)_1^\tau;-a_{n+1},\ldots,-a_p\end{matrix}\right.\right] \\ &= \frac{1}{\omega} G_{p+\tau,q+\sigma}^{m+\nu,n+\mu}\left[\frac{\eta}{\omega} \left| \begin{matrix}a_1,\ldots,a_n;(-d_\beta)_1^\tau;a_{n+1},\ldots,a_p\\ b_1,\ldots,b_m;(-c_\alpha)_1^\sigma;b_{m+1},\ldots,b_q\end{matrix}\right.\right],\end{aligned} \tag{A.29}$$

provided $1 \le n \le p \le q$, $1 \le m \le q$, $1 \le \nu \le \sigma \le \tau$, $1 \le \mu \le \tau$, $\eta \neq 0$, $\omega \neq 0$ and the following conditions are fulfilled also:

$$\begin{aligned}&\Re(b_j + d_\beta) > -1, \quad j = 1,\ldots,m; \quad \beta = 1,\ldots,\mu,\\ &\Re(a_i + c_\alpha) > -1, \quad i = 1,\ldots,n; \quad \alpha = 1,\ldots,\nu,\\ &(a_i - b_j) \quad \text{is not a positive integer for} \quad i = 1,\ldots,n; \quad j = 1,\ldots,m,\\ &(c_\alpha - d_\beta) \quad \text{is not a positive integer for} \quad \alpha = 1,\ldots,\nu; \quad \beta = 1,\ldots,\mu,\\ &\delta > 0, \ |\arg\eta| < \delta\pi, \quad \rho > 0, \quad |\arg\omega| < \rho\pi.\end{aligned} \tag{A.30}$$

Among the many other situations, the case $\delta = 0$, $|\arg\eta| = 0$ (or $\rho = 0$, $\arg\omega = 0$) is possible too, under some additional conditions (see [272, I, p. 164]).

To prove (A.29) we replace one of the F-functions, say $G_{\sigma,\tau}^{\mu,\nu}$, by its definition (A.7) through a contour integral, namely:

$$\text{L.H.S. of (A.29)} = \int_0^\infty G_{p,q}^{m,n}\left[\eta z \left| \begin{matrix}(a_i)\\(b_j)\end{matrix}\right.\right] \frac{1}{2\pi i}\int_{\mathcal{L}} (\omega z)^s$$

$$\times \frac{\prod_{\beta=1}^{\mu}\Gamma(d_\beta - s)\prod_{\alpha=1}^{\nu}\Gamma(1-c_\alpha+s)}{\prod_{\beta=\mu+1}^{\tau}\Gamma(1-d_\beta+s)\prod_{\alpha=\nu+1}^{\sigma}\Gamma(c_\alpha - s)} ds = \frac{1}{2\pi i}\int_{\mathcal{L}} \frac{\prod_{\beta=1}^{\mu}\Gamma(d_\beta - s)\prod_{\alpha=1}^{\nu}\Gamma(1-c_\alpha+s)}{\prod_{\beta=\mu+1}^{\tau}\Gamma(1-d_\beta+s)\prod_{\alpha=\nu+1}^{\sigma}\Gamma(c_\alpha - s)}$$

$$\times \omega^s ds \int_0^\infty z^s G_{p,q}^{m,n}\left[\eta z \left| \begin{matrix}(a_i)\\(b_j)\end{matrix}\right.\right] dz,$$

where, due to the absolute convergence of both integrals under conditions (A.30), the change of their order is permissible and the inner integral in the last line is denoted, in what follows, by $\mathcal{I}$. Then $\mathcal{I}$ can be evaluated as a Mellin transformation of the $G_{p,q}^{m,n}$-function according to formula (A.25), viz.:

$$\mathcal{I} = \frac{\eta^{-(s+1)}\prod_{j=1}^{m}\Gamma(b_j+s+1)\prod_{i=1}^{n}\Gamma(1-a_j-s-1)}{\prod_{j=m+1}^{q}\Gamma(1-b_j-s-1)\prod_{i=n+1}^{p}\Gamma(a_j+s+1)}$$

$$= \frac{\eta^{-s}}{\eta}\frac{\prod_{i=1}^{n}\Gamma(-a_j-s)\prod_{j=1}^{m}\Gamma(1-(-b_j)+s)}{\prod_{i=n+1}^{p}\Gamma(1-(-a_i)+s)\prod_{j=m+1}^{q}\Gamma(-b_j-s)}.$$

Thus, we obtain that

$$\text{the L.H.S. of (A.29)} = \frac{1}{\eta}\frac{1}{2\pi i}\int_{\mathcal{L}}\left(\frac{\omega}{\eta}\right)^s$$

$$\times \frac{\prod_{i=1}^{n}\Gamma((-a_j)-s)\prod_{i=1}^{\mu}\Gamma(d_i-s)\prod_{j=1}^{m}\Gamma(1-(-b_j)+s)\prod_{j=1}^{\nu}\Gamma(1-c_j+s)}{\prod_{i=\mu+1}^{\tau}\Gamma(1-d_i+s)\prod_{i=n+1}^{p}\Gamma(1-(-a_i)+s)\prod_{j=\nu+1}^{\sigma}\Gamma(-c_j-s)\prod_{j=m+1}^{q}\Gamma((-b_j)-s)}$$

$$= \frac{1}{\eta}G_{q+\sigma,p+\tau}^{n+\mu,m+\nu}\left[\frac{\omega}{\eta}\left| \begin{matrix} -b_1,\ldots,-b_m; & c_1,\ldots,c_\sigma; & -b_{m+1},\ldots,-b_q \\ -a_1,\ldots,-a_n; & d_1,\ldots,d_\tau; & -a_{n+1},\ldots,-a_p\end{matrix}\right.\right],$$

which is the second line of (A.29). The third line of this formula follows by using property (A.15).

By specializing the parameters in (A.28), (A.29) we can obtain many important integrals involving the various special functions of mathematical physics, among which the Stieltjes, Hankel, Y-, K- and other integral transforms of the G- and ${}_pF_q$-functions (see e.g. Luke [272, I, p. 164-166]). More results about integral transforms involving generalized hypergeometric functions, asymptotics of their kernels, invertibility of the defining Mellin-Barnes type integrals can be seen in Marichev [275]-[278].

B. Some other auxiliary results for the G-functions

Here we give some results on the G-functions which cannot be encountered in the manuals on the special functions, or at least in the form we use them in our considerations.

It is well known that the Gauss hypergeometric function ${}_2F_1(a,b;c;z)$ with $c \neq 0,-1,-2,\ldots, \Re(c-a-b) > 0$ is defined by means of the absolutely convergent hypergeometric series on the unit circle $|z| = 1$. Moreover, its value at the point $z = 1$ can by expressed by the Euler formula (cf. [106, I, 2.1.3, (10)]):

$$ {}_2F_1(a,b;c;z) = \frac{\Gamma(c)}{\Gamma(a)\ \Gamma(c-b)} \int_0^1 \frac{t^{b-1}(1-t)^{c-b-1}}{(1-tz)^a} dt, \tag{B.1} $$

giving for $z = 1$, $\Re c > \Re b > 0$, $\Re(c-a-b) > 0$:

$$ {}_2F_1(a,b;c;z) = \frac{\Gamma(c)\Gamma(c-a-b)}{\Gamma(c-a)\Gamma(c-b)}. \tag{B.2} $$

For the generalized ${}_{q+1}F_q$-function with an arbitrary $q > 1$ expressions for its value at the point $z = 1$ have been found only for some particular values of the parameters $a_0, a_1, \ldots, a_q$ and $b_1, \ldots, b_q$ satisfyng the general convergence condition $\Re\left(\sum_{k=1}^{q} b_k - \sum_{j=0}^{q} a_j\right) > 0$. These representations are known as the theorems of Saalschütz, Dixon, Watson, Whipple, Dougall, etc., concerning the cases with $q = 2,3,4,6$ (see [106, I, §4.4]). They lead also to expressions for the values of $G^{q+1,1}_{q+1,q+1}$-functions on the unit circle (especially at the point $z = -1$) but only for the corresponding particular choice of the order q and the other parameters of these functions.

An analogous problem stands also for another class of G-functions having singularities on the unit circle, say at $z = 1$. Let us consider the function $G^{q,1}_{q+1,q+1}\left[z \left| \begin{matrix} a_0, a_1, \ldots, a_q \\ b_1, \ldots, b_q, b_{q+1} \end{matrix} \right.\right]$ with $a_0 - b_{q+1} = 1$. As is pointed out by Marichev [280, p. 396], this is a $G^{m,n}_{p,q}$-function with $m + n - p = 0$ and for $\nu^* = \sum_1^q (a_k - b_k) > 0$ it is continuous at the singular point $z = 1$ but "in the general case such a problem (i.e. to find the limit value of $G^{q,1}_{q+1,q+1}(z)$ at $z = 1$) ... is not solved". Here we propose a formula expressing the required value in terms of Γ-functions, under the above conditions, as is usual for these special cases.

Lemma B.1. *Let* $a_0 = b_{q+1} + 1 = \alpha$ *with an arbitrary* α *and let the other parameters of* *the* $G^{q,1}_{q+1,q+1}$*-function satisfy the conditions:* $\Re a_k > \Re b_k > \Re\alpha - 1$, $k = 1, \ldots, q$. *Then,*

$$G^{q,1}_{q+1,q+1}\left[1\left|\begin{matrix}\alpha, (a_j)_1^q \\ (b_k)_1^q, \alpha - 1\end{matrix}\right.\right] = \prod_{k=1}^{q} \frac{\Gamma(b_k - \alpha + 1)}{\Gamma(a_k - \alpha + 1)}. \tag{B.3}$$

Proof. According to definition (A.7),

$$G^{q,1}_{q+1,q+1}\left[1\left|\begin{matrix}\alpha, (a_j)_1^q \\ (b_k)_1^q, \alpha - 1\end{matrix}\right.\right] = \frac{1}{2\pi i}\int_{\mathcal{L}} \left(\prod_{k=1}^{q} \frac{\Gamma(b_k - s)}{\Gamma(a_k - s)}\right) \frac{\Gamma(1 - \alpha + s)}{\Gamma(1 + (1 - \alpha + s))} ds,$$

where $\mathcal{L} = \mathcal{L}_{-i\infty, i\infty} = \{s : \ \Re s = \sigma\}$ is the path of integration i) with a suitably chosen σ such that $\Re b_k > \sigma > \Re\alpha - 1$, $k = 1, \ldots, q$. This integral is absolutely convergent since: $\delta = 0$, $\arg z = \arg 1 = 0$, $q + 1 = 0 + 1$ and $\Re\nu = \Re(\sum_1^{q+1} b_k - \sum_0^q a_j) = \Re\sum_1^q(b_k - a_k) + \alpha - 1 - \alpha < 0 - 1 = -1$. The choice of the abscissa σ ensures that $\mathcal{L}$ leaves to the right all the poles of $\Gamma(b_k - s)$, $k = 1, \ldots, q$, and to the left all the poles of $\Gamma(1 + \alpha + s)$. Indeed, the condition $\Re b_k > \sigma > \Re\alpha - 1$ implies that $\Re(b_k + l) > \sigma > \Re(\alpha - 1 - l')$ for $l = 0, 1, 2, \ldots,\ \ l = 0, 1, 2, \ldots,\ \ k = 1, \ldots, q$. Since $\dfrac{\Gamma(1 - \alpha + s)}{\Gamma(1 + (1 - \alpha + s))} = \frac{1}{1 - \alpha + s}$, then

$$G[1] = \frac{1}{2\pi i}\int_{\mathcal{L}} \left(\prod_{k=1}^{q} \frac{\Gamma(b_k - s)}{\Gamma(a_k - s)}\right) \frac{ds}{1 - \alpha + s}.$$

If we denote $\lambda_k = a_k - b_k > 0$, $k = 1, \ldots, q$, we obtain

$$\begin{aligned}\frac{\Gamma(b_k - s)}{\Gamma(a_k - s)} &= \frac{1}{\Gamma(\lambda_k)} \frac{\Gamma(b_k - s)\Gamma(\lambda_k)}{\Gamma(a_k - s)} = \frac{1}{\Gamma(\lambda_k)} B(\lambda_k, b_k - s) \\ &= \int_0^1 \frac{(1 - t_k)^{\lambda_k - 1}}{\Gamma(\lambda_k)} t_k^{(b_k - s) - 1} dt_k.\end{aligned}$$

We replace this result in the contour integral for $G[1]$ and get

$$\begin{aligned}G[1] &= \frac{1}{2\pi i}\int_{\sigma - i\infty}^{\sigma + i\infty} \left(\prod_{k=1}^{q} \int_0^1 \frac{(1 - t_k)^{\lambda_k - 1}}{\Gamma(\lambda_k)} t_k^{(b_k - s) - 1} dt_k\right) \frac{ds}{1 - \alpha + s} \\ &= \prod_1^q \int_0^1 \frac{(1 - t_k)^{\lambda_k - 1}}{\Gamma(\lambda_k)} t_k^{b_k - \alpha} dt_k \frac{1}{2\pi i}\int_{\sigma - i\infty}^{\sigma + i\infty} \frac{\left(\prod_{k=1}^{q} t_k\right)^{-1 + \alpha - s}}{1 - \alpha + s} ds,\end{aligned}$$

where the change of the order of the integrals is permissible due to the absolute convergence of the contour integrals and the uniform convergence of the B-integrals for

$\Re a_k > \Re b_k > \sigma$. After the substitution $t = \prod_{k=1}^{q} t_k$, $0 \le t \le 1$, the inner integral takes the form

$$\frac{1}{2\pi i}\int_{\sigma-i\infty}^{\sigma+i\infty} \frac{t^{-(1-\alpha+s)}}{1-\alpha+s} ds = \frac{1}{2\pi i}\int_{\sigma_0-i\infty}^{\sigma_0+i\infty} \frac{t^{-s'}}{s'} ds' = \frac{1}{2\pi i}\int_{\sigma_0-i\infty}^{\sigma_0+i\infty} \frac{\Gamma(s')}{\Gamma(s'+1)} t^{-s'} ds',$$

where $s' = s + 1 - \alpha$ and $\sigma_0 = \sigma + 1 - \alpha > 0$. According to [276, p. 130, 1 (1)], $\mathfrak{M}(s') = \frac{\Gamma(s')}{\Gamma(s'+1)}$ is the Mellin transform of the Heaviside function

$$\varphi(t) = H(1-t) = \begin{cases} 0, & t > 1, \\ 1, & t \le 1 \end{cases}.$$

Therefore, by virtue of the inversion formula for the Mellin transform:

$$\varphi(t) = \frac{1}{2\pi i}\int_{\sigma_0-i\infty}^{\sigma_0+i\infty} \mathfrak{M}\left(s'\right) t^{-s'} ds',$$

the inner integral gives, for $0 \le t \le 1$, nothing but $\varphi(t) \equiv 1$. So, finally we have

$$G^{q,1}_{q+1,q+1}\left[1 \left| \begin{matrix} \alpha, (a_j)_1^q \\ (b_k)_1^q, \alpha - 1 \end{matrix} \right.\right] = \prod_{k=1}^{q} \int_0^1 \frac{(1-t_k)^{\lambda_k - 1}}{\Gamma(\lambda_k)} t_k^{b_k - \alpha} dt_k$$

$$= \prod_{k=1}^{q} \frac{\Gamma(\lambda_k)\Gamma(b_k - \alpha + 1)}{\Gamma(\lambda_k)\Gamma(a_k - \alpha + 1)} = \prod_{k=1}^{q} \frac{\Gamma(b_k - \alpha + 1)}{\Gamma(a_k - \alpha + 1)},$$

what was to be proved.

In particular, if $\alpha = 0$, then

$$G^{q,1}_{q+1,q+1}\left[1 \left| \begin{matrix} 0, (a_j)_1^q \\ (b_k)_1^q, -1 \end{matrix} \right.\right] = \prod_{k=1}^{q} \frac{\Gamma(b_k + 1)}{\Gamma(a_k + 1)}, \tag{B.3$'$}$$

provided $\Re a_k > \Re b_k > -1$, $k = 1, \dots, q$.

As a consequence of this formula we obtain the following integral.

Lemma B.2. *If $\Re a_k > \Re b_k > -1$, $k = 1, \dots, q$, then the following integral formula holds:*

$$\int_0^1 G^{q,0}_{q,q}\left[\zeta \left| \begin{matrix} (a_j)_1^q \\ (b_k)_1^q \end{matrix} \right.\right] d\zeta = \prod_{k=1}^{q} \frac{\Gamma(b_k + 1)}{\Gamma(a_k + 1)}. \tag{B.4}$$

Proof. Formally, formula (B.4) can be derived from the Riemann-Liouville integral (A.23) when $m = n = p$, $n = 0$, $\alpha = 1$, under the conditions $\Re b_k > -1$, $k = 1, \ldots, q$,

$$\int_0^z G_{q,q}^{q,0}\left[\omega z \,\middle|\, \begin{matrix}(a_j)\\(b_k)\end{matrix}\right] d\zeta = zG_{q+1,q+1}^{q,1}\left[\omega z \,\middle|\, \begin{matrix}0,(a_j)\\(b_k),-1\end{matrix}\right]$$

$$= \frac{1}{\omega}G_{q+1,q+1}^{q,1}\left[\omega z \,\middle|\, \begin{matrix}1,(a_j+1)_1^q\\(b_k+1)_1^q,0\end{matrix}\right], \tag{B.5}$$

namely, if we take $\omega = z = 1$. Unfortunately (B.5), as a corollary of (A.23), is valid only for $|\omega z| < 1$.

Therefore we shall use another approach.

First we apply to the function $G_{q,q}^{q,0}(\zeta)$ one of the so-called *multiplication theorems* (theorems concerning expansions of G-functions in series, see [106, I, §5.5, (4)]). To this end, we set

$$\zeta = \lambda\eta, \quad \text{where} \quad \lambda = 2,\ \eta = \frac{\zeta}{2}.$$

Then, $n = 0 < p = q$ and $\Re\lambda = 2 > \frac{1}{2}$ and the above-mentioned formula is valid, namely:

$$G_{q,q}^{q,0}(\zeta) = G_{q,q}^{q,0}\left[\lambda\eta \,\middle|\, \begin{matrix}(a_j)_1^q\\(b_k)_1^q\end{matrix}\right]$$

$$= \lambda^{a_q-1}\sum_{j=0}^{\infty}\frac{1}{j!}\left(\frac{1}{\lambda}-1\right)^j G_{q,q}^{q,0}\left[\frac{\zeta}{2} \,\middle|\, \begin{matrix}a_1,\ldots,a_{q-1},a_q-j\\(b_k)_1^q\end{matrix}\right].$$

After a term-by-term integration of this absolutely convergent series in limits from 0 to 1, using formula (B.5) for $\omega = \frac{1}{2}$, $z = 1$ (that is, for $|\omega z| < 1$) we find:

$$\int_0^1 G_{q,q}^{q,0}\left[\zeta \,\middle|\, \begin{matrix}(a_j)_1^q\\(b_k)_1^q\end{matrix}\right] d\zeta = \lambda^{a_q-1}\sum_{j=0}^{\infty}\frac{1}{j!}\left(\frac{1}{\lambda}-1\right)^j \int_0^1 G_{q,q}^{q,0}\left[\omega\zeta \,\middle|\, \begin{matrix}(a_j)_1^{q-1},a_q-j\\(b_k)_1^q\end{matrix}\right] d\zeta$$

$$= \lambda^{a_q-1}\sum_{j=0}^{\infty}\frac{1}{j!}\left(\frac{1}{\lambda}-1\right)^j \lambda G_{q+1,q+1}^{q,1}\left[\omega \,\middle|\, \begin{matrix}1;(a_j+1)_1^{q-1};a_q-j+1\\(b_k+1)_1^q;0\end{matrix}\right]$$

$$= \lambda^{(a_q+1)-1}\sum_{j=0}^{\infty}\frac{1}{j!}\left(\frac{1}{\lambda}-1\right)^j \lambda G_{q+1,q+1}^{q,1}\left[\omega \,\middle|\, \begin{matrix}1;(a_j+1)_1^{q-1};a_q-j+1\\(b_k+1)_1^q;0\end{matrix}\right]$$

$$= G_{q+1,q+1}^{q,1}\left[\lambda\omega \,\middle|\, \begin{matrix}1,(a_j+1)_1^q\\(b_k+1)_1^q,0\end{matrix}\right] =$$

$$= G_{q+1,q+1}^{q,1}\left[1 \,\middle|\, \begin{matrix}1,(a_j+1)_1^q\\(b_k+1)_1^q,0\end{matrix}\right] = G_{q+1,q+1}^{q,1}\left[1 \,\middle|\, \begin{matrix}0,(a_j)_1^q\\(b_k+1)_1^q,-1\end{matrix}\right],$$

where the same multiplication formula is applied once again. Thus the final result follows from Lemma B.1.

The next results present *some new differential formulae for Meijer's G-function .*

Lemma B.3. (Kiryakova [196], [202]). *Let $\eta_1 \geq 0$, $\eta_2 \geq 0, \dots, \eta_q \geq 0$ be arbitrary integer numbers. Consider the differential operator $D^{(\eta)}$, defined by means of the following polynomial of the Euler differential operator $\delta = z\frac{d}{dz}$:*

$$D^{(\eta)} = \left[\prod_{i=1}^{q} D_{\eta_i}\right] = \left[\prod_{i=1}^{q}\prod_{s=0}^{\eta_i - 1}\left(z\frac{d}{dz} - b_i - s\right)\right]. \tag{B.6}$$

Then, Meijer's $G^{m,n}_{p,q}$-function ($q \geq 1$) satisfies the differential relation

$$D^{(\eta)} = G^{m,n}_{p,q}\left[z\,\middle|\,\begin{matrix}(a_j)_1^p\\(b_k)_1^q\end{matrix}\right] = \left(\prod_{i=1}^{q}\varepsilon_i^{\eta_i}\right) G^{m,n}_{p,q}\left[z\,\middle|\,\begin{matrix}(a_j)_1^p\\(b_k+\eta_k)_1^q\end{matrix}\right], \tag{B.7}$$

where $z \neq 0$ is a complex variable,

$$\varepsilon_i = \begin{cases} -1, & i = 1,\dots,m\\ +1, & i = m+1,\dots,q\end{cases}$$

and if some $\eta_i, i = 1,\dots,q$, are equal to zero, then the corresponding factors D_{η_i} are lacking in the differential operator $D^{(\eta)}$, (B.6).

Proof. The following formulas or slight generalizations of them can be found in different variants in the literature (see e.g. [106, I], [272, I, p. 151], [286, p. 9], [370, p. 621]):

$$\begin{aligned}
\frac{d}{dz}\left\{z^{-b_1}G^{m,n}_{p,q}\left[z\,\middle|\,\begin{matrix}(a_j)_1^p\\(b_k)_1^q\end{matrix}\right]\right\} &= -z^{-1-b_1}G^{m,n}_{p,q}\left[z\,\middle|\,\begin{matrix}(a_j)_1^p\\b_1+1,b_2,\dots,b_q\end{matrix}\right], \quad m \geq 1,\\
\frac{d}{dz}\left\{z^{-b_q}G^{m,n}_{p,q}\left[z\,\middle|\,\begin{matrix}(a_j)_1^p\\(b_k)_1^q\end{matrix}\right]\right\} &= -z^{-1-b_q}G^{m,n}_{p,q}\left[z\,\middle|\,\begin{matrix}(a_j)_1^p\\b_1,\dots,b_{q-1},b_q+1\end{matrix}\right], \quad m < q.
\end{aligned}$$

Formula (A.17) is a concise form of them. Taking into account that $z^{b_i+1}\frac{d}{dz}z^{-b_i} = (z\frac{d}{dz} - b_i)$ and the symmetry of the G-function with respect to the parameters $b_1,\dots,b_m$ and to $b_{m+1},\dots,b_q$, we can rewrite these formulas in the form

$$\left(z\frac{d}{dz} - b_i\right)G^{m,n}_{p,q}\left[z\,\middle|\,\begin{matrix}(a_j)_1^p\\(b_k)_1^q\end{matrix}\right] = \varepsilon_i G^{m,n}_{p,q}\left[z\,\middle|\,\begin{matrix}(a_j)_1^p\\b_1,\dots,b_{i-1},b_i+1,b_{i+1},\dots,b_q\end{matrix}\right]$$

for each $i = 1\dots,q$. Analoguosly,

$$\begin{aligned}
&\left[z\frac{d}{dz} - (b_i+1)\right]\left(z\frac{d}{dz} - b_i\right)G^{m,n}_{p,q}[z]\\
&= \varepsilon_i\left[z\frac{d}{dz} - (b_i+1)\right]G^{m,n}_{p,q}\left[z\,\middle|\,\begin{matrix}(a_j)_1^p\\b_1,\dots,b_{i-1},b_i+1,b_{i+1},\dots,b_q\end{matrix}\right]\\
&= \varepsilon_i^2 G^{m,n}_{p,q}\left[z\,\middle|\,\begin{matrix}(a_j)_1^p\\b_1,\dots,b_{i-1},b_i+2,b_{i+1},\dots,b_q\end{matrix}\right].
\end{aligned}$$

Operating in this manner $(\eta_i - 1)$-times (for $\eta_i > 1$), we obtain:

$$D_{\eta_i} G_{p,q}^{m,n}[z] = \left[z\frac{d}{dz} - (b_i + \eta_i - 1)\right] \dots \left[z\frac{d}{dz} - (b_i + 1)\right] \left(z\frac{d}{dz} - b_i\right) G_{p,q}^{m,n}\left[z\left|\begin{matrix}(a_j)_1^p\\(b_k)_1^q\end{matrix}\right.\right]$$
$$= \varepsilon_i^{\eta_i} G_{p,q}^{m,n}\left[z\left|\begin{matrix}(a_j)_1^p\\b_1,\dots,b_{i-1},b_i+\eta_i,b_{i+1},\dots,b_q\end{matrix}\right.\right].$$

This procedure can be repeated subsequently for each $i = 1, \dots, q$ and this leads to the differential relation

$$\left\{\prod_{i=1}^q \prod_{s=0}^{\eta_i - 1}\left(z\frac{d}{dz} - b_i - s\right)\right\} G_{p,q}^{m,n}\left[z\left|\begin{matrix}(a_j)_1^p\\(b_k)_1^q\end{matrix}\right.\right] = \left(\prod_{i=1}^q \varepsilon_i^{\eta_i}\right) G_{p,q}^{m,n}\left[z\left|\begin{matrix}(a_j)_1^p\\(b_k+\eta_k)_1^q\end{matrix}\right.\right],$$

that is, (B.7). Note also, that operators D_{η_i} can be written in the form

$$D_{\eta_i} = z^{b_i+\eta_i}\left(\frac{d}{dz}\right)^{\eta_i} z^{-b_i}.$$

The proof is over.

In other words, this lemma shows that *applying the differential operator*

$$D^{(\eta)}\left[\prod_{i=1}^q \prod_{s=0}^{\eta_i-1}\left(z\frac{d}{dz} - b_i - s\right)\right] = \prod_{i=1}^q\left[z^{b_i+\eta_i}\left(\frac{d}{dz}\right)^{\eta_i} z^{-b_i}\right] \tag{B.6$'$}$$

of Bessel type (see Chapter 3) *to the G-function, we are able to increase each of its parameters $b_1, \dots, b_q$ by an arbitrary integer $\eta_i \ge 0$, $i = 1, \dots, q$.* Similar considerations based on formula (A.17$'$) and concerning the possibility for decreasing each of the parameters $a_1, \dots, a_p$ by given natural numbers $\mu_1, \dots, \mu_p$, are valid too, namely:

Lemma B.4. (Kiryakova [202]). *Let $p \ge 1$, $\mu_1 \ge 0, \dots, \mu_p \ge 0$ be arbitrary integers and*

$$D'^{(\mu)} = \left[\prod_{i=1}^p D'_{\mu_i}\right] = \left[\prod_{i=1}^p \prod_{s=0}^{\mu_i-1}\left(z\frac{d}{dz} + 1 - a_i + s\right)\right]. \tag{B.8}$$

Then, the following differential relation holds:

$$D'^{(\mu)} G_{p,q}^{m,n}\left[z\left|\begin{matrix}(a_j)_1^p\\(b_k)_1^q\end{matrix}\right.\right] = \left(\prod_{i=1}^p \varepsilon'_i\right) G_{p,q}^{m,n}\left[z\left|\begin{matrix}(a_j-\mu_j)_1^p\\(b_k)_1^q\end{matrix}\right.\right] \tag{B.9}$$

for $z \ne 0$ and

$$\varepsilon'_i = \begin{cases} +1, & i = 1, \dots, n \\ -1, & i = n+1, \dots, p. \end{cases}$$

Differential formulas (B.7), (B.9) have a series of useful applications, some of which we shall discuss briefly now.

Corollary B.5. *Let $m = q \geq 1$, $n = p = 0$ and $\eta_1 = \cdots = \eta_q = \eta \geq 0$. Then, for the function $G(z) = G_{0,q}^{q,0}\left[z|(b_k)_1^q\right]$ formula (B.7) gives:*

$$\left\{\prod_{i=1}^{q}\left(z\frac{d}{dz} - b_i - \eta + 1\right)\ldots\left(z\frac{d}{dz} - b_i - 1\right)\left(z\frac{d}{dz} - b_i\right)\right\} G(z)$$
$$= (-1)^q G_{0,q}^{q,0}\left[z|(b_k + \eta)_1^q\right] = (-1)^q z^\eta G(z). \tag{B.10}$$

A quite special case of (B.10), when $\eta = 1$ and $b_q = 0$, is the following differential formula for the function $\Phi(z) = G[z|b_1, \ldots, b_{q-1}, 0]$, viz.:

$$B^*\Phi(z) = (-1)^q z\Phi(z), \tag{B.11}$$

where

$$B^* = \left[\prod_{i=1}^{q}\left(z\frac{d}{dz} - b_i\right)\right] = z\frac{d}{dz}z^{b_{q-1}+1}\frac{d}{dz}\ldots z^{-b_2+b_1+1}\frac{d}{dz}z^{-b_1}.$$

According to [75], [196], [201], etc., the G-function $\Phi(z)$ coincides with the *kernel-function of the so-called Obrechkoff integral transform* (cf. Section 3.9; also [66]-[69], [74]-[76]). So, formula (B.11) is nothing but the *differential equation for this kernel-function*, found by Obrechkoff himself in another way (see [339]).

Corollary B.6. (Kiryakova [196]). *For the Meijer's G-function with $m = p = q \geq 1$, $n = 0$ and $z^{-1} = \left(\frac{t}{x}\right)^\beta$, $\beta > 0$, Lemma B.3 gives the differential relation*

$$G_{m,m}^{m,0}\left[\frac{1}{z}\middle|\begin{matrix}(a_k)_1^m\\(b_k)_1^m\end{matrix}\right] = \left[\prod_{i=1}^{m}\prod_{j=0}^{\lambda_i-1}\left(z\frac{d}{dz} + a_i + j\right)\right] G_{m,m}^{m,o}\left[\frac{1}{z}\middle|\begin{matrix}(a_k + \lambda_k)_1^m\\(b_k)_1^m\end{matrix}\right] \tag{B.12}$$

with some integers $\lambda_1 \geq 0, \ldots, \lambda_m \geq 0$.

Thus, with the application of a suitable differential operator, we are able to increase the upper parameters of a $G_{m,m}^{m,0}$-function. This property turns to be quite useful solving the problem of inverting the generalized fractional integrals, introduced in Chapter 1. Namely, the formal inversion formula leads to an integral whose kernel is a $G_{m,m}^{m,0}$-function with upper parameters a_k less than or equal to the corresponding lower parameters b_k, $k = 1, \ldots, m$. To make this integral convergent, it is necessary to replace this kernel by another $G_{m,m}^{m,0}$-function whose upper parameters exceed the lower ones, i.e. with $a'_k = a_k + \lambda_k \geq b_k$, $k = 1, \ldots, m$. This can be done successfully, if one uses formula (B.12) with the λ_k's depending on the differences $\delta_k = b_k - a_k \geq 0$, $k = 1, \ldots, m$, in the following manner:

$$\lambda_k = \begin{cases}[\delta_k] + 1, & \text{if } \delta_k \text{ is not integer,}\\ \delta_k, & \text{if } \delta_k \text{ is integer,}\end{cases} \quad k = 1, \ldots, m.$$

For more details see Section 1.5.

Corollary B.7. (Kiryakova [198], [202]). *Another special case of (B.7) is a differential relation concerning the generalized hypergeometric functions*

$$ {}_pF_q(a_1,\dots,a_p;b_1,\dots,b_q;-z) = \frac{\prod\limits_{k=1}^{q}\Gamma(b_k)}{\prod\limits_{j=1}^{p}\Gamma(a_j)} G^{1,p}_{p,q+1}\left[z \,\middle|\, \begin{matrix} (1-a_j)_1^p \\ 0,(1-b_k)_1^q \end{matrix} \right]. $$

In particular, for the hyper-Bessel functions ${}_0F_q(-z)$ *of order* $q \geq 1$ *(for more details see Section C), we obtain*

$$ \left[\prod_{i=1}^{q} \prod_{j=1}^{\eta_i} (z\frac{d}{dz} + b_i + j - 1) \right] \; {}_0F_q\left((b_k+\eta_k)_1^q; -z \right) = \left[\prod_{i=1}^{q} \frac{\Gamma(b_i+\eta_i)}{\Gamma(b_i)} \right] \; {}_0F_q\left((b_k)_1^q; -z \right) \tag{B.13} $$

for integers $\eta_i \geq 0$, $i = 1,\dots,q$.

For an application see Corollary 4.1.7.

C. Examples of generalized hypergeometric functions and Meijer's G-function

First of all, let us note that *the generalized hypergeometric function* (A.1) *can be expressed in terms of Meijer's G-function* in the following way (see [106, I, §5.6]):

$$ \begin{aligned} {}_pF_q(a_1,\dots,a_p;b_1,\dots,b_q;z) &= \frac{\prod\limits_{j=1}^{q}\Gamma(b_j)}{\prod\limits_{j=1}^{p}\Gamma(a_j)} G^{1,p}_{p,q+1}\left[-z \,\middle|\, \begin{matrix} 1-a_1,\dots,1-a_p \\ 0,1-b_1,\dots,1-b_q \end{matrix} \right] \\ &= \frac{\prod\limits_{j=1}^{q}\Gamma(b_j)}{\prod\limits_{j=1}^{p}\Gamma(a_j)} G^{p,1}_{q+1,p}\left[-\frac{1}{z} \,\middle|\, \begin{matrix} 1,b_1,\dots,b_q \\ a_1,\dots,a_p \end{matrix} \right]. \end{aligned} \tag{C.1} $$

So, *all the particular cases of* ${}_pF_q$*-functions, mentioned below, have analogous representations by means of the G-functions too.* We start with the following three simplest and well known examples.

Gauss Hypergeometric Function

$$ \begin{aligned} F(a,b;c;z) = {}_2F_1(a,b;c;z) &= \sum_{k=0}^{\infty} \frac{(a)_k(b)_k}{(c)_k}.\frac{z^k}{k!} \\ &= \frac{\Gamma(c)}{\Gamma(a)\Gamma(b)} G^{1,2}_{2,2}\left[-z \,\middle|\, \begin{matrix} 1-a,1-b \\ 0,1-c \end{matrix} \right] \end{aligned} \tag{C.2} $$

($c \neq 0,-1,-2,\dots$; $|z|<1$, or: $z=1$ and $\Re(c-a-b)>0$, or $|z|=1$, $z\neq 1$ and $0 \geq \Re(c-a-b) > -1$).

Confluent Hypergeometric Functions

Kummer function

$$\Phi(a;c;z) = {}_1F_1(a;c;z) = \sum_{k=0}^{\infty} \frac{(a)_k}{(c)_k}\frac{z^k}{k!}$$
$$= \frac{\Gamma(c)}{\Gamma(a)} G^{1,1}_{1,2}\left[-z \,\middle|\, \begin{matrix} 1-a \\ 0, 1-c \end{matrix}\right], \quad c \neq 0, -1, -2, \ldots; \tag{C.3}$$

Tricomi function

$$\Psi(a,c;z) = z^{-a}\,{}_2F_0\left(a, a+1-c, -\frac{1}{z}\right) = \frac{1}{\Gamma(a)\Gamma(a+1-c)} G^{2,1}_{1,2}\left[z \,\middle|\, \begin{matrix} 1-a \\ 0, 1-c \end{matrix}\right], \tag{C.4}$$
$$z \in \mathbb{C}\backslash\{-\infty, 0\};$$

Whittaker function

$$W_{k,\mu}(z) = \exp\left(-\frac{z}{2}\right) z^{\mu+\frac{1}{2}}\Psi\left(\frac{1}{2} - k + \mu, 2\mu+1; z\right)$$
$$= \frac{\exp\left(-\frac{z}{2}\right)}{\Gamma\left(\frac{1}{2} - k + \mu\right)\Gamma\left(\frac{1}{2} - k - \mu\right)} G^{2,1}_{1,2}\left[z \,\middle|\, \begin{matrix} 1+k \\ \frac{1}{2}+\mu, \frac{1}{2}-\mu \end{matrix}\right] = W_{k,-\mu}(z); \tag{C.5}$$

Bessel Functions

Bessel functions (of first kind and order ν)

$$J_\nu(z) = \frac{\frac{z}{2}}{\Gamma(\nu+1)}\,{}_0F_1\left(\nu+1; -\frac{z^2}{4}\right) = \sum_{k=0}^{\infty} \frac{(-1)^k\left(\frac{z}{2}\right)^{2k+\nu}}{k!\Gamma(\nu+k+1)}$$
$$= \left(\frac{z}{2}\right) G^{1,0}_{0,2}\left[\frac{z^2}{4} \,\middle|\, 0, -\nu\right]; \tag{C.6}$$

Modified Bessel function of first kind

$$I_\nu(z) = \frac{\frac{z}{2}}{\Gamma(\nu+1)}\,{}_0F_1\left(\nu+1; \frac{z^2}{4}\right) = \left(\frac{z}{2}\right) G^{1,0}_{0,2}\left[-\frac{z^2}{4} \,\middle|\, 0, -\nu\right]; \tag{C.7}$$

Lommel function $s_{\mu,\nu}(z)$ and Struve function $H_\nu(z)$

$$s_{\mu,\nu}(z) = \frac{z^{\mu+1}}{(\mu-\nu+1)(\mu+\nu+1)}\,{}_1F_2\left(1; \frac{\mu-\nu+3}{2}, \frac{\mu+\nu+3}{2}; -\frac{z^2}{4}\right)$$
$$= 2^{\mu-1}\Gamma\left(\frac{\mu-\nu+1}{2}\right)\Gamma\left(\frac{\mu+\nu+1}{2}\right) G^{1,1}_{1,3}\left[\frac{z^2}{4} \,\middle|\, \begin{matrix} \frac{(\mu+1)}{2} \\ \frac{\mu+1}{2}, \frac{\nu}{2}, -\frac{\nu}{2} \end{matrix}\right], \tag{C.8}$$

$$H_\nu(z) = \frac{s_{\nu,\nu}(z)}{\pi 2^{\nu-1}\left(\frac{1}{2}\right)_\nu}$$
$$= \frac{2}{\sqrt{\pi}\Gamma\left(\nu+\frac{3}{2}\right)}\left(\frac{z}{2}\right)^{\nu+1} {}_1F_2\left(1; \frac{3}{2}+\nu, \frac{3}{2}; -\frac{z^2}{4}\right) \tag{C.8$'$}$$
$$= G^{1,1}_{1,3}\left[\frac{z^2}{4}\,\middle|\,\begin{matrix}\frac{(\nu+1)}{2}\\ \frac{\nu+1}{2}, \frac{\nu}{2}, -\frac{\nu}{2}\end{matrix}\right].$$

Also, the following

ELEMENTARY FUNCTIONS

can be considered as ${}_pF_q$- or G-functions, namely:

$$(1\pm z)^{-\alpha} = {}_0F_1(\alpha; ; \mp z) = {}_2F_1(\alpha, \beta; \beta; \mp z) = \frac{1}{\Gamma(\alpha)} G^{1,1}_{1,1}\left[\mp z\,\middle|\,\begin{matrix}1-\alpha\\ 0\end{matrix}\right], \tag{C.9}$$
$$|z|<1, \quad \Re\alpha > 0,$$

while

$$(1\pm z)^{-\alpha} = (1\pm z)^{\delta-1} = \Gamma(\delta) G^{1,0}_{1,1}\left[\mp z\,\middle|\,\begin{matrix}\delta\\ 0\end{matrix}\right], \quad |z|<1 \tag{C.9}$$

for $\Re\alpha < -1$, $\delta = 1-\alpha$, that is, $\Re\delta > 0$.

More generally,

$$z^\gamma (1-z)^{\delta-1} = \Gamma(\delta)\, G^{1,0}_{1,1}\left[z\,\middle|\,\begin{matrix}\gamma+\delta\\ \gamma\end{matrix}\right] \tag{C.10}$$
$$|z|<1, \quad \Re\delta > 0,$$

which includes the *power functions* ($\delta = 1$)

$$z^\gamma = z^\gamma H(1-z) = G^{1,0}_{1,1}\left[z\,\middle|\,\begin{matrix}\gamma+1\\ \gamma\end{matrix}\right], \quad |z|<1, \tag{C.11}$$

with $H(z)$ standing for the Heaviside unit function;

$$\frac{z^\gamma}{1+az^\alpha} = a^{-\frac{\gamma}{\alpha}} G^{1,1}_{1,1}\left[az^\alpha\,\middle|\,\begin{matrix}\frac{\gamma}{\alpha}\\ \frac{\gamma}{\alpha}\end{matrix}\right]. \tag{C.12}$$

Let us mention *the exponential function:*

$$e^z = {}_0F_0(/, /, z) = {}_1F_1(\alpha, \alpha, z) = G^{1,0}_{0,1}\left[-z|0\right],$$

and more generally,

$$z^\beta e^{-\eta z^\alpha} = \eta^{-\frac{\beta}{\alpha}} G^{1,0}_{0,1}\left[\eta z^\alpha | \frac{\beta}{\alpha}\right]; \tag{C.13}$$

the trigonometric functions

$$\sin z = z\ {}_0F_1\left(/;\frac{3}{2};-\frac{z^2}{4}\right) = \sqrt{\pi}G^{1,0}_{0,2}\left[\frac{z^2}{4}\middle|\frac{1}{2},0\right] \tag{C.14}$$

$$\cos z = {}_0F_1\left(/;\frac{1}{2};-\frac{z^2}{4}\right) = \sqrt{\pi}G^{1,0}_{0,2}\left[\frac{z^2}{4}\middle|0,\frac{1}{2}\right]; \tag{C.15}$$

the logarithmic functions

$$\ln(1+z) = z\ {}_2F_1(1,1;2;-z) = G^{1,2}_{2,2}\left[z\middle|\begin{matrix}1,1\\1,0\end{matrix}\right], |z|<1, \tag{C.16}$$

$$\ln\left(\frac{1+z}{1-z}\right) = 2z\ {}_2F_1\left(\frac{1}{2},1;\frac{3}{2};z^2\right) = G^{1,2}_{2,2}\left[-z^2\middle|\begin{matrix}1,\frac{1}{2}\\\frac{1}{2},0\end{matrix}\right], |z|<1,$$

and some other elemenary functions, like:

$$\arcsin z = z\ {}_2F_1\left(\frac{1}{2},\frac{1}{2};\frac{3}{2};z^2\right) = \frac{1}{2\sqrt{\pi}}G^{1,2}_{2,2}\left[-z^2\middle|\begin{matrix}1,1\\\frac{1}{2},0\end{matrix}\right], |z|<1, \tag{C.17}$$

$$\arctan z = z\ {}_2F_1\left(\frac{1}{2},1;\frac{3}{2};-z^2\right) = \frac{1}{2}G^{1,2}_{2,2}\left[z^2\middle|\begin{matrix}1,\frac{1}{2}\\\frac{1}{2},0\end{matrix}\right],$$

$$\left[\frac{1}{2}\left(1+\sqrt{1-z}\right)\right]^{1-2a} ={}_2 F_1\left(a,a-\frac{1}{2};2a;z\right) = \frac{(2a-1)}{\sqrt{\pi}2^{2-2a}}G^{1,2}_{2,2}\left[-z\middle|\begin{matrix}1-a,\frac{3}{2}-a\\0,1-2a\end{matrix}\right],$$

$$\left(1+\sqrt{z}\right)^{-2a} + \left(1+\sqrt{z}\right)^{-2a} = 2\ {}_2F_1\left(a,a+\frac{1}{2};\frac{1}{2};z\right) = \frac{2^{2a}}{\Gamma(2a)}G^{1,2}_{2,2}\left[-z\middle|\begin{matrix}1-a,\frac{1}{2}-a\\0,\frac{1}{2}\end{matrix}\right].$$

Many other elementary functions can be shown to be special cases of the G-functions (for details see [370, 8.4], [272], [286], etc.).

Most of the *classical orthogonal polynomials* should be listed here as special cases of the ${}_pF_q$-functions and therefore of the G-functions. Here is a partial list of them:

LAGUERRE POLYNOMIALS

$$\begin{aligned} L_n^{(\alpha)}(z) &= \frac{e^z}{z^\alpha n!}\frac{d^n}{dz^n}\left\{e^{-z}z^{n+\alpha}\right\} = \frac{(1+\alpha)n}{n!}\ {}_1F_1(-n;a+\alpha;z) \\ &= \binom{n+\alpha}{n}\ {}_1F_1(-n;1+\alpha;z), \\ L_n^{(0)}(z) &= L_n(z); \quad \alpha>-1; n=0,1,2,\dots \end{aligned} \tag{C.18}$$

JACOBI POLYNOMIALS

$$\begin{aligned} P_n^{\alpha,\beta}(z) &= \frac{(-1)^n}{2^n n!}(1-z)^{-\alpha}(1+z)^{-\beta}\frac{d^n}{dz^n}\left[(1-z)^{n+\alpha}(1+z)^{n+\beta}\right] = \\ &= \binom{n+\alpha}{n}\ {}_2F_1\left(-n,n+\alpha+\beta+1;\alpha+1;\frac{1-z}{2}\right) \end{aligned} \tag{C.19}$$

Ultraspherical (Gegenbauer) polynomials $\left(P_n^{(\alpha,\beta)} \text{ with } \alpha=\beta=\nu-\frac{1}{2}\right)$

$$\begin{aligned} C_n^\nu(z) &= \frac{(-2)^n(\nu)_n}{n!\ (2\nu+n)_n}(1-z^2)^{\frac{1}{2}-\nu}\frac{d}{dz^n}\left\{(1-z)^{\nu+n-\frac{1}{2}}\right\} \\ &= \frac{2^{2n}(\nu)_n}{(2\nu+n)_n}P_n^{(\nu-\frac{1}{2},\nu-\frac{1}{2})}(z), \qquad (C.20) \\ & |z|<1; \quad \nu>-\frac{1}{2}; \quad n=0,1,2,\dots \end{aligned}$$

Legendre Polynomials and Functions $\left(P_n^{(\alpha,\beta)} \text{ with } \alpha=\beta=0; C_n^\nu \text{ with } \nu=\frac{1}{2}\right)$

$$\begin{aligned} P_n(z) &= \frac{1}{2^n n!}\frac{d^n}{dz^n}\left\{(z^2-1)^n\right\} = (-1)^n P_n^{(0,0)}(z) \\ &= \frac{(-1)^n(2n)!\Gamma\left(n+\frac{1}{2}\right)}{n!2^{-2n}\sqrt{\pi}}C_n^{\frac{1}{2}}(z) \qquad (C.21) \\ &= (-1)^n\ {}_2F_1\left(-n,n+1;1;\frac{1-z}{2}\right); \quad |z|<1, n=-0,1,2,\dots; \end{aligned}$$

more generally, the (generalized) *Legendre functions* are defined as

$$\begin{aligned} P_\nu^\mu(z) &= \frac{1}{\Gamma(1-\mu)}\left(\frac{z+1}{z-1}\right)^{\frac{\mu}{2}}\ {}_2F_1\left(-\nu,\nu+1;1-\mu;\frac{1-z}{2}\right), \\ P_\nu^0(z) &= P_\nu(z), \quad z\in\mathbb{C}\backslash\{-1<z<1\} \qquad (C.21') \end{aligned}$$

Tchebisheff Polynomials $\left(P_n^{(\alpha,\beta)} \text{ with } \alpha=\beta=-\frac{1}{2}; \quad C_n^\nu \text{ with } \nu=0\right)$

$$\begin{aligned} T_n(z) &= \cos(n\arccos x) = \frac{n}{2}C_n^0(z) = \frac{n!\sqrt{\pi}}{\Gamma(n+\frac{1}{2})}P_n^{(-\frac{1}{2},-\frac{1}{2})}(z) \\ &= {}_2F_1\left(-n,n;\frac{1}{2};\frac{1-z}{2}\right), \quad n=0,1,2,\dots \qquad (C.22) \end{aligned}$$

Bessel polynomials

$$\begin{aligned} Y_n(z,a,b) &= \sum_{k=0}^n \frac{(-n)_k(a+n-1)_k}{k!}\left(-\frac{z}{b}\right)^k \\ &= {}_2F_0\left(-n,a+n-1;\ ;-\frac{z}{b}\right), \quad n=0,1,2,\dots \qquad (C.23) \end{aligned}$$

Hermite polynomials

$$H_n(z) = (-1)^n\exp(z^2)\frac{d^n}{dz^n}\left\{\exp(-z^2)\right\} = 2^{\frac{n}{2}}\exp\left(\frac{z^2}{2}\right)D_n(z\sqrt{2}), \qquad (C.24)$$

where D_ν stands for the

Parabolic Cylinder Functions

$$D_\nu(z) = 2^{\frac{\nu}{2}} \exp\left(-\frac{z^2}{4}\right) \Psi\left(-\frac{\nu}{2}, \frac{1}{2}; \frac{z^2}{2}\right)$$
$$= \frac{2^{\frac{\nu}{2}} \exp\left(-\frac{z^2}{4}\right)}{\Gamma\left(-\frac{\nu}{2}\right)\Gamma\left(-\frac{\nu}{2}+\frac{1}{2}\right)} G^{2,1}_{1,2}\left[z \,\middle|\, \begin{matrix} 1+\frac{\nu}{2} \\ 0, \frac{1}{2} \end{matrix}\right]. \tag{C.25}$$

Among the other special functions, represented by ${}_pF_q$ and G-functions are: the *exponential integrals* and related functions, *elliptic functions* and *complete elliptic integrals, incomplete cylindrical functions*, etc. (recent results on the latter functions can be seen in e.g. in Al-Saqabi [7], Al-Saqabi and Kalla [8], Kalla [165], Kalla et al. [166]-[173]).

Let us mention also the special functions:

Incomplete Gamma- and Beta-Functions:

$$\gamma(a, z) = \int_0^z e^{-t} z^{\alpha-1} dz = \frac{z^a}{a} \, {}_1F_1(a; a+1; -z) \tag{C.26}$$

$$B_z(p, q) = \int_0^z t^{p-1}(1-t)^{q-1} dt = \frac{z^p}{p} \, {}_2F_1(p, 1-q; p+1; z) \tag{C.27}$$

Error Functions

$$\mathrm{Erf}(z) = \frac{2}{\sqrt{\pi}} \int_0^z e^{-t^2} dt = \frac{2z}{\sqrt{\pi}} \, {}_1F_1\left(\frac{1}{2}; \frac{z}{2}; -z^2\right) \tag{C.28}$$

and a number of other functions, related to them.

The following function has arisen (1937-1938), similarly to Meijer's G-function, in attempts to give meaning to the symbol ${}_pF_q$ when $p > q+1$:

MacRobert's E-function

$$E\left(p, a_1, \ldots, a_p;\ q;\ b_1, \ldots, b_q\ : z\right) = G^{p,1}_{q+1,p}\left[z \,\middle|\, \begin{matrix} 1, b_1, \ldots, b_q \\ a_1, \ldots, a_p \end{matrix}\right]. \tag{C.29}$$

Other special functions, not being generalized hypergeometric ${}_pF_q$-functions (A.1), can be represented by Meijer's G-functions too:

Modified Bessel functions

$$z^\mu Y_\nu(z) = (-1)^m 2^\mu G^{2,0}_{1,3}\left[\frac{z^2}{4} \,\middle|\, \begin{matrix} \frac{1}{2}(\mu-\nu-1)-m \\ \frac{1}{2}(\mu+\nu), \frac{1}{2}(\mu-\nu), \frac{1}{2}(\mu-\nu-1)-m \end{matrix}\right], \tag{C.30}$$
$$\text{for } m = 0, \pm 1, \pm 2, \ldots \quad ;$$

Modified Bessel function of third kind (Macdonald's function)

$$K_\nu(z) = \frac{\pi}{2\sin(\nu\pi)}[I_{-\nu}(z) - I_\nu(z)] = \frac{1}{2}G^{2,0}_{0,2}\left[\frac{z^2}{4}\middle|\frac{\nu}{2}, \frac{-\nu}{2}\right]; \tag{C.31}$$

Airy functions (of the first and second kind)

$$Ai(z) = \frac{1}{\pi}\sqrt{\frac{z}{3}}K_{\frac{1}{3}}\left(\frac{2}{3}z^{\frac{3}{2}}\right) = \frac{1}{2\pi\sqrt[6]{3}}G^{2,0}_{0,2}\left[\frac{z^3}{9}\middle|0, \frac{1}{3}\right], \tag{C.32}$$

$$\begin{aligned} Bi(z) &= \sqrt{\frac{z}{3}}\left[I_{-\frac{1}{3}}\left(\frac{2}{3}z^{\frac{3}{2}}\right) + I_{\frac{1}{3}}\left(\frac{2}{3}z^{\frac{3}{2}}\right)\right] \\ &= e^{\frac{\pi i}{6}}Ai\left(e^{\frac{2\pi i}{3}}z\right) + e^{-\frac{\pi i}{6}}Ai\left(e^{-\frac{2\pi i}{3}}z\right) = \frac{2\pi}{\sqrt[6]{3}}G^{2,0}_{2,4}\left[\frac{z^3}{9}\middle|\begin{matrix}\frac{1}{6}, \frac{2}{3} \\ 0, \frac{1}{3}, \frac{1}{6}, \frac{2}{3}\end{matrix}\right]; \end{aligned} \tag{C.33}$$

also, products like

$$z^\mu K_\nu(z),\ z^\mu Y_\mu(z)Y_\nu(z),\ z^\omega J_\nu(z)K_\mu(z),\ z^\omega I_\nu(z)K_\mu(z) \quad \text{etc.};$$

are representable as G-functions (see [106, I], [286], [370]).

WHITTAKER FUNCTION

$$W_{k,n}(z) = \exp\left(\frac{z}{2}\right)G^{2,0}_{1,2}\left[z\middle|\begin{matrix}1-k \\ \frac{1}{2}+m, \frac{1}{2}-m\end{matrix}\right]. \tag{C.34}$$

Various other generalizations of the Bessel functions towards the number of indices, like Bessel-Maitland functions, di-, n-, hyper-Bessel functions, etc. are G-functions under some additional conditions and will be considered in Sections D and E below.

D. Generalizations of the Bessel functions

Let us mention first the so-called

BESSEL-CLIFFORD FUNCTIONS

$$C_\nu(z) = z^{-\frac{\nu}{2}} J_\nu\left(2\sqrt{z}\right) = \sum_{r=0}^{\infty} \frac{(-1)^r z^r}{\nu!\Gamma(\nu+r+1)} \tag{D.1}$$

which are *entire functions* of z. For their properties and related functions $D_\nu(z)$, $E_\nu(z)$, $A_\nu^{(i)}(z) = C_\nu(z) \pm i D_\nu(z)$, $i = 1, 2, \ldots$, one can consult the handbooks on special functions and the detailed study of Hayek [126].

Generalizations of the Bessel functions $J_\nu(z)$ with one more additional index μ *have been misnamed in the literature as Bessel-Maitland functions* (in the name of E. M. (Maitland) Wright [511]-[512]) see also Stankovich [484]-[485], Gajic and Stankovich [117], Krätzel [239], Pathak [358]-[360] and the books on special functions (e.g. [276], [468], [369]), namely:

BESSEL-MAITLAND, or WRIGHT FUNCTIONS

$$J_\nu^{(\mu)}(z) = \sum_{k=0}^{\infty} \frac{(-z)^k}{k!\Gamma(1+\nu+\mu k)}, \quad \mu > -1. \tag{D.2}$$

In general, $J_\nu^{(\mu)}(z)$ are not G-functions but the more general H-functions of Fox (see Section E). However, for rational $\mu = \frac{p}{q}$, according to Pathak [360, p. 49], these functions reduce to Meijer's G-functions, viz.

$$J_\nu^{(\mu)}(z) = (2\pi)^{\frac{p-q}{2}} q^{\frac{1}{2}} p^{-\nu-\frac{1}{2}} \times G^{q,0}_{0,p+q}\left[\frac{z^q}{q^q p^p}\middle| 0, \frac{1}{q}, \frac{2}{q}, \ldots, \frac{q-1}{q}; \left(1-\frac{1+\nu}{p}\right)\left(1-\frac{2+\nu}{p}\right), \ldots, \left(1-\frac{p+\nu}{p}\right)\right].$$

D.i. Hyper-Bessel functions and generalized trigonometric functions

In 1953 Delerue [60] introduced for the first time generalizations of the Bessel functions $J_\nu(z)$ with a vector index $\nu = (\nu_1, \ldots, \nu_n)$. Later, these functions were investigated also by Klučantčev [218]-[219], Marichev [276], Adamchik [1], Adamchik and Lizarev [3]-[4], Adamchik and Marichev [5], Dimovski and Kiryakova [80]-[81], Kiryakova [196]-[198], [210], Kiryakova and Spirova [213], etc.

Definition D.1. The function

$$\begin{aligned} J^{(n)}_{\nu_1,\ldots,\nu_n}(z) &= \frac{\left(\frac{z}{n+1}\right)^{\nu_1+\cdots+\nu_n}}{\Gamma(\nu_1+1)\ldots\Gamma(\nu_n+1)}\, {}_0F_n\left((\nu_k+1)_1^n; -\left(\frac{z}{n+1}\right)^{n+1}\right) \\ &= \frac{\left(\frac{z}{n+1}\right)^{\nu_1+\cdots+\nu_n}}{\Gamma(\nu_1+1)\ldots\Gamma(\nu_n+1)} j^{(n)}_{\nu_1,\ldots,\nu_n}(z), \end{aligned} \tag{D.3}$$

where

$$j^{(n)}_{\nu_1,\dots,\nu_n}(z) = {}_0F_n\left((\nu_k+1)_1^n; -\left(\frac{z}{n+1}\right)^{n+1}\right)$$
$$= \sum_{k=1}^{\infty} \frac{(-1)^k \left(\frac{z}{n+1}\right)^{k(n+1)}}{(1+\nu_1)_k \dots (1+\nu_n)_k \, k!}, \quad |z| < \infty, \tag{D.4}$$

is said to be a *hyper-Bessel function of order* n with indices $\nu_1, \dots, \nu_n$ (multiindex $\nu = (\nu_1, \dots, \nu_n)$).

Analogously,

$$I^{(n)}_{\nu_1,\dots,\nu_n}(z) = \frac{\left(\frac{z}{n+1}\right)^{\nu_1+\dots+\nu_n}}{\Gamma(\nu_1+1)\dots\Gamma(\nu_n+1)} i^{(n)}_{\nu_1,\dots,\nu_n} \tag{D.5}$$

with

$$i^{(n)}_{\nu_1,\dots,\nu_n}(z) = {}_0F_n\left((\nu_k+1)_1^n; \left(\frac{z}{n+1}\right)^{n+1}\right) \tag{D.6}$$

is called the *modified hyper-Bessel function of order* n.

When $n = 1$, we obtain the classical Bessel functions, namely:

$$J^{(1)}_\nu(z) = J_\nu(z) = \frac{\left(\frac{z}{2}\right)^\nu}{\Gamma(\nu+1)} {}_0F_1\left(\nu+1; -\left(\frac{z}{2}\right)^2\right),$$
$$I^{(1)}_\nu(z) = I_\nu(z) = \frac{\left(\frac{z}{2}\right)^\nu}{\Gamma(\nu+1)} {}_0F_1\left(\nu+1; \left(\frac{z}{2}\right)^2\right).$$

Sometimes, it is more convenient to deal with the *generalized hypergeometric functions* ${}_0F_n(z)$ only, whose values at the origin $z = 0$ are 1. Following Klučantčev [218] we call these functions $j(z) = j^{(n)}_{\nu_1,\dots,\nu_n}(z)$ and $i(z) = i^{(n)}_{\nu_1,\dots,\nu_n}(z)$, defined by (D.4) and (D.6), *normalized hyper-Bessel functions*. They satisfy the following initial conditions, respectively:

$$j(0) = 1, \; j'(0) = \dots = j^{(n-1)}(0) = 0,$$
$$i(0) = 1, \; i'(0) = \dots = i^{(n-1)}(0) = 0. \tag{D.7}$$

Also, the so-called *Bessel-Clifford functions of n-th order*, closely related to (D.4), (D.6), have been introduced and investigated recently by Hayek and Hernandez [127]-[129]:

$$C^{(n)}_{\lambda_1,\lambda_2,\dots,\lambda_n}(z) = \frac{1}{\Gamma(\lambda_1+1)\dots\Gamma(\lambda_n+1)} {}_0F_n\left((\lambda_k+1)^n; -z\right)$$
$$= z^{-\frac{\lambda_1+\dots+\lambda_n}{n+1}} J^{(n)}_{\lambda_1,\dots,\lambda_n}\left[(n+1) z^{\frac{1}{n+1}}\right] \tag{D.8}$$
$$= \sum_{k=0}^{\infty} \frac{(-1)^k z^k}{\Gamma(\lambda_1+k+1)\dots\Gamma(\lambda_n+k+1)\, k!}.$$

These entire functions ([129]) generalize the Bessel-Clifford functions (D.1) and the Bessel-Clifford functions of third order (satisfying a third order differential equation of

Bessel type and following from (D.8) for $n = 2$), see Hayek and Hernandez [127]-[128]:

$$C_{\mu,\nu}(z) = \frac{z^{\mu+\nu}}{3^{\mu+\nu}\Gamma(\mu+1)\Gamma(\nu+1)} {}_0F_2\left(\mu+1, \nu+1; -\frac{z^3}{2}7\right).$$

The following generalizations of the usual trigonometric functions $\cos z$, $\sin z$, $\operatorname{ch} z$, $\operatorname{sh} z$ are also hyper-Bessel functions.

Definition D.2. For integer $n \geq 1$, the functions

$$k_{1,n}(z) = \cos_n(z) = \sum_{j=0}^{\infty} \frac{(-1)^j z^{nj}}{(nj)!} = {}_0F_{n-1}\left(\left(\frac{k}{n}\right)_1^{n-1}; -\left(\frac{z}{n}\right)^n\right) \tag{D.9}$$

and

$$k_{i,n}(z) = \sin_{n,n-i+1}(z) = \sum_{j=0}^{\infty} \frac{(-1)^j z^{nj+i-1}}{(nj+i-1)!}, \quad i = 2, \dots, n, \tag{D.10}$$

are said to be *(generalized) trigonometric functions of order* n. In particular, $\cos_n(z)$ is a cosine-function of order n and $\sin_{n,1}(z), \dots, \sin_{n,n-1}(z)$ are the generalized sine-functions.

It is interesting to note that when $n = 1$ functions (D.9), (D.10) reduce to the exponential function: $k_{1,1}(z) = \exp(-z)$. For $n = 2$ we obtain the usual trigonometric functions:

$$k_{1,2}(z) = \cos_2(z) = \cos(z); \quad k_{2,2}(z) = \sin_{2,1}(z) = \sin(z).$$

Some properties of the generalized trigonometric functions, analogous to the rules of trigonometry, can be found in Erdélyi at al. [108], Klučantčev [218] and other recent papers. We state here a few results, characterizing functions (D.9), (D.10) as solutions of special cases hyper-Bessel differential equations.

Lemma D.3. *The functions* $y(z) = k_{i,n}(z)$, $i = 1, \dots, n$, *are solutions of the ordinary differential equations of* n*-th order:*

$$\frac{d^n}{dz^n} y(z) = -y(z), \tag{D.11}$$

satisfying the initial conditions

$$\frac{d^{(l-1)}y}{dz^{l-1}}(0) = \delta_{i,l} = \begin{cases} 1, & i = l \\ 0, & i \neq l \end{cases}, \quad l = 1, 2, \dots, n. \tag{D.12}$$

Definition D.4. The n functions

$$h_{i,n}(z) = \sum_{j=0}^{\infty} \frac{z^{nj+i-1}}{(nj+i-1)!}, \quad i = 1, \dots, n \tag{D.13}$$

are said to be *(generalized) hyperbolic functions of order* n. In particular,

$$\begin{aligned} h_n(z) = h_{1,n}(z) &= \sum_{j=0}^{\infty} \frac{z^{nj}}{(nj)!} \\ &= {}_0F_{n-1}\left(\left(\frac{k}{n}\right)_1^{n-1}; \left(\frac{z}{n}\right)^n\right) \end{aligned} \tag{D.14}$$

is the hyperbolic analogue of the $\cos_n(z)$ function (D.9).

For

$$\begin{aligned} n = 1: &\quad h_{1,1}(z) = \exp z, \\ n = 2: &\quad h_{1,2}(z) = \operatorname{ch} z, \quad h_{2,2}(z) = \operatorname{sh} z. \end{aligned}$$

Lemma D.5. *The generalized hyperbolic functions* $y(z) = h_{i,n}(z)$ *are solutions of the initial value problem*

$$\begin{aligned} &\frac{d^n}{dz^n} y(z) = y(z), \\ &\frac{d^{l-1}y}{dz^{l-1}}(0) = \delta_{i,l} = \begin{cases} 1, & i = l \\ 0, & i \neq l \end{cases}, \quad l = 1, 2, \ldots, n. \end{aligned} \tag{D.15}$$

It is seen from the series representations (D.9), (D.10), (D.13) that the generalized trigonometric and hyperbolic functions are hypergeometric series ${}_0F_{n-1}$ with particularly chosen parameters, i.e. they are special cases of the hyper-Bessel functions (D.4), (D.6) of order $(n-1)$. This is like the situation where the usual $\cos(z)$ and $\sin(z)$ are special cases of the so-called *spherical Bessel functions* (with index $\nu = n + \frac{1}{2}$, where n is an integer).

For instance:

Lemma D.6. *The cosine function (D.9) of order* $n \geq 2$ *is a normalized hyper-Bessel function of order* $(n-1) \geq 1$ *with indices* $\nu_k = -\frac{k}{n}$, $k = 1, \ldots, n-1$, *namely:*

$$\begin{aligned} \cos_n(z) &= j_{-\frac{1}{n}, -\frac{2}{n}, \ldots, -\frac{n-1}{n}}(z) \\ &= {}_0F_{n-1}\left(\left(\frac{k}{n}\right)_1^{n-1}; -\left(\frac{z}{n}\right)^n\right). \end{aligned} \tag{D.16}$$

Proof.

$$
{}_0F_{n-1}\left(\left(\frac{k}{n}\right)_1^{n-1};-\left(\frac{z}{n}\right)^n\right)=\sum_{j=0}^{\infty}\frac{1}{\prod\limits_{k=1}^{n-1}\left(\frac{k}{n}\right)_j}\cdot\frac{(-1)^j z^{nj}}{n^{nj}j!}
$$

$$
=\sum_{j=0}^{\infty}\frac{\Gamma(j)\prod\limits_{k=1}^{n-1}\Gamma\left(\frac{k}{n}\right)}{\prod\limits_{i=0}^{n-1}\Gamma\left(j+\frac{i}{n}\right)}\cdot\frac{(-1)^j z^{nj}}{j!n^{nj}}.
$$

Since by the Gauss-Legendre *multiplication formula* (see [106, I]):

$$
\prod_{i=0}^{n-1}\Gamma\left(j+\frac{i}{n}\right)=(2\pi)^{\frac{n-1}{2}}\,n^{\frac{1}{2}-nj}\Gamma(nj),
$$

$$
\prod_{k=1}^{n-1}\Gamma\left(\frac{k}{n}\right)=(2\pi)^{\frac{n-1}{2}}\,n^{-\frac{1}{2}},
$$

then,

$$
{}_0F_{n-1}\left(\left(\frac{k}{n}\right)_1^{n-1};-\left(\frac{z}{n}\right)^n\right)=\sum_{j=0}^{\infty}\frac{(j-1)!}{j!n}\cdot\frac{(-1)^j z^{nj}}{(nj-1)!}
$$

$$
=\sum_{j=0}^{\infty}\frac{(-1)^j z^{nj}}{(nj)!}=\cos_n(z).
$$

There is an interesting generalization of the well-known assertion that the Bessel functions are reduced to elementary functions (or linear combinations of them) if and only if their indeces ν are "semi-integer", i.e. $\nu=n+\frac{1}{2}$, $n=0,\pm1,\pm2,\ldots$. These are the so-called *spherical Bessel functions.*

Lemma D.7. (Klučantčev [218]) *The hyper-Bessel functions (D.3), (D.4):*

$$
J^{(n-1)}_{\nu_1,\ldots,\nu_{n-1}}(z),\quad j^{(n-1)}_{\nu_1,\ldots,\nu_{n-1}}(z)
$$

can be reduced to linear combinations of elementary functions if and only if their multi-indeces are "semi-integer", i.e. of the form

$$
(\nu_1,\nu_2,\ldots,\nu_{n-1})=\left(\eta_1+\frac{1}{n},\eta_2+\frac{2}{n},\ldots,\eta_{n-1}+\frac{n-1}{n}\right)\tag{D.17}
$$

with some integers $\eta_1,\eta_2,\ldots,\eta_{n-1}$.

The proof follows from the fact that the *(generalized) trigonometric functions* (D.9), (D.10) *of order n are elementary functions.* Lemma D.7 gives us motivation for the following dfiniiton.

Definition D.8. (Kiryakova [196]) The hyper-Bessel functions (D.3), (D.4) with multi-indices as in (D.17) are said to be *spherical hyper-Bessel functions*:

$$J^{(n-1)}_{\eta_1+\frac{1}{n},\eta_2+\frac{2}{n},\ldots,\eta_{n-1}+\frac{n-1}{n}}(z). \tag{D.18}$$

In Chapter 4 (Section 4.3.i) we establish *generalized Rodrigues formulas*, representing functions (D.18) by means of differential operators (polynomials of $x\frac{d}{dx}$) of the generalized cosine functions. These new formulas are analogues of the representations of $J_{n+\frac{1}{2}}(z)$ by means of $\frac{d^n}{(dz^2)^n}\left\{\frac{\cos z}{z}\right\}$, $n = 0, 1, 2, \ldots$ (see (4.0.2) and [106, II]).

D.ii. Di- and n-Bessel functions

In 1980 Exton [111] generalized the Bessel function $J_\nu(z)$ in the following way

Definition D.9. The function

$$\begin{aligned} A_\nu(z) = A_{\nu,2}(z) &= \sum_{j=0}^{\infty} \frac{(-1)^j}{\{j!\}^2\,\{\Gamma(\nu+1+j)\}^2}\left(\frac{z}{2}\right)^{\nu+2j} \\ &= \left(\frac{z}{2}\right)^{\nu} \sum_{j=0}^{\infty} \frac{1}{\{\Gamma(\nu+1+j)\}^2\, j!}\cdot\frac{\left(-\left(\frac{z}{2}\right)^2\right)^j}{j!} \end{aligned} \tag{D.19}$$

is said to be a *di-Bessel function of Exton.*

He studied the corresponding generating functions, relations to the generalized hypergeometric functions, calculated some integrals of $A_\nu(z)$ and proved that as solutions of self-adjoint differential equations, these functions are orthogonal with respect to weight x^ν, $\nu > 0$. Recent results related to (D.19) belong also to Sarabia [435] and Gonzalez [121]-[122].

Later, in 1984 Agarwal [6] further generalized the functions $J_\nu(z)$ and (D.19) to $(2n-1)$-tuple indeces and studied the properties analogous to those in [111].

Definition D.10. For $n \geq 2$, the function

$$\begin{aligned} A_{\nu,n}(z) &= \sum_{j=0}^{\infty} \frac{(-1)^j}{\{j!\}^n\,\{\Gamma(\nu+1+j)\}^n}\left(\frac{z}{2}\right)^{\nu+2j} \\ &= \left(\frac{z}{2}\right)^{\nu} \sum_{j=0}^{\infty} \frac{1}{\{\Gamma(\nu+1+j)\}^n\,(j!)^{n-1}}\cdot\frac{\left(-\left(\frac{z}{2}\right)^2\right)^j}{j!} \end{aligned} \tag{D.20}$$

is said to be an *n-Bessel function of Agarwal.*

It should be noted that some of the authors recently working with the n-Bessel functions *have not observed their relation with the hyper-Bessel functions of Delerue* of 1953. That is why, we explicitly give this obvious relation here.

Lemma D.11. (Kiryakova [196]). *The n-Bessel functions (D.20) are nothing but Delerue's hyper-Bessel functions of odd order* $(2n-1)$ *(up to a substitution* $x = 2^{\frac{n-1}{n}} n z^{\frac{1}{n}}$*), viz.*

$$A_{\nu,n}(z) = J^{(2n-1)}_{\underbrace{\nu,\nu,\ldots,\nu}_{n},\underbrace{0,\ldots,0}_{n-1}}(x) = J^{(2n-1)}_{\underbrace{\nu,\nu,\ldots,\nu}_{n},\underbrace{0,\ldots,0}_{n-1}}\left(2^{\frac{n-1}{n}} n z^{\frac{1}{n}}\right). \tag{D.21}$$

Proof. Obviously,

$$\begin{aligned} A_{\nu,n}(z) &= \frac{\left(\frac{z}{2}\right)^{\nu}}{\{\Gamma(\nu+1)\}^{n}}\, {}_0F_{2n-1}\left(\underbrace{\nu+1,\ldots,\nu+1}_{n \text{ times}},\underbrace{1,1,\ldots,1}_{(n-1)\text{ times}};\ -\left(\frac{z}{2}\right)^{2}\right) \\ &= \frac{\left(\frac{x}{2n}\right)^{n\nu+(n-1)0}}{\{\Gamma(\nu+1)\}^{n}\{\Gamma(1)\}^{n-1}}\, {}_0F_{2n-1}\left(\underbrace{\nu+1,\ldots,\nu+1}_{n \text{ times}},\underbrace{1,1,\ldots,1}_{(n-1)\text{ times}};\ -\left(\frac{x}{2n}\right)^{2n}\right) \\ &= J^{(2n-1)}_{\underbrace{\nu,\ldots,\nu}_{n \text{ times}},\underbrace{1,1,\ldots,1}_{(n-1)\text{ times}}}\left(n 2^{\frac{n-1}{n}} z^{\frac{1}{n}}\right) \\ &= \frac{\left(\frac{z}{2}\right)^{\nu}}{\{\Gamma(\nu+1)\}^{n}}\, j^{(2n-1)}_{\underbrace{\nu,\ldots,\nu}_{n \text{ times}},\underbrace{1,1,\ldots,1}_{(n-1)\text{ times}}}\left(n 2^{\frac{n-1}{n}} z^{\frac{1}{n}}\right). \end{aligned}$$

In particular, the *di-Bessel functions* are:

$$A_{\nu}(z) = A_{\nu,2}(z) = J^{(3)}_{\nu,\nu,0}(x) = J^{(3)}_{\nu,\nu,0}\left(2^{\frac{3}{2}}\sqrt{z}\right) \tag{D.22}$$

and the *tri-Bessel functions*:

$$A_{\nu,3}(z) = J^{(5)}_{\nu,\nu,\nu,0,0}(x) = J^{(5)}_{\nu,\nu,\nu,0,0}\left(2^{\frac{2}{3}}\ 3\ z^{\frac{1}{3}}\right), \tag{D.23}$$

etc.

From Lemmas D.7, D.11 one obtains the following corollary.

Corollary D.12. *The n-Bessel functions are not spherical hyper-Bessel functions and therefore, they cannot be represented by linear combinations or (purely) differential operators of elementary functions.*

Due to (D.21), a series of results for the n-Bessel functions can be obtained from these on the hyper-Bessel functions; see Chapters 3 and 4.

Concerning the *recent theory of Bessel functions and hyper-Bessel operators* (for the earlier history see Watson [507]), it is interesting to mention some articles and books from the point of view of Galois groups (e.g. Duval [97]), of operator algorithms and of matrix calculus (see Bondarenko [34]), etc. Bessel functions for purely imaginary order have been considered also (e.g. Dunster [96]).

E. Fox's H-functions and special cases

Though the ${}_pF_q$ and G-functions are quite general in nature, there still exist examples of special functions which do not form their particular cases. Such functions are: Wright's generalized hypergeometric functions, Wright's generalized Bessel (Bessel-Maitland) functions, Mittag-Leffler functions, generalized parabolic cylinder functions etc. (see Section E.ii). The H-functions generalize the G-functions in quite a natural way and possess almost similar properties. More details, integrals and integral equations involving H-functions, their multivariable analogues and their relations to the existing elementary and special functions can be seen in: Fox [113], Braaksma [40] and in the books: Mathai and Saxena [287], Srivastava, Gupta and Goyal [467], Srivastava and Kashyap [468], Prudnikov, Brychkov and Marichev [370], Srivastava and Bushman [464]-[465], etc.

Here we propose only some basic definitions, properties and examples of the H-functions, having in mind that their proofs and further extensions can be obtained in a way quite similar to that used for the G-functions.

E.i. Definition and basic properties of the H-functions

Like the G-functions, the H-functions are defined and investigated mainly via their representations by *Mellin-Barnes type integrals*

$$f(z)=\frac{1}{2\pi i}\int\limits_{\gamma-i\infty}^{\gamma+i\infty}\frac{\Gamma(a_1+A_1s)\dots\Gamma(a_m+A_ms)\,\Gamma(b_1-B_1s)\dots\Gamma(b_n-B_ns)}{\Gamma(c_1+C_1s)\dots\Gamma(c_p+C_ps)\,\Gamma(d_1-D_1s)\dots\Gamma(d_q-D_qs)}z^s\,ds, \tag{E.1}$$

the latter being studied first by Pincherle (1888), Barnes (1908) and Mellin (1910). Their asymptotic expansions can be found also in Dixon and Ferrar (1936), Erdélyi et al. [106, I, 1.19].

In an attempt to unify and extend the existing results on symmetrical Fourier kernels, Fox [113] has defined the H-function in terms of a general Mellin-Barnes integral (E.1). He has also investigated the most general Fourier kernel associated with this function and obtained its asymptotic behaviour for large values of z. Later, ss the most general special function, this kernel was found to be very useful in the characterization of probability density functions and for obtaining solutions of certain dual integral equations, in fractional calculus and integral transforms, in queuing theory and related stohastic proceses in various problems of mathematical physics, etc.

Definition E.1. By *Fox's H-function* we mean a generalized hypergeometric function, represented by the Mellin-Barnes type integral

$$\begin{aligned} H^{m,n}_{p,q}(z) &= H^{m,n}_{p,q}\left[z\,\middle|\,\begin{matrix}(a_1,A_1),\dots,(a_p,A_p)\\(b_1,B_1),\dots,(b_q,B_q)\end{matrix}\right]\\ &= H^{m,n}_{p,q}\left[z\,\middle|\,\begin{matrix}(a_j,A_j)_1^p\\(b_k,B_k)_1^q\end{matrix}\right]=\frac{1}{2\pi i}\int\limits_{\mathfrak{L}}\mathcal{H}^{m,n}_{p,q}(s)z^s\,ds,\end{aligned} \tag{E.2}$$

where $\mathfrak{L}$ is a suitable contour in $\mathbb{C}$ (as for example i), ii), iii) in (A.7)); orders $(m, n; p, q)$ are non negative integers such that $0 \le m \le q$, $0 \le n \le q$; parameters A_j, $j = 1, \ldots, p$, and B_k, $k = 1, \ldots, q$, are *positive* and a_j, $j = 1, \ldots, p$; b_k, $k = 1, \ldots, q$, are arbitrary complex numbers such that

$$A_j\,(b_k + l) \neq B_k\,(a_j - l' - 1)\,; \quad l,\ l' = 0, 1, 2, \ldots\ ; \quad j = 1, \ldots, p,\ k = 1, \ldots, q,$$

and the integrand in (E.2) has the form

$$\mathcal{H}^{m,n}_{p,q}(s) = \frac{\prod\limits_{k=1}^{m} \Gamma(b_k - B_k s) \prod\limits_{j=1}^{n} \Gamma(1 - a_j + A_j s)}{\prod\limits_{k=1}^{m} \Gamma(1 - b_k + B_k s) \prod\limits_{j=1}^{n} \Gamma(a_j - A_j s)}. \tag{E.3}$$

Further, the following notation is used (cf. (A.10)):

$$\begin{aligned} \rho &= \prod_{j=1}^{p} A_j^{-A_j} \prod_{k=1}^{q} B_k^{B_k}, \\ \Delta &= \sum_{k=1}^{q} B_k - \sum_{j=1}^{p} A_j, \ \delta = m + n - \frac{p+q}{2} \\ \Omega &= \sum_{j=1}^{n} A_j - \sum_{j=n+1}^{p} A_j + \sum_{k=1}^{m} B_k - \sum_{k=m+1}^{q} B_k, \\ \nu^* &= \sum_{k=1}^{q} b_k - \sum_{j=1}^{p} a_j + \frac{p-q}{2} + 1 = \nu + \frac{p-q}{2} + 1. \end{aligned} \tag{E.4}$$

In general, Braaksma [40] has shown that the H-function makes sense and defines an analytic function of z in both cases:

$$\Delta > 0,\ 0 < |z| < \infty; \qquad \text{or} \qquad \Delta = 0,\ 0 < |z| < \rho \tag{E.5}$$

and this does not depend on the choice of $\mathfrak{L}$. More exactly, nowadays one can state the *following conditions, each one of them combined with a suitable contour* $\mathfrak{L}$, *ensuring the convergence of integral* (E.2) (e.g. [370, 8.3]):

i) $\mathfrak{L} = \mathfrak{L}_{\gamma - i\infty, \gamma + i\infty}$; $\Omega > 0,\ |\arg z| < \Omega\frac{\pi}{2}$; or $\Omega \ge 0,\ |\arg z| = \Omega\frac{\pi}{2},\ \gamma\Delta < -\Re\nu^*$;
ii) $\mathfrak{L} = \mathfrak{L}_{+\infty}$; $\Delta > 0,\ 0 < |z| < \infty$; or $\Delta = 0,\ 0 < |z| < \rho$;
or $\Delta = 0,\ \Omega \ge 0,\ |z| = \rho,\ \Re\nu^* < 0$;
iii) $\mathfrak{L} = \mathfrak{L}_{-\infty}$; $\Delta < 0,\ 0 < |z| < \infty$; or $\Delta = 0,\ |z| > \rho$;
or $\Delta = 0,\ \Omega \ge 0,\ |z| = \rho,\ \Re\nu^* < 0$;
(E.5′)

It is easy to observe that the H-function reduces to a G-function (A.7) if all $A_j = B_k = 1$:

$$H^{m,n}_{p,q}\left[z\left|\begin{matrix}(a_j,1)_1^p\\(b_k,1)_1^q\end{matrix}\right.\right] = G^{m,n}_{p,q}\left[z\left|\begin{matrix}(a_j)_1^p\\(b_k)_1^q\end{matrix}\right.\right] \tag{E.6}$$

and also, if $A_j = B_k = c$, $j = 1, \dots, p$; $k = 1, \dots, q$:

$$H^{m,n}_{p,q}\left[z\left|\begin{matrix}(a_1,c),\dots,(a_p,c)\\(b_1,c),\dots,(b_q,c)\end{matrix}\right.\right] = \frac{1}{c}G^{m,n}_{p,q}\left[z^{\frac{1}{c}}\left|\begin{matrix}(a_j)_1^p\\(b_k)_1^q\end{matrix}\right.\right]. \tag{E.6$'$}$$

More generally the same reduction holds if all the A_j and B_k are positive *rational numbers*. Then, $H^{m,n}_{p,q}(z)$ is representable as $G^{M,N}_{P,Q}(\gamma z^r)$, where r is the L.C.M. of the A_j's, B_k's, namely (see e.g. [360, (2.2)], [370, 8.3.2, (22)]):

$$\begin{aligned} H^{m,n}_{p,q}\left[z\left|\begin{matrix}(a_j,A_j)_1^p\\(b_k,B_k)_1^q\end{matrix}\right.\right] &= \left\{ r(2\pi)^{\frac{1}{2}\left(2m+2n-p-q-\sum_1^m\beta_k-\sum_1^n\alpha_j+\sum_{m+1}^q\beta_k+\sum_{n+1}^p\alpha_j\right)} \right\} \\ &\times \prod_{k=1}^q \beta_k^{b_k-\frac{1}{2}} \prod_{j=1}^p \alpha_j^{\frac{1}{2}-a_j} G^{M,N}_{P,Q}\left[\gamma z^r\left|\begin{matrix}e_1,\dots,e_p\\f_1,\dots,f_q\end{matrix}\right.\right], \end{aligned} \tag{E.7}$$

where: $\alpha_j = rA_j$, $\beta_k = rB_k$; $\gamma = \prod_{k=1}^q \beta_k^{-\beta_k} \prod_{j=1}^p \alpha_j^{\alpha_j}$;

$$M = \sum_1^m \beta_k,\ N = \sum_1^n \alpha_j,\ P = \sum_1^p \alpha_j,\ Q = \sum_1^q \beta_k;$$

$$e_j = \left\{\left(1-\frac{1-a_j}{\alpha_j}\right),\dots,\left(1-\frac{\alpha_j-a_j}{\alpha_j}\right)\right\}_{j=1}^n, \left\{\frac{a_j}{\alpha_j},\dots,\frac{a_j+\alpha_j-1}{\alpha_j}\right\}_{j=n+1}^p$$

$$f_k = \left\{\frac{b_k}{\beta_k},\dots,\frac{b_k+\beta_k-1}{\beta_k}\right\}_{k=1}^m, \left\{\left(1-\frac{1-b_k}{\beta_k}\right),\dots,\left(1-\frac{\beta_k-b_k}{\beta_k}\right)\right\}_{k=m+1}^q.$$

Basic properties:
The H-function is *symmetric* in the pairs (a_1, A_1), ..., (a_n, A_n), likewise in (a_{n+1}, A_{n+1}), ..., (a_p, A_p); (b_1, B_1), ..., (b_m, B_m) or (b_{m+1}, B_{m+1}), ..., (b_q, B_q). *Reduction formulas* similar to (A.13), (A.13$'$) hold, for example, if $n \geq 1$, $q > m$:

$$\begin{aligned} &H^{m,n}_{p,q}\left[z\left|\begin{matrix}(a_1,A_1),(\alpha_2,A_2),\dots,(a_p,A_p)\\(b_1,B_1),\dots,(b_{q-1},B_{q-1}),(a_1,A_1)\end{matrix}\right.\right] \\ &= H^{m,n-1}_{p-1,q-1}\left[z\left|\begin{matrix}(a_2,A_2),\dots,(a_p,A_p)\\(b_1,B_1),\dots,(b_{q-1},B_{q-1})\end{matrix}\right.\right]. \end{aligned} \tag{E.8}$$

Also, similarly to (A.14)-(A.15):

$$z^\sigma H^{m,n}_{p,q}\left[z\left|\begin{matrix}(a_j,A_j)_1^p\\(b_k,B_k)_1^q\end{matrix}\right.\right] = H^{m,n}_{p,q}\left[z\left|\begin{matrix}(a_j+\sigma A_j,A_j)_1^p\\(b_k+\sigma B_k,B_k)_1^q\end{matrix}\right.\right], \tag{E.9}$$

$$H_{p,q}^{m,n}\left[z\left|\begin{matrix}(a_j,A_j)_1^p\\(b_k,B_k)_1^q\end{matrix}\right.\right]=H_{q,p}^{n,m}\left[\frac{1}{z}\left|\begin{matrix}(1-b_k,B_k)_1^q\\(1-a_j,A_j)_1^p\end{matrix}\right.\right]. \tag{E.10}$$

A generalization of the *multiplication formula* (A.16) (see [287, p. 5, (1.2.7)] or [370, 8.3.2, (12)]) holds in the form (see denotations in (E.4) and (A.16)):

$$H_{p,q}^{m,n}\left[z\left|\begin{matrix}(a_j,A_j)_1^p\\(b_k,B_k)_1^q\end{matrix}\right.\right]=(2\pi)^{-(k-1)\delta}k^{\nu^*}H_{kp,kq}^{km,kn}\left[\left(zk^{-\Delta}\right)^k\left|\begin{matrix}(\Delta(k,a_j),A_j)_1^p\\(\Delta(k,b_k),B_k)_1^q\end{matrix}\right.\right]. \tag{E.11}$$

Fox's H-function can be represented also by means of a power series ([370, 8.3.2, (3)-(4)]) or by a series of the form (A.11).

The same properties as (A.12), (A.12′) hold, for example, for $m=0$ and under conditions i), ii):

$$H_{p,q}^{0,n}\left[z^k\left|\begin{matrix}(a_j,A_j)_1^p\\(b_k,B_k)_1^q\end{matrix}\right.\right]\equiv 0, \tag{E.12}$$

(see [370, 8.3.2, (2)]) and similarly for $n=0$, i), iii). In particular, for the $H_{m,m}^{m,0}$-function, frequently used in Chapter 5 as a kernel-function, the latter property takes the form:

$$H_{m,m}^{m,0}\left[z\left|\begin{matrix}(a_k,C_k)_1^m\\(b_k,C_k)_1^m\end{matrix}\right.\right]\equiv 0 \quad \text{in} \quad |z|>1, \tag{E.13}$$

since $A_k=B_k=C_k$, $k=1,\ldots,m$, $\Omega=\Delta=\delta=0$, $\rho=1$, $\nu^*=\nu+1$ and condition iii) is satisfied with $\mathfrak{L}=\mathfrak{L}_{+\infty}:\Delta=0,|z|>1$.

Several differential formulas of the form (A.17)-(A.18) can be found in the literature as well as the following formula ([287, p. 6, (1.3.5)], [370, 8.3.2, (20)]):

$$\begin{aligned}&\prod_{j=1}^{\eta}\left(z\frac{d}{dz}-c_j\right)z^\alpha H_{p,q}^{m,n}\left[z^\sigma\omega\left|\begin{matrix}(a_j,A_j)_1^p\\(b_k,B_k)_1^q\end{matrix}\right.\right]\\&\quad=z^\alpha H_{p+\eta,q+\eta}^{m,n+\eta}\left[z^\sigma\omega\left|\begin{matrix}(c_i-d,\sigma)_1^\eta,(a_j,A_j)_1^p\\(b_k,B_k)_1^q\,(c_i-\alpha+1,\sigma)_{1}\,\eta\end{matrix}\right.\right],\quad \sigma>0,\end{aligned} \tag{E.14}$$

used in Chapter 5.

Unlike the case with the G-function and the differential equation (A.19), there *does not exist a differential equation* satisfied by the H-function in the general case. However, in view of relation (E.7), one can write down the *corresponding differential equation* satisfied by an H-function (E.2) with *rational parameters* $A_j=\frac{\alpha_j}{r}$, $B_k=\frac{\beta_k}{r}$, where α_j, β_k and r are positive integers, $j=1,\ldots,p$, $k=1,\ldots,q$ and the same notation is used as in (E.7) (see [360, (3.2)]):

$$\left[(-1)^{P-M-N}\gamma z^r\prod_{j=1}^{P}\left(\frac{1}{r}z\frac{d}{dz}-e_j+1\right)-\prod_{k=1}^{Q}\left(\frac{1}{r}z\frac{d}{dz}-f_k\right)\right]H(z)=0. \tag{E.15}$$

Therefore, from (E.15) we can derive the linear differential equations satisfied by special cases of H-functions with rational additional parameters A_j, B_k.

Asymptotic expansions. According to Braaksma [40, p. 278], if $\Delta > 0$:

$$H_{p,q}^{m,n}(z) = \mathcal{O}\left(|z|^{\gamma}\right), \quad |z| \to 0 \tag{E.16}$$

with $\gamma = \min\limits_{1 \le k \le m} \Re\left(\frac{b_k}{B_k}\right)$; and

$$H_{p,q}^{m,n}(z) = \mathcal{O}\left(|z|^{\gamma^*}\right), \quad |z| \to \infty \tag{E.17}$$

with $\gamma^* = \min\limits_{1 \le j \le m} \Re\left(\frac{a_j - 1}{A_j}\right)$; also the H-function can vanish exponentially for large values of z, if for example,

$$\Delta > 0, \ \varepsilon = \Delta + \frac{p-q}{2} > 0, \ |\arg z| < \varepsilon \frac{\pi}{2},$$

and then,

$$H_{p,q}^{m,n}(z) = \mathcal{O}\left(\exp\left[-\Delta\left(\frac{|z|}{\rho}\right)^{\frac{1}{\Delta}}\right] |z|^{\frac{\varepsilon + \frac{1}{2}}{\Delta}}\right), \quad |z| \to \infty. \tag{E.17$'$}$$

It is interesting to note that *in the case* $p = q$, the H-function like the G-function has along with the singular points $z = 0, \infty$ also a third singularity at the point $z = (-\rho)^{p-m-n}$. The asymptotic behaviour as $|z| \to \rho$ (through values $|z| < \rho$ or $|z| > \rho$) seems to have not been established explicitly in the general case.

However, following Marichev's approach ([279]-[280], also see [370, 8.2.2, (60)-(61)]), in special cases one can find an explicit asymptotic formula for $|z| \to \rho$. In particular, the H-function (E.12) with $n = 0$, $p = q$, $A_j = B_j = C_j$, $j = 1, \ldots, m \Rightarrow \rho = 1$, has three regular singular points $z = 0, 1, \infty$ and near the third singularity $z = 1$ we can evaluate its behaviour as follows:

$$H_{m,m}^{m,0}\left[z \,\middle|\, \begin{matrix} (a_j, C_j)_1^m \\ (b_j, C_j)_1^m \end{matrix}\right] \sim \frac{(1-z)^{-\nu^*}}{\Gamma(1-\nu^*)}, \qquad \text{as } z \to 1, \ |z| < 1, \tag{E.18}$$

where according to (E.4):

$$-\nu^* = \sum_{j=1}^{m} (b_j - a_j) - 1 < -1, \quad \text{if} \ \ a_j > b_j, \ j = 1, \ldots, m.$$

Integrals of H-functions

The *Riemann-Liouville fractional integral of order* $\alpha > 0$:

$$R^{\alpha}\left\{z^{\lambda} H_{p,q}^{m,n}\left[\eta z^h \,\middle|\, \begin{matrix} (a_j, A_j)_1^p \\ (b_k, B_k)_1^q \end{matrix}\right]\right\} = z^{\lambda+\alpha} H_{p+1,q+1}^{m,n+1}\left[\eta z^h \,\middle|\, \begin{matrix} (-\lambda, h), (a_j, A_j)_1^p \\ (b_k, B_k)_1^q, (-\lambda-\alpha, h) \end{matrix}\right]; \tag{E.19}$$

The *Mellin transform*

$$\int_0^\infty z^{s-1} H_{p,q}^{m,n}\left[\eta z \,\middle|\, \begin{matrix}(a_j, A_j)_1^p \\ (b_k, B_k)_1^q\end{matrix}\right] ds = \eta^{-s} \mathcal{H}_{p,q}^{m,n}(-s), \tag{E.20}$$

see (E.3).

If we formally put $s = 1$ in (E.20), or more precisely follow the proof of Lemma B.2, we obtain an auxiliary integral for H-function (E.13), quite similar to (B.4) and useful for the considerations in Chapter 5, viz.:

AUXILIARY INTEGRAL

$$\int_0^1 H_{m,m}^{m,0}\left[\zeta \,\middle|\, \begin{matrix}(a_k, C_k)_1^m \\ (b_k, C_k)_1^m\end{matrix}\right] d\zeta = \prod_{k=1}^m \frac{\Gamma(b_k + C_k)}{\Gamma(a_k + C_k)}, \tag{E.21}$$

if $\Re a_k > \Re b_k > -C_k$, $k = 1, \dots, m$. The above integral is a special case of a more general *integral formula, involving a product of two H-functions* and analogous to integral formulas (A.28)-(A.29), namely ([370, 2.25, (1)]):

$$\begin{aligned}&\int_0^\infty x^{\alpha-1} H_{u,v}^{s,t}\left[\sigma x \,\middle|\, \begin{matrix}(c_i, C_i)_1^u \\ (d_l, D_l)_1^v\end{matrix}\right] H_{p,q}^{m,n}\left[\omega x^r \,\middle|\, \begin{matrix}(a_j, A_j)_1^p \\ (b_k, B_k)_1^q\end{matrix}\right] dx \\ &\quad = \sigma^{-\alpha} H_{p+v,q+n}^{m+t,n+s}\left[\frac{\omega}{\sigma^r} \,\middle|\, \begin{matrix}(a_j, A_j)_1^n, (1 - d_l - \alpha D_l, r D_l)_1^v, (a_j, A_j)_{n+1}^p \\ (b_k, B_k)_1^m, (1 - c_i - \alpha C_i, r C_i)_1^u, (b_k, B_k)_{m+1}^q\end{matrix}\right].\end{aligned} \tag{E.21$'$}$$

The Laplace transform and other integrals of the H-function can follow from the above formula as a special case.

Detailed study of *convolutional integral equations involving H-functions* as kernels can be found in Srivastava [459], Srivastava and Bushman [463]-[465]. For *H-functions of several variables* and related integral equations and transforms see Srivastava and Goyal [466], Srivastava, Gupta and Goyal [467], Nguen and Yakubovich [317].

E.ii. Special cases of the H-functions: functions of Mittag-Leffler and Bessel-Maitland (Wright) functions, Wright's generalized hypergeometric functions

Naturally, all the elementary and special functions mentioned in Section C, being Meijer's G-functions, are special cases of the H-functions too. Other examples are:

MITTAG-LEFFLER FUNCTION AND SOME GENERALIZATIONS

One of the best-known examples for an *entire function of order* $\rho > 0$ *and type* 1 is the Mittag-Leffler function

$$E_\alpha(z) = \sum_{k=0}^\infty \frac{z^k}{\Gamma(1 + k\alpha)}, \quad \alpha > 0, \quad \left(\rho := \frac{1}{\alpha}\right), \tag{E.22}$$

or alternatively denoted by

$$E_\rho(z;1)=\sum_{k=0}^{\infty}\frac{z^k}{\Gamma\left(1+\frac{k}{\rho}\right)},\qquad \rho>0. \tag{E.22$'$}$$

This is also a typical example of a commonly used special function which does not follow as a particular case of the G-functions (except for the case of rational $\rho=\frac{1}{\alpha}>0$) but is an H-function ([468, p. 42]):

$$E_\alpha(z)=H^{1,1}_{1,2}\left[-z\left|\begin{matrix}(0,1)\\(0,1),(0,\alpha)\end{matrix}\right.\right]. \tag{E.23}$$

The useful properties of functions (E.22)-(E.22$'$) are easily transferred to its generalization $E_\rho(z;\mu)$ with an arbitrary complex parameter μ, allowing greater freedom and applications.

Definition E.2. The entire function defined by means of the power series

$$E_\rho(z;\mu)=\sum_{k=0}^{\infty}\frac{z^k}{\Gamma\left(\mu+\frac{k}{\rho}\right)},\qquad \rho>0,\quad \mu\in\mathbb{C}, \tag{E.24}$$

is said to be a *Mittag-Leffler type function*, or briefly a *Mittag-Leffler (M.-L.) function.*

Now, we have the representation ([468, p. 42]):

$$E_\rho(z;\mu)=z^{\rho(\mu+1)}H^{1,1}_{1,2}\left[-z\left|\begin{matrix}(0,1)\\(0,1),\left(1-\mu,\frac{1}{\rho}\right)\end{matrix}\right.\right], \tag{E.25}$$

following from the contour integral representation (see [276, p. 101]):

$$E_\rho(z;\mu)=\frac{1}{2\pi i}\int_{\mathfrak{L}}\frac{\Gamma(s)\Gamma(1-s)}{\Gamma\left(\mu-\frac{s}{\rho}\right)}(-z)^{-s}ds. \tag{E.26}$$

It is seen then, that only for rational $\rho=\frac{m}{n}>0$ can the Mittag-Leffler function be represented by means of a G-function (cf. (E.7)):

$$E_{\frac{m}{n}}(z;\mu)=mn^{\frac{1}{2}-\mu}(2\pi)^{\frac{n+1}{2}-m}z^{\frac{m}{n}(\mu-1)}G^{1,1}_{1,2}\left[\frac{(-z)^m}{n^n}\left|\begin{matrix}\left(1-\frac{k}{m}\right)_1^m\\\left(\frac{k-1}{m}\right)_1^m,\left(1+\frac{1-\mu-k}{n}\right)_1^n\end{matrix}\right.\right]. \tag{E.25$'$}$$

Function (E.24) has been studied recently in detail by Džrbashjan [98]-[102], Džrbashjan and Bagyan [103]. Its *special cases* are:

$$E_1(z;1)=\exp z,\ E_1(z;2)=\frac{e^z-1}{z},\ E_{\frac{1}{2}}(z;1)=\operatorname{ch}\sqrt{z},\ E_{\frac{1}{2}}(z;2)=\frac{\operatorname{sh}\sqrt{z}}{z},\quad \text{etc.}$$

Similarly to the differential relation

$$\frac{d}{dz}\left(e^{\alpha z}\right) = \alpha e^{\alpha z}, \qquad \text{relating} \quad D = \frac{d}{dz} \quad \text{and} \quad e^z = E_1(z;1),$$

the Mittag-Leffler function with arbitrary ρ, μ satisfies a *more general differential relation*

$$D_{\rho,\mu} E_\rho(\alpha z;\mu) = \alpha E_\rho(\alpha z;\mu), \quad \alpha \neq 0 \tag{E.27}$$

where $D_{\rho,\mu}$ is the fractional order $\left(\frac{1}{\rho} > 0\right)$ differential operator (2.2.15), considered in Section 2.2:

$$D_{\rho,\mu}\left\{\sum_{k=0}^{\infty} a_k z^k\right\} = \sum_{k=1}^{\infty} a_k \frac{\Gamma\left(\mu + \frac{k}{\rho}\right)}{\Gamma\left(\mu + \frac{k-1}{\rho}\right)} z^{k-1},$$

which is said to be a *Džrbashjan-Gelfond-Leontiev differentiation* and can be written also as a generalized Erdélyi-Kober fractional derivative (1.6.7):

$$D_{\rho,\mu} f(z) = D_\rho^{\mu-1,\frac{1}{\rho}} \frac{f(z)}{z} = \left\{ z^{1-\mu}\left(\frac{d}{dz}\right)^{\frac{1}{\rho}} z^{\mu-1} f\left(z^{\frac{1}{\rho}}\right)\right\}_{z\to z^\rho}.$$

The following integral relation is important ($\mu > 0$, $\alpha > 0$, see [101, p. 120]):

$$\frac{1}{\Gamma(\alpha)} \int_0^z (z-\zeta)^{\alpha-1} E_\rho\left(\lambda \zeta^{\frac{1}{\rho}};\mu\right) \zeta^{\mu-1} d\zeta = z^{\mu+\alpha-1} E_\rho\left(\lambda \zeta^{\frac{1}{\rho}};\mu+\alpha\right), \tag{E.28}$$

since it shows the *relation between Mittag-Leffler functions with different parameters μ and $\mu+\alpha$* via the Riemann-Liouville integral of order $\alpha > 0$. In particular,

$$E_\rho(z;\mu) = \Phi\left\{E_\rho(z;1)\right\}, \tag{E.29}$$

where the "transmutation" Φ is the Erdélyi-Kober fractional integral

$$\Phi f(z) = \frac{1}{\Gamma(\mu-1)} \int_0^1 (1-\sigma)^{(\mu-1)-1} f\left(z\sigma^{\frac{1}{\rho}}\right) d\sigma = I_\rho^{0,\mu-1} f(z).$$

The following *convoluitonal type integral formula* turned out to be very useful in finding a convolution $\left(\underset{\rho,\mu}{*}\right)$ of the Džrbashjan-Gelfond-Leontiev integration operator $l_{\rho,\mu}$ (2.2.16)-(2.2.16′) in Dimovski and Kiryakova [77]-[78], namely:

$$\begin{aligned} &\int_0^z \zeta^{\alpha-1} E_\rho\left(\lambda\zeta^{\frac{1}{\rho}};\alpha\right)(z-\zeta)^{\beta-1} E_\rho\left(\mu(z-\zeta)^{\frac{1}{\beta}};\beta\right) d\zeta \\ &\qquad = \frac{z^{\alpha+\beta-1}}{\lambda-\mu}\left\{\lambda E_\rho\left(\lambda z^{\frac{1}{\rho}};\alpha+\beta\right) - \mu E_\rho\left(\mu z^{\frac{1}{\rho}};\alpha+\beta\right)\right\}. \end{aligned} \tag{E.30}$$

The Laplace transform of the Mittag-Leffler function is:

$$\mathfrak{L}\left\{z^{\mu-1}E_\rho\left(z^{\frac{1}{\rho}};\mu\right);s\right\}=\left(1-\frac{1}{s^{\frac{1}{\rho}}}\right)^{-1}. \tag{E.31}$$

Then, a generalization of the Mittag-Leffler functions can be considered (see Gupta [123]): $E^m_{\frac{1}{\rho},\lambda}(z)$ with $m>0$ integer, $\lambda>0$, which has a more general Laplace transform:

$$\mathfrak{L}\left\{E^m_{\frac{1}{\rho},\lambda}\left(z^{\frac{1}{\rho}}\right);s\right\}=\frac{1}{s^{\lambda-1}}\left(1-\frac{1}{s^{\frac{1}{\rho}}}\right)^{-m}. \tag{E.31$'$}$$

Another generalization of $E_\rho(z;\mu)$ has been considered by Imanaliev and Weber [148] together with corresponding results (integral rpresentations, asymptotic formulas, etc.) in view of Džrbashjan's theory, namely:

$$E^m_\rho(z;\mu)=\sum_{k=0}^{\infty}C^k_{k+m}\frac{z^k}{\Gamma\left(\mu+\frac{k}{\rho}\right)}\quad\text{with}\quad C^k_{k+m}=\frac{(k+m)!}{k!m!}, \tag{E.32}$$

where $\rho>\frac{1}{2}$, $\mu\in\mathbb{C}$, $m=0,1,\ldots$ and $E^0_\rho(z;\mu)=E_\rho(z;\mu)$.

Just as the di-Bessel and hyper-Bessel functions generalize the Bessel functions towards increasing the number of indices, similar generalizations have been studied also for the Mittag-Leffler functions $E_\rho(z;\mu)$. For example, Džrbashjan [100] introduced *such a generalization with two-dimensional indices* ρ, μ:

$$\Phi_{\rho_1,\rho_2}(z;\mu_1,\mu_2)=\sum_{k=0}^{\infty}\frac{z^k}{\Gamma\left(\mu_1+\frac{k}{\rho_1}\right)\Gamma\left(\mu_2+\frac{k}{\rho_2}\right)}, \tag{E.33}$$

where $0<\rho_1<\infty$, $0<\rho_2<\infty$ and $-\infty<\mu_1<+\infty$, $-\infty<\mu_2<+\infty$. This is an example of an entire function of

$$\text{order}\quad\rho=\frac{\rho_1\rho_2}{\rho_1+\rho_2}\qquad\text{and type}\quad\sigma=\left(\frac{\rho_1}{\rho}\right)^{\frac{\rho}{\rho_1}}\left(\frac{\rho_2}{\rho}\right)^{\frac{\rho}{\rho_2}}.$$

Special cases are:

$$E_\rho(z;\mu)=\Phi_{\rho,\infty}(z;\mu,1)$$

and the Bessel functions

$$J_\nu(z)=\left(\frac{z}{2}\right)^\nu\Phi_{1,1}\left(-\frac{z^2}{4};1,\nu+1\right).$$

Wright's (Maitland) generalized hypergeometric and Bessel functions

The next step in generalizing the special functions is to consider functions analogous to the ${}_pF_q$ functions (A.1).

Definition E.3. *Wright's generalized hypergeometric functions* are defined by means of series and represented as H-functions as follows (cf. (C.1)):

$$\begin{aligned} {}_p\Psi_q\left[\begin{matrix}(a_1,A_1),\ldots,(a_p,A_p)\\(b_1,B_1),\ldots,(b_q,B_q)\end{matrix};z\right] &= \sum_{k=0}^{\infty}\frac{\Gamma(a_1+kA_1)\ldots\Gamma(a_p+kA_p)}{\Gamma(b_1+kB_1)\ldots\Gamma(b_q+kB_q)}\frac{z^k}{k!} \\ &= H^{1,p}_{p,q+1}\left[-z\left|\begin{matrix}(1-a_1,A_1),\ldots,(1-a_p,A_p)\\(1-b_1,B_1),\ldots,(1-b_q,B_q)\end{matrix}\right.\right]. \end{aligned} \tag{E.34}$$

Naturally,

$$\begin{aligned} {}_p\Psi_q\left[\begin{matrix}(a_1,1),\ldots,(a_p,1)\\(b_1,1),\ldots,(b_q,1)\end{matrix};z\right] &= \frac{\prod_{i=1}^{q}\Gamma(b_i)}{\prod_{j=1}^{p}\Gamma(a_j)}\,{}_pF_q(a_1,\ldots,a_p;\ b_1,\ldots,b_q;\ z) \\ &= \frac{\prod_{i=1}^{q}\Gamma(b_i)}{\prod_{j=1}^{p}\Gamma(a_j)}\,G^{1,p}_{p,q+1}\left[-z\left|\begin{matrix}1-a_1,\ldots,1-a_p\\1-b_1,\ldots,1-b_q\end{matrix}\right.\right]. \end{aligned}$$

As a special case, when $p=0$, $q=1$, $b_1=\nu+1$, $B_1=\mu$, we obtain a generalization of the Bessel function (C.6):

$$J_\nu(z) = \left(\frac{z}{2}\right)^\nu \sum_{k=0}^{\infty}\frac{\left(-\frac{z^2}{4}\right)^k}{\Gamma(\nu+k+1)k!} = \frac{\frac{z}{2}}{\Gamma(\nu+1)}\,{}_0F_1\left(\nu+1;\ \frac{z^2}{4}\right),$$

namely:

$$J^\mu_\nu(z) := {}_0\Psi_1\left[(\nu+1,\mu);\ -z\right] = H^{1,0}_{0,2}\left[z|(0,1),\ (-\nu,\mu)\right]. \tag{E.35}$$

This is the so-called *Wright generalized Bessel function*, misnamed also as the *Bessel-Maitland function* ([511]-[512]; the second name of E. M. Wright is Maitland), defined by the power series:

$$J^\mu_\nu(z) = \sum_{k=0}^{\infty}\frac{(-z)^k}{\Gamma(\nu+k\mu+1)k!};\quad J^1_\nu(z) = J_\nu(z). \tag{E.36}$$

Again, if $\mu=\frac{p}{q}$ is a rational parameter, then $J^\mu_\nu(z)$ can be represented as a G-function and can be shown to satisfy the following differential equation of order $(p+q)$:

$$\left[(-z)^q q^{-q}p^{-p} - \prod_{j=1}^{q+p}\left(\frac{1}{q}z\frac{d}{dz}-d_j\right)\right]J^\mu_\nu(z) = 0,$$

$$d_j = \left\{\frac{j}{q}\right\}_1^{q-1},\ \left\{\left(1-\frac{j+\nu+1-q}{p}\right)\right\}_q^{q+p-1}.$$

This equation, written in the alternative form:

$$(-1)^q z^{\frac{\nu}{\mu}} J_\nu^\mu(z) = \left(\mu z^{1-\frac{1}{\mu}} \frac{d}{dz}\right)^p z^{\frac{\nu+p}{\mu}} \left(\frac{d}{dz}\right)^q J_\nu^\mu(z), \tag{E.37}$$

was obtained by Wright [511].

Sometimes, this function is denoted also by $\varphi(\mu, \nu+1;\ -z) = J_\nu^\mu(z)$, where

$$\varphi(\mu,\nu;\ z) = \sum_{k=0}^{\infty} \frac{z^k}{\Gamma(\nu + k\mu)k!}. \tag{E.36$'$}$$

Integral transforms of Hankel type (3.10.6) involving Wright functions $J_\nu^\mu(z)$ or $\varphi(\mu,\nu;\ z)$ instead of Bessel functions $J_\nu(z)$ have been considered by different authors, for example Stankovic [484]-[485], Gajic and Stankovic [117], Krätzel [239].

There exist also further generalizations like the *generalized Bessel-Maitland functions* (Pathak [358]-[359]):

$$J_{\nu,\lambda}^\mu(z) = \left(\frac{z}{2}\right)^{\nu+2\lambda} \frac{(-1)^k \left(\frac{z}{2}\right)^{2k}}{\Gamma(\nu + k\mu + \lambda + 1)\Gamma(\lambda + k + 1)}, \quad \mu > 0. \tag{E.38}$$

In particular,

$$J_{\nu,0}^1(z) = J_\nu(z); \quad J_{\nu,0}^\mu(z) = \left(\frac{z}{2}\right)^\nu J_\nu^\mu\left(\frac{z^2}{4}\right);$$

$$J_{\nu,\lambda}^1(z) = \frac{2^{2-2\lambda-\nu}}{\Gamma(\lambda)\Gamma(\nu+\lambda)} S_{2\lambda+\nu-1,\nu}(z), \quad \text{the Lommel function (C.8).}$$

A corresponding *generalization of the Hankel transform* (3.10.6) has the form (Pathak [358]-[359])

$$\mathfrak{J}_{\nu,\lambda}^\mu \{f(z);\eta\} = \int_0^\infty \sqrt{\eta z} J_{\nu,\lambda}^\mu(\nu z) f(z) dz, \quad \eta > 0, \tag{E.39}$$

which turns into the Hankel transform for $\mu = 1$, $\lambda = 0$; and if $\nu = \pm\frac{1}{2}$, into the *cos-Fourier* (*sin-Fourier*) transforms (3.10.1); and for $\lambda = 0$, into the Maitland transform involving $J_\nu^\mu(z)$.

Open problem E.4. We have seen in Chapter 4 that the hyper-Bessel functions of Delerue (D.3), (D.5) and the generalized hypergeometric functions ${}_pF_q$ (C.1) can be represented as generalized (multiple) fractional integrals (derivatives) of some basic elementary functions. In these cases, the corresponding (differ)integral operators involve the Meijer's $G_{m,m}^{m,0}$-functions as kernels. On this basis, a new classification and properties of the afore-mentioned functions have been derived.

Now, the same problem, yet unresolved, arises for Wright's $J_\nu^{(\mu)}$- and ${}_p\Psi_q$-functions, namely: *they should be representable as generalized fractional differintegrals involving*

$H^{m,0}_{m,m}$-*functions* as kernels (in the sense of Section 5.1), *of some basic elementary functions.* Once such a theory is developed, *find suitable classifications and new useful representations of Wright's functions* (E.34), (E.35), (E.36), (E.38).

Note. Some fractional integral relations for Wright's hypergeometric functions can be seen, for instance, in Dotsenko [92].

— || —

SOME OF THE AUTHOR'S RESULTS GIVEN IN APPENDIX ARE PUBLISHED IN: Kiryakova [196], [198], [202], Dimovski and Kiryakova [80]-[81].

References

Adamchik, V. S. (1986). On the solution of the hyper-Bessel differential equation (in Russian). *Diff. Uravnenia* **22**, No 4, 697-700.

Adamchik, V. S. (1986). *Special Cases of Functions of Hypergeometric Type and Their Applications* (in Russian). Ph.D. Thesis, Belorussian State University, Minsk.

Adamchik, V. S. and Lizarev, A. D. (1986). Asymptotic expansions of solutions of hyper-Bessel differential equations (in Russian). *Dokladi AN Bel. SSR* **30**, No 5, 393-396.

Adamchik, V. S. and Lizarev, A. D. (1987). Asymptotic solutions of differential equations of hypergeometric type (in Russian). *Diff. Uravnenia* **23**, No 9, 1475-1483.

Adamchik, V. S. and Marichev, O. I. (1984). A fundamental system of solutions of the operator $L_n = \left(z\frac{d}{dz}\right)^n + z$ (in Russian). *Diff. Uravnenia* **20**, No 6, 1082-1083.

Agarwal, A. K. (1984). A generalization of di-Bessel function of Exton. *Indian J. Pure and Appl. Math.* **15**(2), 139-148.

Al-Bassam, M. A. (1961). Some properties of Holmgren-Riesz transform in two dimentions. *Annali della Scuola Normale Super. di Pisa, Sci. Fis. e Mat.-Ser. III*, **15**, Fasc. I-II, 2-24.

Al-Bassam, M. A. (1962). Some properties of Holmgren-Riesz transform in two dimentions. *Annali della Scuola Normale Super. di Pisa, Sci. Fis. e Mat.-Ser. III*, **16**, Fasc. I, 75-90.

Al-Bassam, M. A. (1962). On functional representation by a certain type of generalized power series. *Annali della Scuola Normale Super. di Pisa, Sci. Fis. e Mat.-Ser. III*, **16**, Fasc. IV, 351-366.

Al-Bassam, M. A. (1966). H-R transform equation of Laguerre type. *Bull. College of Sci. (Baghdad, Iraq)*, **9**, 181-184.

Al-Bassam, M. A. (1970). On the existence of series solution of differential equations of generalized order. *Portugaliae Math.*, **29**, Fasc. 1, 5-11.

Al-Bassam, M. A. (1973). Some operational properties of the H-R operator. *Annali della Scuola Normale Super. di Pisa, Classe di Sci.*, **27**, Fasc. III, 491-506.

[13] Al-Bassam, M. A. (1976). On some differential and integro-differential equations associated with Jacobi's differential equation. *J. für die reine und angew. Math.* **288**, 211-217.

[14] Al-Bassam, M. A. (1982). On applications of fractional calculus to differential equations of special functions. In: *Ramanujan Int. Symp. on Anal., Dec. 1987*, Pune, 243-275.

[15] Al-Bassam, M. A. (1985). Application of fractional calculus to differential equations of Hermite's type. *Indian J. Pure Appl. Math.* **16**(9). 1009-1016.

[16] Al-Bassam, M. A. (1985). Some applications of fractional calculus to differential equations. In: *Fractional Calculus* (Ed-s: McBride, A. C. and Roach, G. F.), Proc. Workshop held in Glasgow, 1984. Pitman Publ. Ltd., London, 1-11.

[17] Al-Bassam, M. A. (1985). Some applications of generalized calculus to differential and integro-differential equations. In: *Math.Anal. and its Appl-s*, Proc. Internat. Conf. held in Kuwait, 1985. Pergamon Press, Oxford, 61-76.

[18] Al-Bassam, M. A. (1986). On generalized power series and generalized operational calculus and its application. In: *Nonlinear Analysis* (Ed. Rassias, M.). World Sci. Publ., Singapore, 51-88.

[19] Al-Bassam, M. A. (1986). On fractional analysis and its applications. In: *Modern Analysis and its Applications* (Ed. Manocha, H. L.). Prentice-Hall of India Private Ltd., New Delhi,269-308.

[20] Al-Bassam, M. A. (1989). A unified class of differential equations. *Math. Japonica*, **34**, No 4, 513-533.

[21] Al-Bassam, M. A. (1990). *Fractional Calculus and Fractional Differential Equations* (collected works), vol I, II. Kuwait University, Kuwait.

[22] Al-Bassam, M. A. and Kalla, S. L. (1989). On orthogonal polynomials associated with differential equations of Laguerre type. *Serdica (Bulg. Math. Publ.)*, **15**, 217-222.

[23] Al-Saqabi, B. N. (1987). A generalization of elliptic-type integrals. *Hadronic Journal*, **10**, 331-337.

[24] Al-Saqabi, B. N. and Kalla, S. L. (1988/1990). Some results on incomplete cylindrical functions. *Anales Acad. Cie. Argentina* **40**, 217-224.

[25] Askey, R. (1975). *Orthogonal Polynomials and Special Functions.* SIAM, Philadelphia.

[26] Bainov, D. and Shopolov, N. (1967). On a differential equation of n-th order (In Russian). *C. R. Acad. Bulg. Sci.*, **20**, No 11, 1125-1126.

[27] Barnsley, M. (1988). *Fractals Everywhere.* Acad. Press. San Diego (California).

[28] Berger, N. and Handelsman, R. (1975). Asymptotic evaluation of fractional integral operators with applications. *SIAM J. Math. Anal.*, **6**, No 5, 766-773.

[29] Berkel, C. A. M. Van (1992). *Integral Transformations and Spaces of Type S.* Ph.D. Thesis, Techn. Univ. of Eindhoven (Netherlands).

[30] Bernardi, S. D. (1969). Convex and starlike univalent functions. *TAMS*, **135**, 429-446.

Betancor, J. J. (1987). The Hankel-Schwartz transform for functions of compact support. *Rendiconti Mat. delle sue Applicazioni (Roma)* (ser. VII) **7**, No 3-4, 399-409.

Betancor, J. J. (1990). A generalized Hankel transformation of distributions and a Cauchy problem involving the Bessel-type operator $B_{\alpha_0,\alpha_1,\alpha_2} = x^{\alpha_0} D x^{\alpha_1} D x^{\alpha_2}$. *Jñanabha* **20**, 111-136.

Biernacki, M. (1960). Sur l'integrale des fonctions univalentes. *Bull. Acad. Polon. Sci., Ser. Math. et Phys.*, **8**, No 1, 29-34.

Bondarenko, B. A. (1990). *Operator Algorithms in Differential Equations* (in Russian). Fan, Tashkent.

Bora, S. and Saxena, R.K. (1971). On fractional integration. Publ. Institut Math. Beograd **11** (25), 19-22.

Botashev, A. I. (1962). On the theory of operator calculus (in Russian). *Issled. po Integro-Diff. Uravneniam v Kirgizii* **2**, 297-304.

Botashev, A. I. (1965). On the application of integral transforms to the construction of an operational calculus for Ditkin's operator (in Russian). *Issled. po Integro-Diff. Uravneniam v Kirgizii* **3**, 85-92.

Bozhinov, N. (1988). *Convolutional Representations of Commutants and Multipliers.* Publ. House Bulg. Acad. Sci., Sofia.

Bozhinov, N. and Dimovski, I. (1978). Boundary value operational calculi for linear differential operators of second order. *C. R. Acad. Bulg. Sci.*, **31**, No 7, 815-818.

Braaksma, B. L. J. (1963). Asymptotic expansions and analytic continuations for a class of Barnes integrals. *Compositio Math.* **15**, 239-341.

Braaksma, B. L. J. (1975). In memoriam C. S. Meijer. *Niew Archief Wiskunde* (3) **23**, 95-104.

de Branges, L. (1976). A proof of the Bieberbach conjecture. *Acta Math.* **154**, 137-152.

Bragg, L. (1976). The Riemann-Liouville integral and parameter shifting in a class of linear abstract Cauchy problems. *SIAM J. Math. Anal.* **7**, No 1, 1-12.

Brychkov, Yu. A., Glaeske, H.-J. and Marichev, O. I. (1983). Factorization of convolutional type integral transforms (in Russian). *Itogi Nauki i Techniki. Ser. Mat. Analiz* **21**. VINITI, Moscow.

Brychkov, Yu. A.,Prudnikov, A. P. and Shishov, V. (1979). Operational calculus (in Russian). *Itogi Nauki. Ser. Mat. Analiz* **16**. VINITI, Moscow.

Butzer, P. L., Dychkoff, H., Gorlich, E. and Stens, R. L. (1977). Best trigonometric approximation, fractional order derivatives and Lipschitz classes. *Canad. J. Math.* **29**, 781-793.

Butzer, P. L. and Westphal, U. (1975). An access to fractional differentiation via fractional difference quotients. In: *Fractional Calculus and Its Applications (Proc. Internat. Conf. Held in New Haven, 1974)*= Lecture Notes in Math. **457**. Springer-Verlag, New York, 116-145.

[48] Campos, L. M. B. C. (1984). On a concept of derivative of complex order with applications to special functions. *IMA J. Appl. Math.* **33**, 109-133.

[49] Campos, L. M. B. C. (1985-1986). On rules of derivation with complex order for analytic and branched functions. *Portugaliae Math.* **43**, Fasc. 3, 347-376.

[50] Campos, L. M. B. C. (1986). On a systematic approach to some properties of special functions. *IMA J. Appl. Math.* **36**, 191-206.

[51] Campos, L. M. B. C. (1986). On extensions of Laurent's theorem in the fractional calculus with applications to the generation of higher transcendental functions. *Bull. Math. Beograd* **38**,375-390.

[52] Carlson, B. C. and Shaffer, D. B. (1984). Starlike and prestarlike hypergeometric functions. *SIAM J. Math. Anal.* **15**, 737-745.

[53] Carroll, R. (1979). *Transmutation and Operator Differential Equations.* Notas de Matematica **67**. North-Holland, Amsterdam.

[54] Carroll, R. (1979). Transmutation and separation of variables. *Applicable Anal.* **8**, 253-263.

[55] Carroll, R. (1979). Some remarks on transmutation. *Applicable Anal.* **9**, 291-294.

[56] Carroll, R. (1981). A survey of some recent results in transmutation. In: *Spectral Theory of Diff. Operators.* North-Holland, Amsterdam, 81-92.

[57] Carroll, R. (1982). Elliptic transmutation (I). *Proc. Royal Soc. Edinburgh* **91A**, 315-334.

[58] Carroll, R. and Gilbert, J. (1981). Scattering techniques in transmutation and some connection formulas for special functions. *Proc. Japan Acad.* **57**, 34-37.

[59] Carroll, R. and Gilbert, J. (1981). Some remarks on transmutations, scattering theory and special functions. *Math. Annalen* **258**, 39-54.

[60] Delerue, P. (1953). Sur le calcul symbolique à n variables et fonctions hyperbesséliennes (II). *Annales Soc. Sci. Bruxelles, ser. 1*, No 3, 229-274.

[61] Delsarte, J. (1971). *Oeuvres de Jean Delsarte*, v. I, II. Gothier-Villar, Paris.

[62] Delsarte, J. and Lions, J. L. (1957). Transmutations d'opérateurs différentiels dans le domaine complexe. *Commentarii Math. Helvetici* **32**, 113-128.

[63] Dimkov, G. (1986). On a class of polynomials, II. In: *Proc. Complex Analysis and Applications, Varna 1985.* Sofia, 177-183.

[64] Dimovski, I. (1966). Operational calculus for a class of differential operators. *C. R. Acad. Bulg. Sci.* **19**, No 12, 1111-1114.

[65] Dimovski, I. (1968). On an operational calculus for a differential operator. *C. R. Acad. Bulg. Sci.* **21**, No 6, 513-516.

[66] Dimovski, I. (1974). On a Bessel-type integral transformation due to N. Obrechkoff. *C. R. Acad. Bulg. Sci.* **27**, No 1, 23-26.

[67] Dimovski, I. (1974). A transform approach to operational calculus for the general Bessel-type differential operator. *C. R. Acad. Bulg. Sci.* **27**, No 2, 155-158.

[68] Dimovski, I. (1975). Isomorphism of the quotient fields generated by Bessel-type differential operators. *Math. Nachrichten* **67**, 101-107.

Dimovski, I. (1975). Foundations of operational calculi for the Bessel-type differential operators. *Serdica (Bulg. Math. Publ.)* **1,** 51-63.

Dimovski, I. (1976). *A Convolutional Method in Operational Calculus.* (in Bulgarian), Dr. Sci. Thesis, Sofia.

Dimovski, I. (1979). The convolutional method in operational calculus . In: *Proc. Conf Generalized Functions and Operational Calculus, Varna 1975,* Sofia, 69-88.

Dimovski, I. (1981). Convolutional representation of the commutant of Gel'fond-Leont'ev integration operator. Acad. Bulg. Sci. **34,** No 12, 1643-1645.

Dimovski, I. (1982). *Convolutional Calculus.* Publ. House Bulg. Acad. Sci., Sofia. (Second edition: (1990). *Convolutional Calculus.* Kluwer Acad. Publ. (Ser. Math. and Its Appl-s), East European Ser. **43,** Dordrecht, Boston, London.)

Dimovski, I. and Kiryakova, V. (1980). On an integral transformation, due to N. Obrechkoff. *Lecture Notes in Math.* **798**= *Proc. Conf. on Anal. Functions, Kozubnik 1979.* Springer-Verlag, Berlin, 141-147.

Dimovski, I. and Kiryakova, V. (1981). Complex inversion formulae for the Obrechkoff transform. *Pliska (Studia Math. Bulg.)* **4**, 110-116

Dimovski, I. and Kiryakova, V. (1981). Relation between the Laplace and Obrechkoff transform (in Russian). In: *Proc. Generalized Functions and Their Appl-s in Math. Physics, Moscow 1980.* Moscow, 225-231.

Dimovski, I. and Kiryakova, V. (1983). Convolution and commutant of Gel'fond-Leont'ev operator of integration. In: *Proc. Constructive Function Theory, Varna 1981.* Sofia, 288-294.

Dimovski, I. and Kiryakova, V. (1984). Convolution and differential property of the Borel-Dzhrbashian transform. In: *Proc. Complex Anal. and Appl-s, Varna 1981.* Sofia 148-156.

Dimovski, I. and Kiryakova, V. (1985). Transmutations, convolutions and fractional powers of Bessel-type operators via Meijer's G-function. In: *Proc. Complex Anal. and Appl-s, Varna 1983.* Sofia, 45-66.

Dimovski, I. and Kiryakova, V. (1986). Generalized Poisson transmutations and corresponding representations of hyper-Bessel functions. *C. R. Acad. Bulg, Sci.* **39,** No 10, 29-32

Dimovski, I. and Kiryakova, V. (1987). Generalized Poisson representations of hypergeometric functions ${}_pF_q, p < q$ using fractional integrals. In: *Proc.* 16th *Spring Conf. Union Bulg. Math. 1987.* Sofia, 205-212.

Dimovski, I. and Kiryakova, V. (1993). Discrete operational calculi for two-sided sequences. *The Fibonacci Quart.* To appear.

Dimovski, I. and Vasilev, M. (1989). New representations of commutants of powers of the differential operator. In: *Proc. Complex Anal. and Appl-s, Varna 1987.* Sofia, 137-149.

[84] Dinh, Anh Hoang (1991). Factorization of some G-transformation in weighted L_p-spaces and its application to solving integral equations (in Russian). *DAN Byel.SSR*. To appear.

[85] Ditkin, V. A. (1957). On the theory of operator calculus (in Russian). *Dokladi AN SSSR* **116**, No 1, 15-17.

[86] Ditkin, V. A. and Prudnikov, A. P. (1962). An operational calculus for the Bessel operators (in Russian). *Zhournal Vych. Mat. i Mat. Fiziki* **2**, 997-1018.

[87] Ditkin, V. A. and Prudnikov, A. P. (1963). On the theory of operational calculus, generated by the Bessel equation (in Russian). *Zhournal Vych. Mat. i Mat. Fiziki* **3**, No 2, 223-238.

[88] Ditkin, V. A. and Prudnikov, A. P. (1967). Integral transforms (in Russian). *Itogi Nauki, Ser. Mat. Analiz*, 1966. VINITI, Moscow.

[89] Ditkin, V. A. and Prudnikov, A. P. (1971). The operational calculus of a class of generalized Bessel operators. Preprint of a talk at: *Symposium on Operational Calculus and Generalized Functions, Dubrovnik 1971.*

[90] Ditkin, V. A. and Prudnikov, A. P. (1974). *Integral Transforms and Operational Calculus* (in Russian). Nauka, Moscow.

[91] Ditkin, V. A. and Prudnikov, A. P. (1975). *Operational Calculus* (in Russian). Nauka, Moscow.

[92] Dotsenko, M. R. (1991). On some applications of Wright's hypergeometric function. *C. R. Acad. Bulg. Sci.* **44**, No 6, 13-16.

[93] Dryanov, D. (1982). Average modulus of smoothness of fractional index and fractional order derivatives. *C. R. Acad. Bulg. Sci.* **35**, No 12, 1635-1637.

[94] Dryanov, D. (1983). Average modulus of smoothness of fractional index and applications. *C. R. Acad. Bulg. Sci.* **36**, No 1, 41-43.

[95] Dryanov, D. (1985). Equivalence between fractional average modulus of smoothness and fractional one-sided K-functional. *C. R. Acad. Bulg. Sci.* **38**, No 12, 1609-1612.

[96] Dunster, T. M. (1990). Bessel functions of purely imaginary order, with an application to second-order linear differential equations having a large parameter. *SIAM J. Math. Anal.* **21**, No 4, 995-1018.

[97] Duval, Anne (1989). Biconfluence et groupe de Galois. *Publ. IRMA (Lille)* **18**, No 1, 1-11.

[98] Džrbashjan, M. M. (1952). On the integral representation and uniqueness of some classes of entire functions (in Russian). *Dokladi AN SSSR* **85**, No 1, 29-32.

[99] Džrbashjan, M. M. (1954). On the integral representation of functions, continuous on several rays, or: a generalization of the Fourier integral (in Russian). *Izv. AN SSSR* **18**, 427-448.

[100] Džrbashjan, M. M. (1960). On the integral transformations generated by the generalized Mittag-Leffler function (in Russian). *Izv. AN Arm. SSR* **13**, No 3, 21-63.

[101] Džrbashjan, M. M. (1966). *Integral Transforms and Representations of Functions in the Complex Domain* (in Russian). Nauka, Moscow.

Džrbashjan, M. M. (1974). On the closeness of the system of Mittag-Leffler type (in Russian). *Dokladi AN SSSR* **219**, No 6, 1302-1305.

Džrbashjan, M. M. and Bagyan, R. A. (1975). On the integral representation and measures, associated with functions of Mittag-Leffler type (in Russian). Izv. *AN Arm. SSR* 10, No 6, 483-508.

Erdèlyi, A. (1940). On fractional integration and its applications to the theory of Hankel transforms. *Quart. J. Math. Oxford* **11**, 293-303.

Erdèlyi, A. (1951). On some functional transformations. *Univ. Politech. Torino Rendiconti Sem. Mat.* **10**, 217-234.

Erdèlyi, A. et al. (1953). *Higher Transcendental Functions*, vol. I, II. McGraw-Hill, New York.

Erdèlyi, A. et al. (1954). *Tables of Integral Transforms*, vol. I, II. McGraw-Hill, New York.

Erdèlyi, A. et al.(1955). *Higher Transcendental Functions*, vol III. McGraw-Hill, New York.

Erdèlyi, A. and Kober, H. (1940). Some remarks on Hankel transforms. *Quart. J. Oxford* **11**, 212-221.

Erdèlyi, A. and Sneddon, I. (1962). Fractional integration and dual integral equations. *Canad. J. Math.* **14**, 685-693.

Exton, H. (1980). The di-Bessel function. *Indian J. Pure and Appl. Math.* **11** (7), 856-862.

Fage, M. K. and Nagnibida,N. I. (1987). *The Problem of Equivalence of Ordinary Differential Operators* (in Russian). Nauka, Novosibirsk.

Fox, Ch. (1961). The G- and H-functions as symmetric Fourier kernels. *Trans. Amer. Math. Soc.* **98**, 395-429.

Fox, Ch. (1963). Integral transforms based upon fractional integration. *Proc. Cambridge Phil. Soc.* **59**, 63-71.

Fox, Ch. (1965). A formal solution of certain dual integral equations. *Trans. Amer. Math. Soc.* **119**, 389-398.

Fox, Ch. (1971). Solving integral equations by L and L^{-1} operators. *Proc. Amer. Math. Soc.* **29**, 299-306.

Gajic, L. and Stankovic (1976). Some properties of Wright's function. *Publ. l'Institut Math. Beograd, Nouv. Ser.* **20** (34), 91-98.

Galue, L. and Kiryakova, V. (1993). Further results on a family of generalized radiation integrals. *Radiation Phys. & Chem.*, To Appear.

Galue, L., Kiryakova, V. S. and Kalla, S. L. (1993). Solution of dual integral equations by fractional calculus. *Math. Balkanica (N. S.)* **7**, To appear.

Gelfond, A. O. and Leontiev, A. F. (1951). On a generalization of the Fourier series (in Russian). *Mat. Sbornik* **29**(71), 477-500.

Gonzalez, J. M. (1990). Operational calculus for an integral transformation of Watson type and an integrodifferential operator. In: *Fractional Calculus and Its*

Appl-s (Proc. Intern. Conf. Held in Tokyo, 1989), Nihon Univ., 1990; 191-200.

[122] Gonzalez, J. M. (1992). Nuevas Propriedades de la función di-Bessel de Exton. *Rev. Tec. Ing., Univ. Zulia* **15**, No 2, 135-141.

[123] Gupta, S. D. (1973). On certain integral equations. *Math. Balkanica* **3**, 115-117.

[124] Hadamard, J. (1892). Essai sur l'etude des fonctions données par leur développment de Taylor. *Math. Pures and Appl., ser. 4*, **8**, 101-186.

[125] Hardy, G. H., Littlewood, J. E. and Polya, G. (1934). *Inequalities.* Cambridge Univ. Press, London.

[126] Hayek, N. (1966-67). Estudio de la ecuacion diferencial $xy'' + (\nu + 1)y' + y = 0$ y de sus aplicaciones. *Collectanea Math.* **18**, 57-174.

[127] Hayek, N. and Hernandez, V. (1991). Sobre las funciones de Bessel-Clifford de tercer orden. *Actas del* XII *C.E.D.Y.A., Oviedo'91*, 135-140.

[128] Hayek, N. and Hernandez, V. (1992). Representaciones integrales de las funciones de Bessel-Clifford de tercer orden. *Rev. Academia de Ciencias. Zaragoza* **47**, 51-60.

[129] Hayek, N. and Hernandez, V. (1992). On a class of functions connected with the hyper-Bessel functions. *Jñanabha*, **23**, To appear.

[130] Heading, J. (1957). The Stokes phenomenon and certain n-th order differential equations, I. *Proc. Cambridge Phil. Soc.* **53**, part 2, 399-418.

[131] Heading, J. (1960). The Stokes phenomenon, II. *Proc. Cambridge Phil. Soc.* **56**, part 4, 419-441.

[132] Heading, J. (1962) *An Introduction to Phase-Integral Methods.* John Wiley, New York.

[133] Heading, J. (1974). The Stokes phenomenon and generalized contiguous transformations of some generalized hypergeometric functions. *Proc. Cambridge Phil. Soc.* **76**, 423-442.

[134] Hearsh, R. (1975). The method of transmutations. In: Lecture Notes in Math. **446**, *Spinger-Verlag, New York.*, 264-282.

[135] Hohlov, Yu. (1987). Hadamard's products, hypergeometric functions and linear operators in the class of univalent functions (in Russian). *Dokladi AN Ukr.SSR, Ser. A.*, No 7, 25-27.

[136] Hohlov, Yu. (1989). Convolutional operators preserving univalent functions. *Pliska (Studia Math. Bulg.)* **10**, 87-92.

[137] Hriptun, M. D. (1961). About one differential equation of 3rd order (in Russian). In: *Issled. po Sovremennim Problemam Teorii Funkcii Compl. Peremennogo*, Moscow, 490-499.

[138] Hriptun, M. D. (1975). Multiplication theorems for the solutions of generalized differential equation of Bessel of mth order (in Russian). *Diff. Uravnenia* **18**, No 2, 287-293.

[139] Hristova, M. (1984). A general form of the linear operators defined in spaces of analytic functions and commuting with the generalized Hardy-Littlewood operator (in Bulgarian). *Godishnik VUZ. Pril. Matematika.* **20**, No 2, 67-74.

Hristova, M. (1986). Some analytic characteristics of operators, commuting with the fixed power of the generalized Hardy-Littlewood operator (in Bulgarian). *Nauchni Trudove VII "K. Marx"* **1**, 149-159.

Hristova, M. (1992). On a commutant of a general linear operator of differentiation type. *Mathematica Balkanica (N.S.)* **6**, No 4, 333-340.

Hubbell, J. H. (1963). A power-series buildup factor formulation. Application to rectangular and off-axis disk source problems. *J. Res. Nat. Bur. Stand.* **67c**, 291-306.

Hubbell, J. H., Bach, R. L. and Lamkin, J. C. (1960). Radiation field from a rectangular source. *J. Res. Nat. Bur. Stand.* **64c**, 121-138.

Ignatov, Z. G. and Kaishev, V. K. (1987). Some properties of generalized B-splines. *Proc. Conf. Constr. Function Theory, Varna 1987.* Sofia, 233-241.

Ignatov, Z. G. and Kaishev, V. K. (1987). On the computations of distributions of serial correlation coefficients through B-splines. *Proc. 2nd Internat. Tampere Conf. in Ststistics, Finland 1987.* Depart. Math. Sci. Univ. Tampere, Tampere, 479-489.

Iliev, L. (1979). *Zeros of Entire Functions* (in Bulgarian). Publ. House of Bulg. Acad. of Sci., Sofia.

Iliev, L. (1987). *Laguerre Entire Functions.* Publ. House Bulg. Acad. of Sci., Sofia.

Imanaliev, M. I. and Weber, V. K. (1980). On a generalization of functions of Mittag-Leffler type and its application (in Russian). *Issled. po Integro-Diff. Uravneniam v Kirgizii* **13**, 49-59.

Ivanov, K. (1983). On the rates of convergence of two moduli of functions. *Pliska (Studia Math. Bulg.)* **5**, 97-104.

Jakubowski, Z. (1983). On typical-real functions. *Annales Polonici Math.* **40**, 151-160.

Jakubowski, Z. (1985). On the properties of integral operators. *Bull. Australian Math. Soc.* **32**, No 1, 55-68.

Jakubowski, Z. (1989). On some classes of analytic functions. *Proc. Complex Anal. and Appl-s, Varna 1987.* Sofia, 241-249.

Kabe, D. G. (1958). Some applications of Mayer's G-function to distribution problems in statistics. *Biometrica* **45**, 578-580.

Kaishev, V. K. (1991). A Gaussian cubature formula for the computation of generalized B-splines and its application to serial correlation. *Contemporary Math.* **115**, 219-237.

Kamke, E. (1976). *A Handbook on Ordinary Differential Equations* (in Russian). Nauka, Moscow.

Kalla, S. L. (1966). Some theorems of fractional integration. *Proc. Nat. Acad. Sci. India, Sect. A*, **36**, 1007-1012.

Kalla, S. L. (1969). Integral operators involving Fox's H-function, I. *Acta Mexicana Cie. Tech.* **3**, 117-122.

Kalla, S. L. (1969). Integral operators involving Fox's H-function, II. *Acta Mexicana Cie. Tech.* **7**, 72-79.

[159] Kalla, S. L. (1969). Fractional integration operators involving hypergeometric functions, II. *Acta Mexicana Cie. Tech.* **3**, 1-5.

[160] Kalla, S. L. (1970). Fractional integration operators involving hypergeometric functions, I. *Univ. Nac. Tucuman, Rev. Ser.* **A20,** 93-100.

[161] Kalla, S. L. (1970). On operators of fractional integration, I. *Mat. Notae* **22**, 89-93.

[162] Kalla, S. L. (1973). A generalization of the Sonine-Mehler integral. *Univ. Nac. Tucuman, Rev. Ser.* **23**, 111-112.

[163] Kalla, S. L. (1976). On operators of fractional integration, II. *Mat. Notae* **25**, 29-35.

[164] Kalla, S. L. (1980). Operators of fractional integration. In: *Proc. Conf. Analytic Functions, Kozubnik 1979 = Lecture Notes of Math.* **798**, 258-280.

[165] Kalla, S. L. (1986). Generalized elliptic-type integrals. In: *Proc. Complex Anal. and Appl-s, Varna 1985.* Sofia, 295-310.

[166] Kalla, S. L. and Al-Saqabi, B. (1985). A functional relation involving the Ψ -function. *Rev. Tec. Ing., Univ. Zulia* **8**, No 2, 31-35.

[167] Kalla, S. L. and Al-Saqabi, B. (1986). On a generalized elliptic-type integral. *Rev. Brasileira Fis.* **26**, No 1, 145-156.

[168] Kalla, S. L. and Al-Saqabi, B. (1988). Summation of certain infinite series with di-gamma functions. *C. R. Acad. Bulg. Sci.* **41**, No 11, 15-17.

[169] Kalla, S. L. and Al-Saqabi, B. and Conde, S. (1987). Some results related to radiation-field problems. *Hadronic J.* **10**, No 3, 221-230.

[170] Kalla, S. L., Conde, S. and Hubbell, J. H. (1986). Some results on generalized elliptic-type integrals. *Appl. Anal.* **22**, 273-287.

[171] Kalla, S. L. and Galue, L. (1989). Integral operators involving incomplete cylindrical functions. In: *Proc. Complex Anal. and Appl-s, Varna 1987.* Sofia, 250-263.

[172] Kalla, S. L., Galue, L. and Kiryakova, V. (1991). Some expansions related to a family of generalized radiation integrals. *Math. Balkanica (N. S.)* **5**, No 3, 190-202.

[173] Kalla, S. L., Leubner, C. and Hubbell, J. H. (1987). Further results on generalized elliptic-type integrals. *Appl. Anal.* **25**, 269-274.

[174] Kalla, S. L. and Kiryakova, V. S. (1990). A generalized fractional calculus dealing with H-functions. In: *Fractional Calculus and Its Appl-s (Proc. Intern. Conf. Held in Tokyo, 1989)*, Nihon Univ., 1990; 62-69.

[175] Kalla, S. L. and Kiryakova, V. S. (1990). An H-function generalized fractional calculus based upon composition of Erdélyi-Kober operators in L_p. *Math. Japonica* **35**, No 6, 1151-1171.

[176] Kalla, S. L. and Ross, B. (1985). The development of functional relations by means of fractional operators. In: *Fractional Calculus. Proc. Workshop held in Glasgow, 1984 = Res. Notes in Math.* **138**, Pitman, London, 32-43.

[177] Kalla, S. L. and Saxena, R. K. (1969). Integral operators involving hypergeometric functions. *Math. Z.* **108**, 231-234.

Kalla, S. L. and Saxena, R. K. (1974). Integral operators involving hypergeometric functions - II. *Univ. Nac. Tucumán, Rev. Ser.* **A24**, 31-36.

Kaljabin, G. A. (1980). *Theory of Spaces of Functions of Generalized Smoothness and Some Applications* (in Russian). Author's Summary of a Dr. Sci. Thesis. AN SSSR, Moscow.

Kaljabin, G. A. (1985). A characterization of certain function spaces by means of generalized differences. *Soviet Math. Dokladi* **32**, No 2, 583-586.

Kaljabin, G. A. (1988). Characterization of Bessov-Lizorkin-Treibel type spaces by means of generalized differences (in Russian) *Trudy Mat. Instituta AN SSR* **181**, 95-116.

Kaljabin, G. A. and Lizorkin, P. I. (1987). Spaces of functions of generalized smoothness. *Math. Nachr.* **133**, 7-32.

Kapoor, G. P. and Patel, J. (1986). Univalence of Gelfond-Leontiev derivatives of functions defined by gap power series. *Rendiconti Mat. delle sue Appl.* (Roma) (VII) **6**, No 4, 491-502.

Karanikoloff, H. (1949). Sur une équation différentielle envisagée par Kümmer. *C. R. Acad. bulg Sci.* **2**, No 1.

Khachatrjan, I. G. (1978). On transmutation operators for differential equations of higher orders (in Russian). *Izv. Akad. Nauk Arm. SSR* **13**, No 3, 213-237.

Khachatrjan, I. G. (1979). On the existence of a transmutation operator for differential equations of higher orders, preserving the asymptotics of the solutions (in Russian). *Izv. Akad. Nauk Arm. SSR* **14**, No 6, 424-445.

Khachatrjan, I. G. (1980). Transmutation operators for boundary-value problems (in Russian). *Izv. AN Arm. SSR* **15**, No 1, 30-44.

Kilbas, A. A. (1978). Potential-type operators with power-logarithmic kernels in a weighted Hölder space (in Russian). *Izv. AN Bel. SSR, ser. fiz.-mat.* No 2, 29-37.

Kilbas, A. A. (1990). Asymptotic expansions for fractionals integrals and applications. In: *Fractional Calculus and Its Appl-s (Proc. Intern. Conf. Held in Tokyo, 1989)*, Nihon Univ., 1990; 70-79.

Kiryakova, V. S. (1975). *Differential Property and Inversion Formulas for the Obrechkoff transform* (in Bulgarian). M. Sc. Thesis, Sofia University.

Kiryakova, V. (1982). An explanation of Stokes phenomenon by the method of transmutations. In: *Proc. Diff. Equations and Appl-s, Rousse 1981.* Rousse, 349-353.

Kiryakova, V. (1984). An application of Meijer's G-function to Bessel-type operators. In: *Proc. Constructive Function Theory, Varna 1984.* Sofia, 457-462.

Kiryakova, V. (1984). Integral transformation for Bessel-type differential operators via Meijer's G-function. In: *Proc. Complex Anal. and Appl. to Partial Diff. Equations*, Halle, 67-71.

Kiryakova, V. (1986). On operators of fractional integration involving Meijer's G-function. *C. R. Acad. Bulg. Sci.* **39**, No 10, 25-28.

[195] Kiryakova, V. (1986). New integral representations of the generalized hypergeometric functions. *C. R. Acad. Bulg. Sci.* **39**, No 12, 33-36.

[196] Kiryakova, V. (1986). *Generalized Operators of Fractional Integration and Differentiation and Their Applications* (in Bulgarian). Ph.D. Thesis, Sofia Univ, Sofia.

[197] Kiryakova, V. (1986). Applications of the generalized Poisson transformation to solving hyper-Bessel differential equations (in Bulgarian). *Godishnik VUZ. Pril. Matematika* **22**, No 4, 129-140.

[198] Kiryakova, V. (1987). Generalized fractional derivative representations of hypergeometric functions ${}_pF_q$ *with* $p < q$. *Proc.* 16th *Spring Conf. Union Bulg. Math. 1987,* Sofia, 228-235.

[199] Kiryakova, V. (1987). Multiple Erdélyi-Kober fractional differintegrals and their uses in univalent, starlike and convex function theory. *Annual Univ. Sofia "Kl. Ohridski"* **81**. To appear.

[200] Kiryakova, V. (1988). Generalized fractional derivative representations of hypergeometric functions ${}_pF_q$ for $p = q$ and $p = q + 1$. In: *Proc. Constructive Functoin Theory, Varna 1987.* Sofia, 260-267.

[201] Kiryakova, V. (1988). A generalized fractional calculus and integral transforms. In: *Proc. GFCA, Dubrovnik 1987.* Plenum Publ. Co., New York, 205-217.

[202] Kiryakova, V. (1988). A new differential formula for Meijer's G-function. *C. R. Acad. Bulg. Sci.* **41**, No 5, 27-30.

[203] Kiryakova, V. (1988). Fractional integration operators involving Fox's $H^{m,0}_{m,m}$-function. *C. R. Acad. Bulg. Sci.* **41**, No 11, 11-14.

[204] Kiryakova, V. (1988). Generalized $H^{m,0}_{m,m}$-function fractional integration operators in some classes of analytic functions. *Bull. Math. (Beograd)* **40**, No 3-4, 259-266.

[205] Kiryakova, V. (1989). An application of the generalized operators of fractional integration to dual integral equations involving Meijer's G-function. *Pliska (Studia Math. Bulg.)* **10**, 93-107.

[206] Kiryakova, V. (1989). Convolutions of Erdélyi-Kober fractional integration operators. In: *Proc. Complex Anal. and Appl., Varna 1987.* Sofia, 273-283 .

[207] Kiryakova, V. (1989). Explicit solution to a generalized (multiple) Abel integral equation via the H-function. In: *Equadiff 7, Praha 1989 (Enlarged abstracts).* Praha, 153-155.

[208] Kiryakova, V. (1990). On a class of generalized operators of fractional integration. In: *Proc. Jubillee Sess. Devoted to Acad. Chakalov, Samokov 1986*, Sofia, 79-87.

[209] Kiryakova, V. (1990). Poisson and Rodrigues type fractional differential formulas for the generalized hypergeometric functions ${}_pF_q$. *Atti. Sem. Mat. Fis. Univ.Modena.* **39**, 311-322.

[210] Kiryakova, V. (1992). Solving hyper-Bessel differential equations by means of Meijer's G-functions, I: Two alternative approaches. *Reports Strathclyde Univ., Math. Dept.*, No 19, 36p.

] Kiryakova, V. S. and McBride, A. C. (1992). On solving hyper-Bessel differential equations by means of Meijer's G-functions, II: The non homogeneous case. *Reports Strathclyde Univ., Math. Dept.*, No 20, 22p.

] Kiryakova, V. S. and McBride, A. C. (1993). Explicit solution of the non homogeneous hyper-Bessel differential equation. *C. R. Acad. Bulg. Sci.* **46**. To appear.

] Kiryakova, V. and Spirova, S. (1989). Representations of the solutions of hyper-Bessel differential equations via Meijer's G-function. In: *Proc. Complex Anal. and Appl-s, Varna 1987*. Sofia, 284-297.

] Kiryakova, V. S. and Srivastava, H. M. (1993). Generalized (multiple) Riemann-Liouville fractional differintegrals and their use in univalent function theory. In: *Analysis, Geometry and Groups: a Riemann Legacy Volume*. Hadronic Press, Inc., Florida. To appear.

] Kiryutenko, Yu. A. (1981). Operators commuting with the integration in the space of functions, analytic in simplyconnected domains (in Russian). *Mat. Zametki* **29**, No 3, 409-419.

] Kljuchantzev, M. (1975). On the construction of r-even solutions of singular differential equations (in Russian). *Dokladi AN SSSR* **224**, No 5, 1000-1008.

] Kljuchantzev, M. (1976). Integrals of fractional order and singular boundary-value problems (in Russian). *Diff. Equations* **12**, No 6, 983-990.

3] Kljuchantzev, M. (1983). Singular differential operators with $(r-1)$ parameters and Bessel functions with vector indices (in Russian). *Sibirskii Mat. Zhournal* **24**, No 3, 47-62.

9] Kljuchantzev, M. (1987). An introduction to the theory of $(\nu_1, \ldots, \nu_{r-1})$-transforms (in Russian). *Mat. Sbornik* **132**, No 2, 167-181.

0] Kober, H. (1940). On a fractional integral and derivative. *Quart. J. Math. Oxford* **11**, 193-211.

1] Koh, E. L. (1976). A Mikusinski calculus for the Bessel operator B_ν. In: *Proc. Diff. Eqs = Lecture Notes in Math.* **564**, Springer-Verlag, Berlin-Heidelberg-New York, 291-300.

2] Koh, E. L. (1982). A direct extension of Meller's calculus. *Internal J. Math. & and Math. Sci.* **5**, No 4, 785-791.

3] Koh, E. L., Deeba, E. Y. and Ali, M. A. (1987). The Meijer transformation of generalized functions. *Internat. J. Math. and Math. Sci.* **10**, No 2, 267-286.

4] Kohno, M. (1979). An extended Airy function of the first order. *Hiroshima Math. J.* **2**, No 2, 473-489.

5] Komatsu, H. (1966). Fractional powers of operators, I. *Pacific J. Math.* **19**, 285-346.

6] Komatsu, H. (1967). Fractional powers of operators, II: Interpolation spaces. *Pacific J. Math.* **21**, 89-111.

7] Komatsu, H. (1969). Fractional powers of operators, III: Negative powers. *Journal Math. Soc. Japan* **21**, No 2, 205-220.

8] Komatsu, H. (1969). Fractional powers of operators, IV: Potential operators. *Journal Math. Soc. Japan* **21**, No 2, 221-228.

[229] Komatsu, H. (1970). Fractional powers of operators, V: Dual operators . *J. Fac. Sci. Univ. Tokyo, Sect I,* **17**, Part. 1&2, 373-396.

[230] Komatu, Y. (1980). A one-parameter family of operators defined on analytic functions in a circle. In: *Proc. Conf. on Anal. Functions, Kozubnik 1979 = Lecture Notes in Math.* **798**, Springer-Verlag, Berlin, 292-300.

[231] Komatu, Y. (1986). On oscillation in a class of analytic functions related to fractional integration. In: *Proc. Complex Anal. and Appl-s, Varna 1985.* Sofia, 327-334.

[232] Komatu, Y. (1989). On distortion properties of operators defined on analytic functions. In: *Proc. Complex Anal. and Appl-s, Varna 1987.* Sofia, 398-304.

[233] Koornwinder, T. H. (1975). A new proof of a Paley-Wiener theorem for the Jacobi transform. *Arkiv för Math.* **13**, 145-159.

[234] Koprinski, S. (1975). *Operational Methods for Some Generalized Heat Equations (in Bulgarian).* Ph. D. Thesis. Sofia.

[235] Krätzel, E. (1965). Eine Verallgemeineiung der Laplace- und Meijer- Transformation. *Wiss. Z. Friedrich-Shiller-Univ. Math.-naturwiss Reihe* **14**, No 5, 369-381.

[236] Krätzel, E. (1965). Die Faltung der L-Transformation. *Wiss. Z. Friedrich-Shiller-Univ. Math.-naturwiss Reihe* **14**, No 5, 383-390.

[237] Krätzel, E. (1967). Differentiationssätze der L-Transformation und Differentialgleichungen nach dem Operator. *Math. Nachr.* **35**, No 1-2, 105-114.

[238] Krätzel, E. (1967). Übertragung der Post-Widderschen Umkehrformel der Laplace-Transformationen auf die L-Transformation. *Math. Nachr.* **35**, No 5-6, 295-304.

[239] Krätzel, E. (1979). Integral transformations of Bessel-type. In: *Proc. Conf. Generalized Functoins and Operational Calculus, Varna 1975.* Sofia, 148-155.

[240] Krein, S. G., Petunin, Yu. I. and Semenov, E. M. (1978). *Interpolation of Linear Operators* (in Russian). Nauka, Moscow.

[241] Kummer, E. (1839). Sur l'équation $\frac{d^n y}{dx^n} = x^n y$. *Journal de Crelle* **19**, 286.

[242] Kuyosying, S. (1990). Fractals: A new dimension in mathematics. In: *Fractional Calculus and Its Appl-s (Proc. Intern. Conf. Held in Tokyo, 1989)*, Nihon Univ., 1990; 270-280.

[243] Lamb, W. (1984). Fractional powers of operators defined on a Fréchet space. *Proc. Edinburgh Math. Soc.* **27**, 165-180.

[244] Lamb, W. (1985). A distributional theory of fractional calculus. *Proc. Royal Soc. Edinburgh* **99A**, 347-357.

[245] Lamb, W. (1986). Fourier multipliers on spaces of distributions. *Proc. Edinburgh Math. Soc.* **29**, 309-327.

[246] Lamb, W. and McBride, A. C. (1988). On relating two approaches to fractional calculus. *J. Math. Anal. and Appls* **132**, No 2, 590-610.

[247] Lapidus, M. L. (1991). Fractional drum, inverse spectral problems for elliptic operators and a partial resolution of the Weyl-Berry conjecture. *TAMS* **325**, No 2, 465-529.

Lapidus, M. L. (1991). Can one hear the shape of a fractal drum? Partial resolution of the Weyl-Berry conjecture. *Geom. Anal. and Computer Graphics* (MSRI, Berkaley, 1988) **17**, 119-126.

Lapidus, M. L. (1991). Spectral and fractal geometry: From the Weyl-Berry conjecture for the vibrations of fractal drums to the Riemann Zeta-function. In: *Diff. Eq-s and Math. Phys.* (Proc. Conf. held in Alabama, 1990; ed. Bennewitz, C.). Acad. Press, New York, 151-182.

Lapidus, M. L. et Maier, H. (1991). Hypothèse de Riemann, cordes fractales vibrantes et conjecture de Weyl-Berry modifiée. *C. R. Acad. Sci., Paris* **313**, Sér. I, 19-24.

Lapidus, M. L. and Pomerance, C. (1993). The Riemann Zeta-function and the one-dimensional Weyl-Berry conjecture for fractal drums. *Proc. London Math. Soc. (3)* **66**, No 1, 41-69.

Lavoie, J. L., Osler, T. J. and Tremblay, R. (1976). Fractional derivatives and special functions. *SIAM Review* **18**, No 2, 240-268.

Lavoie, J. L., Tremblay, R. and Osler, T. J. (1975). Fundamental properties of fractional derivatives via Pochhammer integrals. In: *Proc. Fractional Calculus and Its Applications, New Haven 1974 = Lecture Notes in Math.* **457**. Springer-Verlag, New York, 323-356.

Letnikov, A. V. (1868). The theory of differentiation of arbitrary power (in Russian). *Mat. Sbornik* **3**, 1-68.

Letnikov, A. V. (1872). On the explanation of basic ideas in the theory of differentiation with an arbitrary power (in Russian). *Mat. Sbornik* **6**, No 1, 413-445.

Libera, R. (1965). Some classes of regular univalent functions. *Proc. Amer. Math. Soc.* **16**, 755-758.

Lighthill, M. J. (1964). *Introduction to Fourier Analysis and Generalized Functions*, Cambridge Univ. Press, London.

Linchouk, N. E. (1985). Representation of the commutant of the Gelfond-Leontiev operator of generalized integration (In Russian). *Izv. Vyssh. Uchebn. Zaved. Mat. 1985*, No 5, 72-74.

Lions, J. L. (1956). Quelque applications d'opératuers de transmutation. In: *Qolloq. Internat. du CNRS* **LXXXXI**, Paris, 125-137.

Lions, J. L. (1956). Opérateurs de Delsarte et problèmes mixtes. *Bull. Soc. Math. France* **84**, 9-95.

Lions, J. L. (1959). Opérateurs de transmutation singuliers et equations d'Euler-Poisson-Darboux généralisées. *Rend. Seminario Math. Fis. Milano*, **28**, 3-16.

Love, E. R. (1967). Some integral equations involving hypergeometric functions. *Proc. Edinburgh Math. Soc.* **15**, No 3, 169-198.

Love, E. R. (1971). Fractional derivatives of imaginary order. *J. London Math. Soc., Ser. 2*, **3**, 241-259.

Love, E. R. (1972). Two index law for fractional integrals and derivatives. *J. Austr. Math. Soc.* **14**, No 4, 385-410.

[265] Love, E. R. (1985). A third index law for fractional integrals and derivatives. In: *Proc Fractional Calculus, Workshop held in Glasgow 1984 = Research Notes in Math.* **138**, Pitman Publ., Boston etc., 63-74.

[266] Love, E. R. (1990). Two theorems on Riesz-type fractional integrals. In: *Fractiona Calculus and Its Appl-s (Proc. Intern. Conf. Held in Tokyo, 1989)*, Nihon Univ., 1990; 85-93.

[267] Love, E. R. and Clements, D. L. (1976). A transformation of Cooke's treatment of som triple integration equations. *J. Austr. Math. Soc.* **19** *(Ser. B)*, part 3, 259-288.

[268] Lowndes, J. S. (1981). On some generalizations of Riemann-Liouville and Weyl fractiona integrals and their applications. *Glasgow Math. J.* **22**, part 2, 173-180.

[269] Lowndes, J. S. (1985). On two new operators of fractional integration. In: *Proc. Fractional Calculus, Workshop held in Glasgow 1984 = Research Notes in Math.* **138**, Pitman Publ., Boston etc., 87-98.

[270] Luchko, Yu. F. and Yakubovich, S. B. (1993). Operational calculus for the generalized fractional differential operator and applications. *Math. Balkanica (N.S.)* **7**. To appear.

[271] Luchko, Yu. F. and Yakubovich, S. B. (1993). Convolution of the generalized fractional integration operators. In: *Complex Analysis and Generalized Functions, Varna'91* (Proc. Int. Conf.). To appear.

[272] Luke, Y. L. (1969). *The Special Fractions and Their Approximations,* 1,2. Acad. Press, New York and London.

[273] Mandelbrot, B. B. (1983). *The Fractal Geometry of Nature.* W. H. Freeman and Co., New York.

[274] Marichev, O. I. (1974). Volterra equation of Mellin convolutional type with a Horn function in the kernel (in Russian). *Izv. AN BSSR, ser. fiz.-mat. nauk*, No 1, 128-129.

[275] Marichev, O. I. (1978). Integral operators with special functions in the kernels, generalizing operators of integration of complex order (in Russian). *Vesci Academi Navuk Bel. SSR, ser. fiz.-mat. navuk*, No 1, 38-44.

[276] Marichev, O. I. (1978). *Method of Calculation of Integrals of Special Functions* (in Russian). Nauka i Technika, Minsk.

[277] Marichev, O. I. (1981). On the evaluation of integration transforms of hypergeometric functions (in Russian). In: *Generalized Functions and Their Applications in Mathematical Physics, Moscow 1980.* Moscow, 323-331.

[278] Marichev, O. I. (1982). On the conditions for invertibility of Mellin-Barnes type integrals (in Russian). *Dokladi AN BSSR* **26**, No 3, 205-208.

[279] Marichev, O. I. (1983). Asymptotics of functions of hypergeometric type (in Russian). *Vesci Acad. Navuk BSSR*, No 4, 18-25.

[280] Marichev, O. I. (1984). On the representation of Meijer's G-function in the vicinity of the singular unity. In: *Proc. Complex Anal. and Appl-s, Varna 1981.* Sofia, 383-398.

Marichev, O. I. (1990). Compositions of fractional integrals and derivatives with power weights. In: *Fractional Calculus and Its Appl-s (Proc. Intern. Conf. Held in Tokyo, 1989)*, Nihon Univ., 1990; 94-99.

Marichev, O. I. and Tuan, Vu K. (1985). Compositional structure of some convolutional type integral transforms (in Russian). In: *Dokladi Razsh. Zased. Seminara Inst. Prikl. Mat. I. N. Vekua, Tbilisi* **1**, No 1, 139-142.

Marichev, O. I. and Tuan, Vu K. (1985). Some Volterra equations with Appell's function F_1 in the kernel (in Russian). In: *Nauchnye Trudy Jubil. Seminara po Kraevym Zadacham, Posv. Akad. Gahovu.* Izd. Univ., Minsk, 169-172.

Marichev, O. I. and Tuan, Vu K. (1986). The factorization of a G-function transform in two spaces of functions. In: *Proc. Complex Anal. and Appl-s, Varna 1985.* Sofia, 418-433.

Martic, B. (1973). A note on fractional integration. *Publ. l'Institut Math. Beograd* **16**(30), 111-113.

Mathai, A. M. and Saxena, R. K. (1973). *Generalized Hypergeometric Functions with Applications in Statistics and Physical Sciences. (Lecture Notes in Math.* **348**). Springer-Verlag, Berlin-Heidelberg-New York.

Mathai, A. M. and Saxena, R. K. (1978). *The H-function with Applications in Statistics and Other Disciplines.* Wiley Eastern Ltd., New Dehli-Bangalore-Bombay.

McBride, A. C. (1979). *Fractional Calculus and Integral Transforms of Generalized Functions.* Pitman, London.

McBride, A. C. (1982). Fractional powers of a class of ordinary differential operators. *Proc. London Math. Soc.(III)* **45**, 519-546.

McBride, A. C. (1987). *Semigroups of Linear Operators: An Introduction.* Pitman Res. Notes in Math. Ser. **156**. Harlow-New York.

McBride, A. C. and Roach, G. F. (1985). *Fractional Calculus = Res. Notes in Math.* **138** (Proc. Workshop held at Ross Priory, Univ. of Strathclyde, 1984.) Pitman Publ., Boston-London-Melbourne.

McBride, A. C. and Spratt, W. Y. (1991). On the range of invertibility of a class of Mellin multiplier transforms, I. *J. Math. Anal. and Appl-s* **156**, No 2, 568-587.

McBride, A. C. and Spratt, W. Y. (1991). On the range of invertibility of a class of Mellin multiplier transforms, III. *Canadian J. Math.* **43**, No 6, 1323-1338.

McBride, A. C. and Spratt, W. Y. (1992). A class of Mellin multipliers. *Canadian Math. Bull.* **35**, No 2, 252-260.

Meijer, G. S. (1936). Über Whittakersche bezw. Besselsche funktionen and deren produkte. *Nieuw Arch. Wisk.(2)* **18**, 10-39.

Meijer, G. S. (1941). Integraldarstellungen für Whittakersche funktionen und ihre produkte, I-II. *Nederl. Akad. Wetensch. Proc.* **44**, 435-441, 559-605.

Meijer, G. S. (1946). On the G-function, I-VIII. *Nederl. Akad. Wetensch. Proc.* **49**, 227-237, 344-356, 457-469, 632-641, 765-772, 936-943, 1063-1072, 1165-1175.

[298] Meller, N. A. (1960). On the operational calculus for the operator $B = t^{-\alpha}\frac{d}{dt}t^{\alpha+1}\frac{d}{dt}$ (in Russian). *Vychisl. Mat.*, No 6, 161-168.

[299] Meller, N. A. (1963). On some applications of the operational calculus to the problem of mathematical analysis (in Russian). *Zhournal Vych. Mat. i Mat. Fisiki* **3**, No 1, 71-78.

[300] Mendez, J. M.(1988). On the Bessel transformation of arbitrary order. *Math. Nachr.* **136**, 233-239.

[301] Mendez, J. M.(1988). A mixed Parseval equation and the generalized Hankel transformations. *Proc. AMS* **102**, No 3, 619-624.

[302] Mikolas, M. (1959). Differentiation and integration of complex order of function represented by trigonometrical series and generalized zeta-functions. *Acta Math. Acad. Sci. Hungarian* **10**, No 1-2, 77-124.

[303] Mikolas, M. (1975). On recent trends in the development, theory and applications of fractional calculus. In: *Fractional Calculus and Appl-s (Proc. Conf. Held in New Heven, 1974) = Lecture Notes in Math.* **457**, Springer-Verlag, New York, 357-375.

[304] Mikolas, M. (1990). A new method of summations based on fractional integration and generalized zeta-functions. In: *Fractional Calculus and Its Appl-s (Proc. Intern. Conf. Held in Tokyo, 1989)*, Nihon Univ., 1990; 106-109.

[305] Mikolas, M. (1990). Historical remarks about the works of M. Riesz in the theory of generalized integro-differential operators and their applications. In: *Fractional Calculus and Its Appl-s (Proc. Intern. Conf. Held in Tokyo, 1989)*, Nihon Univ., 1990; 110-111.

[306] Mikusinski, J. (1983,1987). *Operational Calculus*, v. I, II. Polish Sci. Publ. and Pergamon Press, Warszawa.

[307] Mikusinski, J. and Ryll-Nardzewski (1953). Une théoreme sur le produit de composition de fonctions de plusieurs variables. *Studia Math.* **13**, No 1, 62-68.

[308] Millington, G. (1969). Stokes phenomenon. *Radio Science* **24**, No 2, 95-115.

[309] Miller, K. S. (1975). The Weyl fractional calculus. In: *Fractional Calculus and Its Applications (Proc. Conf. Held in New Haven, 1974) = Lecture Notes in Math.* **457**, Springer-Verlag, New York, 80-91.

[310] Misra, A. P. (1973). Applications fractional derivative operators to Rodrigues' type of formulas for polynomial sets. *Mat. Balkanica* **3**, 358-362.

[311] Mocanu, P. T. (1986). On starlikeness of the Libera transform. *Mathematica* **28**(51), No 2, 153-155.

[312] Nahoushev, A. M. (1969). Some boundary-value problems for hyperbolic equations and equations of mixed type (in Russian). *Diff. Uravnenia* **5**, 44-59.

[313] Nahoushev, A. M. (1969). A new boundary value problems for a degenerate hyperbolic equation. *Soviet. Math. Dokladi* **10**, 935-938.

[314] Nahoushev, A. M. (1988). On the theory of fractional calculus. *Diff. Uravnenia* **24**, No 2, 313-324.

] Nahoushev, A. M. and Salathidinov, M. S. (1988). On the composition of differintegral operators with different initial data (in Russian). *Dokladi AN SSSR* **299**, No 6, 1313-1316.

] Napalkov, V. V. (1978). On the operators, commuting with differentation in spaces of functions of several variables (in Russian). *Mat. Zametki* **24**, No 6, 829-838.

] Nguyen Th. Hai and Yakubovich, S. B. (1992). *The Double Mellin-Barnes type Integrals and Their Applications to Convolution Theory.* World Sci., Singapore-London, etc.

] Nikiforov, A. F. and Uvarov, V. B. (1975). On a new approach to the theory of special functions (in Russian). *Mat. Sbornik* **98**, No 4, 578-590.

] Nikiforov, A. F. and Uvarov, V. B. (1984). *Special Functions in Mathematical Physics* (in Russian). Nauka, Moscow.

] Nikolic-Despotovic, D. and Kiryakova, V. (1988). Abelian theorems for the Obrechkoff transforms. *Rev. Res. Faculty Sci. (Univ. Novi Sad), Mat. Ser.* **18**, No 2, 45-60.

] Nishimoto, K. (1976). Fractional derivative and integral, I. *J. Coll. Eng. Nihon Univ.* **B-17**, 11-19.

] Nishimoto, K. (1983). On a half order's integral equation $f.z = f_{-\frac{1}{2}}$ (Volterra type) and simultaneous half order's differintegral equations. *J. Coll. Eng. Nihon Univ. Ser.* **B-24**, 7-13.

] Nishimoto, K. (1984). Tables of fractional differintegrations of elementary functions. *J. Coll. Eng. Nihon Univ. Ser.* **B-25**, 41-46.

] Nishimoto, K. (1984). *Fractional Calculus.* Descartes Press, Koriyama (Japan).

] Nishimoto, K. (1985). Applications of fractional calculus to the solution of linear second order differential equations of Fuchs type. In: *Fractional Calculus (Proc. Workshop held at Ross Priory, Univ. Strathclyde, 1984). = Res. Notes in Math.* **138**. Pitman, Boston-London-Melbourne, 140-153.

] Nishimoto, K. (1987). *Fractional Calculus, II.* Descartes Press. Koriyama (Japan).

] Nishimoto, K. (1988). Applications of fractional calculus to a third order linear ordinary differential equation of Fuchs type. *J. Coll. Eng. Nihon Univ. Ser.* **B-29**, 21-28.

] Nishimoto, K. (1988). Application of fractional calculus to a fourth order linear ordinary differential equation of Fuchs type. *J. Coll. Eng. Nihon Univ. Ser.* **B-29**, 29-36.

] Nishimoto, K. (1990). *Fractional Calculus and Its Applications (Proc. Internat. Conf. Held at the Center of Nihon Univ., Tokyo 1989).* Nihon Univ., (Koriyama, Japan).

] Nishimoto, K. (1991). *An Essence of Nishimoto's Fractional Calculus (Calculus of the* 21st *Century).* Descartes Press. Koriyama (Japan).

] Nishimoto, K. (1992). Power functions in fractional calculus of Nishimoto and that of Lacroix and Riemann-Liouville. *J. Fract. Calculus* **2**, November, 11-25.

[332] Nishimoto, K. and Kalla, S. (1988). Applications of a fractional calculus to a third order linear ordinary differential equation. *Serdica (Bulg. Math. Publ.)* **14**, No 4, 385-391.

[333] Nishimoto, K. and Kalla, S. (1989). Use of fractional calculus to solve certain linear differential equatoins. *J. Coll. Eng. Nihon Univ. Ser.* **B-30**, 23-28.

[334] Nishimoto, K., Owa, Sh. and Srivastava, H. M. (1984). Solutions to a new class of fractional differential equations. *J. Coll. Eng. Nihon Univ. Ser.* **B-25**, 75-78.

[335] Nishimoto, K. and Srivastava, H. M. (1989). Certain classes of infinite series summable by means of fractional calculus. *J. Coll. Eng. Nihon Univ. Ser.* **B-30**, March, 97-106.

[336] Nunokawa, M. and Owa, Sh. (1990). An application of fractional calculus for a certain analytic functions. In: *Fractional Calculus and Its Appl-s (Proc. Intern. Conf. Held in Tokyo, 1989)*, Nihon Univ., 1990; 175-178.

[337] Obrechkoff, N. (1950-52). On some properties of the real functions, defined on the real half-line (in Bulgarian). *Godishnik FMM Sofia Univ. "Kl. Ohridski"* **47**, No 1, II, 109-134.

[338] Obrechkoff, N. (1954). On some integral representations of functions in conection with differential equations (in Bulgarian). *Izvestija Mat. Institut (Sofia)* **1**, No 2, 3-33.

[339] Obrechkoff, N. (1958). On some integral representations of real functions on the real half-line. *Izvestija Mat. Institut (Sofia)* **3**, No 1, 3-28.

[340] Oldham, K. B. and Spanier, J. (1974). *The Fractional Calculus.* Acad. Press, New York-London.

[341] Olver, F. W. J.(1974). *Introduction to Asymptotics and Special Functions.* Acad. Press, New York- London.

[342] Osipov, S. I. (1963). On the expansion of the polynomial of the operator $B\alpha$ (in Russian). *Zhournal Vych. Mat. i Mat. Fiziki* **3**, No 1, 191-194.

[343] Osipov, S. I. (1964). An expansion of the polynomial of the Bessel operators (in Russian). *Zhournal Vych. Mat. i Mat. Fiziki* **4**, 149-155.

[344] Owa, S. (1990). On certain generalization subclasses of analytic functions involving fractional calculus. In: *Fractional Calculus and Its Appl-s (Proc. Intern. Conf. Held in Tokyo, 1989)*, Nihon Univ., 1990; 179-184.

[345] Owa, S. and Ahuja, O. P. (1986). A class of functions defined by using the Hadamard product. *Hokkaido Math. Journal* **15**, No 2, 217-232.

[346] Owa, S., Altintas, O. and Pashkuleva, D. (1987). Notes on certain subclasses of analytic functions with negative coefficients. *J. Fac. Sci. & Techn. Kinki Univ.* **23**, 11-18.

[347] Owa, S. and Nishimoto, K. (1984). A note on a class of convex functions. *J. Coll. Eng. Nihon Univ. Ser.* **B-25**, 53-56.

[348] Owa, S. and Obradovic, M. (1986). A remark on Rucheweyh derivatives. *Bull. Soc. Royale Sci. Liège*, 55e année, No 2, 279-284.

Owa, S. and Obradovic, M. (1986). Certain subclasses of Basilevic functions of type α. *Internat. J. Math. & Math. Sci.* **9**, No 2, 347-359.

Owa, S., Saigo, M. and Furuto, D. M. (1991). On a certain fractional operators. *Proc. Japan Acad.* **67**, Ser. A, No 5, 151-153.

Owa, S. and Srivastava, H. M. (1986). Some applications of the generalized Libera integral operators. *Proc. Japan Acad.* **62**, Ser. A, No 4, 125-128.

Owa, S. and Srivastava, H. M. (1987). Univalent and starlike generalized hypergeometric functions. *Canadian J. Math.* **39**, No 5, 1057-1077.

Owa, S. and Srivastava, H. M. (1987). Some inequalities for starlike and conve functions of type α. *Boll. della Unione Mat. Italiana (7)*, **1**-A, 221-227.

Owa, S., Srivastava, H. M. and Shen, C. Y. (1988). Applications of certain linear operator defined by a Hadamard product of convolution. *Utilitas Math.* **33**, 173-181.

Parashar, B. P. (1967). Domain and range of fractional integration operators. *Math. Japonicae* **12**, No 2, 141-145.

Paris, R. B. (1985). On the asymptotic expansions of solutions of an nth order linear differential equation with power coefficients. *Proc. Royal Irish Acad.* **85A**, No 2, 201-220.

Paris, R. B. and Wood, A. D. (1987). Results, old and new on the hyper-Bessel equation. *Proc. Royal Soc. Edinburgh* **106A**, 259-265.

Pathak, R. S. (1966). Certain convergence theorems and asymptotic properties of a generalization of Lommel and Maitland transformations. *Proc. Nat. Acad. Sci. India* **A-36**, No 1, 81-86.

Pathak, R. S. (1966-67). An inversion formula for a generalization of the Lommel and Maitland transformations . *J. Sci. Banaras Hindu Univ.* **17**, No 1, 65-69.

Pathak, R. S. (1972). A general differential equation satisfied by a special function. *Progress of Math.* **6**, No 1-2, 46-50.

Pathak, R. S. (1981). Abelian theorems for the G-transformation. *J. Indian Math.* **45**, 243-249.

Pathak, R. S. and Pandey, J. N. (1979). The G-transform of generalized functions. *Rocky Mountain J. of Math.* **9**, No 2, 307-325.

Penzel, F. (1986). *Zur Theorie verallgemeinerter Integralgleichungen.* Dissertation-Doktors der Naturwissenschaften. Darmstadt.

Popov, V. (1983). Function spaces, generated by the averaged moduli of smoothness. *Pliska* **5**, 132-143.

Przeworska-Rolewicz, D. (1973). Algebraic theory of right invertible operators. *Studia Math.* **48**, 129-144.

Prudnikov, A. P. (1962). On the theory of operational calculus (in Russian). *Dokladi AN SSSR* **142**, No 4, 794-797.

Prudnikov, A. P. (1990). On Watson transforms. In: *Fractional Calculus and Its Appl-s (Proc. Intern. Conf. Held in Tokyo, 1989)*, Nihon Univ., 1990; 185-190.

[368] Prudnikov, A. P. and Brychkov, Yu. A. and Marichev, O. I. (1981). *Integral and Series. Elementary Functions* (in Russian). Nauka, Moscow.

[369] Prudnikov, A. P. and Brychkov, Yu. A. and Marichev, O. I. (1983). *Integral and Series. Special Functions* (in Russian). Nauka, Moscow.

[370] Prudnikov, A. P. and Brychkov, Yu. A. and Marichev, O. I. (1986). *Integral and Series. Additional Chapters* (in Russian). Nauka, Moscow.

[371] Prudnikov, A. P. and Brychkov, Yu. A. and Marichev, O. I. (1989). Evaluation of integrals and Mellin transforms (in Russian). *Itogi Nauki i Tehniki. Ser. Math. Analiz* **27**, VINITI, Moscow, 3-146.

[372] Raichinov, I. (1963). Sur une théorème de G. Polya. *Publ. Inst. Math. Beograd* **2**, 141-144.

[373] Raichinov, I. (1970). On the linear operators, commuting with the operation of integration (in Russian). *Mat. Analiz i ego Prilozhenia,* Rostov-na-Donu **2**, 63-72.

[374] Raichinov, I. (1970). Linear operators, acting in spaces of analytic functions and commuting with a fixed power of integration (in Bulgarian). *Godishnik VTUZ. Matematika* **6**, No 2, 25-32.

[375] Raichinov, I. (1970). On the character of the mapping, realized by linear operators, commuting with a fixed power of integration (in Bulgarian). *Godishnik VTUZ. Matematika* **6**, No 2, 33-42.

[376] Raichinov, I. (1971). Sufficient conditions for biuniqueness of the mapping of linear operators, commuting with the fixed power of the integration operator (in Bulgarian). *Godishnik VTUZ. Matematika* **7**, No 4, 7-18.

[377] Raichinov, I. (1972). On the non trivial invariant subspaces of the integration operator (in Russian). *Natura (Univ. de Plovdiv "P. Hilendarskii")* **5**, No 1, 23-25.

[378] Raichinov, I. (1978). Linear operators defined in spaces of complex functions of many variables and commuting with the operators of integration. *Serdica (Bulg. Math. Publ.)* **4**, 316-323.

[379] Raichinov, I. (1979). On some commutation properties of the algebra of the linear operators , acting in spaces of analytic functions, I (in Bulgarian). *Godishnik VUZ. Pril. Matematika* **15**, No 3, 27-40.

[380] Raichinov, I. (1981). On some commuting properties of the algebra of the linear operators, acting in spaces of analytic functions, II (in Bulgarian). *Godishnik VUZ. Pril. Matematika* **17**, No 1, 61-71.

[381] Raichinov, I. (1981). On the question of minimally commuting elements of the algebras of linear operators, acting in spaces of analytic functions (in Bulgarian). *Godishnik VUZ. Pril. Matematika* **17**, No 4, 91-99.

[382] Raichinov, I. and Raichinov, R. (1983). On the linear operators , acting in spaces of analytic functions and commuting with the Euler operator (in Russian). *D. R. Acad. Bulg. Sci.* **36**, No 8, 1025-1026.

Raichinov, I. and Raichinov, R. (1990). On linear operators, acting in spaces of analytic functions and commuting with the Euler operators. *Pliska (Studia Math. Bulg.).* To appear.

Raina, R. K. (1979). On the Weyl fractional differintegration. *Indian J. Pure and Appl. Math.* **10**, No 1-2, 37-41.

Raina R. K. (1984). On composition of certain fractional integral operators. *Indian J. Pure and Appl. Math.* **15**, No 5, 509-516.

Raina, R. K. (1984). On the multiple Weyl fractional integral of a general system of polynomials. *Bolletino U.M. Italia (6)* **3-A**, 283-287.

Raina, R. K. (1984). A theorem connecting the two-dimensional Weyl operator and H-function transform. *J. Korean Math. Soc.* **21**, No 1, 71-78.

Raina, R. K. (1985). A note on the fractional derivatives of a general system of polynomials. *Indian J. Pure and Appl. Math.* **16**, No 7, 770-774.

Raina, R. K. and Kalla, S. L. (1984). On moments of probability distribution functions and H-function transform. *Rev. Téc. Ing., Univ. Zulia* **7**, No 1, 59-62.

Raina, R. K. and Kiryakova, V. S. (1983). On the Weyl fractional operator of two dimensions. *C. R. Acad. Bulg. Sci.* **36**, No 10, 1273-1276.

Raina, R. K. and Koul, C. L. (1977). Fractional derivatives of the H-functions. *Jñanabha* **7**, 97-105.

Raina, R. K. and Koul, C. L. (1979). On Weyl fractional calculus. *Proc. Amer. Math. Soc.* **73**, No 2, 188-192.

Raina, R. K. and Koul, C. L. (1981). On Weyl fractional calculus and H-function transform. *R. Kyungpook Math. J.* **21**, No 2, 275-279.

Raina, R. K. and Saigo, M. (1993). A note on fractional calculus operators involving Fox's H-function on space $F_{p,\mu}$. In: *Recent Advances in Fractional Calculus* (ed. by R. N. Kalia). Global Publ. Co. To appear.

Rodriguez, J. (1989). Operational calculus for the generalized Bessel operator. *Serdica (Bulg. Math. Publ.)* **15**, 179-186.

Rodriguez, J. (1989). Operational calculus through the convolutional method for the Bessel operator. *Pure and Appl. Math. Sci.* **29**, No 1-2, 71-81.

Rodriguez, J. (1990). Associated operational calculus for the Bessel-Clifford operator. *Jñanabha* **20**, 79-91.

Rodriguez, J. (1993). Una nueva convolucion para el calculo operacional del operator $B_{-\nu}$. *Revista Tecnica Fac. Ing., Univ. Zulia* **16**, To appear.

Rooney, P. G. (1972). On the ranges of certain fractional integrals. *Canadian J. Math.* **24**, 1190-1216

Rooney, P. G. (1973). A technique for studying the boundedness and extendability of certain types of integrals. *Canadian J. Math.* **25**, 1090-1102.

Rooney, P. G. (1979). On the range of the Hankel transformation. *Bull. London Math. Soc.* **11**, 45-48.

[402] Rooney, P. G. (1982). Multipliers for the Mellin transformation. *Canadian Math. Bull.* **25**, No 3, 257-262.

[403] Rooney, P. G. (1983). On integral transformations with G-function kernels. *Proc. Royal Soc. Edinburgh* **93A**, 265-297.

[404] Ross, B. (ed.). (1975). *Fractional Calculus and Its Applications (Proc. Internat. Conf. Held in New Haven, 1974)*= Lecture Notes in Math. **457**. Springer-Verlag, New York.

[405] Ross, B. (1975). *A Chronological Bibliography of Fractional Calculus with Commentary.* Preprint, 15p.

[406] Ross, B. (1975). A brief history and exposition of the fundamental theory of fractional calculus. In: Ross, B. (ed.) = [404].

[407] Ross, B. (1978). A use for a derivative of complex order in the fractional calculus, or: What, indeed, is $\frac{d^{3-\frac{1}{2}i}}{dx^{3-\frac{1}{2}i}}$ and what can we do with it? *Indian J. Pure and Appl. Math.* **9**, No 4, 400-406.

[408] Ross, B. (1983). Serendipity in Mathematics, or: How one is led to discover that $\sum_{n=1}^{+\infty} 1.3\ldots(2n-1)/n.2^n.n! = \ln 4$. *Amer Math. Monthly* **90**, No 8, 562-566.

[409] Ross, B. and Kalla, S. L. (1985-86). The method of fractional operators applied to summation. *Real Analysis Exchange* **11**, 271-282.

[410] Ross, B. and Sachdeva, B. K. (1990). The continuity property $D^0 f(x) = f(x)$, its proof, usefulness and applications. In: *Fractional Calculus and Its Appl-s (Proc. Intern. Conf. Held in Tokyo, 1989)*, Nihon Univ., 1990; 201-206.

[411] Ruscheweyh, S. (1975). New criteria for univalent functions. *Proc. Amer. Math. Soc.* **49**, 109-115.

[412] Ruscheweyh, S. (1982). *Convolutions in Geometric Function Theory.* Les presses de l'Univ. Montréal (Canada).

[413] Rusev, P. (1979). On the representations of analytic functions by series of Laguerre polynomials. *Soviet Math. Dokladi* **20**, No 6, 1221-1223.

[414] Rusev, P. (1984). *Analytic Functions and Classical Orthogonal Polynomials.* Publ. House Bulg. Acad. of Sci., Sofia.

[415] Saigo, M. (1978). A remark on integral operators involving the Gauss hypergeometric functions. *Math. Reports Coll. Gen. Educ., Kyushu Univ.* **11**, No 2, 135-143.

[416] Saigo, M. (1979). A certain boundary value problem for the Euler-Darboux equation, I. *Math. Japonica* **24**, No 4, 377-385.

[417] Saigo, M. (1980). A certain boundary value problem for the Euler-Darboux equation, II. *Math. Japonica* **25**, No 2, 211-220.

[418] Saigo, M. (1980). On the Hölder continuity of the generalized fractional integrals and derivatives. *Math. Reports Call. Gen. Educ., Kyushu Univ.* **12**, No 2, 55-62.

Saigo, M. (1981). A certain boundary value problem for the Euler-Darboux equation, III. *Math. Japonica* **26**, No 1, 103-119.

Saigo, M. (1985). A generalization of fractional calculus. In: *Proc. Workshop Held in Glasgow, 1984 = Res. Notes in Math.* **138**, Pitman Adv. Publ. Program, Boston, London, Melbourne, 188-198.

Saigo, M. (1987). A generalization of fractional calculus and boundary value problems for Euler-Darboux equation. In: *Diff. Equations and Appl-s (Proc. Conf. Held in Rousse, 1985).* Rousse, 907-913.

Saigo, M. and Glaeske, H.-J.(1990). Fractional calculus operators involving the Gauss function on spaces $F_{p,\mu}$ and $F'_{p,\mu}$. *Math. Nachr.* **147**, 285-306.

Saigo, M. and Raina, R. K. (1988). Fractional calculus operators associated with a general class of polynomials. *Fukuoka Univ. Sci. Reports* **18**, No 1, 15-22.

Saigo, M. and Raina, R. K. (1991). On the fractional calculus operator involving Gauss's series and its applications to certain statistical distortions. *Rev. Techn. Ing., Univ. Rulia* **14**, No 1, 53-62.

Saigo, M., Saxena, R. K. and Ram, J. (1992). On the fractional calculus operator associated with the H-function. *Ganita Sendesh* **6**, No 1, 36-47.

Saigo, M. and Yakubovich, S. B. (1991). On the theory of convolution integrals for G-transforms. *Fukuoka Univ. Sci. Reports* **21**, No 2, 181-193.

Saitoh, H. (1990). Properties of certain multivalent functions and an application of fractional calculus. In: *Fractional Calculus and Its Appl-s (Proc. Intern. Conf. Held in Tokyo, 1989)*, Nihon Univ., 1990; 215-220.

Sakaljuk, K. D.(1960). A generalized Abel integral equation (in Russian). *Dokladi AN SSSR* **131**, No 4, 748-751.

Sakaljuk, K. D.(1962). A generalization of the Abel integral equation (in Russian). *Uchebnye Zapiski Kishinev. Univ.* **50**, 95-102.

Samko, S. G. (1968). On a generalized Abel equation and operator of fractional integration (in Russian). *Diff. Uravnenia* **4**, No 2, 298-314.

Samko, S. G. (1971). On operators of potential type (in Russian). *Dokladi AN SSSR* **196**, No 2, 299-301.

Samko, S. G. (1973). On the space $I^{\alpha}(L_p)$ of fractional integrals and operators of potential type (in Russian). *Izv. AN Arm. SSR, Mat.* **8**, No 5, 359-383.

Samko, S. G. (1985). One-dimensional and multi-dimensional integral equations of first kind with a weak singularity in the kernel (in Russian). In: *Nauchnye Trudy Jubilee Sem. po Kraevym Zadacham, posv. Akad. Gahovu.* Izd. Univ. Minsk, 103-115.

Samko, S. G., Kilbas, A. A. and Marichev, O. I. (1987). *Integrals and Derivatives of Fractional Orders and Some of Their Applications* (in Russian). Nauka i Tehnika, Minsk. [To appear in English , Gordon and Breach Sci. Publ-s, New Yokrk, London].

Sarabia, J. (1987). Sobre la funcion de Exton (di-Bessel). *Rev. Tec. Ing., Univ. Zulia* **10**, No 1, 65-73.

[436] Saxena, R. K. (1967). On the formal solution of certain dual integral equations involving H-functions. *Proc. Cambridge Phil. Soc.* **63**, 171-178.

[437] Saxena, R. K. (1967). On the formal solution of dual integral equations . *Proc. Amer. Math. Soc.* **18**, 1-8.

[438] Saxena, R. K. and Kiryakova, V. S. (1992). On a relation between the two-dimensional H-transforms in terms of Erdélyi-Kober operators . *Math. Balkanica (N.S.)* **6**, No 2, 133-140.

[439] Saxena, R. K., Kiryakova, V. S. and Davie, O. P. (1993). A unified approach to certain fractional integration operators . *Math. Balkanica (N. S.)* **7**. To appear.

[440] Saxena, R. K. and Kumbhat, R. K. (1974). Integral operators involving H-functions. *Indian J. Pure Appl. Math.* **5**, 1-6.

[441] Saxena, R. K. and Ram, J. (1990). On the two-dimensional Whittaker transform. *Serdica (Bulg. Math. Publ.)* **16**, 27-30.

[442] Saxena, R. K. and Ram, J. (1990). On certain multidimensional generalized Kober operators. *Collectanea Math.* **41**, No 1, 27-34.

[443] Sekine, T. (1990). On distortion theorems of fractional calculus. In: *Fractional Calculus and Its Appl-s (Proc. Intern. Conf. Held in Tokyo, 1989)*, Nihon Univ., 1990; 232-237.

[444] Sekine, T., Owa, S. and Nishimoto, K.(1986). An application of the fractional calculus. *J. Coll. Eng. Nihon Univ. Ser.* **B-27**, 31-37.

[445] Sendov, B. (1967). *Approximation with Respect to the Hausdorff Distance* (in Russian). Dr. Sci. Thesis, Mat. Institut AN SSSR, Moscow.

[446] Sendov, B. (1982). On the constants of Whitney. *C. R. Acad. Bulg. Sci.* **35**, No 4, 431-434.

[447] Sendov, B. (1983). A modified Steklov function (in Russian). *C. R. Acad. Bulg. Sci.* **36**, No 3, 315-317.

[448] Sendov, B. and Popov, V. (1983,1988). *Averaged Moduli of Smoothness* (in Bulgarian). Publ. House BAS, Sofia. [in English; A Wiley Intersci. Publ. Series of Texas, Monographs and Tracts. Chichester, New York, Brisbane, Toronto, Singapore].

[449] Shulev, K. Z. (1976). On the operator of fractional integration and differentiation of an analytic function (in Bulgarian). *Godishnik VUZ. Pril. Matematika* **12**, No 3, 97-109.

[450] Slater, I. J. (1966). *Generalized Hypergeometric Functions.* Cambridge Univ. Press, London and New York.

[451] Sneddon, I. N. (1951). *Fourier Transforms.* McGraw-Hill, New York.

[452] Sneddon, I. N. (1966). *Mixed Boundary Value Problem in Potential Theory.* North-Holand Publ. Co., Amsterdam.

[453] Sneddon, I. N. (1972). *The Use of Integral Transforms.* McGraw-Hill, New York.

[454] Sneddon, I. N. (1975). The use in mathematical analysis of Erdélyi-Kober operators and some of their applications. In: *Fractional Calculus and Its Applications*

(Proc. Internat. Conf. Held in New Haven, 1974)= Lecture Notes in Math. **457**. Springer-Verlag, New York, 37-79.

Sneddon, I. N. (1979). *The Use of Operators of Fractional Integration in Applied Mathematics.* RWN - Polish Sci. Publishers, Warszawa-Poznan.

Sonine, N. Ya. (1872). On the differentiation of an arbitrary exponent (in Russian). *Mat. Sbornik* **6**, No 1, 1-38.

Sprinkhuizen-Kuiper, I. G. (1979). A fractional integral operator corresponding to a negative power of a certain second order differential operator. *J. Math. Anal. Appl.* **72**, No 2, 674-702.

Srivastava, H. M. (1968). Fractional integration and inversion formulae associated with the generalized Whittaker transform. *Pacific J. Math.* **26**, 375-377.

Srivastava, H. M. (1972). A class of integral equations involving the H-function as kernel. *Proc. Kongl. Nederl. Akad. Weten.* **A75**, No 3, 212-220.

Srivastava, H. M. (1980). Quelques functions generatrices pour une class de polynomes definies per une formule generale du type de Rodrigues. *C. R. Acad. Paris Ser.* **A-B 290**, A993-A996.

Srivastava, H. M. (1982). Some biorthogonal polynomials suggested by the Laguerre polynomials. *Pacific J. Math.* **98**, No 1, 235-250.

Srivastava, H. M. (1983). The Weyl fractional integral of a general class of polynomials. *Boll. Unione mat. Italiana* **132**, No 1, 219-228.

Srivastava, H. M. and Buschman, R. G. (1973). Composition of fractional integral operators involving Fox's H-function. *Acta Maxicana Ci. Tecn.* **7**, 21-28.

Srivastava, H. M. and Buschman, R. G. (1977). *Convolution Integral Equations With Special Functions Kernels.* Wiley Eastern, New Delhi.

Srivastava, H. M. and Buschman, R. G. (1992). *Theory and Applications of Convolution Integral Equations.* Kluwer Acad. Publ. (Ser. Math. and Its Appl-s, vol. **79**). Dordrecht, Boston, London.

Srivastava, H. M. and Goyal, S. P. (1985). Fractional derivatives of the H-functions of several variables. *J. Math. Anal. and Appl.* **112**, 641-651.

Srivastava, H. M., Gupta, K. C. and Goyal, S. P. (1982). *The H-function of One and Two Variables with Applications.* South Asian Publishers, New Delhi, Madras.

Srivastava, H. M. and Kashyap, B. R. K. (1981). *Special Functions in Queuing Theory and Related Stohastic Processes.* Acad. Press, New York, London and San Francisco.

Srivastava, H. M., Lavoie, J.-L. and Tremblay, R. (1979). The Rodrigues type representations for a certain class of special functions. *Ann. Math. Pure. Appl.* (4)**119**, 9-24.

Srivastava, H. M. and Manocha, H. L. (1984). *A Treatise on Generating Functions.* Halsted Press, John Wiley and Sons, New York, Brisbane and Toronto.

Srivastava, H. M. and Owa, S. (1986). A certain one parameter additive family of operators defined on analytic functions. In: *J. Math. Anal. and Appl.* **118**, No. 1, 80-87.

[472] Srivastava, H. M. and Owa, S. (1986). A new class of analytic functions with negative coefficients. *Comm. Math. Univ. Sancti Pauli* **35**, No 2, 175-188.

[473] Srivastava, H. M. and Owa, S. (1987). Some characterizations and distortion theorems involving fractional calculus, generalized hypergeometric functions , Hadamard products, linear operators and certain subclasses of analytic functions. *Nagoya Math. J.* **106**, 1-28.

[474] Srivastava, H. M. and Owa, S. (1987). Some applications of fractional calculus operators to certain analytic and multivalent funcitons. *J. Math. Anal. and Appl.* **122**, No 1, 187-196.

[475] Srivastava, H. M. and Owa, S. (ed-s) (1989). *Univalent Functions, Fractional Calculus and Its Applications.* John Wiley and Sons, Halsted Press, New York.

[476] Srivastava, H. M. and Owa, S. (ed-s) (1992). *Current Topics in Analytic Function Theory.* World Scientific, Singapore etc.

[477] Srivastava, H. M., Owa, S. and Ahuja, O. P. (1988). A new class of analytic functions associated with the Ruscheweyh derivatives. *Proc. Japan Acad.* **64**, Ser A, No 1, 17-20.

[478] Srivastava, H. M., Owa, S. and Nishimoto, K. (1984). A note on a certain class of fractional differintegral equations. *J. Coll. Eng. Nihon Univ. Ser.* **B-25**, 69-73.

[479] Srivastava, H. M., Owa, S. and Nishimoto. K. (1985). Some fractional differintegral equations. *J. Math. Anal. and Appl.* **106**, No 2, 360-366.

[480] Srivastava, H. M. and Saigo, M. (1987). Multiplication of fractional calculus operators and boundary value problems involving the Euler-Darboux equation . *J. Math. Anal. and Appl.* **121**, No 2, 325-369.

[481] Srivastava, H. M., Saigo M. and Owa, S. (1988). A class of distortion theorems involving certain operators of fractional calculus. *J. Math. Anal. and Appl.* **131**, No 2, 412-420.

[482] Srivastava, H. M., Sekine, T., Owa, S. and Nishimoto, K. (1986). Fractional derivatives and fractional integrals of certain subclasses of starlike and convex functions. *J. Coll. Eng. Nihon Univ. Ser.* **B-27**, 35-46.

[483] Srivastava, R. (1990). The multidimensional Weyl fractional integral of certain classes of polynomials . In: *Fractional Calculus and Its Appl-s (Proc. Intern. Conf. Held in Tokyo, 1989)*, Nihon Univ., 1990; 262-269.

[484] Stankovic, B. (1959). O nulama jedne funcije operatorskog racuna. *Annuaire Fac. des Lettres et Sci. a Novi Sad* **4**, 379-381.

[485] Stankovic, B. (1970). On the function of E. M. Wright. *Publ. de l'Institut Mathematique, Nouvelle Ser.* **10**(24), 113-124.

[486] Takev, M. D. (1992). On the Hausdorff dimension of completed graphs. *C. R. Acad. Bulg. Sci.*, **45**, No 7, 13-15.

[487] Timan, A. F. (1960). *Theory of the Approximation of Functions of a Real Variable* (in Russian). FizMatGiz, Moscow.

Tkachenko, V. A. (1977). On operators commuting with generalized differentiation in spaces of analytic functions with a given indicator of growth. *Math. USSR Sbornik* **31**, No 3, 389-407.

Tkachenko, V. A. (1977). Equations of convolution type in spaces of analytic functions. *Math. USSR Izvestja* **11**, No 2, 361-374.

Tkachenko, V. A. (1979). Spectral expansions in spaces of analytic functions (in Russian). *Izvestja AN SSSR, Ser. Mat.* **43**, No 3, 654-713.

Tkachenko, V. A. (1979). On operators, commuting with generalized integration in spaces of analytic functions (in Russian). *Mat. Zametki* **25**, No 2, 271-282.

Tkachenko, V. A. (1980). Spectral theory in spaces of analytic functionals for operators, generated by a multiplication by an independent variable (in Russian). *Mat. Sbornik* **112**, No 3, 421-466.

Tricomi, F. G. (1957). *Lectures on Partial Differential Equations* (in Russian). Inostr. Literatura, Moscow.

Trimeche, K. (1986). Mean-periodic functions associated with a singular differential operator in the complex domain. In: *Proc. Complex Anal. and Appl-s, Varna 1985.* Sofia, 715-719.

Tuan, Vu K. (1986). Generalized integral transformations of convolutional type in some spaces of functions. In: *Proc. Complex Anal. and Appl-s, Varna 1985.* Sofia, 720-735.

Tuan, Vu K. (1987). *Integral Transforms and Their Compositional Structure* (in Russian). Summary of a Dr. Sci. Thesis, Belorussian State Univ., Minsk.

Tuan, Vu K. and Marichev, O. I. and Yakubovich, S. B. (1986). Composition structure of integral transformations. *Soviet Math. Dokladi* **33**, No 1, 166-170.

Ufljand, Ja. S. (1977). *Method of Dual Equations in the Problems of Mathematical Physics* (in Russian). Nauka, Leningrad.

Uspensky, J. V. (1927). On the development of arbitrary functions in series of Hermite's and Laguerre's polynomials. *Ann. Math.* (2) **28**, 593-619.

Vasilev, M. V. (1985). Commutants of powers of the differentiation operator in spaces of holomophic functions. In: *Proc. Complex Anal. and Appl-s, Varna 1983.* Sofia, 239-251.

Vasilev, M. V. (1990). The commutant of a generalization of the backward shift operator and its powers. *C. R. Acad. Bulg. Sci.* **43**, No 7, 19-22.

Virchenko, N. A. (1984). On some associated dual integral equations (in Russian). *Ukrain. Mat. Zhournal* **36**, No 2, 139-142.

Virchenko, N. A. (1986). Dual integral equations involving hypergeometric functions ${}_2F_1(a,b,c;-xt)$, ${}_3F_2(\nu,\alpha_1,\alpha_2;\mu+\nu;-xt)$ (in Russian). *Dokladi AN USSR, Ser. A., Fiz. Mat. i Techn. Nauki*, No 2, 3-5.

Virchenko, N. A. (1989). *Dual (Triple) Integral Equations* (in Russian). Vyshcha Shkola, Kiev.

Virchenko, N. A. and Makarenko, L. G. (1975). On some dual integral equations (in Russian). *Ukrain. Mat. Zhournal* **27**, No 6, 790-794.

[506] Volterra, V. (1959). *Theory of Functionals and of Integral and Integro-Differential Equations.* Dover Publ., Inc., New York.

[507] Watson, G. N. (1944). *A Treatise on the Theory of Bessel Functions.* Cambridge Univ. Press, Cambridge (second edition).

[508] Weinstein, A. (1953). Generalized axially symmetric potential theory. *Bull. AMS* **59**, 20.

[509] Weinstein, A. (1955). The generalized radiation problem and the Euler-Poisson-Darboux equation. *Summa Brazil Math.* **3**, 125-147.

[510] Widder, D. V. (1946). *The Laplace Transform.* Princeton Univ. Press, Princeton.

[511] Wright, E. M. (1933). On the coefficients of power series having exponential singularities. *J. London Math. Soc.* **8**, 71-79.

[512] Wright, E. M. (1940). The generalized Bessel functions of order greater than one. *Quart. J. Math. Oxford, ser.* **2**, 36-48.

[513] Yakubovich, S. B. (1990). On a constructive method for building integral convolutions (in Russian). *DAN Byel.SSR* **34**, No 7, 588-591.

[514] Yakubovich, S. B. (1991). On the integral convolutions of Laplace for G-functions (in Russian). *Vestsi AN Byel.SSR (Ser. Fiz.-Mat.)*, No 6, 11-16.

[515] Yakubovich, S. B. and Luchko, Yu. F. (1991). Generalization of the Leibniz rule on integral convolutions (in Russian). *DAN Byel.SSR* **35**, No 2, 111-115.

[516] Yakubovich, S. B. and Luchko, Yu. F. (1991). The generalizations of an integral analog of the Leibniz rule on the G-convolutions. *Extracta Math.* **6**, No 2-3, 119-122.

[517] Yakubovich, S. B. and Nguyen Th, Hai (1991). Integral convolutions for H-transforms (in Russian). *Izv. Vysh. Uch. Zaved. Mat.* **8**, 72-79

[518] Zemanian, A. H. (1966). Some Abelian theorems for the distributional Hankel and K-transformations. *SIAM J. Appl. Math.* **14**, No 6, 1255-1265.

[519] Zemanian, A. H. (1968). *Generalized Integral Transformations.* Intersci. Publishers, New York.

Citation index

LIST OF ARTICLES, REFERRED TO IN INTRODUCTION

LIST OF ARTICLES, REFERRED TO IN CHAPTER 1

Section 1.4: [191] [257] [289] [309] [384] [385] [386] [387] [388] [390] [393] [415] [434] [438] [439] [441] [452] [455] [483]

Section 1.5: [164] [202] [289] [415] [420] [421] [453] [454] [455]

Section 1.6: [64] [65] [66] [67] [68] [69] [79] [101] [117] [120] [146] [147] [156] [157] [158] [159] [160] [161] [162] [163] [164] [194] [196] [201] [202] [208] [239] [272] [276] [289] [291] [355] [360] [387] [390] [404] [412] [415] [416] [417] [418] [419] [420] [421] [457] [484] [485] [511] [512]

List of articles, referred to in Chapter 2

[104] [109] [161] [164] [220] [268] [269] [289] [399] [400] [452] [454] [455]

Section 2.1: [64] [68] [69] [70] [71] [72] [73] [206] [270] [271] [306] [317]

Section 2.2: [72] [73] [77] [78] [98] [99] [100] [101] [102] [103] [120] [146] [147] [183] [196] [258] [370] [434] [490]

Section 2.3: [72] [73] [77] [78] [83] [112] [139] [140] [141] [215] [258] [316] [372] [373] [374] [375] [376] [377] [378] [379] [380] [381] [382] [383] [488] [489] [490] [491] [492] [500]

Section 2.4: [72] [73] [77] [78] [98] [99] [100] [101] [107] [196] [206] [488] [489] [492]

List of articles, referred to in Chapter 3

[64] [65] [66] [67] [68] [69] [70] [71] [157] [158] [159] [160] [161] [164] [177] [178] [262] [289] [290] [291] [329] [339] [355] [404] [415] [416] [417] [418] [419] [420] [421] [434] [457]

Section 3.1: [26] [64] [65] [66] [67] [68] [69] [70] [71] [73] [106] [184] [216] [217] [218] [272] [342] [343] [365]

Section 3.2: [5] [36] [70] [73] [74] [87] [190] [272] [365]

Section 3.3: [5] [32] [36] [37] [60] [61] [64] [65] [66] [67] [68] [69] [70] [71] [73] [74] [75] [76] [85] [86] [87] [89] [121] [126] [127] [128] [129] [190] [192] [193] [196] [201] [221] [222] [234] [235] [236] [237] [238] [239] [272] [286] [289] [290] [298] [299] [312] [313] [320] [339] [342] [343] [395] [396] [397] [398] [457] [508] [509] [519]

Section 3.4: [1] [5] [6] [26] [34] [60] [68] [71] [75] [79] [80] [106] [111] [122] [127] [128] [129] [132] [133] [137] [138] [184] [190] [191] [192] [193] [196] [197] [210] [216] [217] [218] [241] [289] [339] [356] [357] [435] [494]

Section 3.5: [38] [39] [43] [53] [54] [55] [56] [57] [58] [59] [61] [62] [68] [69] [70] [71] [73] [79] [134] [185] [186] [187] [216] [217] [234] [259] [260] [261] [299] [306]

Section 3.6: [1] [2] [5] [64] [65] [68] [69] [70] [73] [75] [76] [79] [80] [127] [128] [129] [190] [191] [192] [193] [212] [213] [243] [244] [245] [246] [288] [289] [290] [292] [293] [294]

Section 3.7: [60] [79] [80] [81] [107] [126] [127] [128] [129] [130] [131] [132] [133] [155] [196] [218] [219] [224] [308] [341] [356] [357]

Section 3.8: [26] [38] [39] [72] [79] [80] [106] [196] [272] [286] [287] [289] [291] [307]

Section 3.9: [1] [37] [64] [65] [66] [67] [68] [69] [70] [74] [75] [76] [79] [85] [86] [88] [89] [90] [91] [106] [107] [190] [192] [196] [235] [236] [237] [238] [239] [298] [337] [338] [339] [361] [366] [403] [453] [510] [518]

Section 3.10: [31] [32] [36] [74] [75] [76] [79] [80] [86] [87] [107] [109] [190] [191] [192] [193] [196] [197] [201] [210] [211] [212] [213] [219] [223] [234] [235] [236] [237] [238] [300] [301] [401] [451] [453] [510] [518]

List of articles, referred to in Chapter 4

[25] [106] [108] [252] [272] [286] [287] [318] [319] [341] [367] [368] [369] [414] [450] [452] [467] [468] [507]

Section 4.1: [6] [60] [70] [71] [106] [111] [218] [252] [272]

Section 4.2: [25] [81] [106] [107] [196] [198] [200] [209] [252] [286] [352] [369]

Section 4.3: [106] [107] [198] [200] [202] [209] [252] [272] [310] [369] [468]

Section 4.4: [106] [113] [244] [245] [272] [276] [288] [339] [361] [362] [368] [403] [495] [496] [497] [515] [516]

Section 4.5: [9] [80] [81] [163] [164] [195] [196] [198] [200] [202] [209]

List of articles, referred to in Chapter 5

Section 5.1: [125] [157] [158] [161] [163] [164] [174] [175] [203] [204] [399] [425] [434] [440] [463]

Section 5.2: [84] [174] [175] [203] [207] [287] [288] [289] [290] [291] [307] [363] [428] [429] [430] [431] [432] [433] [434] [464] [465] [506]

Section 5.3: [110] [115] [116] [119] [205] [267] [276] [286] [434] [436] [437] [452] [455] [498] [502] [503] [504] [505]

Section 5.4: [64] [65] [66] [67] [68] [69] [70] [71] [73] [79] [196] [206] [239] [270] [271] [317] [513] [514] [515] [516] [517]

Section 5.5: [30] [33] [42] [52] [63] [104] [124] [135] [136] [150] [151] [152] [164] [183] [199] [200] [204] [209] [220] [230] [231] [232] [254] [255] [256] [311] [321] [322] [323] [324]

[325] [326] [327] [328] [329] [330] [331] [332] [333] [334] [335] [336] [344] [345] [346] [347] [348] [349] [350] [351] [352] [353] [354] [370] [404] [411] [412] [415] [420] [427] [434] [443] [444] [456] [471] [472] [473] [474] [475] [476] [477] [478] [479] [480] [481] [482]

Section 5.6: [28] [29] [44] [104] [114] [119] [174] [175] [196] [199] [201] [203] [204] [205] [207] [214] [257] [270] [271] [276] [284] [288] [317] [384] [385] [386] [387] [388] [389] [390] [391] [392] [393] [403] [438] [439] [441] [442] [458] [459] [465] [495] [496] [510] [515] [516] [517] [519]

Section 5.7: [27] [38] [45] [46] [47] [65] [68] [69] [70] [71] [73] [82] [88] [93] [94] [95] [118] [119] [142] [143] [144] [145] [149] [153] [154] [166] [168] [169] [170] [172] [174] [175] [179] [180] [181] [182] [196] [199] [201] [203] [204] [205] [207] [214] [225] [226] [227] [228] [229] [242] [243] [244] [245] [246] [247] [248] [249] [250] [251] [273] [286] [287] [288] [289] [291] [292] [293] [294] [304] [306] [329] [364] [390] [402] [404] [408] [409] [424] [434] [438] [439] [445] [446] [447] [448] [468] [486] [487]

List of articles referred to in Appendix

[106] [108] [272] [276] [286] [287] [368] [369] [370] [467] [468]

Section A: [41] [106] [107] [272] [275] [276] [277] [278] [279] [280] [286] [295] [296] [297] [370] [450] [468]

Section B: [1] [66] [67] [68] [69] [74] [75] [76] [106] [196] [198] [201] [202] [272] [276] [280] [286] [339] [370]

Section C: [7] [8] [106] [165] [166] [167] [168] [169] [170] [171] [172] [173] [272] [286] [370]

Section D: [1] [3] [4] [5] [6] [34] [60] [80] [81] [96] [97] [106] [108] [111] [117] [121] [122] [126] [127] [128] [129] [196] [197] [198] [210] [213] [218] [219] [239] [276] [360] [369] [435] [468] [484] [485] [507] [511] [512]

Section E: [40] [77] [78] [92] [98] [99] [100] [101] [102] [103] [106] [113] [117] [123] [148] [239] [276] [279] [280] [287] [317] [358] [359] [360] [370] [459] [463] [464] [465] [466] [467] [468] [484] [485] [511] [512]

List of open problems, stated in the notes

Open problems *NoNo* 3.7.12, 3.10.1; 5.4.5, 5.5.7, 5.7.1, 5.7.2, 5.7.3; E.4.